Mechanics of
Solids and
Structures

Mechanics of Solids and Structures

P P Benham

MSc(Eng), PhD, DSc(Eng), DIC, FIMechE, FPRI
Professor and Head of Aeronautical Engineering
Queen's University, Belfast

and

The late F V Warnock

OBE, MSc(Eng), PhD, FRCScI, FIMechE
Formerly Professor and Head of Mechanical Engineering
Queen's University, Belfast

S I Units

PITMAN

PITMAN BOOKS LIMITED
39 Parker Street, London WC2B 5PB

PITMAN PUBLISHING INC
1020 Plain Street, Marshfield, Massachusetts 02050

Associated Companies
Pitman Publishing Pty Ltd, Melbourne
Pitman Publishing New Zealand Ltd, Wellington
Copp Clark Ltd, Toronto

© P P Benham and F V Warnock 1973, 1976

First published in Great Britain 1973
Revised and reprinted 1976
Reprinted 1978, 1981

Text set in 10/11 pt. Monotype Times New Roman, printed and bound
in Great Britain at The Pitman Press, Bath

ISBN 0 273 36186 4 (Cased edition)
ISBN 0 273 36191 0 (Paperback edition)

Preface

The adoption in Europe of the International System of Units (SI) has necessitated the extensive revision of existing textbooks. One such is our book on *Mechanics of Solids and Strength of Materials*, published in 1965. Since then there have been changes in university and H.N.C. syllabuses and the introduction of C.E.I. and H.N.D. examinations. This change over to SI units has also presented the authors with the opportunity of updating material and presentation where required. The principal area of revision has been in the first eleven chapters, in which the basic principles of structural and solid mechanics have to be introduced and thoroughly understood. As the new title implies, more emphasis has been placed on elementary structural analysis and its integration with solid mechanics in demonstrating statical determinacy and indeterminacy, superposition and the basic principles of equilibrium, compatibility and stress–strain behaviour.

The division of the original book into three parts, namely (1) the theoretical analysis of force, stress and displacement, (2) the mechanical properties of materials, and (3) the experimental analysis of stress and strain, was well received and has been retained.

The first eleven chapters in Part 1 are concerned with forces and displacements in statically determinate and indeterminate structures, the analysis of uniaxial stress and strain due to various forms of loading such as thermal, pressure, bending and torsion. The concepts of strain energy and elastic stability are also introduced in this part of the text.

In Chapters 12 and 13 a detailed analysis is made of the elastic stress and strain relationships in terms of equilibrium and compatibility of displacements in a complex stress system. There are relatively few practical problems involving complex stress that can be solved by the above principles without recourse to the mathematical theory of elasticity; however, some useful examples (thick-walled cylinders, rotating discs, etc.) are discussed and solved in Chapter 14.

The design of machines and structures is generally such that stresses and strains are kept within the elastic range, and in order to do this a criterion of yielding under complex stress is needed. Various theories of yielding are discussed in Chapter 15.

An important new addition is in Chapter 16 dealing with the bending of thin circular plates and shells, which is now included in C.E.I. and most degree syllabuses.

v

The importance of stress concentration, i.e. a sharp rise in stress due to irregularities in the shape of a component, cannot be overemphasized, particularly in relation to fracture by fatigue. Chapter 17 deals with various aspects of stress concentration. The final chapter in Part 1 introduces some elementary concepts of plasticity, namely deformations beyond the elastic limit, in bending, torsion and axially-symmetrical components. Although materials generally work in their elastic range, the engineer needs to have some idea of the consequences of plastic deformation, whether by design or accident. Viscoelasticity is also introduced in this chapter in order to assist in the growing use of plastics as engineering materials.

There are many worked examples in all chapters in this first part, and at the end of the book a number of problems have been included, many of which have been taken from past examination papers of universities and various professional institutions.

Theoretical stress analysis is of little use if the engineer has not also the necessary knowledge of how the material of a machine or structure will behave in service. The mechanical properties of materials have been an extensive field for research for the past 150 years or more, and there are still numerous problems to be understood and solved. There are a number of excellent textbooks dealing with this subject in detail; however, the engineering student needs initially to be put in the picture so that he can achieve full benefit from a laboratory course. Part 2 of this book is therefore aimed at presenting the main features and some established facts about mechanical properties, and also the very necessary aspect of how these properties are studied in the laboratory by mechanical testing. There are six chapters (19 to 24) which are devoted to a discussion of topics such as tension, compression, fatigue, creep and brittle fracturie, etc. Chapter 24 introduces the properties and testing of non-metalltic materials, to remind the student that metals do not have a complete monopoly in engineering and that there is an important place for non-metallics.

At the end of each chapter in this part of the book a short bibliography has been given for students requiring further reading of a more detailed nature.

The final part of the book consists of three chapters devoted to an introduction to various techniques of experimental stress and strain analysis. Solutions for the stresses in some engineering problems by theoretical means may be very lengthy and complex and occasionally impossible without many simplifying assumptions. It is in these circumstances, and also for checking of other theoretical solutions, that experimental techniques have been developed and are now widely used in industry. Most university and technical college laboratory work now includes instruction and experiments in the use of the electrical resistance strain gauge, photo-elasticity and other techniques. Chapters 25 to 27 will provide the necessary introduction to these methods and references for more detailed reading if required.

There are two Appendices: one gives alternative sign conventions in bending theory, and the other develops relationships for the properties of areas.

It should perhaps be pointed out that the three separate parts of the book can and probably should be studied concurrently rather than consecutively. In fact it has been occasionally necessary to refer the reader from one part of the book to another in order to make the best use of the material available.

Some use has been made of data, diagrams and photographs from other published literature and manufacturers' catalogues, and although individual reference has been made where appropriate in the text, the authors also wish at this point to make grateful acknowledgement to all persons and organizations concerned.

1973   P. P. BENHAM
                                                               F. V. WARNOCK

Contents

8 Elastic Strain Energy

9 Theory of Torsion

10 Statically Indeterminate Beams and Frames

Appendixes

Problems

List of Plates

(*At end of book*)

Principal Notation

α	angle, coefficient of thermal expansion
β	angle
γ	shear strain
δ	deflection, displacement
ε	direct strain
η	efficiency, viscosity
θ	angle, angle of twist, co-ordinate
λ	wavelength of light
ν	Poisson's ratio
ρ	radius of curvature, density, resistivity
σ	direct stress
τ	shear stress
ϕ	angle, co-ordinate, stress function
ω	angular velocity
Θ	body force
A	area
C	complementary energy
D	diameter
E	Young's modulus of elasticity
F	force, photoelastic fringe value
G	shear or rigidity modulus of elasticity
H	force
I	second moment of area
J	polar second moment of area
K	bulk modulus of elasticity, fatigue strength factor, strain gauge factor, stress concentration factor
L	length
M	bending moment
N	number of stress cycles, speed of rotation
P	force, product moment of area
Q	shear force
R	body force, electrical resistance, photoelastic relative retardation, radius of curvature, reaction force, stress ratio
S	cyclic stress
T	temperature, torque
U	strain energy
V	volume
W	weight, load
X	body force
Y	body force
Z	body force, section modulus

a	area, distance
b	breadth, distance
c	distance
d	depth, diameter
e	eccentricity
e	base of Napierian logarithms
g	gravitational constant
h	distance, eccentricity
j	number of joints
k	radius of gyration
l	length
m	mass, modular ratio, number of members
n	number: of coils in a spring, photoelastic fringes
p	pressure
q	shear flow
r	co-ordinate, radius
s	length
t	thickness, time
u	displacement in the x or r direction
v	deflection, displacement in the y or θ direction, velocity
w	displacement in the z direction, load intensity, weight per unit volume
x	co-ordinate, distance
y	co-ordinate, distance
z	co-ordinate, distance

It should be noted that a number of these symbols have also been used to denote constants in various equations.

Part 1

Theoretical Analysis of Stress and Strain

Chapter 1
Statically Determinate Frames and Beams

1.1. Structural and solid-body mechanics are concerned with analysing the effects of applied loads. These are *external* to the material of the structure or body, and result in *internal* reacting forces, together with deformations and displacements, conforming to the principles of Newtonian mechanics. It follows that a proper understanding of the subject must begin from a familiarity with the principles of statics, the cornerstone of which is equilibrium.

1.2. REVISION OF STATICS

A particle is in a state of equilibrium if the resultant force and moment acting on it are zero, and hence according to Newton's law of motion it will have no acceleration and will be at rest. This hypothesis can be extended to clusters of particles that interact with each other with equal and opposite forces but have no overall resultant. Thus it is evident that solid bodies, structures or any subdivided part will be in equilibrium if the resultant of all external forces and moments is zero. This may be expressed mathematically in the following six equations which relate to Cartesian co-ordinate axes x, y and z.

$$\left.\begin{array}{l} \Sigma F_x = 0 \\ \Sigma F_y = 0 \\ \Sigma F_z = 0 \end{array}\right\} \tag{1.1}$$

where F_x, F_y and F_z represent the components of force vectors in the co-ordinate directions.

$$\left.\begin{array}{l} \Sigma M_x = 0 \\ \Sigma M_y = 0 \\ \Sigma M_z = 0 \end{array}\right\} \tag{1.2}$$

where M_x, M_y and M_z are components of moment vectors caused by the external forces acting about the axes x, y, z.

The above six equations are the necessary and sufficient conditions for equilibrium of a body.

If the forces all act in one plane, say $z = 0$, then

$$\Sigma F_z = \Sigma M_x = \Sigma M_y = 0$$

3

are automatically satisfied and the equilibrium conditions to be satisfied in a two-dimensional system are

$$\left.\begin{matrix} \Sigma F_x = 0 \\ \Sigma F_y = 0 \\ \Sigma M_z = 0 \end{matrix}\right\} \tag{1.3}$$

Forces and moments are vector quantities and may be resolved into components, that is to say a force or a moment of a certain magnitude and direction may be replaced and exactly represented by two or more components of different magnitudes and in different directions.

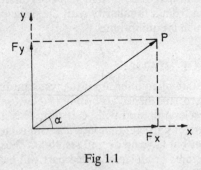

Fig 1.1

Considering firstly the two-dimensional case shown in Fig. 1.1, the force P may be replaced by the two components F_x and F_y provided that

$$\left.\begin{matrix} F_x = P \cos \alpha \\ F_y = P \sin \alpha \end{matrix}\right\} \tag{1.4}$$

If the force P were arbitrarily oriented with respect to three axes x, y, z as in Fig. 1.2 then it could be replaced or represented by the following components:

$$\left.\begin{matrix} F_x = P \cos \alpha \\ F_y = P \cos \beta \\ F_z = P \cos \gamma \end{matrix}\right\} \tag{1.5}$$

A couple or moment vector about an axis can similarly be resolved into a representative system of component moment vectors about other axes, as shown in Fig. 1.3 and represented by the following equations:

$$\left.\begin{matrix} M_x = M \cos \alpha \\ M_y = M \cos \beta \\ M_z = M \cos \gamma \end{matrix}\right\} \tag{1.6}$$

It is sometimes more convenient to replace a system of applied forces by a resultant which of course must have the same effect as those forces. Considering a two-dimensional case as illustrated in Fig. 1.4 then the most general solution is obtained by choosing any point A through which

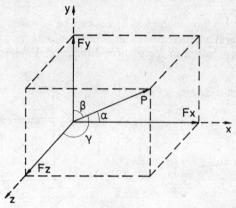

Fig 1.2

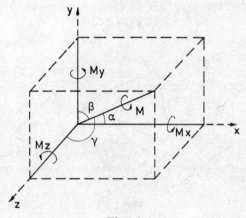

Fig 1.3

the resultant can act. Then the total force components in the co-ordinate directions are

$$F_x = \Sigma F_x \atop F_y = \Sigma F_y \Big\}$$ (1.7)

and the resultant force is given by

$$R = \sqrt{(F_x{}^2 + F_y{}^2)}$$ (1.8)

However, this is not sufficient in itself since the moment due to the forces must be represented. This is done by having a couple acting about A such that

$$\bar{M} = \Sigma M_2 \tag{1.9}$$

In general, then, any system of forces can be replaced by a resultant force through and a couple about any chosen point.

The equivalent solution for a three-dimensional system of forces is

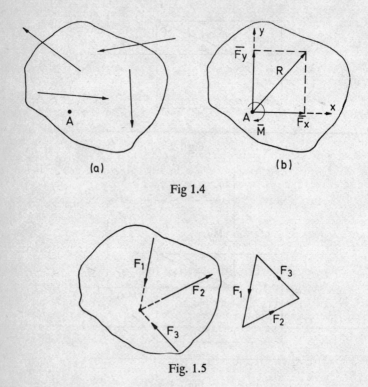

(a)   (b)

Fig 1.4

Fig. 1.5

similarly a couple and a resultant force whose direction is parallel to the axis of the couple.

One of the most useful constructions in force analysis is termed the *triangle of forces*. If a body is acted on by three forces then, for equilibrium to exist, these must act through a common point or else there will exist a couple about the point causing the body to rotate. In addition the magnitude and direction of the three force vectors must be such as to form a closed triangle as shown in Fig. 1.5.

1.3. TYPES OF STRUCTURAL AND SOLID BODY COMPONENTS

Structures are made up of series of members of regular shape that have a particular function for load carrying. The shape and function are, through usage, implied in the name attached to the member. The first group is concerned with carrying loads parallel to a longitudinal axis. Examples are shown in Fig. 1.6. A member which prevents two parts of a structure from moving apart is subjected to a pull at each end, or tensile force, and is termed a *tie* (*a*). Conversely a slender member which prevents parts of a structure moving toward each other is under compressive force and is termed a *strut* (*b*). A vertical member which is perhaps not too slender and supports some of the mass of the structure is called a *column* (*c*). A *cable* (*d*) is a generally recognized term for a flexible string under tension which connects two bodies. It cannot supply resistance to bending action.

One of the most important of structural members is that which is frequently supported horizontally and carries transverse loading. This is known as a *beam* (*e*), a common special case of which is termed a *cantilever* (*f*), where one end is fixed and provides all the necessary support. A *beam-column* (*g*), as the name implies, combines the separate functions already described. The *arch* (*h*) has the same function as the beam or beam-column, but is curved in shape.

The filling-in and carrying of load over an area or space are achieved by *flat slabs or plates* (*i*), by *panels* (*j*) and also by *shells* (*k*), which are curved versions of the former.

The transmission of torque and twist is achieved through a member which is frequently termed a *shaft* (*l*).

The members described above can have a variety of cross-sectional shapes depending on the particular type of loading to be carried. Some typical cross-sections are illustrated in Fig. 1.7. They are made generally by rolling steel and rolling or extruding aluminium alloy.

1.4. TYPICAL JOINTS IN STRUCTURES

The separate members of a structural framework are joined together by bolting, riveting or welding, two examples of which are shown in Fig. 1.8. Now, if these joints were ideally stiff, when the members of the framework were deformed under load, the angles between the members at the joint would not change. This would also imply that the joint was capable of transmitting a couple. However, calculations for a complete structure on this basis would become rather involved and tedious.

It is found in practice that there is some degree of rotation between members at a joint due to the elasticity of the system. Furthermore, it has been shown that it is not unreasonable, for purposes of

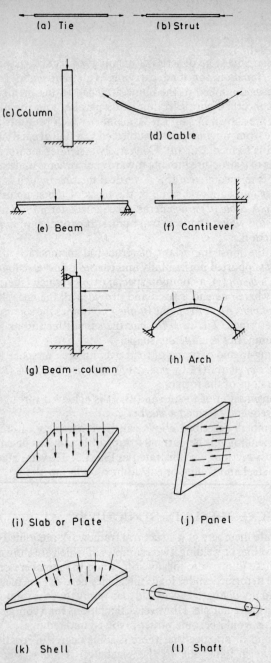

(a) Tie   (b) Strut

(c) Column   (d) Cable

(e) Beam   (f) Cantilever

(g) Beam - column   (h) Arch

(i) Slab or Plate   (j) Panel

(k) Shell   (l) Shaft

Fig 1.6

calculation, to assume that these joints may be represented by a simple ball and socket or pin in a hole. Even with this arrangement, which of course cannot transmit a couple or bending moment (other than by friction which is ignored), deformations of the members are relatively

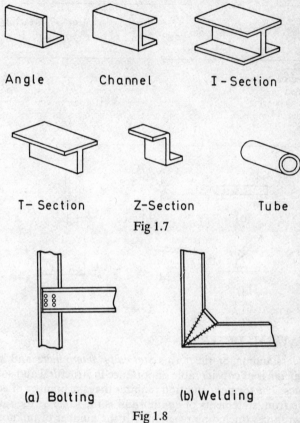

Angle   Channel   I-Section

T- Section   Z-Section   Tube

Fig 1.7

(a) Bolting   (b) Welding

Fig 1.8

small. Consequently changes in angle at the joints are also small, which is why this approximation is not unreasonable when applied to the actual joints. Thus it is common practice when calculating the forces in the members of a framework to assume that all joints are pinned.

1.5. TYPES OF SUPPORT FOR FRAMEWORKS, BEAMS AND COLUMNS

The applied loading on a framework, beam or column is transmitted to the supports which will provide the required reacting forces to maintain overall equilibrium. Examples of supports of various kinds suitable to react to loading in a plane (two dimensions) are shown in Fig. 1.9. In the accompanying table the possible displacement and reacting force components are indicated.

		Displacement			Reacting force		
		x	y	θ	R_z	R_y	M_z
(a)	Fixed or built-in	—	—	—	√	√	√
(b)	Pinned	—	—	√	√	√	—
(c)	Linear bearing	—	√	—	√	—	√
(d)	Roller	√	—	√	—	√	—
(e)	Free	√	√	√	—	—	—

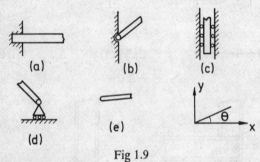

Fig 1.9

1.6. STATICAL DETERMINACY

The understanding of the terms *statically determinate* and *statically indeterminate* is of considerable importance in structural and solid-body mechanics. The former condition requires that the number of equations available from statements of equilibrium is the same as the number of unknown forces (including reactions). If the number of unknown reactions or internal forces in the structure or component is greater than the number of equilibrium equations available, then the problem is said to be *statically indeterminate*. Additional equations have to be found by considering the displacement or deformation of the body.

The above statements are quite general and apply throughout this text.

1.7. STIFFNESS OF FRAMES

The principles discussed above will now be considered in relation to plane and space frames. There are three classes of frame or truss, in concept, although one is not of practical interest:

(a) *Under-stiff*. If there are more equilibrium equations than unknown forces or reactions the system is unstable and is not a structure but a mechanism.

(b) *Just-stiff*. This is the statically determinate case. If any member is removed then a part or the whole of the frame will collapse.

(c) *Over-stiff*. This is the statically indeterminate case. There is at least one member more than is required for the frame to be just stiff.

Some examples are given in Fig. 1.10.

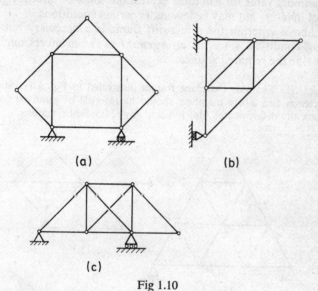

(a)

(b)

(c)

Fig 1.10

It is useful to express the three cases in the form of mathematical criteria. Let the number of joints, including support points, in a frame be j, the number of members, m, and the number of reactions, r. Now, for a space frame there are three equilibrium equations applicable to each joint, namely $\Sigma F_x = 0$, $\Sigma F_y = 0$, $\Sigma F_z = 0$; hence there are $3j$ equations to determine $m + r$ unknown forces and reactions, and the statically determinate case is represented by

$$m + r = 3j \qquad (1.10)$$

When $m + r < 3j$ the members form a mechanism, and for $m + r > 3j$ the frame is over-stiff, or redundant, and therefore statically indeterminate.

There are six conditions for overall equilibrium; therefore the minimum value for r, when using the above criteria to allow for any general loading system, is six. Certain arrangements of loading may result in less than six reactions being required.

For frames lying only in one plane there are only two equilibrium equations at each joint and so the relationships comparable with the above are

$$\left.\begin{array}{c} m + r < 2j \\ m + r = 2j \\ m + r > 2j \end{array}\right\} \tag{1.11}$$

The minimum value for r in these expressions must be three for general forms of loading, but may be less under certain conditions.

The above criterion for a just-stiff frame is a necessary but not a sufficient condition, since the *arrangement* of the members might still not provide the required stiffness.

Example 1.1. Examine the plane frames illustrated in Fig. 1.11. State the class of each and where members should be inserted or removed to make each statically determinate. Also indicate any redundant reactions.

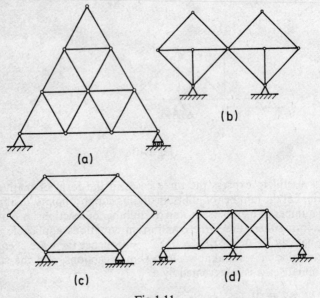

Fig 1.11

(a) $m = 18$, $r = 3$, $j = 10$: hence $m + r > 2j$ and the frame is over-stiff or redundant. Any member may be removed from the central hexagon structure. None of the members may be removed from the apexes.

(b) $m = 14$, $r = 4$, $j = 9$: hence $m + r = 2j$ and the frame is just stiff and statically determinate.

(c) $m = 8$, $r = 3$, $j = 6$: hence $m + r < 2j$, which constitutes a mechanism. To make statically determinate either position fix the right-hand support or insert a member between any pair of unconnected joints.

(d) $m = 15$, $r = 4$, $j = 8$: hence $m + r > 2j$, which is redundant both in members and reactions. Hence remove either the left or right support and a diagonal member from each square.

1.8. FREE BODY DIAGRAMS

When commencing to analyse any force system acting on a component or structure it is essential firstly to have a diagram showing the forces acting. If the structure or part of it is separated from its surroundings and the appropriate reactions, required to maintain equilibrium, are inserted then a diagram of this system is called a *free-body* diagram. Examples of this are shown in Fig. 1.12.

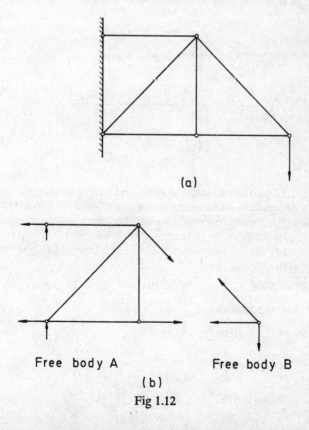

(a)

Free body A   Free body B

(b)

Fig 1.12

1.9. DETERMINATION OF FORCES IN A PLANE PIN-JOINTED FRAME

(a) Resolution at Joints

At any joint in a plane frame two equilibrium equations apply, namely

$$\Sigma F_x = 0 \quad \text{and} \quad \Sigma F_y = 0 \quad \text{(for a frame lying in the } z\text{-plane)}$$

hence only two unknown forces can be determined. The method therefore entails making a free-body diagram centred on each joint at which there are only two unknowns. The forces are then resolved in the x and y directions so that the above two equations can be applied. It will probably be necessary at first to determine support reactions by considering equilibrium of the whole frame. This generally gives at least one joint where there is a known reaction and only two members having unknown forces from which to start the analysis.

Example 1.2. Determine the magnitude and the type of the forces in the members of the plane pin-jointed frame shown in Fig. 1.13.

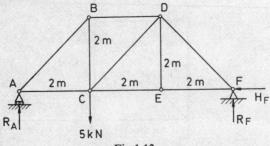

Fig 1.13

In Fig. 1.13, in order to save the space and having to draw the complete frame twice, once on its supports and a second time as a free body away from the supports with the reactions in place, the possible reactions have been placed directly on to the configuration diagram.

Considering the equilibrium of the whole frame and taking moments $F(\Sigma M_z = 0)$,

$$6R_A - 5000 \times 4 = 0 \quad \text{so that} \quad R_A = 3333\,\text{N}$$

For horizontal equilibrium,

$$\Sigma F_x = 0 \quad \text{gives} \quad H_F = 0$$

For vertical equilibrium,

$$\Sigma F_y = 0 \quad \text{gives} \quad R_A + R_F - 5000 = 0$$

Hence $R_F = 1667\,\text{N}$

The only joints at which there are two unknowns are A and F. Start say at A and *assume initially that each member is subjected to tension which is designated as positive* and indicated by arrowheads as shown in Fig. 1.14.

Tension   Compression

Fig 1.14

Joint A

The free-body diagram is as shown in Fig. 1.15.

$$\Sigma F_y = 0 \quad \text{gives} \quad F_{AB} \sin 45° + 3333 = 0$$

Hence $F_{AB} = -4713\,\text{N}$

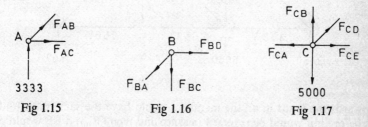

Fig 1.15   Fig 1.16   Fig 1.17

The negative sign shows that in this case the wrong assumption was made and that AB is in compression and not tension.

$$\Sigma F_x = 0 \quad \text{gives} \quad F_{AB} \cos 45° + F_{AC} = 0$$

Hence $F_{AC} = 4713 \cos 45° = 3333\,\text{N}$

The next step *has* to be at joint B (rather than C).

Joint B (Fig. 1.16)

$$\Sigma F_y = F_{BA} \cos 45° + F_{BC} = 0$$
$$F_{BC} = 4713 \cos 45° = 3333\,\text{N}$$
$$\Sigma F_x = F_{BD} - F_{BA} \cos 45° = 0$$
$$F_{BD} = -4713 \cos 45° = -3333\,\text{N} \quad \text{(compression)}$$

Joint C (Fig. 1.17)

There are now only two unknowns at this joint, F_{CD} and F_{CE}.

$$\Sigma F_y = F_{CB} - 5000 + F_{CD} \cos 45° = 0$$
$$F_{CD} = (-3333 + 5000)\sqrt{2} = 2356\,\text{N}$$
$$\Sigma F_x = F_{CA} - F_{CE} - F_{CD} \cos 45° = 0$$
$$F_{CE} = 3333 - 2356 \cos 45° = 1667\,\text{N}$$

Continuing the above process for joints E and D gives

$$F_{ED} = 0 \qquad F_{EF} = 1667\,\text{N} \qquad F_{DF} = 2356\,\text{N}$$

The final force distribution in the frame is shown in Fig. 1.18.

As the force in DE is zero one might ask whether that member is re-quired. With this particular applied load, in fact, DE could be removed quite safely, since the two members CE and EF really act as one con-tinuous member CEF in tension. However, if the applied load at C acted

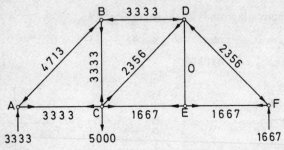

Fig 1.18

upwards the loads in all the members would have the same magnitude as before but would be reversed in sense and then CE and EF would be in compression. While ED was in position even without any force in it, it would keep CE and EF in line, whereas if ED were removed the joint E would move vertically due to the slightest misalignment and result in collapse.

(b) Method of Sections

If the forces are required in only a few members of a frame with many joints then unnecessary effort might be involved in using method (a)

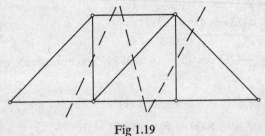

Fig 1.19

above. Instead, the frame is cut by a section through *not more than three members* as illustrated in several ways in Fig. 1.19. Treat either part of the sectioned frame as a free body and insert force vectors at the cut ends of members in the tensile sense as in Fig. 1.20. Equilibrium for the free body requires that

$$\Sigma F_x = 0 \quad \Sigma F_y = 0 \quad \Sigma M_z = 0$$

Hence $\Sigma F_y = 3333 - F_{BC} = 0$

$F_{BC} = 3333\,\text{N}$

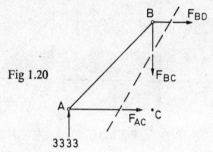

Fig 1.20

3333

Moments about A give

$$\Sigma M_z = 2F_{BD} + 2F_{BC} = 0$$
$$F_{BD} = -3333\,\text{N} \quad \text{(compression)}$$

Finally,

$$\Sigma F_x = F_{BD} + F_{AC} = 0$$
$$F_{AC} = 3333\,\text{N}$$

This procedure is continued for other sections through the frame as may be required.

(c) Graphical Method

A useful alternative to the analytical method (a) above is achieved through the graphical construction of a force polygon.

If forces acting at a point are in equilibrium then the vector diagram is a closed polygon. The polygon can only be built up on the basis of not more than two unknowns at any joint. This often means that reactions have first to be determined from the equilibrium conditions for the whole frame.

The next step is to employ Bow's notation and letter each of the areas bounded by members and the spaces between forces or reactions outside the frame. Choose a suitable force scale to give a reasonable size of polygon. (Some experience is needed to be able to prejudge the likely shape and size of a polygon and usually a rough preliminary sketch is helpful.) Start the polygon at a joint where there is a known force or forces and not more than two unknowns. Proceeding mentally in a clockwise (or anticlockwise) path across the known force from one lettered space to the next, draw a line parallel to the direction of the force and of length appropriate to the force scale chosen. Then insert the space letters on each end of the line to give the correct sense.

Consider the frame previously analysed by other methods, as shown in

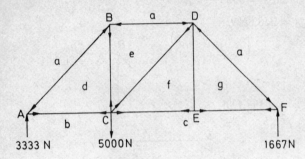

Fig 1.21

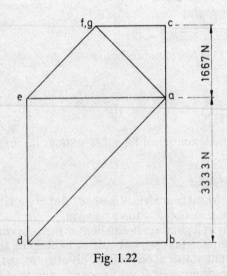

Fig. 1.22

Fig. 1.21, where the reactions are known and the space letters have been inserted.

For joint A a line is marked *ba* and drawn vertically upwards to represent the reaction of 3333 N. Next draw from *a* and *b* lines parallel to AB and AC to meet at *d*. The sense of the force in the member is given by reading the letters (clockwise), for example *a* to *d*, which from the polygon gives a force towards A; or *d* to *b*, which represents a force away from A. The appropriate arrowheads may then be inserted at each end of those members and one can next move to joint B, where there are now only two unknowns. The above procedure is then repeated, which adds another triangle to the diagram, and so on for each joint until the polygon is complete for the whole frame as illustrated in Fig. 1.22. The arrowheads show whether a member is subjected to tension or compression.

(d) Method of Tension Coefficient

This method was proposed by Southwell in 1920 as a convenient way of solving for the forces in space frames, and it is of course also applicable to plane frames. It amounts to a shorthand notation for resolution at joints.

The *tension coefficient* for a member is defined as the force, F, in the member divided by its length, L. Consider the member AB shown in

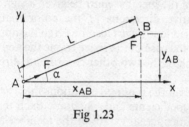

Fig 1.23

Fig. 1.23; the resolved components of the force F_{AB} in the co-ordinate directions x and y will be $F_{AB} \cos \alpha$ and $F_{AB} \sin \alpha$. Now, $\cos \alpha$ and $\sin \alpha$ can be expressed as the ratio of the length of the member projected onto the x and y axes divided by the length, L_{AB}, of the member; thus

$$\cos \alpha = \frac{x_{AB}}{L_{AB}} \quad \text{and} \quad \sin \alpha = \frac{y_{AB}}{L_{AB}}$$

Therefore

$$F_{AB} \cos \alpha = F_{AB} \frac{x_{AB}}{L_{AB}} = t_{AB} x_{AB} \tag{1.12}$$

and

$$F_{AB} \sin \alpha = F_{AB} \frac{y_{AB}}{L_{AB}} = t_{AB} y_{AB} \tag{1.13}$$

where t_{AB} is the *tension coefficient* for the member.

Since the majority of configuration diagrams give dimensions which are related to the co-ordinate directions, it is simpler to express resolved components of force in terms of the tension coefficient and the projected length, rather than sines and cosines of angles which might need to be found from tables. All members are assumed to be in tension initially and hence have a positive coefficient. If in the solution a tension coefficient turns out to be negative then this shows that the member is actually in compression. There is no need in this method to work out reactions at supports in advance as they can simply be included as unknowns in the equilibrium equations.

The first step is to choose a set of reference co-ordinate axes and directions and then to insert all the necessary reactions at the support points in the co-ordinate directions. The arrowhead direction is quite arbitrary. Now commencing at any joint and considering one co-ordinate direction, write down the equilibrium equation. This will consist of the sum of the products of the tension coefficient, t, and the projected length, say x, (for that co-ordinate direction) for each member at that joint plus any reaction or applied force at that point. It is important to remember that, taking the origin of the co-ordinate axes referenced at the joint, the projected lengths of the member *must* be given a sign appropriate to the positive and negative directions of the co-ordinate axes. Similarly, reactions and applied loads *must* also be given the appropriate sign relative to the axes. In the case of the space frame the above procedure must be repeated for each of the two other co-ordinate directions and then in turn for each joint.

Having obtained all the required equations these are solved for all the unknown tension coefficients and reactions. Each tension coefficient is then multiplied by the actual length of the member to give the force in the member. It will perhaps now be evident that, for clarity and ease of checking, a tabular system of solution is most desirable. The method will now be illustrated in the following example of a space frame.

Example 1.3. Use the method of tension coefficients to determine the reactions and the forces in the space frame shown in Fig. 1.24.

Eqn. No.	Joint/direction	Equilibrium equation
1	A − x	$-H_A + 2t_{AD} = 0$
2	A − y	$4t_{AD} + R_A = 0$
3	A − z	$-S_a = 0$
4	B − x	$-H_B - 2t_{BD} + 8t_{BE} = 0$
5	B − y	$R_B + 4t_{BD} = 0$
6	B − z	$S_B + 3t_{BD} + 3t_{BE} = 0$
7	C − x	$-H_C - 2t_{CD} + 8t_{CE} = 0$
8	C − y	$R_C + 4t_{CD} = 0$
9	C − z	$-S_C - 3t_{CD} - 3t_{CE} = 0$
10	D − x	$-20 - 2t_{AD} + 2t_{BD} + 2t_{CD} + 10t_{DE} = 0$
11	D − y	$-4t_{DA} - 4t_{DB} - 4t_{DC} - 4t_{DE} = 0$
12	D − z	$3t_{DC} - 3t_{DB} = 0$
13	E − x	$-8t_{EB} - 8t_{EC} - 10t_{ED} = 0$
14	E − y	$4t_{ED} - 50 = 0$
15	E − z	$-3t_{EB} + 3t_{EC} = 0$

Member	Tension coefficient	Length, m	Force, kN	Reaction, kN
AD	+20	4·48	+89·6	$H_A = 40$
BD	−16·25	5·39	−87·5	$R_A = -80$
CD	−16·25	5·39	−87·5	$S_A = 0$
BE	−7·8	8·55	−66·6	$H_B = -30$
CE	−7·8	8·55	−66·6	$R_B = 65$
DE	+12·5	10·78	+134·8	$S_B = 25·4$
				$H_C = -30$
				$R_C = 65$
				$S_C = 25·4$

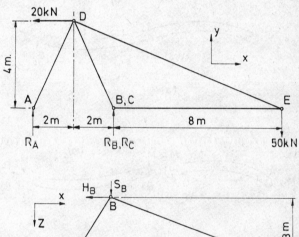

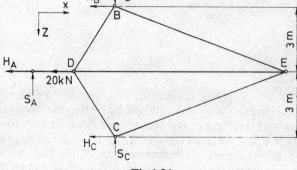

Fig 1.24

1.10. STATICS OF BENDING

In the statical analysis of frameworks the members were only subjected to axial force, namely tension or compression. The next step is to consider the effect of transverse loads acting on slender members. The deformation that results is termed *bending* and is of course very common in structures and machines—floor joists, railway axles, aeroplane wings, leaf springs, etc. External applied loads which cause bending give rise to internal reacting forces. These have to be determined before it is possible to calculate stress and deflection.

1.11. FORCE AND MOMENT EQUILIBRIUM DURING BENDING

A slender curved bar is shown in Fig. 1.25(a) subjected to various transverse loads. In general, to maintain equilibrium a force and a couple will be required at each end of the bar. The force can be resolved into two

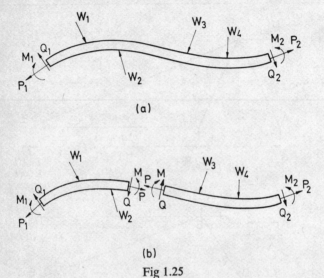

(a)

(b)

Fig 1.25

components, one perpendicular and the other parallel to each end cross-section. The internal forces are obtained by cutting the bar to give two free bodies and inserting at the cut sections the necessary forces and moments for equilibrium, as shown in Fig. 1.25(b). The couple, M, is termed a *bending moment*, the transverse force, Q, is called a *shear force*, and P is a *longitudinal force*. The most important stresses and deflections that occur during bending are due to M, rather than Q or P.

1.12. FORMS OF LOADING AND SUPPORT

The transverse externally applied load on a beam or bar can take one of two forms, concentrated or distributed. The former is illustrated in Fig. 1.26(a) in which the load acts on the surface of the beam along a line perpendicular to the longitudinal axis. This, of course, is an idealization, and in practice a concentrated load will cover a very short length of the beam.

A load which is distributed is shown in Fig. 1.26(*b*) and occupies a length of beam surface. The load intensity is always taken as constant across the beam thickness but may be uniformly or non-uniformly distributed along part or the whole length of the beam. In practice the particular conditions of force and displacement at beam supports may vary considerably. Theoretical solutions of beam problems generally

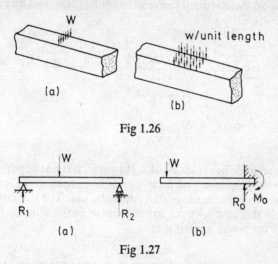

Fig 1.26

Fig 1.27

employ two simplified forms of support. These are termed simply *supported* and *built-in* or fixed. The former is illustrated in Fig. 1.27(*a*) in which the beam rests on knife-edges or rollers. When the beam is bent under load the support reaction is only a transverse force and there is no restraining couple. Hence the deflection at the support is zero and the beam is free to take up a slope dictated by the applied load. The built-in support shown in Fig. 1.27(*b*) reacts with a transverse force and a couple. Thus both deflection and slope are fully restrained. The particular example illustrated of a beam built-in at one end and free at the other is termed a *cantilever*.

The number and type of supports also has a further important bearing on a beam solution by making it either *statically determinate* or *statically indeterminate*. In the former the support reactions can be found simply from force and moment equilibrium equations. This applies, for example, to beams on two simple supports or one built-in support and no other. The two equilibrium equations are insufficient to find the reactions at the supports of a statically indeterminate beam owing to the presence of redundant forces. In this case it is necessary to consider also the deflection of the beam in order to obtain additional equations to solve for the reactions.

1.13. SIGN CONVENTION

It is important that the analysis of internal forces in bending shall be consistent within and between various problems, and this is best achieved by adopting a sign convention for bending moment, shear force, distance, etc. There is no standardized convention for bending, and quite a variety

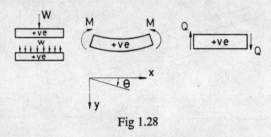

Fig 1.28

may be found in different textbooks. However, it is not important which convention one chooses and finds easiest to remember so long as one remains consistent in the use of that particular one. The sign convention* to be used in this text, without any attempt at justification, is illustrated in Fig. 1.28 for positive quantities.

1.14. SHEAR FORCE AND BENDING MOMENT DIAGRAMS

Both stresses and deflections during bending are directly related to S.F. (shear force) and B.M. (bending moment); it is therefore desirable to know the distribution along the member and hence where maximum or minimum values occur. To this end S.F. and B.M. are computed for a number of cross-sections along the beam and diagrams plotted showing the distribution and magnitude.

1.15. CANTILEVER CARRYING A CONCENTRATED LOAD AT THE FREE END

This example is illustrated in Fig. 1.29(a). The first step is to choose any section XX at a distance x from the left-hand end, cut the beam at this position and draw the free-body diagram, as in Fig. 1.29(b), inserting the required internal forces and moments in the positive sense according to

---

* Two alternative conventions are given in Appendix 1 with a summary of the various bending equations to be developed.

the sign convention. It is advisable not to mentally balance the free body and put on the forces in what seems the "right" sense.

Vertical equilibrium for the free body gives

$$W + Q_x = 0$$

where Q_x denotes the shearing force at a distance x from the left end; hence

$$Q_x = -W$$

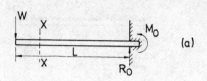

(a)

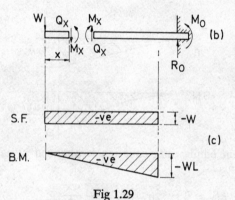

(b)

(c)

Fig 1.29

It is fairly obvious that this value of shear force is obtained at whatever distance the beam is cut between $x = 0$ and $x = L$.

Taking moments about the mid-point of section XX will give a moment equilibrium equation

$$Wx + M_x = 0 \quad \text{or} \quad M_x = -Wx$$

This is a linear variation from $M = 0$ at $x = 0$ to $M = -WL$ at $x = L$.

The reactions at the built-in end are, from the above,

$$R_0 = W \quad \text{and} \quad M_0 = -WL$$

These could of course be found at the beginning simply from considering overall equilibrium of the beam.

The S.F. and B.M. diagrams are as shown in Fig. 1.29(c).

1.16. CANTILEVER CARRYING A UNIFORMLY DISTRIBUTED LOAD

Following the above procedure, vertical equilibrium of the free body in Fig. 1.30 is obtained as

$$wx + Q_x = 0 \quad \text{so that} \quad Q_x = -wx$$

The shear-force diagram is therefore a linear variation from 0 to $-wL$ as shown.

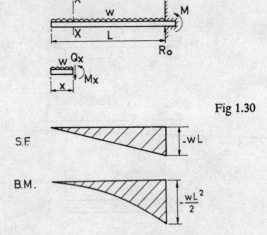

Fig 1.30

The moment equilibrium equation is obtained by taking moments about XX; thus

$$wx\frac{x}{2} + M_x = 0 \quad \text{or} \quad M_x = -\frac{wx^2}{2}$$

which gives a parabolic shape of B.M. diagram varying from $M = 0$ at $x = 0$ to $M = -wL^2/2$ at $x = L$.

The support reactions are therefore

$$R_0 = wL \quad \text{and} \quad M_0 = -\frac{wL^2}{2}$$

1.17. SIMPLY SUPPORTED BEAM CARRYING A UNIFORMLY DISTRIBUTED LOAD

In this case, Fig. 1.31, the reaction at the left-hand end has to be determined before the values of S.F. can be expressed. From vertical equilibrium and symmetry,

$$R_1 = R_2 = +\frac{wL}{2}$$

For the equilibrium of the free-body section,

$$+Q_x - \frac{wL}{2} + wx = 0$$

Shear Force

$$Q_x = -wx + \frac{wL}{2}$$

This equation shows that the S.F. is zero at mid-span and equal to the reactions, $wL/2$, at $x = 0$ and L.

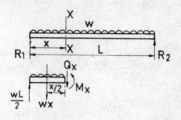

Fig 1.31

Moment equilibrium gives

$$M_x - \frac{wL}{2}x + wx\frac{x}{2} = 0$$

Bending Moment

$$M = \frac{wLx}{2} - \frac{wx^2}{2}$$

This gives a diagram which is parabolic, being zero at each support and having a maximum value of $wL^2/8$ at $x = L/2$.

1.18. TWO CONCENTRATED LOADS ON A SIMPLY SUPPORTED BEAM (Fig. 1.32)

The left reaction is first required and as the beam is statically determinate the reactions can be found from the following two equilibrium equations:

Vertical

$$-R_A - R_D + W_1 + W_2 = 0$$

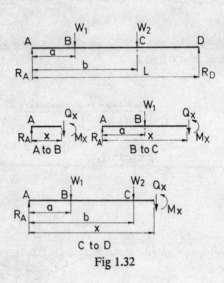

Fig 1.32

Moments

$$-R_A L + W_1(L - a) + W_2(L - b) = 0$$

$$R_A = \frac{W_1(L - a) + W_2(L - b)}{L}$$

The two concentrated loads cause mathematical discontinuities and so separate free-body diagrams and equilibrium equations have to be expressed for each different section, as shown in Fig. 1.32 and in the following equations:

Vertical Equilibrium

AB $0 < x < a$ $+Q - R_A = 0$ $Q = +R_A$

BC $a < x < b$ $+Q - R_A + W_1 = 0$

$Q = +R_A - W_1$

CD $b < x < L$ $+Q - R_A + W_1 + W_2 = 0$

$Q = +R_A - W_1 - W_2$

Moment Equilibrium

AB $0 < x < a$ $M - R_A x = 0$ $M = R_A x$

BC $a < x < b$ $M - R_A x + W_1(x - a) = 0$

$M = R_A x - W_1(x - a)$

CD $b < x < L$ $M - R_A x + W_1(x - a) + W_2(x - b)$

$M = R_A x - W_1(x - a) - W_2(x - b)$

The diagrams resulting from the above equations are shown in Fig. 1.33. It should be noted that

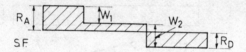

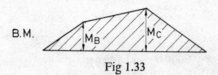

Fig 1.33

(a) The S.F. diagram *changes* in value at a support or concentrated load by the amount of the reaction or load.

(b) The B.M. is *always zero* at the ends of a beam which are either unsupported or on a simple support.

1.19. SIMPLY SUPPORTED BEAM WITH AN APPLIED COUPLE

Apart from transverse loads, bending can be caused by a couple applied to the beam at a cross-section, as shown in Fig. 1.34. The reactions at the supports are obtained from moment equilibrium of the whole beam from which $RL = \bar{M}$, or $R = \bar{M}/L$. The shear force is constant along the length of beam and of value $Q = -\bar{M}/L$.

For $x < a$ $M + Rx = 0$; hence $M = -\dfrac{\bar{M}x}{L}$

For $x > a$ $M + Rx - \bar{M} = 0$; hence $M = \bar{M} - \dfrac{\bar{M}x}{L}$

$$= \frac{\bar{M}}{L}(L - x)$$

The S.F. and B.M. diagrams were derived from these equations.

Example 1.4.   Sketch the S.F. and B.M. diagrams for the cantilever shown in Fig. 1.35.

Even though there is no loading on AB and therefore no shear force or bending moment, the distance x will still be measured from A rather than B.

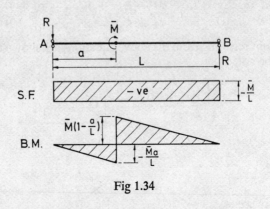

Fig 1.34

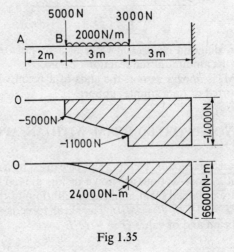

Fig 1.35

Shear Force

$$0 < x < 2 \qquad Q = 0$$

$$2 < x < 5 \qquad Q = -5000 - 2000(x - 2)$$

$$5 < x < 8 \qquad Q = -5000 - 6000 - 3000 = -14000\,\text{N}$$

Bending Moment

$$0 < x < 2 \qquad M = 0$$

$$2 < x < 5 \qquad M = -5000(x - 2) - 2000(x - 2)\frac{(x - 2)}{2}$$

$$= -1000(x - 2)(x + 3) \, \text{N-m (parabolic distribution)}$$

$$5 < x < 8 \qquad M = -5000(x - 2) - 6000(x - 3\tfrac{1}{2})$$
$$- 3000(x - 5)$$

$$= -14\,000x + 46\,000 \, \text{N-m (linear distribution)}$$

The diagrams and principal values are shown in Fig. 1.35.

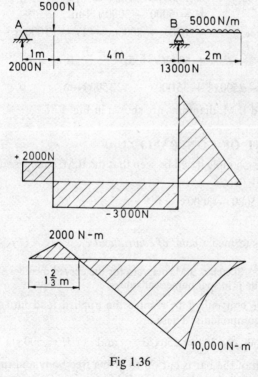

Fig 1.36

Example 1.5. Sketch the S.F. and B.M. diagrams for the simply supported beam shown in Fig. 1.36.

Equilibrium of the Whole Beam

$$-R_A - R_B + 5000 + 10\,000 = 0$$

Moments about B

$$-5R_A + (5000 \times 4) - 5000 \times 2 \times 1 = 0$$
$$R_A = 2000 \, \text{N} \qquad R_B = 13\,000 \, \text{N}$$

Shear Force

$$0 < x < 1 \qquad +Q - 2000 = 0 \qquad Q = +2000 \, \text{N}$$
$$1 < x < 5 \qquad +Q - 2000 + 5000 = 0 \qquad Q = -3000 \, \text{N}$$
$$5 < x < 7 \qquad +Q - 2000 + 5000 - 13\,000 + 5000(x - 5) = 0$$
$$Q = -5000x + 35\,000 \, \text{N}$$

Bending Moment

$$0 < x < 1 \qquad -2000x + M = 0 \qquad M = 2000x \, \text{N-m}$$
$$1 < x < 5 \qquad -2000x + 5000(x - 1) + M = 0$$
$$M = 5000 - 3000x \, \text{N-m}$$
$$5 < x < 7 \qquad -2000x + 5000(x - 1) - 13\,000(x - 5)$$
$$+ 5000(x - 5)\frac{(x - 5)}{2} + M = 0$$

$$M = -2500x^2 + 35\,000x - 122\,500 \, \text{N-m}$$

The S.F. and B.M. diagrams are shown in Fig. 1.36.

1.20. POINT OF CONTRAFLEXURE

In the above example it will be seen that the B.M. changes sign and has a zero value at

$$M = 5000 - 3000x = 0$$
$$x = 1\tfrac{2}{3} \, \text{m}$$

This point is termed a *point of contraflexure*.

Example 1.6. Sketch a B.M. diagram for the curved member shown in Fig. 1.37 giving the principal numerical values.

It is more convenient to resolve the applied load into vertical and horizontal components. These are:

$$V = 500 \cos 30° = 250\sqrt{3} \qquad \text{and} \qquad H = 500 \sin 30° = 250$$

A portion of the bar is cut off to give a free body and the forces and couple are inserted as shown. The B.M. required to balance H is

$$250 \times 2(1 - \cos \theta)$$

and to balance V,

$$-250\sqrt{3} \times 2 \sin \theta$$

Hence
$$M = 500(1 - \cos \theta) - 500\sqrt{3} \sin \theta$$

At $\theta = 0$ $M = 0$
          $\theta = \pi/2$ $M = -365 \text{ N-m}$
          $\theta = \pi$ $M = +1000 \text{ N-m}$

Also, if $M = 0 = 500(1 - \cos \theta - \sqrt{3} \sin \theta)$, then

$$\cos \theta = -\tfrac{1}{2} \quad \text{and} \quad \theta = 120°$$

which gives a point of contraflexure.

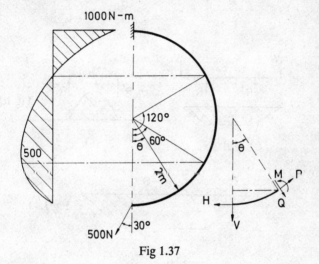

1000 N-m

120°

60°

θ

500

2m

θ

M

H

Q

V

500N   30°

Fig 1.37

When $dM/d\theta = 0$,
$$\sin \theta - \sqrt{3} \cos \theta = 0 \qquad \theta = 60°$$

$$M_{60°} = 500 \left(1 - \tfrac{1}{2} - \sqrt{3} \frac{\sqrt{3}}{2}\right) = -500 \text{ N-m}$$

The B.M. diagram is plotted for convenience along the vertical axis as shown in Fig. 1.37.

1.21. THE PRINCIPLE OF SUPERPOSITION

Consider the beam problem in Fig. 1.38. Although this would not be difficult to analyse for B.M. in its present form, imagine it split up into the two separate cases of Figs. 1.39 (*a*) and (*b*). These have effectively been dealt with in Sections 1.15 and 1.17 respectively, and the B.M. diagrams for the two parts are as shown. Now, bending moment is always a first-order function of applied load, i.e. there are never terms in W^2 or w^2, etc.; therefore at any section along the beam the sum of the bending moments due to the loads acting separately would be exactly the same as the bending moment due to the combined load. Similarly the reactions at the supports due to the separate loads may be summed to give the reactions caused by the combined loading. This very useful technique is

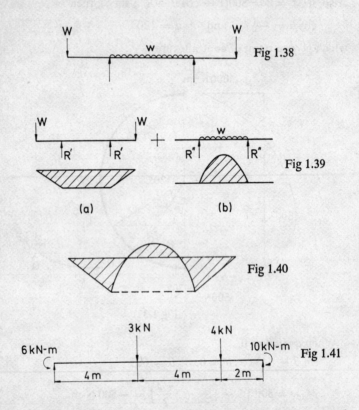

Fig 1.38

Fig 1.39

(a)   (b)

Fig 1.40

6 kN-m   3 kN   4 kN   10 kN-m

4 m   4 m   2 m

Fig 1.41

an application of the *principle of superposition* and will be demonstrated at regular stages throughout this text when analysing force, stress, strain, displacement, etc.

The solution of the present example is obtained by summing the ordinates of bending moment at every section along the beam. However, this may be done more simply graphically than algebraically by placing the two B.M. diagrams together as shown in Fig. 1.40 and the resultant is obtained as the shaded area.

Example 1.7. Use the principle of superposition to determine the B.M. diagram for the beam loaded as shown in Fig. 1.41.

The problem is split into the three separate cases in Figs. 1.42(a), (b) and (c) and the B.M. diagrams can be drawn by inspection from previous experience. The positive diagrams at (a) and (b) are first combined by simply adding the ordinates where the loads are applied and joining up with straight lines. The negative diagram at (c) is then inverted and imposed on the one above, the resultant being obtained from the shaded overlaps at (d).

1.22. RELATIONSHIPS BETWEEN LOADING, SHEAR FORCE AND BENDING MOMENT

Consider the free body, Fig. 1.43, of a small slice of beam of length dx carrying uniform loading w per unit length.

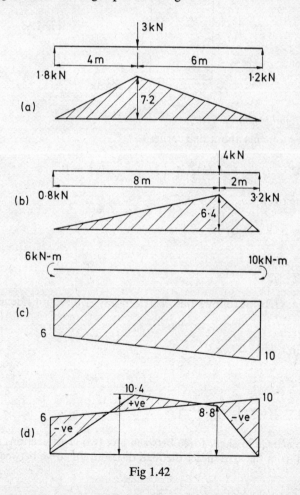

Fig 1.42

Vertical Equilibrium

$$-Q + w\,dx + \left(Q + \frac{dQ}{dx}\,dx\right) = 0$$

$$w + \frac{dQ}{dx} = 0$$

$$w = -\frac{dQ}{dx} \tag{1.14}$$

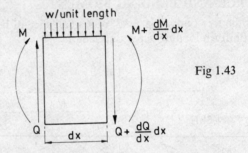

Fig 1.43

Moment Equilibrium

Taking moments about one corner,

$$-M - Q\,dx + w\,dx\frac{dx}{2} + \left(M + \frac{dM}{dx}\,dx\right) = 0$$

Neglecting $(dx)^2$,

$$-Q + \frac{dM}{dx} = 0$$

$$Q = \frac{dM}{dx} \tag{1.15}$$

From eqn. (1.14) it follows that, between any two sections denoted by 1 and 2,

$$\int_1^2 dQ = \int_1^2 - w\,dx$$

or

$$Q_2 - Q_1 = \int_1^2 - w\,dx + A \tag{1.16}$$

Thus the *change* in shear force between any two cross-sections may be obtained from the area under the load distribution curve between those sections.

From eqn. (1.15),

$$\int_1^2 dM = \int_1^2 Q\,dx$$

or

$$M_2 - M_1 = \int_1^2 Q\,dx + B \tag{1.17}$$

Thus the *change* in bending moment between any two sections is found from the area under the shear force diagram between those sections.
Also from eqns. (1.14) and (1.15),

$$w = -\frac{d^2M}{dx^2} \tag{1.18}$$

When $Q = 0$, $dM/dx = 0$, i.e. the S.F. is zero when the *slope* of a B.M. diagram is zero and not simply when M is a maximum, although this also is implied. It is possible to have $M = M_{max}$ when $dM/dx \neq 0$ (see Fig. 1.36).

Example 1.8. Use the relationships developed in Section 1.22 to find the position of zero shear force and the value of the maximum bending moment

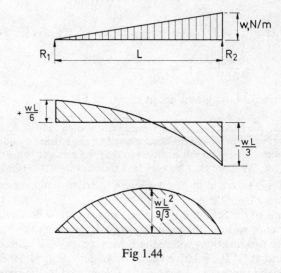

Fig 1.44

for the simply supported beam shown in Fig. 1.44. It carries a distributed load which varies linearly in intensity from zero at the left to w newtons per metre at the right-hand end. Plot the S.F. and B.M. diagrams.

By similar triangles, the load intensity at x from the left end is $(w/L)x$
From eqn. (1.16) the shear force is given by

$$Q = -\int \frac{wx}{L}\,dx + A = -\frac{wx^2}{2L} + A$$

and from eqn. (1.17),

$$M = \int Q\,dx + B = -\frac{wx^3}{6L} + Ax + B$$

Now, $M = 0$ when $x = 0$ and L. Hence

$$B = 0 \quad \text{and} \quad A = \frac{wL}{6} \quad (=R_1)$$

Thus

$$Q = -\frac{wx^2}{2L} + \frac{wL}{6}$$

and when $Q = 0$, $x = L/\sqrt{3}$.

Now,

$$M = -\frac{wx^3}{6L} + \frac{wLx}{6}$$

Since $dM/dx = Q = 0$ is a possible solution in this case for M_{max},

$$M_{max} = -\frac{w}{6L}\left(\frac{L}{\sqrt{3}}\right)^3 + \frac{wL}{6}\frac{L}{\sqrt{3}} = \frac{wL^2}{9\sqrt{3}}$$

The S.F. and B.M. diagrams are as shown in Fig. 1.44.

Example 1.9. A beam carries a uniformly distributed load of 20 kN/m over
the full length of 5 m. It is simply supported at the left-hand end and some
other position. Determine this position so that the greatest value of bending
moment along the beam may be kept to a minimum.

The beam is sketched in Fig. 1.45 with the right-hand support placed
at a distance z from the end.

The forms of the S.F. and B.M. diagrams may now be sketched from
past experience without the need for any calculation at this stage.

There are two positions where the B.M. reaches peak values, between
D and E and at E. The relative values of M_1 and M_2 will depend on the
position of the right-hand support. As E is moved to the right M_1 gets
larger and M_2 smaller, and when moved to the left M_2 increases and M_1
decreases. The optimum position for the least values of M_1 and M_2 is
therefore when $M_1 = M_2$.

At this stage it is useful to remember that, from eqn. (1.17),

$M_1 = $ area A of S.F. diagram

and

$M_2 = $ (area A − area B) of S.F. diagram = area C

Before one can use the above information it is necessary to calculate the reaction R_1. By moments about E,

$$-R_1(5 - z) + 20\frac{(5 - z)^2}{2} - \frac{20z^2}{2} = 0$$

$$R_1 = \frac{250 - 100z}{5 - z}$$

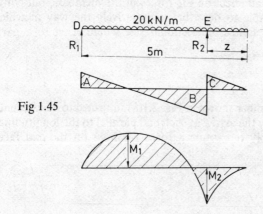

Fig 1.45

Now, triangles A, B and C are all similar, and if area $A =$ area C then the heights of these two triangles must be the same; therefore

$$\frac{250 - 100z}{5 - z} = 20z$$

$$250 - 100z = 100z - 20z^2$$

$$2z^2 - 20z + 25 = 0$$

from which

$$z = 1 \cdot 465 \, \text{m}$$

For completeness the reactions may now be calculated:

$R_1 = 29 \cdot 3$ $R_2 = 70 \cdot 7 \, \text{kN}$

and

$M_1 = M_2 = 21 \cdot 5 \, \text{kN-m}$

Chapter 2
Statically Determinate Stress Systems

2.1. The effects of external applied forces can now be measured in terms of the internal reacting forces in a solid body or the members of a framework, as described in the previous chapter. However, at that stage no mention was made of the *cross-sectional size and shape* of the members. This aspect had no effect on the forces in the members, but conversely one should be able to describe quantitatively the way in which two members of different cross-sectional size would react to a particular value of force.

2.2. STRESS

Consider the member shown in Fig. 2.1(*a*) subjected to an external force, *F*, represented by the arrow at each end parallel to the longitudinal axis. The arrow simply represents a force *resultant* on the end faces and

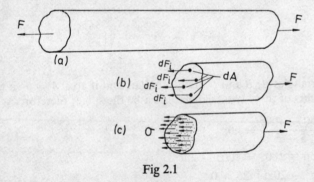

Fig 2.1

obviously the force is not actually applied solely along the line of the arrow. Similarly the internal force reaction does not act along just a single line but is transmitted throughout the bulk of material from grain to grain. If part of the member is cut off to give a free body as in Fig. 2.1(*b*) then equilibrium will be maintained by appropriate components of internal reacting force such as dF_i acting on elements of area dA.

For equilibrium $\Sigma dF_i = F$ and $\Sigma dA = A$ the total cross-sectional area.

The internal *force per unit area* is

$$\text{Lt } dA \to 0 \quad \frac{dF_i}{dA}$$

and this is termed *stress*. In general stress varies in magnitude and direction, so that dF_i/dA is in fact the stress at a point.

(i) Direct Stress

In the simple case in Fig. 2.1 the *average* direct stress is F/A and will be denoted by the symbol σ as in Fig. 2.1(*c*). Direct or normal stress acts perpendicular to a plane and when acting outwards from the plane is termed *tensile stress* and given a *positive* sign. Stress acting towards a plane is termed *compressive stress* and is *negative* in sign. In order to denote the direction of a stress with respect to co-ordinate axes, a suffix notation is used, so that σ_x, σ_y, σ_z represent the components of direct or normal stress in the x, y and z-directions as shown in Fig. 2.2(*a*).

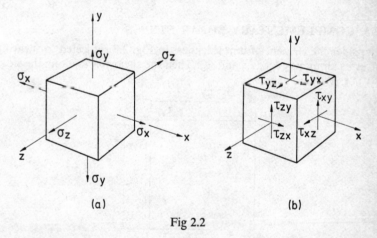

Fig 2.2

(ii) Shear Stress

When a person who is running wishes to slow down he applies his "brakes". This is achieved through mounting pressure between the soles of his shoes and the ground and thus increased frictional force parallel to the ground. This concept of a force applied tangential or parallel to a surface is termed a shear force (see also Section 1.11). If internal reacting shear force is expressed as a force per unit area then it is termed a *shear stress*. It also acts parallel to any associated plane within the material and is denoted by the symbol τ. A double suffix notation is required to define shear stresses with respect to co-ordinate axes. The first suffix gives the direction of the normal to the plane on which the stress is acting, and the second suffix indicates the direction of the shear stress component. Thus τ_{xy} is a shear stress acting on the yz-plane (the normal in the x direction) and pointing in the y direction. The sign convention associated with shear stress is defined in Chapter 13 at which stage it is of importance. There are twelve possible shear stress components in a three-dimensional stress system as indicated in Fig. 2.2(b) (those on the obscured faces have been omitted for clarity) which reduce to six independent components as explained later.

(iii) Hydrostatic Stress

This is a special state of direct stress which should be mentioned now although its importance will become more evident at later stages. Hydrostatic stress may be represented by the stress set up in a body immersed at a great depth in a fluid. The external applied pressure being equal at all points round the body gives rise to internal reacting compressive force and hence compressive stress equal in all directions, i.e. $\sigma_x = \sigma_y = \sigma_z = \sigma$.

2.3. COMPLEMENTARY SHEAR STRESS

Consider the element of unit thickness in Fig. 2.3 subjected to shearing stresses along its edge, τ_{xy} and τ_{yx}. Then the shear *forces* along the sides

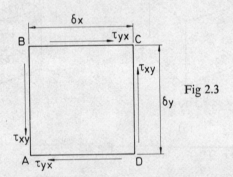

Fig 2.3

AB and CD are $(\tau_{xy} \times 1 \times dy)$ and along AD and BC are $(\tau_{yx} \times 1 \times dx)$. Now, to maintain rotational equilibrium of the element, the above *forces* must balance out, which may be expressed by taking moments about a z-axis through the centre of the element:

$$2(\tau_{xy} \times 1 \times dy) \times \frac{dx}{2} - 2(\tau_{yx} \times 1 \times dx) \times \frac{dy}{2} = 0$$

or

$$\tau_{xy} = \tau_{yx}$$

These are termed *complementary shear stresses*.

Thus a shear stress on one plane is *always* accompanied by a complementary shear stress of the same sign and magnitude on a perpendicular plane.

2.4. A STATICALLY DETERMINATE STRESS SYSTEM

If the stresses within a body can be calculated purely from the conditions of equilibrium of the applied loading and the internal forces then the problem is said to be statically determinate. There are very few examples of this nature; however, they do give further illustration of the application of equilibrium, and the remainder of the chapter will be devoted to solutions in this category.

2.5. ASSUMPTIONS AND APPROXIMATIONS

Exact solutions for stress, displacement, etc., in real engineering problems are not always mathematically possible and even those that are possible can involve lengthy computation and advanced mathematical techniques which are not necessarily justifiable. This is because we seldom know the exact conditions of applied loading on a component or structure for its expected working life, and the materials used are not wholly predictable in behaviour. It therefore becomes necessary and desirable in most engineering problems to make some simplifying approximations and assumptions which, while not changing the basic nature of the problem, will allow a simpler solution and an answer which is not too far from the truth. It is important, however, that any assumptions or approximations are clearly stated at the start so that the reader may assess the validity of the answer in respect of what might be the exact solution.

Some of the problems to follow are in general not statically determinate, but with some realistic geometrical limitations, they can be solved purely from equilibrium conditions to give answers which although not exact are reasonably accurate.

2.6. TIE BAR AND STRUT OR COLUMN

These are the simplest examples of statically determinate stress situations, since the equilibrium condition is simply that the external force at the ends of the member must be balanced by the internal force, which is the average stress multiplied by the cross-sectional area.

The two cases are illustrated in Fig. 2.4. For the bar at (*a*), subjected to tension,

$$\sigma_x A = F \quad \text{or} \quad \sigma_x = \frac{F}{A} \text{ (tensile stress)}$$

and for the bar at (b), under compressive force,

$$\sigma_x A = -F \quad \text{or} \quad \sigma_x = -\frac{F}{A} \text{ (compressive stress)}$$

In the diagrams the bars have been cut at any arbitrary section to show the appropriate stresses on each part.

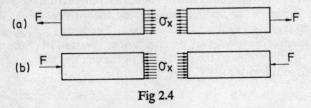

Fig 2.4

Example 2.1. A double-acting hydraulic cylinder has a piston 250 mm in diameter and a piston rod 75 mm in diameter. The water pressure is 7 MN/m² on one side of the piston and 300 kN/m² on the other side, and on the return stroke the pressures are interchanged. Determine the maximum stress in the rod.

The two loading situations are shown in Fig. 2.5.

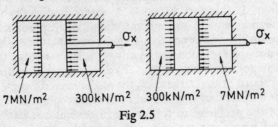

Fig 2.5

The first step is to calculate the areas on which pressures and stresses are acting.

$$\text{Full piston area} = \frac{\pi}{4} \times 0.25^2 = 0.0491 \text{ m}^2$$

$$\text{Rod area} = \frac{\pi}{4} \times 0.075^2 = 0.0044 \text{ m}^2$$

Reduced piston area $= 0.0447 \text{ m}^2$

In the first case the equilibrium equation is

$$0.0044\sigma_x + (0.0491 \times 7) - (0.0447 \times 0.3) = 0$$

Hence

$$\sigma_x = -75 \text{ MN/m}^2$$

On the other stroke the equilibrium condition is

$$0{\cdot}0044\sigma_x + (0{\cdot}0491 \times 0{\cdot}3) - (0{\cdot}0447 \times 7) = 0$$

and

$$\sigma_x = +67{\cdot}8 \, \text{MN/m}^2$$

Therefore the maximum stress set up in the rod is $75 \, \text{MN/m}^2$ in compression.

2.7. THIN RING OR CYLINDER ROTATING

If a cylinder or ring is rotating at constant velocity then an inward radial component of force is required to provide the centripetal acceleration on each element of material. This inward force may be resolved into the tangential direction at each end of a typical element as shown in Fig. 2.6(b), which is thus subjected to circumferential tensile stress. This may

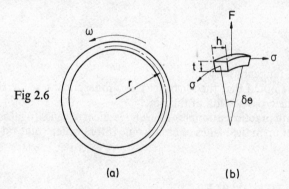

Fig 2.6

(a)   (b)

be determined from a consideration of dynamical equilibrium. Alternatively we can reduce this dynamical situation to a static one by applying D'Alembert's principle, in which forces to accelerate masses are replaced by equal and opposite static forces. In this case the centripetal force is replaced by, what is often termed, a centrifugal force acting radially outwards, and we can then apply statical equilibrium.

Consider the rim of a wheel (no spokes) or a slice through a thin-walled cylinder each rotating about a central axis at velocity ω. If this problem is to be statically determinate then the diameter of the ring or cylinder must be large, say >10 times the cross-sectional dimensions of the rim. It is then possible to assume a uniform distribution of stress over the cross-section in the circumferential direction, and in the radial and axial directions the stresses can be taken as zero or negligible.

A small element of arc of the ring of cross-sectional area A $(=th)$ rotating at uniform velocity, ω, is shown in Fig. 2.6(b). The forces acting on the element are the radial inertia force, F, and the circumferential tensile stress, σ, acting over the area, A. The resolved component of the radial forces inwards is

$$2\sigma th \sin \frac{\delta\theta}{2} \simeq 2\sigma th \frac{\delta\theta}{2} \text{ for small values of } \delta\theta$$

Mass of element $= \rho th r \, \delta\theta$

where ρ is the mass per unit volume.

Radial centrifugal force, $F = \rho th r \, \delta\theta \, \omega^2 r$

For radial equilibrium,

$$2\sigma th \frac{\delta\theta}{2} - F = 0$$

$$\sigma th \, \delta\theta - \rho th r^2 \omega^2 \, \delta\theta = 0$$

and

$$\sigma = \rho\omega^2 r^2$$
$$= \rho v^2 \tag{2.1}$$

where v is the tangential velocity.

It should be noted that the tensile stress is independent of the shape and area of the cross-section of the rim.

An important practical example of the above effect of centrifugal action is the stress set up in the blades of gas turbine rotors which rotate at very high speed.

2.8. SUSPENDED CABLES

(i) A common form of loading on a cable, for example in suspension bridges, is shown in Fig. 2.7(a). The loading, w, per unit length, is distributed uniformly on a horizontal base, the weight of the cable being neglected. In this particular example, the ends A and B are set at different heights above the lowest point. It is useful for the analysis to cut the cable at O, to insert a reaction at that point and to consider the equilibrium of the right-hand part of the cable. The free-body diagram in Fig. 2.7(b) shows that equilibrium is satisfied by the triangle of forces T_B, T_0 and wx_1. The position of the lowest point O and hence the distance x_1 is not known. The distance x_1 can be determined by equilibrium of moments of the forces for either part of the cable. Thus, taking moments about B,

$$T_0 y_1 - wx_1 \frac{x_1}{2} = 0 \tag{2.2}$$

or, for the left-hand part, moments about A give

$$T_0 y_2 + w(l - x_1)\frac{(l - x_1)}{2} = 0 \qquad (2.3)$$

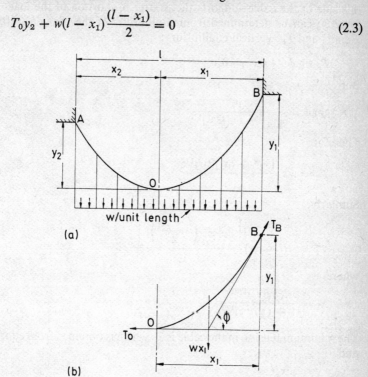

Fig 2.7

Inspection of eqn. (2.2) shows that, for any point on the cable having co-ordinates (x, y) relative to O,

$$y = \frac{wx^2}{2T_0} \qquad (2.4)$$

which is an equation for a parabola.

Returning to eqns. (2.2) and (2.3) and eliminating T_0 gives

$$\frac{y_1}{y_2} = \frac{x_1{}^2}{(l - x_1)^2}$$

from which

$$x_1 = \frac{l(y_1/y_2)^{1/2}}{1 + (y_1/y_2)^{1/2}}$$

If the ends A and B are at the same level then $y_2 = y_1$ and $x_1 = l/2$.

The maximum tension in the cable will naturally occur at end B since this side of the cable supports the greater proportion of the load. The force T_B can be determined from equilibrium of the triangle of forces T_0, wx_1 and T_B. Vertical equilibrium:

$$T_B \sin \phi - wx_1 = 0$$

But

$$\sin \phi = \frac{y_1}{[y_1{}^2 + (x_1/2)^2]^{1/2}}$$

Therefore

$$T_B = wx_1 \frac{[y_1{}^2 + (x_1/2)^2]^{1/2}}{y_1} \tag{2.5}$$

Similarly,

$$T_A = w(l - x_1) \frac{\left[y_2{}^2 + \left(\frac{l - x_1}{2}\right)^2\right]^{1/2}}{y_2} \tag{2.6}$$

where

$$x_1 = \frac{l(y_1/y_2)^{1/2}}{1 + (y_1/y_2)^{1/2}}$$

The minimum tension in the cable is T_0 which is obtained from eqn. (2.2) and

$$T_0 = \frac{wx_1{}^2}{2y_1} \tag{2.7}$$

Any required stress values are determined by dividing the appropriate tension by the cross-sectional area, a, of the cable, since it is assumed that the stress is uniformly distributed across the section and thus $\sigma = T/a$.

(ii) The alternative case to the one above is that in which a wire is freely suspended between two fixed points A and B, and is subjected to uniform loading, w, per unit length along the wire, which may be simply the weight of the wire itself or some additional external load such as snow. The action of the loading on the wire is such as to cause tension to be set up as the internal reaction. The wire cannot sustain any bending, so that the axis of the wire must at all points coincide with the direction of the resulting tension. Referring to the free-body diagram, Fig. 2.8, the horizontal tension at the lowest point O is T_0 and the tension at a point C at a distance s *along* the wire from O is T.

Horizontal equilibrium

$$T \cos \phi = T_0 \tag{2.8}$$

Vertical equilibrium

$$T \sin \phi = ws \qquad (2.9)$$

Eliminating the slope of the wire, ϕ, at C gives

$$\frac{T_0^2}{T^2} + \frac{w^2 s^2}{T^2} = 1$$

or

$$T^2 = T_0^2 + w^2 s^2 \qquad (2.10)$$

which is the complete equilibrium equation for the wire.

w/unit length

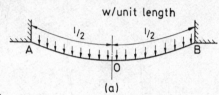

Fig 2.8

(a)

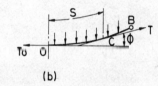

(b)

Eliminating T from eqns. (2.8) and (2.9), an equation for the shape of the wire is obtained:

$$s = \frac{T_0}{w} \tan \phi \qquad (2.11)$$

which describes the common catenary.

It is apparent from eqn. (2.10) that the maximum tension occurs at the ends A and B, and if l is the length of the wire then $T = T_{max}$ at $s = l/2$ and $T_{max} = [T_0^2 + (w^2 l^2/4)]^{1/2}$. Also if a is the cross-sectional area of the wire then

$$\sigma_{max} = \frac{T_{max}}{a} = \frac{1}{a}[T_0^2 + (w^2 l^2/4)]^{1/2} \qquad (2.12)$$

If T_0, which is also equal to the horizontal component of the tension at the ends of the wire, is not known, but the slope ϕ_A or ϕ_B is defined then T_{max} can be found from eqn. (2.9). Thus

$$T_{max} = \frac{wl}{2 \sin \phi_A}$$

2.9. STRESSES IN THIN SHELLS DUE TO PRESSURE OR SELF-WEIGHT

Thin-walled shells are used extensively in engineering in two principal forms: (a) made of reinforced concrete to form part of a civil engineering structure and stressed partly by self-weight and partly by the environment, namely wind, snow, etc.; (b) made of metal and used generally in engineering as storage containers for liquid, powder, gas, etc. Stresses

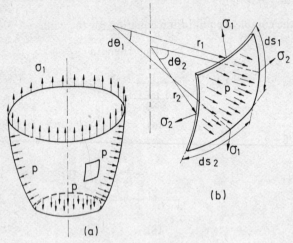

(a)

(b)

Fig 2.9

will arise due to, say, uniform internal liquid or gas pressure, e.g. in a steam boiler, or pressure due to weight of liquids or solids contained.

In general, a shell of arbitrary wall thickness subjected to pressure contains a three-dimensional stress system. The stresses are perpendicular to the thickness and in two principal orthogonal directions tangential to the surface geometry. Each of these stresses has a non-uniform distribution through the thickness of material, and the solution is statically indeterminate. If, however, the wall thickness is less than about $\frac{1}{10}$ of the principal radii of curvature of the shell, the variation of tangential stresses through the wall thickness is small and the radial stress may be neglected. The solution can then be treated as statically determinate.

In the present treatment the following additional simplifying assumptions will be made: that the shell acts as a membrane and does not provide bending resistance so that there are only uniform direct stresses present, which are often called membrane stresses; that the shell is formed by a surface of revolution and there are no discontinuities or sharp bends.

Consider first a general axi-symmetrical shell, Fig. 2.9(a) from which is cut an element bounded by two meridional lines and two lines perpendicular to the meridians as at (b). The notation is as follows:

σ_1 = tensile stress in meridional direction, meridional stress
σ_2 = tensile stress along a parallel circle, hoop stress
r_1 = meridional radius of curvature
r_2 = radius of curvature perpendicular to the meridian
t = wall thickness

The forces on the sides of the element are in the meridional direction $\sigma_1 t \, ds_2$ and in the perpendicular direction $\sigma_2 t \, ds_1$. These forces have components inwards towards their centres of curvature. The radial components are $2\sigma_1 t \, ds_2 \sin (d\theta_1/2)$ and $2\sigma_2 t \, ds_1 \sin (d\theta_2/2)$. For small values of $d\theta$ the total radial force becomes

$$\sigma_1 t \, ds_2 \, d\theta_1 + \sigma_2 t \, ds_1 \, d\theta_2 = \sigma_1 t \, ds_2 \frac{ds_1}{r_1} + \sigma_2 t \, ds_1 \frac{ds_2}{r_2}$$

The radial force due to the pressure, p, is

$$p \, ds_1 \, ds_2$$

Thus, for equilibrium of the element,

$$\sigma_1 t \, ds_2 \frac{ds_1}{r_1} + \sigma_2 t \, ds_1 \frac{ds_2}{r_2} = p \, ds_1 \, ds_2$$

and

$$\frac{\sigma_1}{r_1} + \frac{\sigma_2}{r_2} = \frac{p}{t} \tag{2.13}$$

In general σ_1 and σ_2 are both different and unknown and so another equilibrium equation has to be formulated relevant to the particular problem, in addition to eqn. (2.13), in order to solve for the stresses.

(i) Thin Sphere

The simplest case to which eqn. (2.13) may be applied is the sphere. The symmetry about any axis implies that $\sigma_1 = \sigma_2 = \sigma$, and of course $r_1 = r_2 = r$; therefore the circumferential stress in any direction is $\sigma = pr/2t$.

(ii) Thin Cylinder

This particular problem will be considered from first principles rather than using eqn. (2.13) in order to demonstrate the further use of free body diagrams.

Axial Equilibrium. The force acting on each closed end of the cylinder owing to the internal pressure, p, Fig. 2.10(b), is obtained from the product of the pressure and the area on which it acts. Thus

Axial force $= pr^2\pi$

The part of the vessel shown in the free-body diagram, Fig. 2.10(b), is in axial equilibrium simply under the action of the applied force above and the axial stress, σ_z, in the material, the radial pressure shown having

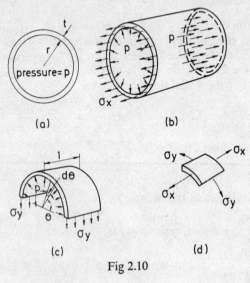

(a)   (b)

(c)   (d)

Fig 2.10

no axial resultant force. The cross-sectional area of material is approximately $2\pi rt$, and therefore the internal force is $\sigma_z \times 2\pi rt$, and for equilibrium,

$$2\pi rt\sigma_z = \pi r^2 p \qquad\qquad (2.14)$$

or

$$\sigma_z = \frac{pr}{2t}$$

Circumferential Equilibrium. Considering the equilibrium of part of the vessel, if the cylinder is cut across a diameter as in the free-body diagram in Fig. 2.10(c) the internal pressure acting outwards must be in equilibrium with the circumferential stress, σ_y, as shown. Consider the unit length of cylinder, and the small length of shell subtending an angle $d\theta$ shown in the diagram. The radial component of force on the element is $p \times 1 \times r\,d\theta$; hence the vertical component is $p \times 1 \times r\,d\theta \sin \theta$. Therefore the total vertical force due to pressure is

$$\int_0^\pi pr \sin \theta \, d\theta = 2pr$$

It is useful to note that the vertical force can also be found by considering the pressure acting on the projected area at the diameter. This fact also shows that the axial force is independent of the shape of the end closures.

The internal force required for equilibrium is obtained from the stress σ_y acting on the two ends of the strip of shell. Hence the internal force is $\sigma_y \times 2 \times t \times 1$. For equilibrium, $2t\sigma_y = 2rp$, or

$$\sigma_y = \frac{pr}{t} \qquad (2.15)$$

Comparing eqns. (2.14) and (2.15) it is seen that the circumferential stress is twice the axial stress. Fig. 2.10(d) shows a small element of the shell subjected to the axial and circumferential (hoop) stresses.

Example 2.2.   A concrete dome is 250 mm thick, has a radius of 30 m and subtends an angle of 120° at the support ring. Calculate the stresses at the support due to self-weight. The density for concrete is $2 \cdot 3 \, \text{Mg/m}^3$. The dome is illustrated in Fig. 2.11.

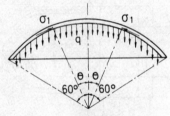

Fig 2.11

Consider the vertical equilibrium of a circular segment containing an angle 2θ; then if q is the force per unit area due to weight of concrete,

$$(2\pi r \sin \theta \times t\sigma_1 \sin \theta) + 2\pi r(r - r \cos \theta)q = 0$$

$$\sigma_1 = -\frac{qr}{t} \frac{1 - \cos \theta}{\sin^2 \theta} = -\frac{qr}{t} \frac{1}{1 + \cos \theta}$$

Since self-weight, unlike applied pressure, does not act radially everywhere, eqn. (2.13) has to be modified to take account of the radially resolved component of weight, giving

$$\frac{\sigma_1}{r_1} + \frac{\sigma_2}{r_2} = -\frac{q}{t} \cos \theta$$

Substituting for σ_1,

$$\sigma_2 = \frac{qr}{t} \left[\frac{1}{1 + \cos \theta} - \cos \theta \right]$$

The stresses at the supports are obtained by putting $\theta = 60°$ and $q = 2\cdot3 \times 9\cdot81 \times 250/1000 = 5\cdot65\,\text{kN/m}^2$.

$$\sigma_1 = -\frac{5\cdot65 \times 30}{0\cdot25} \times \frac{1}{1\frac{1}{2}} = -452\,\text{kN/m}^2$$

$$\sigma_2 = \frac{5\cdot65 \times 30}{0\cdot25}\,(\tfrac{2}{3} - \tfrac{1}{2}) = 113\,\text{kN/m}^2$$

2.10. PROBLEMS IN SHEAR

(i) Pinned Connection

A forked pin joint in, say, a tow bar is illustrated in Fig. 2.12. Tensile force P is transmitted across the joint being carried by shearing action on sections XX and YY of the pin.

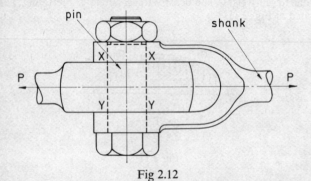

Fig 2.12

Assuming that there is a uniform shear stress τ over the cross-sectional area A of the pin, then

Total internal shear force $= 2\tau A$

For equilibrium,

$$P = 2\tau A$$

so the shear stress at the two sections of the pin is

$$\tau = \frac{P}{2A} \tag{2.16}$$

(ii) Pure Torsion of a Thin Circular Tube

In general the twisting or torsion of solid and hollow circular section members is statically indeterminate and is discussed fully in Chapter 8. However, as with thin-walled shells, if the radius of a circular tube is large (say >10 times) compared with the wall thickness then the stress can reasonably be taken as uniform through the thickness and the problem can be treated as statically determinate.

In Fig. 2.13 a tube of mean radius r and wall thickness t is subjected to opposing torques at each end. The deformation action is that of twisting, and considering the equilibrium of part of the tube, by cutting it at some section away from the end effects and perpendicular to the axis, an internal reacting torque will be required for equilibrium and to maintain the

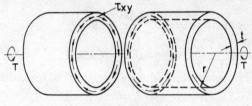

Fig 2.13

twist. This reaction torque will take the form of a "uniform" shearing stress acting on the cut face shown. The shear stress cannot vary around the tube or else this would imply a variation in the complementary longitudinal shear stress and thus a resultant axial force.

Internal torque $= \tau \times 2\pi rt \times r$

For equilibrium,

$$T = 2\pi r^2 t$$

$$\tau = \frac{T}{2\pi r^2 t} \tag{2.17}$$

Example 2.3. A closed-ended cylindrical steel pressure vessel is 2 m in internal diameter. It is made up from 10 mm thick steel sheet formed into two semi-cylinders, which are riveted along longitudinal single-lap seams with 12 mm diameter rivets at 36 mm pitch. The dished ends are also single-lap riveted on to the cylinder at each end and the circumferential joints are made with 8 mm diameter rivets at 30 mm pitch. If the maximum shear stress in any rivet is not to exceed 60 MN/m² calculate the maximum allowable internal pressure and the resulting longitudinal and circumferential stresses in the wall of the cylinder.

Longitudinal Seam

Maximum shear force per rivet $= 60 \times 10^6 \times \dfrac{\pi}{4} \times \dfrac{12^2}{10^6} = 6 \cdot 8\,\text{kN}$

Separating force at joints due pressure over one pitch length
$= 0 \cdot 036 \times 2 \times p$

The equilibrium condition is

$$0.036 \times 2 \times p = 2 \times 6.8$$

so that

$$p = 189 \, \text{kN/m}^2$$

Circumferential Seam

$$\text{Maximum shear force per rivet} = 60 \times 10^6 \times \frac{\pi}{4} \times \frac{8^2}{10^6} = 3.02 \, \text{kN}$$

$$\text{Number of rivets} = \pi \times \frac{2000}{30} = 209$$

$$\text{Total force in shear allowable} = 630 \, \text{kN}$$

This must be in equilibrium with the longitudinal force due to pressure, which is

$$\frac{\pi}{4} \times 2^2 \times p$$

Hence

$$\pi p = 630 \qquad \text{and} \qquad p = 201 \, \text{kN/m}^2$$

Therefore the maximum allowable pressure is $189 \, \text{kN/m}^2$.

$$\text{Circumferential stress, } \sigma_1 = \frac{pr}{t} = \frac{189 \times 1}{0.01} = 18.9 \, \text{MN/m}^2$$

$$\text{Axial stress, } \sigma_2 = \frac{18.9}{2} = 9.45 \, \text{MN/m}^2$$

Example 2.4. It is required to make a large concrete foundation block which, when supporting a compressive load together with the self-weight

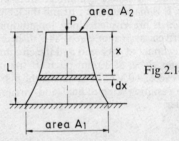

Fig 2.14

will have the same compressive stress at all cross-sections. Determine a suitable profile.

A cross-section through the proposed block is shown in Fig. 2.14.

Let A be the cross-sectional area at the top of the strip and $A + dA$

the area at the base. Since the strip is in equilibrium, then equating upward and downward forces at DE,

Compressive stress × area of section at DE
= (compressive stress × area of section at BC)
+ weight of element BCED

$$\sigma\,(A + dA) = \sigma A + wA\,dx$$

Therefore

$$\frac{dA}{A} = \frac{w}{\sigma}\,dx$$

where w is the weight of unit volume of the block. Integrating each side,

$$\int_{A_2}^{A}\frac{dA}{A} = \int_{0}^{x}\frac{w}{\sigma}dx$$

Therefore

$$\log_{e}\frac{A}{A_2} = \frac{w}{\sigma}x$$

and

$$A = A_2\,e^{(w/\sigma)x}$$

where e is the base of the Napierian logarithms (=2·718). Since

$$A_2 = \frac{P}{\sigma}$$

where P is the applied load, then

$$A = \frac{P}{\sigma}\,e^{(w/\sigma)x}$$

The area of the cross-section A_1 is found by putting $x = L$. Therefore

$$A_1 = \frac{P}{\sigma}\,e^{(w/\sigma)L}$$

Chapter 3
Stress–Strain Relations

3.1. As explained previously a statically indeterminate problem cannot be solved from the conditions of equilibrium alone; additional equations are required to find all the unknowns. These equations are obtained by studying the *geometry of deformation* of the component or structure and the *load–deformation* or *stress–strain relationship* for the material. These topics are dealt with in this chapter, but the detailed application in worked examples is held over to Chapter 5.

3.2. DEFORMATION

Deformations may occur in a material for a number of reasons, such as external applied loads, change in temperature, tightening of bolts, irradiation effects, etc. Bending, twisting, compression, torsion and shear or combinations of these are common modes of deformation. In some materials, e.g. rubber, plastics, wood, the deformations are quite large for relatively small loads, and readily observable by eye. In metals, however, the same loads would produce very small deformations requiring the use of sensitive instruments for measurement.

Stress values do not always provide the limiting factor in design, for although a component may be safe and employ material economically with regard to stress, the deformations accompanying that stress might be dangerous or inconvenient. For example, too high a deflection of an aeroplane wing can result, among other things, in a detrimental change in aerodynamic characteristics. A lathe bed which was not sufficiently rigid would not permit of the required tolerances in machining. A perfectly safe sag in a dance-hall floor might upset the poise of the dancers.

In this and succeeding chapters there will be many problems in which the analysis of displacements will be considered specifically in addition to the determination of stress magnitude.

3.3. STRAIN

As explained in Chapter 2 the effect of a force applied to bodies of different size can be compared in terms of stress, i.e. the force per unit area. Likewise the deformation of different bodies subjected to a particular load is a function of size, and therefore comparisons are made by expressing deformation as a non-dimensional quantity given by the change in dimension per unit of original dimension, or in the case of shear as a change in angle between two initially perpendicular planes. The non-dimensional expression of deformation is termed *strain*.

Direct or Normal Strain

Consider the bar shown in Fig. 3.1 subjected to axial tensile loading. If

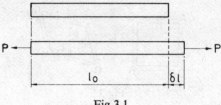

Fig 3.1

the resulting extension of the bar is δl and its original length is l_0, then the direct tensile strain is

$$\varepsilon = \frac{\delta l}{l_0}$$

If two bars identical in material, length l_0 and area were each subjected to a tensile force P then the extension in each would be the same, say δl, and the strain would be $\delta l/l_0$. The bars are now joined end on end and the same tensile force, P, is applied. The overall extension of the combined bar will be $2\delta l$, but since the original length is now $2l_0$ the strain is $2\delta l/2l_0$ or $\delta l/l_0$, i.e. the same as for the separate bars. Hence strain rather than elongation is an appropriate measure of deformation.

Similarly, if the bar had been compressed by an amount δl, then the compressive strain would be

$$\varepsilon = -\frac{\delta l}{l_0}$$

Strain is positive for an increase in dimension and negative for a reduction in dimension.

The same suffix notation is used for strains as for stresses. ε_x is the strain of a line measured in the x-direction, and ε_y the strain of a line in the y-direction.

Shear Strain

An element which is subjected to shear stress experiences deformation as shown in Fig. 3.2. The tangent of the angle through which two adjacent sides rotate relative to their initial position is termed *shear strain*. In many cases the angle is very small and the angle itself is used, expressed in radians, instead of the tangent, so that

$$\gamma = \angle AOB - \angle A'OB' = \phi$$

When $\angle A'OB' < \angle AOB$ then γ is defined as a positive shear strain, and when $\angle A'OB' > \angle AOB$, γ is termed a negative shear strain.

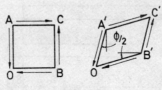

Fig 3.2

Volumetric Strain

The term "hydrostatic stress" was used in Chapter 2 to describe a state of tensile or compressive stress equal in all directions within or external to a body. Hydrostatic stress causes a change in volume of the material which, if expressed per unit of original volume, gives a volumetric strain denoted by ε_v.

Volumetric strain may be expressed in terms of the three co-ordinate linear strains, and for some problems this is a necessary relationship which can be derived as follows. Let a cuboid of material have sides initially of lengths, x, y and z. If under some load system the sides change in length by dx, dy and dz then the new volume is

$$(x + dx)(y + dy)(z + dz)$$

Neglecting products of small quantities,

New volume $= xyz + zy\,dx + xz\,dy + xy\,dz$

Original volume $= xyz$

Change in volume $= zy\,dx + xz\,dy + xy\,dz$

Volumetric strain, $\varepsilon_v = \dfrac{zy\,dx + xz\,dy + xy\,dz}{xyz}$

$$= \frac{dx}{x} + \frac{dy}{y} + \frac{dz}{z}$$

$$= \varepsilon_x + \varepsilon_y + \varepsilon_z \qquad (3.1)$$

i.e. volumetric strain is given by the sum of the three linear co-ordinate strains.

3.4. ELASTIC LOAD–DEFORMATION BEHAVIOUR OF MATERIALS

Studies of material behaviour made by Robert Hooke in 1678 showed that up to a certain limit the extension δl of a bar subjected to an axial tensile loading P was often directly proportional to P, as in Fig. 3.3(a). This behaviour in which $\delta l \propto P$ is known as *Hooke's law*. It is similarly found that for many materials uniaxial compressive load and compressive deformation are proportional up to a certain limit of load. A cylindrical bar which is twisted about its axis by opposing torques applied at each end is also found to have a linear torque–twist relationship

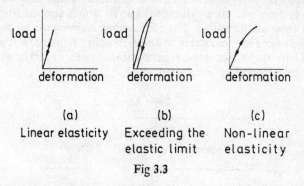

(a)   (b)   (c)

Linear elasticity   Exceeding the   Non-linear
             elastic limit   elasticity

Fig 3.3

up to a certain point. The maximum load up to which Hooke's law is applied is termed the *limit of proportionality*. If in each of the above cases at any particular load the same deformation exists both with increasing and decreasing load, and if after completely unloading the body it returns to exactly its original size, then it is said to exhibit the property of *elasticity*. This behaviour exists only over a certain range of load and deformation, the end point being termed the *elastic limit*. In general, the limit of proportionality is a shade lower than the elastic limit. If the elastic limit is exceeded it is found that some permanent deformation remains after removal of the load, as illustrated in Fig. 3.3(b).

Metals generally obey a linear load–deformation law up to their elastic limits, exhibiting what is termed *linear elasticity*; however, there are some materials, principally non-metallic, which have an elastic range as defined above, but exhibit a non-linear load–deformation relationship, Fig. 3.3(c).

3.5. ELASTIC STRESS–STRAIN BEHAVIOUR OF MATERIALS

If the load in Fig. 3.3(a) is divided by the cross-sectional area of the bar, A, and the extensions on the abscissa are divided by the original length of the bar, l_0, a graph of stress against strain is obtained. Since A and l_0 are constants the stress–strain behaviour is also linear in the elastic range.

The slope of the line is constant and may be expressed as

$$\frac{W}{A}\bigg/\frac{\delta l}{l_0} = \frac{\sigma}{\varepsilon} = E$$

where E is a constant for the material, and is called *Young's modulus of elasticity*. Since ε is non-dimensional, E has the dimensions of stress, i.e. force per unit area. Some typical approximate values of E are given for a few materials in Table 3.1.

A relationship between shear stress and shear strain may be derived from a torsion test on a cylindrical bar in which applied torque and angular twist are measured. The connections between torque and shear stress and twist and shear strain will be derived in Chapter 8. It is sufficient to state here that shear stress is proportional to shear strain within the elastic limit. Hence $\tau/\gamma = $ constant $= G$.

Table 3.1

Material	E GN/m^2	G GN/m^2
Steels	190–200	78–82
Copper	110–120	37–44
Aluminium	69–70	24–26
Glass	50–70	26–32
Plastic (Perspex)	2·8	1·0

The constant of proportionality, G, is known as the *modulus of rigidity*, or *shear modulus*, and has the dimensions of force per unit area. Approximate values of G for various materials are given in Table 3.1.

It can also be demonstrated experimentally that volumetric strain is proportional to hydrostatic stress within the elastic range. The constant relating those two quantities is termed the *bulk modulus* and is denoted by the symbol K. Thus

$$\frac{\sigma}{\varepsilon_v} = K$$

It will be shown in Chapter 13 that the elastic constants, E, G and K are related to one another.

Example 3.1.   Calculate the overall change in length of the tapered rod shown in Fig. 3.4. It carries a tensile load of 10 kN at the free end, and at the step change in section a compressive load of 2 MN/m evenly distributed around a circle of 30 mm diameter. $E = 208\,GN/m^2$.

Firstly consider the general case of a tapered rod fixed at one end and subjected to a tensile load W at the other end as in Fig. 3.5. The mean radius of any arbitrary slice at a distance x from the upper end is

$$r = r_0 - (r_0 - r_1)\frac{x}{L}$$

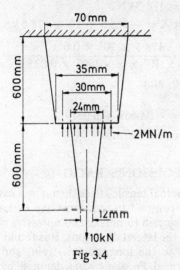

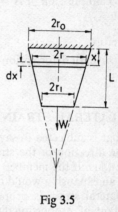

Fig 3.4   Fig 3.5

So the cross-sectional area of the slice is

$$A_x = \pi\left[r_0 - (r_0 - r_1)\frac{x}{L}\right]^2$$

If the slice extends an amount du under load then its strain is

$$\frac{du}{dx} = \frac{W}{A_x}\frac{1}{E}$$

The total extension of the rod is

$$u = \int_0^L \frac{W}{A_x E}dx$$

$$= \frac{W}{\pi E}\int_0^L \frac{dx}{\left[r_0 - (r_0 - r_1)\dfrac{x}{L}\right]^2}$$

$$= \frac{WL}{E\pi r_0 r_1}$$

Returning now to the stepped taper rod in Fig. 3.4, the extension of the lower part will be

$$u_B = \frac{10\,000 \times 0.6}{208 \times 10^9 \times \pi \times 0.024 \times 0.012} = +0.0319\,\text{mm}$$

The compressive load on the upper part will be treated as an axial concentrated load of magnitude

$$2 \times \pi \times 0.03 = 0.06\pi\,\text{MN} = 188.5\,\text{kN}$$

Resultant load acting on part A $= -188.5 + 10 = 178.5\,\text{kN}$

$$\text{Compression of A} = -\frac{178.5 \times 10^3 \times 0.6}{208 \times 10^9 \times \pi \times 0.07 \times 0.035}$$

$$= -0.0669\,\text{mm}$$

Therefore

Overall deformation of rod $= -0.0669 + 0.0319$

$$= -0.035\,\text{mm}$$

3.6. LATERAL STRAIN AND POISSON'S RATIO

If a bar is subjected to say longitudinal tensile stress then it will extend in the direction of the stress and contract in the transverse or lateral directions. If the member were subjected to uniaxial compressive stress then an expansion would occur in the lateral directions. It is found that the lateral strain is proportional to the longitudinal strain, and the constant of proportionality is termed *Poisson's ratio* denoted by the symbol ν. Hence

Lateral strain $= -\nu \times$ Direct strain (due to stress)

For most metals ν is in the range from 0·28 to 0·32. It is important to remember that lateral strain can occur without being accompanied by lateral stress.

3.7. THERMAL STRAIN

The effect of a change of temperature on a piece of material is a small change in size and hence strain. This can, in some circumstances, induce considerable stresses. The dependence of size on temperature variation is measured in terms of the basic quantity known as the *coefficient of linear thermal expansion* per unit temperature per unit length, denoted by α.

A rod of length l_0 has its temperature changed from T_0 to T and the accompanying change in length is

$$\delta l = \alpha l_0 (T - T_0)$$

This may be expressed as a thermal strain:

$$\varepsilon_T = \frac{\delta l}{l_0} = \alpha(T - T_0)$$

Increasing temperature causes expansion and thus a positive strain, while decreasing temperature results in contraction and negative strain. An important feature about this behaviour is that if there is no restraint on the material there can be strain unaccompanied by stress. However, if there is any restriction on free change in size then a *thermal stress* will result.

The total strain in a body experiencing thermal stress may be divided into two components, the strain associated with the stress, ε_σ, and the strain resulting from temperature change, ε_T. Thus

$$\varepsilon = \varepsilon_\sigma + \varepsilon_T$$

Hence

$$\varepsilon = \frac{\sigma}{E} + \alpha(T - T_0)$$

which is a more general form of the simple uniaxial stress–strain law.

3.8. GENERAL STRESS–STRAIN RELATIONSHIPS

Consider an element of material as in Fig. 3.6(a) subjected to a uniaxial stress, σ_x; the corresponding strain system is shown at (b). In the x-direction the strain is ε_x and in the y- and z-directions the strains are $-\nu\varepsilon_x$

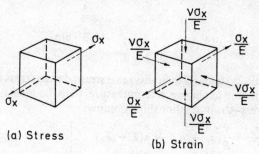

(a) Stress   (b) Strain

Fig 3.6

and $-\nu\varepsilon_x$, respectively. These strains may be written in terms of stress as $\varepsilon_x = \sigma_x/E$ and $\varepsilon_y = \varepsilon_z = -\nu\sigma_x/E$, the negative sign indicating contraction.

The element in Fig. 3.7(a) is subjected to triaxial stresses σ_x, σ_y and σ_z. The total strain in the x-direction is therefore composed of a strain due to σ_x, a lateral strain due to σ_y and a further lateral strain due to σ_z. Using the principle of superposition, the resultant strain in the x-direction is therefore the sum of the separate strains as shown at (b); hence

$$\varepsilon_x = \frac{\sigma_x}{E} - \frac{v\sigma_y}{E} - \frac{v\sigma_z}{E}$$

or

$$\varepsilon_x = \frac{\sigma_x}{E} - \frac{v}{E}(\sigma_y + \sigma_z)$$

Similarly

$$\varepsilon_y = \frac{\sigma_y}{E} - \frac{v}{E}(\sigma_z + \sigma_x)$$

and

$$\varepsilon_z = \frac{\sigma_z}{E} - \frac{v}{E}(\sigma_x + \sigma_y)$$

$$(3.2)$$

There is no lateral strain associated with shear strain; hence the shear-stress/shear-strain relationship is the same for both uniaxial and

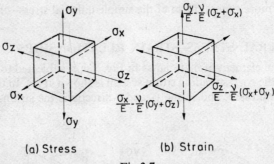

(a) Stress   (b) Strain

Fig 3.7

complex strain systems. If in addition to strain due to stress there is also thermal strain due to change in temperature, then the most general form of the six stress–strain relationships obtains:

$$\varepsilon_x = \frac{\sigma_x}{E} - \frac{v}{E}(\sigma_y + \sigma_z) + \alpha(T - T_0)$$

$$\varepsilon_y = \frac{\sigma_y}{E} - \frac{v}{E}(\sigma_z + \sigma_x) + \alpha(T - T_0)$$

$$\varepsilon_z = \frac{\sigma_z}{E} - \frac{v}{E}(\sigma_x + \sigma_y) + \alpha(T - T_0)$$

$$(3.3)$$

$$\gamma_{xy} = \frac{\tau_{xy}}{G} \qquad \gamma_{yz} = \frac{\tau_{yz}}{G} \qquad \gamma_{zx} = \frac{\tau_{zx}}{G}$$

$$(3.4)$$

3.9. STRAINS IN A STATICALLY DETERMINATE PROBLEM

The stresses in a thin-walled cylinder under internal pressure were found in Chapter 2 as a statically determinate problem, and now that stress–strain relationships have been developed the strains in the cylinder can be found.

From eqns. (2.14) and (2.15) from equilibrium,

$$\sigma_x = \frac{Pr}{2t} \qquad \sigma_y = \frac{Pr}{t}$$

and σ_z is negligible in comparison; therefore, from the stress–strain equations (3.3) the axial strain is

$$\varepsilon_x = \frac{Pr}{2tE} - \frac{vPr}{tE} = \frac{Pr}{2tE}(1 - 2v)$$

and the circumferential or hoop strain is

$$\varepsilon_y = \frac{Pr}{tE} - \frac{vPr}{2tE} = \frac{Pr}{2tE}(2 - v)$$

Taking a value for v of 0·3 it is found that the ratio of the hoop to the axial strain is

$$\frac{\varepsilon_y}{\varepsilon_x} = \frac{1\cdot7}{0\cdot4} = 4\cdot25$$

whereas the ratio for hoop to axial stresses was only 2·0. In the thin sphere there is only circumferential stress and strain:

$$\varepsilon = \frac{\sigma}{E} - \frac{v\sigma}{E}$$

and since

$$\sigma = \frac{Pr}{2t}$$

$$\varepsilon = \frac{Pr}{2tE}(1 - v)$$

3.10. PLASTIC STRESS–STRAIN BEHAVIOUR OF MATERIALS

With relatively few exceptions in the design of structures and machines, stresses and strains are limited to the elastic range of a material as described in Section 3.5. However, it is important to appreciate how a material behaves beyond the elastic range in what is termed the *plastic range* of stress and strain. As there is some elementary analysis involving yielding and plasticity in Chapters 16 and 18, the basic concept will be briefly introduced here. A more detailed treatment will be found in Chapter 19.

When a metal specimen is subjected to uniaxial tension to fracture, the measurements of stress and strain will result in a diagram typically as shown in Fig. 3.8(*a*) or (*b*). When the material has passed through the

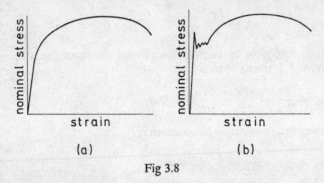

(a)   (b)

Fig 3.8

elastic range and enters the plastic range it is said to be *yielding*. Stress continues to increase with strain, but at a slower rate than in the elastic range, until a maximum value of nominal stress (load divided by original cross-sectional area) is reached which is termed the *tensile strength*. Thereafter the specimen enters a failure range terminating in complete fracture at one cross-section. The plastic range of strain is generally between 50 and 300 times the elastic range of strain, and demonstrates what is described as *ductility*, namely the ability to deform plastically.

3.11. VISCOELASTIC BEHAVIOUR OF MATERIALS

Some materials, notably plastics and rubbers, do not exhibit a linear elastic range like metals, but show an interdependence of stress and strain with time. In addition these materials have a "memory" in the sense that current strain or stress is always dependent on the loading history and after unloading considerable recovery of residual strain can occur at zero load. The above behaviour is termed *viscoelasticity*.

The simplest representation of viscoelasticity is the combined features of a Hookean solid and a Newtonian liquid. The former provides an elastic component, and the latter a viscous component. From the latter, stress is proportional to strain rate, $\dot{\varepsilon}$, and the constant of proportionality is η, the *coefficient of viscosity*. Hence the simplest possible relationship between stress, strain and time for a linear viscoelastic material is

$$\sigma = E\varepsilon + \eta\dot{\varepsilon} \tag{3.5}$$

A simple model representation of this equation is shown in Fig. 3.9, in which a linear elastic spring is placed in parallel with a Newtonian liquid dashpot. When the model (termed a *Voigt–Kelvin solid*) is stressed

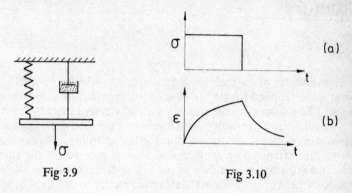

Fig 3.9   Fig 3.10

as in Fig. 3.10(*a*) its strain–time response is expressed by eqn. (3.5) and the curve shown at (*b*). There will be further discussion of viscoelasticity in Chapters 18 and 24.

Chapter 4
Displacements in Statically Determinate Structures

4.1. A knowledge of the manner and the extent to which a structure will deform under applied loading is obviously important. It was shown in Chapter 1 how forces in members of a statically determinate framework could be calculated. Knowing these forces and the load-deformation or stress-strain relationships discussed in the last chapter, it is possible to calculate the extension or compression of each member of the framework. This information can then be used in several methods to estimate the displacement of any joint in the structure relative to, say, the support points.

4.2. CHANGE OF GEOMETRY METHOD

This method simply uses the changes in length of members under load to describe the change in geometry or configuration of the framework, thus giving the displacement of joints. The principle of the method will

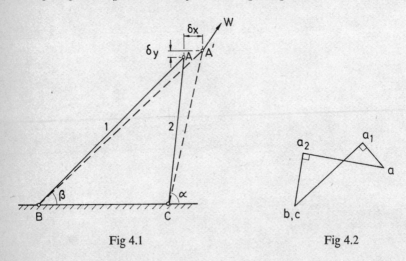

Fig 4.1   Fig 4.2

now be illustrated through the very simple situation of two pin-jointed members loaded as shown in Fig. 4.1. The forces in the members AB and AC can be calculated from two equilibrium equations, and thus, using the stress–strain relationship, the changes in length of the members are

$$\delta L_1 = \frac{P_1 L_1}{A_1 E} \quad \text{and} \quad \delta L_2 = \frac{P_2 L_2}{A_2 E} \tag{4.1}$$

where suffix 1 applies to AB and suffix 2 to AC.

The joint A moves to A' (exaggerated for clarity) under load where the displacements are δx and δy in the co-ordinate directions.

For small deformations the geometrical relationships are approximately

$$\left. \begin{aligned} \delta L_1 &= \delta x \cos \beta + \delta y \sin \beta \\ \delta L_2 &= \delta x \cos \alpha + \delta y \sin \alpha \end{aligned} \right\} \tag{4.2}$$

The slight change in the angles α and β under load is ignored as a second-order quantity.

Using the values of δL_1 and δL_2 in eqns. (4.1) and solving eqns. (4.2) simultaneously, the displacement components, δx and δy, can be determined.

Although this method is quite satisfactory and is also applicable to three-dimensional frames, for any degree of complexity the arithmetic becomes rather tedious and other methods are therefore preferable.

4.3. WILLIOT DIAGRAM

A very convenient graphical construction for determining displacements of frameworks was developed by Williot in 1877.

Referring to Fig. 4.1 again, the actual movement of the members may be regarded as a combination of change in axial length and a swing or rotation about one end. Since displacements are extremely small compared to the length of the members, the rotation component may be treated as a movement perpendicular to the direction of a member.

To draw the displacement diagram, shown in Fig. 4.2, the procedure is as follows. Mark a point b to represent B; then the same point also represents C, since there is no relative movement between B and C. Draw ba_1 to a suitable scale to represent δL_1 parallel to BA and in the correct sense for extension. Now draw a line through a_1 perpendicular to BA which is the line of the rotation component, although its length is not known. Next draw a line ca_2 parallel to CA to represent δL_2 and in the correct sense. Finally, draw a line from a_2 perpendicular to the direction of CA to intersect the previous rotation component. The point of intersection a defines the resultant position of A. The movement of A relative to the "ground" points B and C can then be measured off the diagram.

Example 4.1. Determine the vertical and horizontal displacements of the joint D in the plane pin-jointed framework in Fig. 4.3(*a*). All members have a cross-sectional area of 1000 mm² and modulus of 200 GN/m².

The first step is to calculate the forces in all the members by one of the methods described earlier. Next the change in length of each member is determined and it is convenient to tabulate the foregoing information:

Member	Force W, kN	Length L, m	Change in length, WL/EA, mm
BC	+20	2	+0·2
CD	+10	2	+0·1
FG	−10	2	−0·1
FC	−10√2	2√2	−0·2
GC	+10	2	+0·1
GD	−10√2	2√2	−0·2

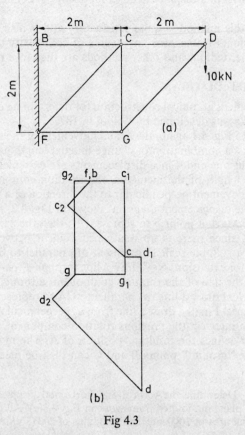

(a)

(b)

Fig 4.3

The displacement diagram (Fig. 4.3(b)) can now be drawn starting with the triangle BCF. BC extends to the right, so that c_1 is drawn to the right of b at a distance of 0·2 mm (to scale). C moves diagonally down to the left, so fc_2 is drawn in the same sense, 0·2 mm (to scale). The perpendiculars to the directions of BC and FC are c_1c and c_2c, respectively, meeting at c. We can now proceed to treat triangle FCG in the same manner arriving at point g. Finally the displacement components of triangle CGD will give the position of D (d). From the completed diagram,

> Vertical displacement of D = 1·266 mm
>
> Horizontal displacement of D = 0·3 mm

4.4. WILLIOT–MOHR DIAGRAM

The Williot diagram can only be drawn if there is a triangle of members with two of its apexes as "ground" points from which to commence. It is evident that there could be many framework configurations which did not satisfy this condition. For this reason Otto Mohr, in 1887, suggested an extension to the Williot method. It is best explained and understood by using a specific example.

Consider the plane frame in Fig. 4.4 which was solved for member forces in Example 1.2. The first step is to determine the change in length of the members. Assuming a constant value of (area A × modulus E) for all members of 200 MN, the numerical values are given in the following table.

Member	Force, N	Length, m	Length change, mm
AB	−4713	2·82	−0·0667
AC	+3333	2·00	+0·0333
BC	+3333	2·00	+0·0333
BD	−3333	2·00	−0·0333
CD	+2356	2·82	+0·0333
CE	+1667	2·00	+0·0167
DE	0	2·00	0
DF	−2356	2·82	−0·0333
EF	+1667	2·00	+0·0167

If now, for example, C is regarded temporarily as a position fixed joint, then a Williot diagram could be drawn starting from A and C. It would give the correct *distortion* of the framework and the *relative* displacements of joint to joint as in Fig. 4.5. However, as the assumption that AC is horizontal is not true, the Williot solution would be for the framework displaced relative to "ground" as shown rather exaggerated in Fig. 4.4. Mohr, then, proposed the required correction to the diagram to restore the real situation where F remains in the horizontal plane.

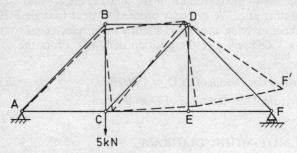

Fig 4.4

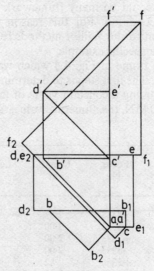

Fig 4.5

This is achieved by giving the frame a rigid-body rotation until F' coincides with F, which means that on the Williot diagram the apparent vertical distance between a and f must be eliminated. The other joints will move in a direction perpendicular (approximately) to lines connecting each joint to A by an amount equal to the distance of the joint from A multiplied by F'F/AF. This statement is represented on the Williot diagram by drawing the frame rotated through 90° (in the opposite sense of F′ moving to F), and to a suitable scale so that when a' is placed on a (since A does not move), f' is horizontally in line with f. The points a', b', c', d', e', f' now become the "ground" points for a, b, c, d, e, f, and the true displacements of the joints in the framework relative to the supports are given by the distances between the corresponding points $a'a$, $b'b$, $c'c$, etc.

4.5. VIRTUAL WORK METHOD

The principle of *virtual work* is one of the most powerful tools in structural analysis. Apart from a brief introduction here, it will be referred to again in Chapter 10 on "Strain Energy". However, for a more detailed treatment the reader is advised to consult one of the texts on structural analysis or energy methods. The principle may be stated as follows: If a system of forces acts on a particle which is in statical equilibrium and the particle is given any virtual displacement then the net work done by the forces is zero.

A virtual displacement is any arbitrary displacement which is mathematically conceived and does not actually have to take place, but must be geometrically possible. The forces must not only be in equilibrium but are assumed to remain constant and parallel to their original lines

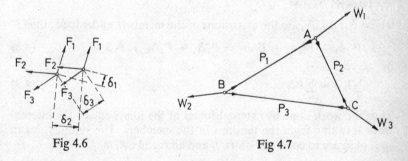

Fig 4.6   Fig 4.7

of action. If these conditions are applied to the particle shown in Fig. 4.6 which is in statical equilibrium then

$$F_1\delta_1 + F_2\delta_2 + F_3\delta_3 = 0 \tag{4.3}$$

We can now proceed to apply eqn. (4.3) to a simple assembly of bars in equilibrium under external loads W_1, W_2, W_3 as illustrated in Fig. 4.7. Let the *tensile* forces in the members be P_1, P_2, P_3.

Consider joint A given a virtual displacement to A′. The displacement of W_1 is δ_1 and the displacements of P_1 and P_2 are Δ_{1A} and Δ_{2A} in directions *opposite* to the sense of the arrowheads. (A member *extends* under tension but the force *in the member* is pulling in the direction opposite to the extension.)

From eqn. (4.3),

$$W_1\delta_1 - P_1\Delta_{1A} - P_2\Delta_{2A} = 0$$

the negative signs being due to the products of force in one direction and displacement in the opposite sense. Hence

$$W_1\delta_1 = P_1\Delta_{1A} + P_2\Delta_{2A} \tag{4.4}$$

For joint B,

$$W_2\delta_2 = P_1\Delta_{1B} + P_3\Delta_{3B} \tag{4.5}$$

and for joint C,

$$W_3\delta_3 = P_3\Delta_{3C} + P_2\Delta_{2C} \tag{4.6}$$

Adding eqns. (4.4), (4.5) and (4.6),

$$W_1\delta_1 + W_2\delta_2 + W_3\delta_3 = P_1(\Delta_{1A} + \Delta_{1B}) + P_2(\Delta_{2A} + \Delta_{2\upsilon})$$
$$+ P_3(\Delta_{3B} + \Delta_{3C}) \tag{4.7}$$

Put $(\Delta_{1A} + \Delta_{1B}) = \Delta_1$

$(\Delta_{2A} + \Delta_{2C}) = \Delta_2$

$(\Delta_{3B} + \Delta_{3C}) = \Delta_3$

where Δ_1, Δ_2, Δ_3 are the extensions of the members under load; then

$$W_1\delta_1 + W_2\delta_2 + W_3\delta_3 = P_1\Delta_1 + P_2\Delta_2 + P_3\Delta_3 \tag{4.8}$$

or

$$\sum_j W\delta = \sum_m P\Delta \tag{4.9}$$

Thus the work done by external forces at the joints equals the internal work resulting from the tensions in the members. The summations in eqn. (4.9) are to cover all joints, j, and all members, m.

4.6. FRAME DISPLACEMENTS USING VIRTUAL WORK

(*a*) Actual Forces, Extensions and Displacements

If a plane frame is subjected to *one* external load only and the displacement of the loaded joint is required *in the direction* of the load, then eqn. (4.9) can be used with all real values. The method will be illustrated in relation to part of Example 4.1. Let the vertical displacement of joint D, which is where the load is applied and in the same direction, be δ. Then

$$\sum_j W\delta = 10\delta$$

and, using numerical values from the table in Example 4.1,

$$\sum_m P\Delta = (20 \times 0\cdot2) + (10 \times 0\cdot1) + (-10 \times -0\cdot1)$$
$$+(-10\sqrt{2} \times -0\cdot2) + (10 \times 0\cdot1) + (-10\sqrt{2} \times -0\cdot2)$$
$$= 12\cdot66\,\text{kN-mm}$$

Therefore

$$10\delta = 12\cdot66 \quad \text{and} \quad \delta = 1\cdot266\,\text{mm}$$

The horizontal displacement of D cannot be found by the above method, since there is no force in the horizontal direction at D to give an external work term.

The above method of using the real forces and extensions cannot be used on a frame carrying several external loads, unless all the displacements at the loaded joints, except one, are known, since there would be several unknown δ's on the left-hand side of eqn. (4.9).

(b) **Actual Extensions and Displacements with Hypothetical Forces**

The previous section demonstrated the use of eqn. (4.9) in which the work terms were composed of the real forces and real displacements. However, the equation is equally valid if the displacements are those actually occurring due to the applied loading, but the load and force terms are completely fictitious, so long as they form an equilibrium system. This feature enables one to determine the displacement of any joint in any direction, whether loaded or not, by the use of a "dummy" load of unit magnitude.

Consider again Example 4.1 and the need to find the horizontal displacement at D. A dummy unit load is inserted horizontally at D. Then the equilibrium system of forces in the members *due to the unit load only* (the 10 kN is removed) can be found; the values are as given in the following table.

Member	Actual extensions, Δ, due to 10 kN load	Hypothetical forces, P', due to dummy load	$P'\Delta$
BC	+0·2	1	+0·2
FC	−0·2	0	0
FG	−0·1	0	0
CG	+0·1	0	0
CD	+0·1	1	+0·1
GD	−0·2	0	0
			$\Sigma = 0\cdot3$

If the real horizontal displacement of D due to the actual load of 10 kN is δ, then the external virtual work is $1 \times \delta$. The internal virtual work is given by the sum of the products of the actual changes in length, Δ, of the members due to the actual load of 10 kN and the hypothetical equilibrium set of forces, P', resulting from the unit dummy load. The simple computation is shown in the right-hand column of the above table. Applying eqn. (4.9) gives

$$1\delta = \Sigma P'\Delta \qquad \text{or} \qquad \delta = 0\cdot3\,\text{mm}$$

Example 4.2. Determine the vertical displacement of joint G of the frame in Example 4.1.

Following the same procedure as above a unit dummy load is placed vertically at G; the corresponding force system *due to the unit load only* is shown in the table below.

Member	Actual extensions Δ	Forces due to unit load P'	$P'\Delta$
BC	+0·2	+1	+0·2
FC	−0·2	−$\sqrt{2}$	+0·2$\sqrt{2}$
FG	−0·1	0	0
CG	+0·1	1	+0·1
CD	+0·1	0	0
GD	−0·2	0	0
			$\Sigma = +0\cdot5828$

Thus

$$1\delta_G = 0\cdot5828 \qquad \text{or} \qquad \delta_G = 0\cdot5828\,\text{mm}$$

Example 4.3. Determine the forces in the members of the plane framework illustrated in Fig. 4.8(*a*) and calculate the vertical deflection at the joint carrying the central load. All members are 3 m in length and 1500 mm² in cross-section. $E = 205$ GN/m².

Firstly a force polygon is constructed for the actual loading on the frame (Fig. 4.8(*b*)). Although we are trying to find the displacement of a joint in the direction of the applied load, because there are three external loads having unknown displacements, eqn. (4.9) cannot be solved using the approach in Section 4.6(*a*). Instead we resort to the method in Section 4.6(*b*) and after removing the 10 kN loads insert a unit vertical

load only at the central joint in the lower span. The appropriate force polygon is then drawn as in Fig. 4.8(c) (to a larger scale than at (b)).
The forces are all tabulated below. Since A, L and E are the same for

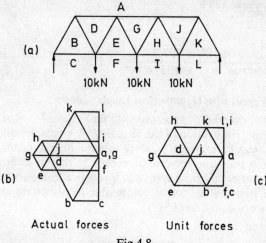

Actual forces   Unit forces

Fig 4.8

all members, the changes in length have one constant of proportionality to the forces and it is therefore wasted effort to include a specific column in the table.

Member	Actual forces P, kN	Unit load forces, P', kN	$\dfrac{PP'}{kN^2}$
AD	−17·3	−0·575	9·950
AG	−23·1	−1·15	26·550
AJ	−17·3	−0·575	9·950
KL	+8·65	+0·289	2·475
IH	+20·2	+0·868	17·700
EF	+20·2	+0·868	17·700
BC	+8·65	+0·289	2·475
AB	−17·3	−0·575	9·950
BD	+17·3	+0·575	9·950
DE	−5·8	−0·575	3·325
EG	+5·8	+0·575	3·325
GH	+5·8	+0·575	3·325
HJ	−5·8	−0·575	3·325
JK	+17·3	+0·575	9·950
AK	−17·3	−0·575	9·950

$$\Sigma = 139\cdot900$$

At this point the factor L/AE required to give changes in length can be included. Therefore

$$1\delta = \frac{L}{AE} \times 139 \cdot 9$$

and

$$\delta = \frac{3 \times 10^3 \times 139 \cdot 9}{1500 \times 10^{-6} \times 205 \times 10^6} = 1 \cdot 363 \text{ mm}$$

(c) Actual Forces with Hypothetical Displacements

Although this chapter is concerned with finding *displacements* of frameworks, the final illustration of the virtual work principle will be devoted to finding *forces* in frameworks, since this will help to confirm the principle. We now take the converse procedure to that in section (b) above and use the true equilibrium system of forces in a framework in conjunction with a hypothetical set of compatible displacements to compose the virtual work equation (4.9).

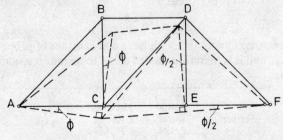

Fig 4.9

Consider the plane frame in Fig. 4.9, which was analysed by various methods in Chapter 1. It is only possible to find one unknown force at a time using eqn. (4.9), so let this be, for example, that in member BD. It is also necessary to include the applied load of 5 kN at C on the left-hand side of eqn. (4.9). If no other forces are to appear in the equation, then the other members must not change in length and so rigid body rotations are given to the other parts of the frame. This results in the hypothetical displaced configuration of Fig. 4.9, from which eqn. (4.9) can be expressed as

$$5000\delta = (P\Delta)_{BD}$$

Let the rigid-body rotation of triangle ABC be ϕ; for small displacements $\sin \phi \approx \phi$ so that $\delta = AC \cdot \phi = 2\phi$.

For the system of displacements to be compatible, triangle CDF rotates through an angle $\phi/2$. The compression of BD can now be expressed as

$$\Delta_{BD} = BC . \phi + DE . \phi/2$$
$$= 2\phi + 2\phi/2$$

Substituting these values in the virtual-work equation,

$$5000 \times 2\phi = P_{BD}(2\phi + 2\phi/2)$$

Hence

$$P_{BD} = 3333 \, N$$

The actual value of ϕ is now seen to be immaterial as it cancels through in the virtual-work equation.

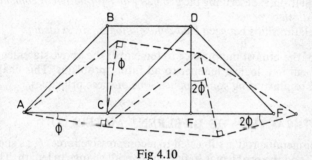

Fig 4.10

To find the force in CE the displacement configuration in Fig. 4.10 is adopted giving

$$5000 . AC . \phi = P_{CE} . BC . \phi + P_{CE} . DE . 2\phi$$

from which

$$P_{CE} = 1667 \, N$$

Summary

The principle of virtual work is expressed in the equation

$$\sum_j W\delta = \sum_m P\Delta$$

and only two conditions are necessary, namely that the force system satisfies the requirements of equilibrium and that the deformation system satisfies all requirements of compatibility. It is not necessary for external loads to be the actual loads, or for the bar forces to be those forces induced by the loads, nor is it necessary for the deflections and bar extensions to be those caused by the loads.

Chapter 5
Statically Indeterminate Stress Systems

5.1. The subject has now been developed sufficiently so that statically indeterminate problems involving uniform direct tensile or compressive stress can be solved. The principles of solution are fundamental to the mechanics of solids and structures and are stated as follows:

1. Equation(s) of *equilibrium of forces* both internal (as a function of stress) and external (applied).
2. Equation(s) describing the *geometry of deformation* or *compatibility of displacements*.
3. Relationships between *load-deformation* or *stress-strain*.

It is of the utmost importance to remember the above statements and apply them in a logical manner to all future problems. This chapter is devoted to illustrating some applications of these principles.

5.2. INTERACTION OF DIFFERENT MATERIALS

(*a*) A bi-metallic rod is subjected to a compressive force, *F*, as shown in Fig. 5.1, and the problem is to find the overall change in length. The two

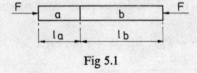

Fig 5.1

materials will be denoted by subscripts *a* and *b*.

Equilibrium
It should be fairly obvious that the same force is carried through each material, so that

$$P_a = P_b = F \tag{5.1}$$

Geometry of Deformation
The overall change in length is the sum of the changes in the two parts of the rod, so that

$$\delta = \delta_a + \delta_b \tag{5.2}$$

82

Stress–Strain Relations

$$\frac{\sigma_a}{\varepsilon_a} = E_a \tag{5.3}$$

$$\frac{\sigma_b}{\varepsilon_b} = E_b \tag{5.4}$$

Solution

Let the cross-sectional area be $A = A_a = A_b$. Then eqns. (5.3) and (5.4) can be re-expressed as

$$\frac{P_a}{A_a} = E_a \frac{\delta_a}{l_a} \tag{5.5}$$

and

$$\frac{P_b}{A_b} = E_b \frac{\delta_b}{l_b} \tag{5.6}$$

Substituting the values of δ_a and δ_b into eqn. (5.2) gives

$$\delta = \frac{P_a l_a}{A_a E_a} + \frac{P_b l_b}{A_b E_b}$$

thus, using eqn. (5.1),

$$\delta = \frac{F}{A}\left(\frac{l_a}{E_a} + \frac{l_b}{E_b}\right)$$

(b) A simple but important variation in principle of the above situation is shown in Fig. 5.2, in which the two materials are arranged concentrically.

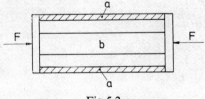

Fig 5.2

Equilibrium

In this case the load is *shared* in some unknown proportions between the two parts so that

$$P_a + P_b = F \tag{5.7}$$

Geometry of Deformation

If the unloaded lengths are initially the same, then they will remain the same under load; hence

$$\delta_a = \delta_b = \delta \tag{5.8}$$

Stress–Strain Relations

Again

$$\frac{\sigma_a}{\varepsilon_a} = E_a \tag{5.9}$$

$$\frac{\sigma_b}{\varepsilon_b} = E_b \tag{5.10}$$

Solution

From eqns. (5.8), (5.9) and (5.10),

$$P_a = E_a A_a \frac{\delta}{l}$$

$$P_b = E_b A_b \frac{\delta}{l}$$

Substituting in the equilibrium equation (5.7),

$$E_a A_a \frac{\delta}{l} + E_b A_b \frac{\delta}{l} = F$$

Thus

$$\delta = \frac{Fl}{E_a A_a + E_b A_b} \tag{5.11}$$

$$P_a = \frac{F E_a A_a}{E_a A_a + E_b A_b} \tag{5.12}$$

and

$$P_b = \frac{F E_b A_b}{E_a A_a + E_b A_b} \tag{5.13}$$

5.3. INTERACTION OF DIFFERENT STIFFNESS COMPONENTS

Situations similar to those in Section 5.2 can arise if an assembly is made of one type of material but certain component parts have different stiffnesses (load per unit deformation of an elastic member) which interact. For example, consider a flexible mounting for a machine of weight W

consisting of a *rigid* (does not bend) rectangular plate supported on four coil springs of stiffness K placed symmetrically with respect to the corners of the plate as shown in Fig. 5.3. After the machine has been located centrally on the plate, the springs are further loaded by means of four bolts having their lower ends fixed firmly into concrete and their upper ends passing freely through the plate. The nuts are tightened until the bolts are stretched by an amount y. The springs have then been compressed by an amount y' due to the combined action of the load W and

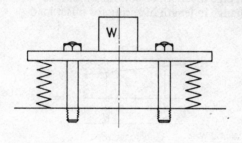

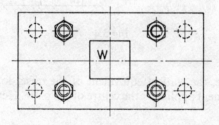

Fig 5.3

the tension in the bolts. Determine how far the machine can be moved along a central line parallel to the long edge of the plate without the change in length of the bolts exceeding half the original extension, y.

The system is symmetrical about a vertical central plane parallel to the long edge of the plate, and it is only necessary to consider the effect of $W/2$ acting on a pair of springs and bolts.

The first step is to consider a likely mode of deformation and the associated equilibrium condition. When the load is moved a distance x away from the centre as in Fig. 5.4, one end of the plate will sink further and the other end will rise. The statical equivalent to the offset load is a central load $W/2$ and a couple about the centre of the plate $Wx/2$. The latter will simply cause a small rotation of the plate and the fall at one end will equal the rise at the other.

Let T = Initial tension in bolts due to extension y
$\quad\;\; F$ = Force exerted by spring due to the compression y'
$\quad \Delta T$ = Change in force in bolt for offset load
$\quad \Delta F$ = Change in force in spring for offset load
$\quad\;\; k$ = Stiffness of a bolt
$\quad\;\; K$ = Stiffness of a spring
$\quad\;\; \delta$ = Change in length of bolt for offset load
$\quad\;\; \delta'$ = Change in length of spring for offset load

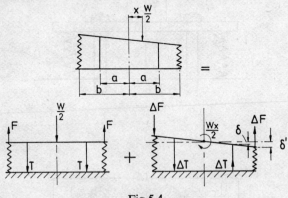

Fig 5.4

Equilibrium

As stated above, the statical equivalent of the offset load is a vertical force and a moment applied at the centre of the plate. Hence, for vertical equilibrium,

$$\frac{W}{2} + 2T - 2F = 0 \tag{5.14}$$

For rotational equilibrium about the centre,

$$-\frac{W}{2}x + 2\Delta Ta + 2\Delta Fb = 0 \tag{5.15}$$

where a and b are as shown in Fig. 5.4.

Load Deformation

For vertical central load,

$$\frac{T}{y} = k \qquad \frac{F}{y'} = K \tag{5.16}$$

For rotation due to the offset load,

$$\frac{\Delta T}{\delta} = k \qquad \frac{\Delta F}{\delta'} = K \qquad (5.17)$$

Geometry

Since the plate is rigid, from similar triangles,

$$\frac{\delta}{\delta'} = \frac{a}{b} \qquad (5.18)$$

The stipulated condition is that $\delta \not> \tfrac{1}{2}y$, or for maximum x,

$$\delta = \tfrac{1}{2}y \qquad (5.19)$$

Solution

There are thus eight equations to solve for the eight unknown quantities. Substituting for ΔT and ΔF from eqns. (5.17) in the moment equation (5.15),

$$-\frac{Wx}{2} + 2k\delta a + 2\,K\delta'b = 0 \qquad (5.20)$$

From eqn. (5.18),

$$\delta' = \frac{b}{a}\delta$$

and for maximum x,

$$\delta = \tfrac{1}{2}y$$

Substituting these values into eqn. (5.20),

$$-\frac{Wx}{2} + kya + \frac{Kyb^2}{a} = 0 \qquad (5.21)$$

Eliminating T and F between eqns. (5.14) and (5.16),

$$k = \frac{Ky' - (W/4)}{y}$$

Substituting for k in eqn. (5.21) and simplifying,

$$x = \frac{2K}{Wa}(a^2y' + b^2y) - \frac{a}{2} \qquad (5.22)$$

Example 5.1. A rigid member AB, the weight of which can be neglected, is supported horizontally at the pin joints A and C and by the column at B as shown in Fig. 5.5(*a*). The stiffness of member CD is 2 kN/m and of the column is 5 kN/m. Calculate the force in CD, which is initially unstressed, when a vertical load of 10 kN is applied at B.

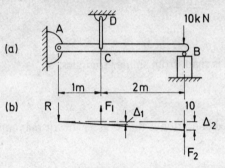

Fig 5.5

Equilibrium

Let the reaction at A be R and the forces in CD and in the column be F_1 and F_2 respectively.

Vertical $\qquad 10 + R - F_1 - F_2 = 0$ $\qquad\qquad\qquad$ (5.23)

Moments $\qquad 3R - 2F_1 = 0$ $\qquad\qquad\qquad\qquad\qquad$ (5.24)

Geometry of Deformation

The deformation is as shown in Fig. 5.5(b), so that

$$\frac{\Delta_1}{\Delta_2} = \frac{1}{3} \qquad\qquad\qquad (5.25)$$

Load-Deformation Relations

$$\frac{F_1}{\Delta_1} = 2 \qquad\qquad\qquad (5.26)$$

$$\frac{F_2}{\Delta_2} = 5 \qquad\qquad\qquad (5.27)$$

From eqns. (5.25), (5.26) and (5.27),

$$F_1 = \frac{2}{15}F_2$$

From eqns. (5.23) and (5.24),

$$F_1 + 3F_2 = 30$$

Hence the force in CD is

$$F_1 = 1{\cdot}275\,\text{kN}$$

5.4. RESTRAINT OF THERMAL STRAIN

Returning to the problem in Section 5.2(*b*), Fig. 5.2, we will remove the compressive force and instead simply change the temperature of the composite member from T_0, when the two materials are stress free, to a final temperature T. Let the two materials be copper and steel denoted by superscripts c and s. The change in temperature will cause a change in length of the assembly, but because of the different values of α (coefficient of thermal expansion) each material will put a restraint on the other and thermal stresses will be induced.

Equilibrium
Since there is no applied external force, the sum of the internal forces in the copper and steel must be zero. Therefore

$$P^c + P^s = 0 \tag{5.28}$$

or

$$\sigma^c A^c + \sigma^s A^s = 0 \tag{5.29}$$

Geometry of Deformation
Since the two materials are initially stress free and their ends are tied together the total strain must be the same for each. Therefore

$$\varepsilon^c = \varepsilon^s \quad \text{or} \quad \varepsilon_\sigma{}^c + \varepsilon_T{}^c = \varepsilon_\sigma{}^s + \varepsilon_T{}^s \tag{5.30}$$

where ε_σ = strain due to stress, ε_T = strain due to temperature change.

Stress–Strain Relation

$$\varepsilon^c = \frac{\sigma^c}{E^c} + \alpha^c(T - T_0) \tag{5.31}$$

$$\varepsilon^s = \frac{\sigma^s}{E^s} + \alpha^s(T - T_0) \tag{5.32}$$

Solution
Equating ε^c and ε^s from eqns. (5.31) and (5.32),

$$\frac{\sigma^c}{E^c} + \alpha^c(T - T_0) = \frac{\sigma^s}{E^s} + \alpha^s(T - T_0)$$

Substituting for, say, σ^s from eqn. (5.29),

$$\frac{\sigma^c}{E^c} + \alpha^c(T - T_0) = -\frac{1}{E^s}\frac{\sigma^c A^c}{A^s} + \alpha^s(T - T_0)$$

$$\sigma^c\left(\frac{1}{E^c} + \frac{A^c}{E^s A^s}\right) = (T - T_0)(\alpha^s - \alpha^c)$$

or

$$\sigma^c = \frac{A^s E^s E^c (T - T_0)(\alpha^s - \alpha^c)}{A^s E^s + A^c E^c}$$

and

$$\sigma^s = -\frac{A^c E^c E^s (T - T_0)(\alpha^s - \alpha^c)}{A^s E^s + A^c E^c}$$

The negative sign for σ^s does not necessarily indicate a compressive stress but simply that it is opposite in sign to σ^c. The type of stress in each material is determined by the numerical values of the quantities, T_0, T, α^s and α^c. For example, put $E^s = 205\,\mathrm{GN/m^2}$, $E^c = 115\,\mathrm{GN/m^2}$, $\alpha^s = 11 \times 10^{-6}$ per deg C, $\alpha^c = 16 \times 10^{-6}$ per deg C. Also let $A^s = 600$ mm^2, $A^c = 1200\,\mathrm{mm^2}$, $T_0 = 20°C$ and $T = 100°C$; then

$$\sigma^c = \frac{\begin{array}{c}600 \times 10^{-6} \times 205 \times 10^9 \times 115 \\ \times 10^9(100 - 20)(11 - 16) \times 10^{-6}\end{array}}{[(600 \times 10^{-6} \times 205 \times 10^9) + (1200 \times 10^{-6} \times 115 \times 10^9)]}$$

$$= -\frac{600 \times 205 \times 115 \times 80 \times 5 \times 10^3}{(600 \times 205) + (1200 \times 115)}$$

$$= -21{\cdot}7\,\mathrm{MN/m^2}$$

and

$$\sigma^s = -\frac{\sigma^c A^c}{A^s} = -\frac{-21{\cdot}7 \times 1200}{600} = +43{\cdot}4\,\mathrm{MN/m^2}$$

Thus for an *increase* in temperature, α^c being greater than α^s, the copper is prevented from expanding as much as if it was free and is put into compression. The steel is forced to expand more than it would if free and is therefore in tension.

Example 5.2. A copper ring having an internal diameter of 150 mm and external diameter of 154 mm is to be shrunk onto a steel ring, of the same width, having internal and external diameters of 140 mm and 150·05 mm respectively.

What change in temperature is required in the copper ring so that it will just slide on to the steel ring?

What will be the uniform circumferential stress in each ring and also the interface pressure when assembled and back at room temperature?

Assume that there is no stress in the width direction.

$$E_s = 205\,\mathrm{GN/m^2}; \qquad E_c = 100\,\mathrm{GN/m^2}; \qquad \alpha_c = 18 \times 10^{-6}\,\text{per deg C}$$

The circumferential length of the copper ring has to be increased by heating till it is fractionally larger than the circumferential length of the steel ring.

$$\text{Minimum required change in circumference} = \pi d_s - \pi d_c$$
$$= \pi \times 0.05$$

$$\text{Change in circumference due to heating} = \pi d_c \times \alpha(T - T_0)$$
$$= \pi \times 150 \times 18$$
$$\times 10^{-6}(T - T_0)$$

Therefore

$$T - T_0 = \frac{0.05}{2700 \times 10^{-6}} = 18.5°C$$

When the assembly has returned to ambient temperature assume that the circumferential stresses in the copper and steel are uniformly distributed over each cross-section.

Equilibrium
Let the width of each ring be w; then

$$(w \times 2)\sigma_c + (w \times 5)\sigma_s = 0$$
$$\sigma_c = -2.5\sigma_s \tag{5.33}$$

Geometry of Deformation
The circumferential strains in the copper and steel must be the same; i.e.

$$\varepsilon_s = \varepsilon_c \tag{5.34}$$

Stress–Strain Relations
$$\varepsilon_s = \sigma_s/E_s \tag{5.35}$$
and
$$\varepsilon_c = \sigma_c/E_c + \alpha_c\Delta T \tag{5.36}$$

since it is only the copper ring that has the thermal strain component.

Solution
From eqns. (5.34), (5.35) and (5.36),

$$\frac{\sigma_s}{E_s} = \frac{\sigma_c}{E_c} + \alpha_c\Delta T$$

Using eqn. (5.33),

$$\sigma_s\left(\frac{1}{E_s} + \frac{2.5}{E_c}\right) = -18 \times 10^{-6} \times 18.5$$

The negative sign is due to ΔT being a reduction in temperature. Substituting for E_s and E_c,

$$\sigma_s = -11.15 \, \text{MN/m}^2$$

and

$$\sigma_c = +27{\cdot}9 \, \text{MN/m}^2$$

The radial pressure at the interface between the two rings may be treated as a thin cylinder under internal or external pressure, so that

$$p = \sigma_c t / r = 27{\cdot}9 \times 2/75 = 745 \, \text{kN/m}^2$$

5.5. VOLUME CHANGES

The following problem analyses the change in volume of a vessel subjected to pressure and makes use of the relationship between hydrostatic stress and volume strain.

Example 5.3. A thin spherical steel shell has a mean diameter of 3 m, a wall thickness of 6 mm, and is just filled with water at 20°C and atmospheric pressure. Find the rise in gauge pressure if the temperature of the water and shell rises to 50°C, and then determine the volume of water that would escape if a small leak developed at the top of the vessel.

Steel: Young's modulus, $E = 200 \, \text{GN/m}^2$
       Coefficient of linear expansion $= 11 \times 10^{-6}$ per deg C
       Poisson's ratio $= 0{\cdot}3$

Water: Bulk modulus, $K = 2{\cdot}2 \, \text{GN/m}^2$
       Coefficient of volumetric expansion $= 0{\cdot}207 \times 10^{-3}$ per deg C

Equilibrium

Let the gauge pressure in the sphere after rise in temperature be P; then from Chapter 2 the equilibrium condition is

$$\sigma = \frac{Pr}{2t} = 125P \tag{5.37}$$

Geometry of Deformation

If there is to be a pressure at all then the water and sphere must remain in overall contact, and hence

Change in volume of sphere = Change in volume of water

or

$$\varepsilon_{v \, sphere} = \varepsilon_{v \, water} \tag{5.38}$$

since the original volume is the same for each.

Stress–Strain Relations

For the water the total volumetric strain is the sum of that due to pressure and that due to thermal strain:

$$\varepsilon_{v\,water} = -(P/K) + \alpha_v(T - T_0) \tag{5.39}$$

For the sphere the total volumetric strain is a function of strain due to stress (from pressure) and thermal strain:

$$\varepsilon_{v\,sphere} = \varepsilon_{v\,stress} + \varepsilon_{v\,thermal} \tag{5.40}$$

Solution

From eqn. (5.39),

$$\varepsilon_{v\,water} = -\frac{P}{2 \cdot 2 \times 10^9} + (0 \cdot 207 \times 10^{-3} \times 30)$$

$$= -(0 \cdot 445P \times 10^{-9}) + (6210 \times 10^{-6})$$

The change in the internal capacity or volume of the sphere is

$$\tfrac{4}{3}\pi(r + \delta r)^3 - \tfrac{4}{3}\pi r^3$$

which gives, neglecting products of the small quantity δr,

$$\tfrac{4}{3}\pi \times 3r^2\,\delta r$$

Expressing this as a volumetric strain,

$$\frac{\tfrac{4}{3}\pi \times 3r^2\,\delta r}{\tfrac{4}{3}\pi r^3} = 3\frac{\delta r}{r}$$

It will now be shown that $\delta r/r$ is the linear or hoop strain in the material of the sphere.

$$\text{Change in circumference} = 2\pi(r + \delta r) - 2\pi r = 2\pi\delta r$$

Therefore

$$\text{Hoop strain} = \frac{2\pi\delta r}{2\pi r} = \frac{\delta r}{r}$$

So that

$$\text{Volumetric strain of vessel} = 3 \times \text{hoop strain}$$

Now, the hoop strain is given by eqn. (3.3):

$$\varepsilon = \frac{\sigma}{E} - \frac{v\sigma}{E} + \alpha(T - T_0)$$

$$= \frac{125P}{200 \times 10^9}(1 - 0 \cdot 3) + (11 \times 10^{-6} \times 30)$$

Therefore the total volumetric strain is

$$\varepsilon_{v\,sphere} = 3\left(\frac{125 \times 0{\cdot}7P}{200 \times 10^9} + 330 \times 10^{-6}\right)$$

Hence, using eqn. (5.38),

$$(-0{\cdot}445P \times 10^9) + (6210 \times 10^{-6}) =$$
$$3\left[\frac{125 \times 0{\cdot}7P}{200 \times 10^9} + (330 \times 10^{-6})\right]$$

from which

$$P = 2{\cdot}97\,\text{MN/m}^2$$

The volume of water which escapes through the leak is simply the difference of the *free* thermal expansions of the water and the vessel, since obviously there is no pressure present to affect the issue.

Volume of water escaping
$$= [(6210 \times 10^{-6}) - (3 \times 330 \times 10^{-6})] \times (4/3)\pi \times 1500^3$$
$$= 7{\cdot}4 \times 10^{-3}\,\text{m}^3$$

5.6. CONSTRAINED MATERIAL

The following example illustrates the use of the three-dimensional stress–strain relationships in a statically indeterminate problem.

A cylindrical block of concrete is encompassed by a close fitting thin steel tube of inside radius r and wall thickness t as shown in Fig. 5.6. If

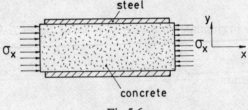

Fig 5.6

the concrete is subjected to a uniform axial compressive stress, σ_x, determine the required ratio of r/t so that the axial strain in the tube is equal to that in the concrete. The moduli and Poisson's ratio for concrete and steel are E_c, ν_c and E_s, ν_s respectively.

The physical nature of the problem is that as the concrete is compressed it expands laterally on to the tube. This action sets up circumferential tensile stress in the tube which in turn results in axial contraction. It is required to make the latter the same as the compression of the concrete by a suitable choice of r/t. Superscripts c and s will be used to denote concrete and steel. Using eqns. (3.3),

$$\text{Axial strain in concrete, } \varepsilon_x{}^c = \frac{\sigma_x{}^c}{E_c} - \frac{2v_c}{E_c}\sigma_y{}^c \qquad (5.41)$$

$(\sigma_y{}^c = \sigma_z{}^c$ from symmetry)

Circumferential strain in concrete,

$$\varepsilon_y{}^c = \frac{\sigma_y{}^c}{E_c}(1 - v_c) - \frac{v_c\sigma_x{}^c}{E_c} \qquad (5.42)$$

$$\text{Circumferential strain in tube, } \varepsilon_y{}^s = \frac{\sigma_y{}^s}{E_s} = \frac{\sigma_y{}^c r}{t E_s} \qquad (5.43)$$

Since the hoop stress, $\sigma_y{}^s$, in the tube is derived from the pressure $\sigma_y{}^c$ exerted by the concrete substituted into eqn. (2.15),

$$\text{Axial strain in tube, } \varepsilon_x{}^s = -v_s\varepsilon_y{}^s = -\frac{v_s\sigma_y{}^s}{E_s} = -\frac{v_s\sigma_y{}^c r}{t E_s} \qquad (5.44)$$

using eqn. (5.43).

The geometry of deformation conditions are that the circumferential strains in the concrete and steel are equal:

$$\varepsilon_y{}^c = \varepsilon_y{}^s$$

and similarly the axial strains are equal:

$$\varepsilon_x{}^c = \varepsilon_x{}^s$$

From eqns. (5.42) and (5.43),

$$\frac{\sigma_y{}^c}{E_c}(1 - v_c) - \frac{v_c\sigma_x{}^c}{E_c} = \frac{\sigma_y{}^c r}{t E_s}$$

$$\sigma_y{}^c\left(\frac{1 - v_c}{E_c} - \frac{r}{t E_s}\right) = \frac{v_c\sigma_x{}^c}{E_c} \qquad (5.45)$$

and from eqns. (5.41) and (5.44),

$$\frac{\sigma_x{}^c}{E_c} - \frac{2v_c}{E_c}\sigma_y{}^c = -\frac{v_s\sigma_y{}^c r}{t E_s}$$

$$\sigma_y{}^c\left(\frac{2v_c}{E_c} - \frac{v_s r}{t E_s}\right) = \frac{\sigma_x{}^c}{E_c} \qquad (5.46)$$

From eqns. (5.45) and (5.46),

$$\frac{r}{tE_s}(1 - v_s v_c) = \frac{1 - v_c}{E_c} - \frac{2v_c^2}{E_c}$$

Hence

$$\frac{r}{t} = \frac{E_s}{E_c} \frac{1 - v_c - 2v_c^2}{1 - v_s v_c} \tag{5.47}$$

Taking $E_s = 15 E_c$, $v_s = 0.29$ and $v_c = 0.25$,

$$\frac{r}{t} = \frac{15[1 - 0.25 - (2 \times 0.25^2)]}{1 - (0.29 \times 0.25)} = \frac{15 \times 0.625}{0.9275} = 10.1$$

Chapter 6
Bending: Stresses

6.1. In Chapter 1 it was shown that, by considering the equilibrium condition of a beam, the externally applied load is related to the shear force and bending moment within the beam. The next stage is to establish the distributions of direct and shear stresses in the beam in terms of bending moment and shear force. The analysis will be developed for a beam of symmetrical cross-section subjected to an external moment at each end causing bending in the plane of symmetry. The beam incurs a bending moment which is constant along the length and is therefore said to be in *pure bending*.

6.2. ASSUMPTIONS IN SIMPLE BENDING THEORY

This is a statically indeterminate problem, and hence all three basic principles stated in the introduction of the preceding chapter will need to be employed. However, the steps will be considered in a different order since at this stage there is insufficient information about the stress distribution to enable an equilibrium equation to be formulated. The approach will be to consider a prismatic beam of symmetrical cross-section subjected to a simple loading condition from which the *geometry of deformation* can be studied and strain distribution determined. The *stress–strain relations* will give the stress distribution which can be related to forces and moments through an *equilibrium condition*.

Before commencing the analysis it is necessary to make some assumptions and these are as follows:

1. Transverse sections of the beam which are plane before bending will remain plane during bending.
2. From consideration of symmetry during bending, transverse sections will be perpendicular to circular arcs having a common centre of curvature.
3. The radius of curvature of the beam during bending is large compared with the transverse dimensions.
4. Longitudinal elements of the beam are subjected only to simple tension or compression, and there is no lateral stress.
5. Young's modulus for the beam material has the same value in tension and compression.

6.3. DEFORMATIONS IN BENDING

(a) Longitudinal

Consideration of the beam subjected to pure bending shown in Fig. 6.1 indicates that the lower surface stretches and is therefore in tension and the upper surface shortens and thus is in compression. Hence there must be an xz-plane in between in which longitudinal deformation is zero. This is termed the *neutral plane*, and a transverse axis lying in the neutral

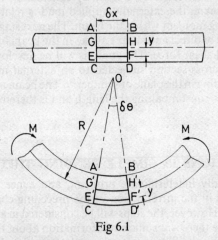

Fig 6.1

plane is the *neutral axis*. Consider the deformations between two sections AC and BD, a distance δx apart, of an initially straight beam. A longitudinal force EF at a distance y below the neutral axis will have initially the same length as the fibre GH at the neutral axis. During bending EF stretches to become E'F', but GH being at the neutral axis is unstrained when it becomes G'H'. Therefore, if R is the radius of curvature of G'H',

$$G'H' = GH = \delta x = R\,\delta\theta$$
$$E'F' = (R + y)\,\delta\theta$$

and the longitudinal strain in fibre E'F' is

$$\varepsilon_x = \frac{E'F' - EF}{EF}$$

But $EF = GH = G'H' = R\,\delta\theta$; therefore

$$\varepsilon_x = \frac{(R + y)\,\delta\theta - R\,\delta\theta}{R\,\delta\theta}$$

Hence

$$\varepsilon_x = \frac{y}{R} \tag{6.1}$$

and since $R = dx/d\theta$, therefore also

$$\varepsilon_x = y\frac{d\theta}{dx} \qquad (6.2)$$

From eqn. (6.1) it will be seen that strain is distributed linearly across the section, being zero at the neutral surface and having maximum values at the outer surfaces. It is important to note here that eqn. (6.1) is entirely independent of the type of material, whether it is in an elastic or plastic state and linear or non-linear in stress and strain.

(b) Transverse

Regarding deformations in the y- and z-directions, it is apparent that changes in length of the beam will result in changes in the transverse dimensions. For example, the fibres in compression will be associated with an increase in thickness, whereas the region in tension will show a decrease in beam thickness. The transverse strains will be

$$\varepsilon_z = \varepsilon_y = -\frac{\nu\sigma_x}{E} \qquad (6.3)$$

and the cross-section of the beam will take up the shape shown in Fig. 6.2. This can be easily demonstrated by bending an eraser. The neutral

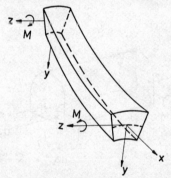

Fig 6.2

surface, instead of being plane, will be curved. This behaviour is termed *anticlastic curvature*. The deformations are extremely small and do not upset the solution for longitudinal strains.

6.4. STRESS–STRAIN RELATIONSHIP

As it has been assumed that $\sigma_y = \sigma_z = 0$ then the stress–strain relationship that is applicable for *linear-elastic bending* is

$$\varepsilon_x = \frac{\sigma_x}{E}$$

and, from eqn. (6.1),

$$\varepsilon_x = \frac{\sigma_x}{E} = \frac{y}{R}$$

or

$$\frac{\sigma_x}{y} = \frac{E}{R} \tag{6.4}$$

Thus bending stress is also distributed in a linear manner over the cross-section, being zero when y is zero, that is at the neutral plane, and being maximum tension and compression at the two outer surfaces where y is maximum, as shown in Fig. 6.3(a).

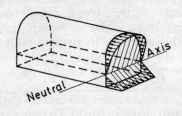

(a)

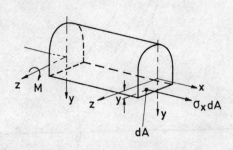

(b)

Fig 6.3

6.5. EQUILIBRIUM OF FORCES AND MOMENTS

(a) Position of the Neutral Plane and Axis

Consider a small element of area, δA, at a distance y from some arbitrary location of the neutral surface, Fig. 6.3(b).

The force on the element in the x-direction is

$$\delta F_x = \sigma_x \delta A$$

Therefore the total longitudinal force on the cross-section is

$$F_x = \int_A \sigma_z \, dA$$

where A is the total area of the section.

Since there is no external axial force in pure bending the internal force resultant must be zero; therefore

$$F_x = \int_A \sigma_z \, dA = 0$$

Using eqn. (6.4) to substitute for σ_z,

$$\frac{E}{R} \int_A y \, dA = 0$$

Since E/R is not zero, the integral must be zero, and as this is the first moment of area about the neutral axis, it is evident that the centroid of the section must coincide with the neutral axis. (The first moment of area of a section about its centroid is zero; see Appendix 2.)

(b) Internal Resisting Moment

Returning to the element in Fig. 6.3(b), the moment of the axial force about the neutral surface is $\delta F_x y$. Therefore the total *internal resisting moment* is

$$\int_A y \, dF_x = \int_A y \sigma_z \, dA$$

This must balance the external applied moment M, so that for equilibrium

$$\int_A y \sigma_z \, dA = M$$

or, substituting for σ_z,

$$M = \frac{E}{R} \int_A y^2 \, dA$$

Now $\int_A y^2 dA$ is the *second moment of area* (see Appendix 2) of the cross-section about the neutral axis and will be denoted by I. Thus

$$M = \frac{EI}{R}$$

or

$$\frac{M}{I} = \frac{E}{R} \tag{6.5}$$

6.6. THE BENDING RELATIONSHIP

Combining eqns. (6.4) and (6.5) gives the fundamental relationship between bending stress, moment and geometry of deformation:

$$\frac{M}{I} = \frac{\sigma}{y} = \frac{E}{R} \tag{6.6}$$

6.7. SECTION MODULUS

For the outer surfaces of the beam,

$$M = \hat{\sigma}_t \frac{I}{y_{t\ max}} = \hat{\sigma}_c \frac{I}{y_{c\ max}}$$

where the subscripts denote tension and compression. The quantities $I/y_{t\ max}$ and $I/y_{c\ max}$ are a function of geometry only; they are termed the *section moduli* and are denoted by Z_t and Z_c. Thus

$$\sigma_{t\ max} = \frac{M}{Z_t} \quad \text{and} \quad \sigma_{c\ max} = \frac{M}{Z_c} \tag{6.7}$$

6.8. THE GENERAL CASE OF BENDING

The bending relationship, eqn. (6.6), is only an exact solution for the case of pure bending; however in practice many beam problems involve bending moment and shearing force which vary along the length. In these cases it has been shown that eqn. (6.6), even if not exact, provides a solution which is quite accurate for engineering design, where cross-sections are clear from support points and concentrated loads.

Example 6.1.   A solid circular bar is built into a wall at one end and carries a concentrated load of 1 kN at a distance of 2 m from the wall. Calculate a suitable diameter so that the maximum tensile bending stress does not exceed 150 MN/m² allowing some margin for safety.

Maximum bending moment $= 2 \times 1 = 2$ kN-m

Let the diameter of the bar be d; then

$$I = \frac{\pi d^4}{64} \quad \text{and} \quad y_{max} = \frac{d}{2}$$

Assuming a safety factor of 2 on the stress,

$$\hat{\sigma} \not> 75 \text{ MN/m}^2$$

Using eqn. (6.6),

$$\frac{2000}{\pi d^4/64} = \frac{75 \times 10^6}{d/2}$$

$$d^3 = \frac{2000 \times 32}{75 \times 10^6 \times \pi} \, m^3$$

$$d = 0{\cdot}065 \, m = 65 \, mm$$

Example 6.2. A tee section bar has dimensions as shown in Fig. 6.4. The bar is used as a simply supported beam of span 1·5 m, the flange being horizontal.

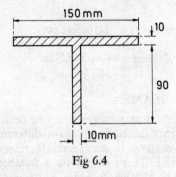

Fig 6.4

Calculate the uniformly distributed load which can be applied if the maximum tensile stress is not to exceed 100 MN/m². What is then the greatest bending stress in the flange?

The first step is to find the position of the centroid, which will also give the neutral axis. Taking first moments of area about the top surface,

$$(240 \times 10)h = (1500 \times 5) + (900 \times 55)$$
$$h = 23{\cdot}8 \, mm$$

To find the second moment of area we can use the parallel axis theorem (see Appendix 2):

$$I_{xx} = \frac{150 \times 10^3}{12} + (150 \times 10 \times 18{\cdot}8^2) + \frac{10 \times 90^3}{12}$$

$$\qquad + (10 \times 90 \times 31{\cdot}2^2)$$

$$\qquad = 2028 \times 10^3 \, mm^4$$

$$M_{max} = \frac{w \times 1{\cdot}5^2}{8} = 0{\cdot}281w \, N\text{-}m$$

The maximum tensile stress occurs on the bottom surface, where

$$y_{max} = 76 \cdot 2 \, \text{mm}$$

Therefore

$$\frac{100 \times 10^6}{0 \cdot 0762} = \frac{0 \cdot 281 w}{2028 \times 10^{-9}}$$

$$w = \frac{202 \cdot 8}{0 \cdot 0762 \times 0 \cdot 281} = 10 \cdot 7 \, \text{kN/m}$$

Since stress is proportional to distance from the neutral axis,

$$\frac{\hat{\sigma}_c}{y_c} = \frac{\hat{\sigma}_t}{y_t}$$

Therefore the greatest stress in the flange is

$$\hat{\sigma}_c = 100 \times 10^6 \times \frac{23 \cdot 8}{76 \cdot 2} = 31 \cdot 3 \, \text{MN/m}^2$$

6.9. COMPOSITE BEAMS

In some circumstances it may be necessary or desirable to construct a beam such that the cross-section contains two different materials. Usually the object is for one material to act as a reinforcement to the other, perhaps weaker, material. There would be a number of reasons (cost, weight, size, etc.) why the whole beam could not be made from the stronger material. The positioning of the reinforcement material might not be symmetrical with respect to the centroid of the cross-section and it could be embedded within or fixed in some manner to the outside of the main bulk material.

The arguments which were applied to the analysis of simple bending of a homogeneous beam also apply to the composite beam since the two materials constrain each other to deform in the same manner, e.g. to an arc of a circle for pure bending.

(a) Symmetrical Sections

Consider a beam cross-section consisting of a central part of, say, plastic or timber with reinforcing plates firmly bonded (no sliding) to the upper and lower surfaces along the length of the beam as shown in Fig. 6.5. The section is symmetrical about the centroid and neutral surface.

Equilibrium

Since there must be no net end load, longitudinal equilibrium gives

$$\int_{Am} \sigma_m \, dA_m + \int_{Ar} \sigma_r \, dA_r = 0 \tag{6.8}$$

where the subscripts m and r refer to the main and reinforcing materials respectively.

Equilibrium of internal and external bending moments gives

$$M_m + M_r = M$$

Hence

$$\int_{Am} \sigma_m y_m \, dA_m + \int_{Ar} \sigma_r y_r \, dA_r = M \tag{6.9}$$

The above equations are independent of whether the material is in an elastic or a plastic condition.

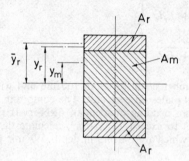

Fig 6.5

Geometry of Deformation
For a linear elastic strain distribution,

$$\frac{1}{R} = \frac{\varepsilon_m}{y_m} = \frac{\varepsilon_r}{y_r} \tag{6.10}$$

These relationships are purely a function of geometry and therefore are independent of the material and its properties.

Stress–Strain Relations

$$\left. \begin{array}{l} \sigma_m = E_m \varepsilon_m \\ \sigma_r = E_r \varepsilon_r \end{array} \right\} \tag{6.11}$$

Solution
From eqns. (6.10) and (6.11), $\sigma_m = y_m E_m / R$ and $\sigma_r = y_r E_r / R$. Substitution in the equilibrium equation (6.9) gives

$$\left(\frac{E_m}{R} \right) \int_{Am} y_m{}^2 \, dA_m + \left(\frac{E_r}{R} \right) \int_{Ar} y_r{}^2 \, dA_r = M$$

but the integrals are the second moments of area of the core and cover plates, I_m and I_r respectively, about the neutral surface. Therefore

$$\frac{E_m I_m + E_r I_r}{R} = M$$

Substituting for $1/R$ gives

$$\frac{\sigma_m}{E_m y_m} = \frac{\sigma_r}{E_r y_r} = \frac{M}{E_m I_m + E_r I_r}$$

or

$$\left.\begin{aligned} \sigma_m &= \frac{M E_m y_m}{E_m I_m + E_r I_r} \\[2mm] \sigma_r &= \frac{M E_r y_r}{E_m I_m + E_r I_r} \end{aligned}\right\} \tag{6.12}$$

Example 6.3. A timber beam of depth 100 mm and width 50 mm is to be reinforced with steel plates on each side. The composite section will be subjected to a maximum bending moment of 6 kN-m. If the maximum stress in the timber is not to exceed 12 MN/m², calculate the required thickness for the steel plates, which are 100 mm in depth. What is the maximum stress in the steel?

$$E_{steel} = 205 \, \text{GN/m}^2; \qquad E_{timber} = 15 \, \text{GN/m}^2.$$

$$I_{timber} = \frac{50 \times 100^3}{12} = 4 \cdot 17 \times 10^6 \, \text{mm}^4$$

$$I_{steel} = \frac{t}{12} \times 100^3 = 0 \cdot 0833 \times 10^6 t \, \text{mm}^4$$

From eqn. (6.12) for limiting stress in the timber,

$$12 \times 10^6 = \frac{6000 \times 0 \cdot 05 \times 15 \times 10^9 \times 10^{12}}{(15 \times 10^9 \times 4 \cdot 17 \times 10^6) + (205 \times 10^9 \times 0 \cdot 0833 \times 10^6 t)}$$

from which $t = 21 \cdot 9$ mm; and since this is the total thickness of steel, the plates are each 10·95 mm thick.

The maximum stress in the steel is

$$\sigma = \frac{6000 \times 0 \cdot 05 \times 205 \times 10^9 \times 10^{12}}{(15 \times 10^9 \times 4 \cdot 17 \times 10^6) + (205 \times 10^9 \times 0 \cdot 0833 \times 10^6 \times 0 \cdot 022)}$$

$$= 975 \, \text{MN/m}^2$$

(b) Equivalent Sections

In beam sections which are symmetrical geometrically, but unsymmetrical with respect to location of the different materials, the neutral axis no longer coincides with the centroid of the section. The problem can still be solved using eqns. (6.8)–(6.11), but the arithmetic is more tedious since the neutral axis has first to be found.

A more convenient approach, which is also valid, is to transform the composite section into an equivalent (from the view of resisting forces and moments) section of only one of the two (or more) materials. The solution is then of course a simple routine.

Consider the beam section shown in Fig. 6.6(a), in which the unknown

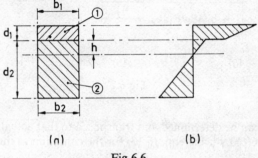

(a)   (b)

Fig 6.6

neutral axis has been placed at a distance h from the interface between the two materials. Assuming no sliding, the strain at the interface must be the same for each material; therefore, at the interface,

$$\varepsilon = \frac{\sigma_1}{E_1} = \frac{\sigma_2}{E_2} \qquad (6.13)$$

Also, rearranging eqns. (6.12),

$$M = \frac{\sigma_1}{h}\left(I_1 + \frac{E_2}{E_1}I_2\right) = \frac{\sigma_2}{h}\left(I_2 + \frac{E_1}{E_2}I_1\right) \qquad (6.14)$$

where I_1 and I_2 are the respective second moments of area about the neutral axis. Now,

$$\frac{E_2}{E_1}I_2 = \frac{E_2}{E_1}\left(\frac{b_2 d_2^{\,3}}{12} + b_2 d_2 \bar{y}_2^{\,2}\right) = b_1'\left(\frac{d_2^{\,3}}{12} + d_2 \bar{y}_2^{\,2}\right)$$

where $b_1' = (E_2/E_1)b_2$, so exactly the same resisting moment will exist if material 2 is replaced by material 1 having the *same* depth d_2 but a new width of $(E_2/E_1)b_2$. This forms the equivalent section shown in Fig. 6.7(a), which is entirely made of material 1. In the converse manner, material 1 can be replaced by material 2 if the width is made $(E_1/E_2)b_1$ to give the equivalent section as in Fig. 6.7(b). The neutral axis can be found quite simply (from either equivalent section) as it now coincides with the centroid. The stress distribution in either equivalent section,

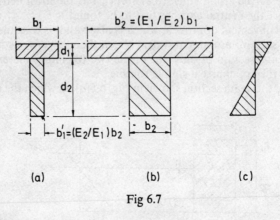

(a)   (b)   (c)

Fig 6.7

Fig. 6.7(c), can be determined and transposed to that actually occurring as in Fig. 6.6(b) by using eqn. (6.13) for the condition at the interface.

Example 6.4. A timber beam of rectangular section 100 mm × 50 mm and simply supported at the ends of a 2 m span has a 30 mm × 10 mm steel strip securely fixed to the top surface as shown in Fig. 6.8(a) to protect the timber from trolley wheels. When the trolley is exerting a force at mid-span of 2 kN, determine the stress distribution at that section. $E_{steel} = 20 E_{wood}$.

Firstly consider the equivalent section made entirely of timber as in Fig. 6.8(b). The position of the centroid and hence of the neutral axis from the top surface is given by

$$\bar{y}(6000 \times 5000) = (6000 \times 5) + (5000 \times 60)$$

$$\bar{y} = 30 \, \text{mm}$$

$$I = \frac{600 \times 10^3}{12} + (6000 \times 25^2) + \frac{50 \times 100^3}{12} + (5000 \times 30^2)$$

$$= 1246 \cdot 6 \times 10^4 \, \text{mm}$$

Maximum bending moment, $M = (2000 \times 2)/4 = 1000 \, \text{N-m}$

At lower surface, $\sigma = \dfrac{1000 \times 0 \cdot 08}{1246 \cdot 6 \times 10^{-8}} = 6 \cdot 4 \, \text{MN/m}^2$

At interface in timber, $\sigma = 6 \cdot 4 \times 20/80 = 1 \cdot 6 \, \text{MN/m}^2$

Since $E_{steel} = 20E_{wood}$, then from eqn. (6.13), $\sigma_{steel} = 20\sigma_{wood}$.

At interface in steel, $\sigma = 1\cdot6 \times 20 = 32\,\mathrm{MN/m^2}$

At top surface, $\sigma = 6\cdot4 \times \dfrac{30}{80} \times 20 = 48\,\mathrm{MN/m^2}$

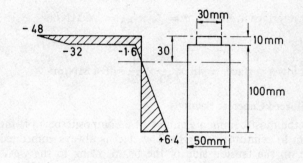

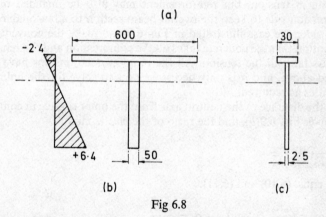

Fig 6.8

The distribution in the equivalent section is shown in Fig. 6.8(*b*) and the actual distribution in Fig. 6.8(*a*). If next the equivalent section is made out of steel then this is represented in Fig. 6.8(*c*).

The centroid is at

$$\bar{y}(300 + 250) = (300 \times 5) + (250 \times 60)$$
$$\bar{y} = 30\,\mathrm{mm}$$

which is the same as before since the proportions are the same. As only width dimensions have been changed by a factor of 20 compared to the previous equivalent section, there is no need to recalculate I, since

$$I_s = \frac{I_w}{20} = \frac{1246\cdot6 \times 10^4}{20} = 623\cdot3 \times 10^3\,\mathrm{mm^4}$$

At top surface, $\sigma = \dfrac{1000 \times 30}{623 \cdot 3 \times 10^{-9}} = 48 \text{ MN/m}^2$

At interface in steel, $\sigma = 48 \times \dfrac{20}{30} = 32 \text{ MN/m}^2$

At interface in timber, $\sigma = 32 \times \dfrac{1}{20} = 1 \cdot 6 \text{ MN/m}^2$

and

At lower surface, $= 48 \times \dfrac{80}{30} \times \dfrac{1}{20} = 6 \cdot 4 \text{ MN/m}^2$

(c) Reinforced Concrete Sections

Perhaps the most common example of a composite beam is the use of steel bars to reinforce concrete. The steel is always embedded in the concrete on the tension side of the beam owing to the weakness of concrete in tension, but reinforcement may also be included on the compression side to keep the overall beam section to a reasonable size.

Consider the case illustrated in Fig. 6.9 and make the conventional assumption that the concrete takes all the compression and the reinforcing bars take all the tension. All the required relationships have been derived above, and it is only necessary now to solve for the unknown quantities as required.

Let the distance of the neutral axis from the outer surface in compression be h, Fig. 6.9(b), and the ratio of the elastic moduli

$$\frac{E_{steel}}{E_{concrete}} = m$$

From eqns. (6.10) and (6.11),

$$\sigma_c = \frac{y_c E_c}{R} \quad \text{and} \quad \sigma_s = \frac{y_s E_s}{R}$$

Substituting the stresses into the equation (6.8) for longitudinal equilibrium, and noting that the stresses in the steel and concrete are of opposite sign, gives

$$-\frac{E_c}{R} \int_{A_c} y_c \, dA_c + \frac{E_s}{R} \int_{A_s} y_s \, dA_s = 0 \qquad (6.15)$$

For the steel reinforcement the tensile stress is considered constant over the cross-sectional area A_s and concentrated at $y = (d - h)$ as in Fig. 6.9(c), so that eqn. (6.15) becomes

$$-E_c \frac{bh^2}{2} + E_s (d - h) A_s = 0 \qquad (6.16)$$

and thus

$$h = \left[\left(\frac{mA_s}{b} \right)^2 + \frac{2mA_s d}{b} \right]^{\frac{1}{2}} - \frac{mA_s}{b} \tag{6.17}$$

which gives the position of the neutral axis.

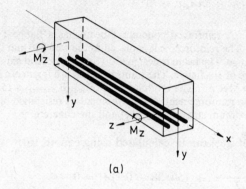

(a)

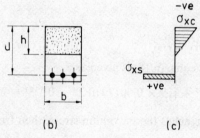

(b)   (c)

Fig 6.9

Substituting for σ_c and σ_s in eqn. (6.9) for equilibrium of moments,

$$\frac{E_c}{R} \int_{A_c} y_c^2 \, dA_c + \frac{E_s}{R} \int_{A_s} y_s^2 \, dA_s = M$$

or

$$\frac{E_c}{R} \frac{bh^3}{3} + \frac{E_s}{R} A_s (d - h)^2 = M$$

Now,

$$\frac{1}{R} = \frac{\sigma_c}{y_c E_c} = \frac{\sigma_s}{y_s E_s}$$

so that substituting for $1/R$ gives

$$\left.\begin{array}{l} \sigma_c = \dfrac{My_c}{\frac{1}{3}bh^3 + mA_s(d-h)^2} \\[4mm] \sigma_s = \dfrac{M(d-h)m}{\frac{1}{3}bh^3 + mA_s(d-h)^2} \end{array}\right\} \tag{6.18}$$

Example 6.5. A reinforced concrete T-beam has a flange 1·5 m wide and 100 mm deep. The reinforcement is placed in the web 380 mm from the upper edge of the flange. The beam is designed so that the neutral axis coincides with the lower edge of the flange. The limits of stress are for steel 110 MN/m² and for concrete 4 MN/m². The modular ratio $E_{steel}/E_{concrete}$ is 15. Calculate (a) the area of the reinforcement, (b) the moment of resistance of the beam, (c) the actual maximum stress in the steel and the concrete.

The area of steel can be calculated using eqn. (6.16):

$$-\frac{1\cdot5 \times 0\cdot1^2}{2} + 15(0\cdot38 - 0\cdot1)A_s = 0$$

whence

$$A_s = 1790 \text{ mm}^2$$

The denominator in eqns. (6.18) is given as

$$\tfrac{1}{3} \times 1\cdot5 \times 0\cdot1^3 + [15 \times 0\cdot00179 (0\cdot38 - 0\cdot1)^2] = 2\cdot61 \times 10^{-3} \text{ m}^4$$

Assuming the steel reaches the maximum stress, then from eqn. (6.18),

$$M = \frac{2\cdot61 \times 10^{-3} \times 110 \times 10^6}{0\cdot28 \times 15} = 68\cdot3 \text{ kN-m}$$

Alternatively,

$$M = \frac{2\cdot61 \times 10^{-3} \times 4 \times 10^6}{0\cdot1} = 104\cdot5 \text{ kN-m}$$

for the concrete to reach its maximum stress.
Therefore

Maximum moment of resistance = 68·3 kN-m

Actual maximum stress in concrete, $\sigma_c = \dfrac{68\cdot3 \times 10^3 \times 0\cdot1}{2\cdot61 \times 10^{-3}}$

$$= 2\cdot62 \text{ MN/m}^2$$

6.10. COMBINED BENDING AND DIRECT STRESSES

A number of situations arise in practice where a member is subjected to a combination of bending and longitudinal load. Problems of this type can be most easily dealt with by superposition of the individual components of stress to give resultant values.

Consider the rectangular-section beam shown in Fig. 6.10, which is

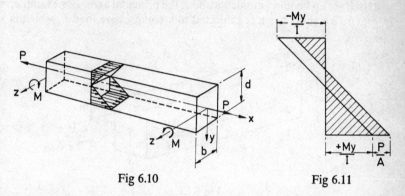

Fig 6.10   Fig 6.11

subjected to bending moments M about the z-axis, and an axial load P in the x direction.

If the end load acted alone, there would be a longitudinal stress

$$\sigma_x = \frac{P}{A}$$

If the moment acted alone, the axial stress would be

$$\sigma_x = \pm \frac{My}{I}$$

By superposition the resultant stress due to P and M is

$$\sigma_x = \frac{P}{A} \pm \frac{My}{I}$$

$$= \frac{P}{bd} \pm \frac{12My}{bd^3} \tag{6.19}$$

The distribution of σ_x over the cross-section is shown in Figs. 6.10 and 6.11 by the shaded portions. An interesting feature is that the neutral surface no longer passes through the centroid of the cross-section since

$$\int_A \sigma_x \, dA \neq 0$$

as in the case of simple bending.

Note that y in eqn. (6.19) is a distance from the centroid of the section and *not* from the new neutral axis.

6.11. ECCENTRIC END LOAD

If the end load P does not act at the centroid of the cross-section then it will itself set up bending moments about the principal axes. For example, in Fig. 6.12 a short beam is subjected to a compressive load P, which is

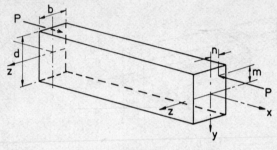

Fig 6.12

eccentric from the z and y axes by amounts m and n respectively. The equivalent equilibrium system is with the load P acting at the centroid and moments Pm and Pn acting about the z and y axes.

The direct stress due to P alone is therefore

$$\sigma_x' = -\frac{P}{A}$$

Due to bending about the y axis,

$$\sigma_x'' = \pm \frac{Pnz}{I_{yy}}$$

Due to bending about the z axis,

$$\sigma_x''' = \pm \frac{Pmy}{I_{zz}}$$

Therefore the resultant longitudinal stress is

$$\sigma_x = \sigma_x' + \sigma_x'' + \sigma_x'''$$

$$= -\frac{P}{A} \pm \frac{Pnz}{I_{yy}} \pm \frac{Pmy}{I_{zz}} \qquad (6.20)$$

$$= \frac{P}{A}\left(-1 \pm \frac{nz}{k_{yy}^2} \pm \frac{my}{k_{zz}^2}\right) \qquad (6.21)$$

where k_{yy} and k_{zz} are the radii of gyration about the relevant axes. For the rectangular section shown in Fig. 6.12, $k^2 = I/A$; therefore

$$k_{yy}{}^2 = \frac{b^2}{12} \quad \text{and} \quad k_{zz}{}^2 = \frac{d^2}{12}$$

and

$$\sigma_x = \frac{P}{bd}\left(-1 \pm \frac{12nz}{b^2} \pm \frac{12my}{d^2}\right) \tag{6.22}$$

where z lies between $\pm b/2$ and y between $\pm d/2$.

In some materials which are strong in compression but weak in tension, such as concrete, it is necessary to limit the eccentricities m and n so that no tensile stress is set up. The condition for no tension is that the compressive stress due to P is greater than or equal to the maximum tensile bending stresses set up by the moments Pm and Pn. Therefore, from eqn. (6.22),

$$\frac{6n}{b} + \frac{6m}{d} \leqslant 1 \tag{6.23}$$

This equation defines the locus of maximum eccentricity as shown by the shaded area in Fig. 6.13. When P is applied within the shaded area

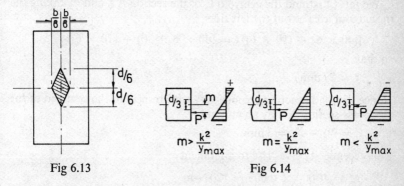

Fig 6.13   Fig 6.14

then no tensile stress will be set up anywhere in the cross-section. The limits on the z and y axes are $\pm b/6$ and $\pm d/6$, which has resulted in what is known as the *middle-third rule* for no tension. Typical distributions for various amounts of eccentricity along *one* principal axis are shown in Fig. 6.14.

Example 6.6.   In a tensile test within the elastic range on a specimen of circular cross-section an extensometer is being used which will only measure deformation on one side of the specimen. Determine how much eccentricity of loading will give rise to a 5% difference between the surface stress derived from the extensometer and the average stress over the cross-section.

Let the average stress be σ; then for a 5% error due to non-axial loading the resultant stress on one edge of the specimen will be 0.95σ, and at the opposite end of the diameter 1.05σ. From eqn. (6.21),

$$0.95\sigma = \sigma\left(1 - \frac{my_{max}}{k_z^2}\right)$$

Therefore

$$m = \frac{0.05k_z^2}{y_{max}}$$

For a circular cross-section $I = Ak^2 = \pi d^4/64$, and hence $k_z^2 = d^2/16$. Therefore

$$m = \frac{0.05d^2/16}{d/2} = 0.00625d$$

Thus in a tensile test on a specimen of 10 mm diameter, the eccentricity of loading must be *less* than 0.063 mm to avoid surface stresses being more than 5% greater than the average direct stress.

Example 6.7.   A slotted machine link 6 mm thick, illustrated in Fig. 6.15(*a*) is subjected to a tensile load of 40 kN acting along the centre line of the end faces. Find the stress distribution for a section through the slot such as AA.

We must first find the centroid C of the section AA and by taking the moment of area about the left side:

$$[(40 \times 6) + (10 \times 6)]\bar{x} = (40 \times 6 \times 20) + (10 \times 6 \times 55)$$

so that

$$\bar{x} = 27 \text{ mm}$$

Hence the load is acting eccentrically with respect to the centroid of the slot cross-section AA by an amount

$$e = 30 - 27 = 3 \text{ mm}$$

This eccentricity gives rise to a moment

$$M = 40\,000 \times 0.003 = 120 \text{ N-m}$$

For bending only, the neutral axis passes through C and the greatest bending stresses occur at B and D; thus $y_{max} = 27$ and 33 mm and the second moment of area is $91.28 \times 10^3 \text{ mm}^4$. Therefore

$$\sigma_B = -\frac{120 \times 0.027}{91.28 \times 10^{-9}} = -35.5 \text{ MN/m}^2$$

$$\sigma_D = +\frac{120 \times 0.033}{91.28 \times 10^{-9}} = +43.4 \text{ MN/m}^2$$

$$\text{Direct stress} = \frac{40\,000}{0.05 \times 0.006} = +133 \text{ MN/m}^2$$

Resultant Stresses

At B $+133 - 35\cdot5 = 97\cdot5\,\text{MN/m}^2$

At D $+133 + 43\cdot4 = 176\cdot5\,\text{MN/m}^2$

At E $+133 + 17\cdot1 = 150\cdot1\,\text{MN/m}^2$

At F $+133 + 30\cdot2 = 163\cdot2\,\text{MN/m}^2$

The distribution of stress is shown in Fig. 6.15(*b*).

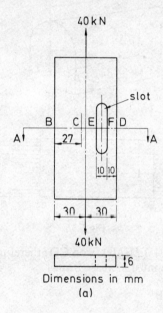

40 kN

slot

B C E F D

A   A

27

10 10

30   30

40 kN

6

Dimensions in mm

(a)

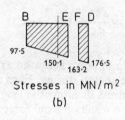

B   E F D

97·5

150·1

163·2   176·5

Stresses in MN/m^2

(b)

Fig 6.15

6.12. SHEAR STRESSES IN BENDING

The presence of shear force indicates that there must be shear stress on transverse planes in the beam. It is not possible to make use of the conditions of geometry of deformation and the stress–strain relationships except in the development of an exact solution. However, from the assumptions about the validity of the bending stress distribution, it is possible to estimate the transverse and longitudinal shear stress distributions in the beam by using only the condition of equilibrium.

Firstly, consider the bending stress distribution in the short section of beam of length dx shown in Fig. 6.16. The bending moment increases

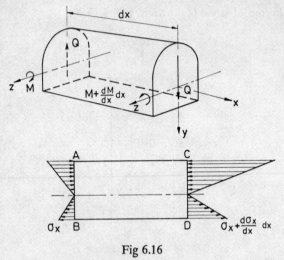

Fig 6.16

from M on AB to $M + (dM/dx)dx$ on CD; therefore the bending stress on any arbitrary fibre must increase from

$$\sigma = \frac{My}{I} \quad \text{on AB} \tag{6.24}$$

to

$$\sigma + \frac{d\sigma}{dx}dx = \left(M + \frac{dM}{dx}dx\right)\frac{y}{I} \quad \text{on CD} \tag{6.25}$$

at each value of y.

Now consider the strip of beam below an xz-plane ABCD, Fig. 6.17, at a distance y_1, from the neutral surface. Each fibre in this strip will have an increase in bending stress $d\sigma$ along the length dx, so that taken over the area A' there will be a resultant axial force equal to

$$\int_{A'} d\sigma\,dA = P \tag{6.26}$$

One way of maintaining equilibrium of the strip is by means of a shear force Q_{yx} acting on the underside of the plane ABCD. Therefore

$$Q_{yx} = \int_{A'} d\sigma\,dA \tag{6.27}$$

Now subtracting eqn. (6.24) from eqn. (6.25) gives

$$d\sigma = dM\frac{y}{I}$$

Hence, from eqn. (6.27),

$$Q_{yx} = \int_{A'} dM\frac{y}{I}\, dA \qquad (6.28)$$

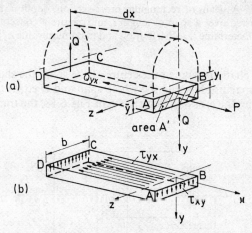

Fig 6.17

If the width of the strip at the plane $y = y_1$ is b and this is small compared with the depth of the beam, then the shear stress on ABCD is almost uniformly distributed and can therefore be expressed as

$$\tau_{yx} = \frac{Q_{yx}}{b\, dx}$$

and combining with eqn. (6.28) gives

$$\tau_{yx} = \frac{dM}{dx}\frac{1}{bI}\int_{A'} y\, dA \qquad (6.29)$$

But, from eqn. (1.15), $dM/dx = Q$ the vertical shear force on the section and the integral is the first moment of area of A' about the neutral surface; therefore eqn. (6.29) becomes

$$\tau_{yx} = Q\frac{A\bar{y}}{bI}$$

Using the principle of complementary shear stresses, Fig. 6.17(b), it is then evident that the vertical shear stress, τ_{yx}, is also given by

$$\tau_{xy} = Q\frac{A\bar{y}}{bI} \qquad (6.30)$$

This solution is only exact for a constant shear force along the beam; however, if the cross-section is small compared with the span, the error introduced by a varying shear force is quite small.

Example 6.8. A beam of rectangular cross-section, depth d, thickness b, is simply supported over a span of length l, and carries a concentrated load W at mid-span. Determine the distribution and maximum value of the transverse shear stress.

Although changing sign at the centre of the span, the shear force Q is constant in magnitude along the whole span and is equal to $W/2$.
Considering the cross-section shown in Fig. 6.18, the transverse shear

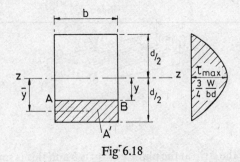

Fig. 6.18

stress on some arbitrary line AB at a distance y from the neutral surface is given by eqn. (6.30), where A' is the shaded area below AB and $\bar{y}$ is the distance of the centroid of A' from the neutral surface. Therefore

$$A'\bar{y} = b\left(\frac{d}{2} - y\right) \times \frac{1}{2}\left(\frac{d}{2} + y\right) = \frac{b}{2}\left[\left(\frac{d}{2}\right)^2 - y^2\right]$$

The vertical shear stress is therefore

$$(\tau_{xy})_{AB} = \frac{Q}{bI} \times \frac{b}{2}\left[\left(\frac{d}{2}\right)^2 - y^2\right]$$

$$= \frac{W}{4I}\left[\left(\frac{d}{2}\right)^2 - y^2\right] \qquad (6.31)$$

The above expression shows that the distribution of vertical shear stress down the depth of the section is parabolic. The shear stress is zero at the outer fibres where $y = \pm d/2$, as it must be since the complementary shear stress in the longitudinal direction must be zero at a free surface. The maximum value is at the neutral surface where $y = 0$; therefore

$$\tau_{xy\ max} = \frac{Wd^2}{16I} = \frac{3W}{4bd}$$

If uniformly distributed the shear stress would be given by the shear force divided by the area, or

$$\tau_{xy\ mean} = \frac{W}{2bd}$$

Hence the maximum shear stress is 1·5 times the mean value.

Example 6.9. A box beam is built up of plate material riveted together as shown in Fig. 6.19. It is simply supported at each end of a 3 m span and carries

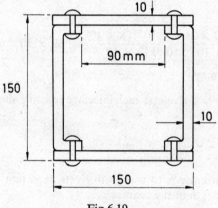

Fig 6.19

a concentrated load of 12 kN at 1 m from one end. Estimate a suitable rivet diameter if the rivets are to be pitched at about 100 mm intervals. The shear stress is not to exceed 50 MN/m² in each rivet.

The maximum shear force will be equal to the larger of the two reactions, $12 \times 2/3 = 8$ kN.

Now, the force tending to shear the rivets is due to the variation of bending stress along the length of the beam and is given by eqn. (6.29) and the next equation if slightly rearranged as follows:

$$\text{Shear force per unit length} = \tau_{yx}b = \frac{QA\bar{y}}{I}$$

If the rivet pitch is p and the cross-sectional area a, then the allowable shear force per unit length is $\hat{\tau}a/p$. Hence

$$\frac{\hat{\tau}a}{p} = \frac{QA\bar{y}}{I}$$

Required area, $a = \dfrac{QA\bar{y}p}{\hat{\tau}I}$

$$I \approx \frac{0 \cdot 15 \times 0 \cdot 15^3}{12} - \frac{0 \cdot 13 \times 0 \cdot 11^3}{12} - (2 \times 0 \cdot 09 \times 0 \cdot 01 \times 0 \cdot 06^2)$$

$$\approx 0 \cdot 2132 \times 10^{-4}\,\mathrm{m}^4$$

At the interface between the outer plates and the side plates where shearing would occur,

$$A\bar{y} = 150 \times 10 \times 70 = 105 \times 10^3\,\mathrm{mm}^3$$

Therefore

$$a = \frac{8000 \times 105 \times 10^{-6} \times 0 \cdot 1}{50 \times 10^6 \times 0 \cdot 2132 \times 10^{-4}} = 7 \cdot 89 \times 10^{-5}\,\mathrm{m}^2$$

$$= 78 \cdot 9\,\mathrm{mm}^2$$

As there are two rivets at each interface resisting shear, the diameter of each is

$$d = \left(\frac{78 \cdot 9}{2} \times \frac{4}{\pi}\right)^{\frac{1}{2}} = 7 \cdot 1\,\mathrm{mm}$$

The best compromise is to use 7 mm rivets at 97 mm pitch, giving 31 pitches in the length of the beam.

6.13. BENDING AND SHEAR STRESSES IN I-SECTION BEAMS

The I-section beam shown in Fig. 6.20(a) is widely used in the construction of buildings, bridges, etc. The shape is efficient to resist both bending and shear. The latter is carried almost entirely by the web, and the flanges are located where the bending stress is highest. This practical I-section is usually idealized for ease of calculation into the rectangular shapes shown at (b). The second moment of area of the section is derived in Appendix 2 in two ways; and one of these gives

$$I = \frac{BD^3 - bd^3}{12}$$

the bending stress distribution being

$$\sigma = \frac{12My}{BD^3 - bd^3}$$ (6.32)

which is illustrated in Fig. 6.21.

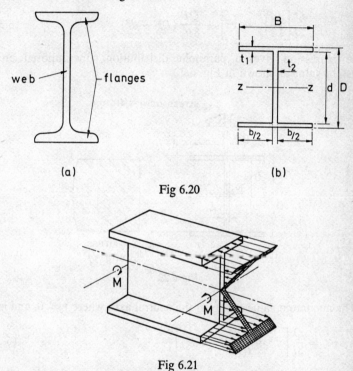

Fig 6.20

Fig 6.21

It can be shown that for a typical I-section the flanges carry approximately 80% of the total bending moment on the cross-section. In designing an I-section against failure consideration must be given, in addition to the strength of the tension flange, to the avoidance of buckling (Chapter 11) in the compression flange.

Shear Stresses in Web and Flanges

The shear stress distributions are rather more complex than in a rectangular section. Firstly, we will examine the distribution of vertical shear τ_{xy} parallel to the axis yy.

For the web, from eqn. (6.30),

$$\tau_{xy} = \frac{Q}{It_1} \int_y^{D/2} y \, dA$$

Because the section has a different width for the flange and the web this integral has to be expressed in two parts:

$$\tau_{xy} = \frac{Q}{It_1}\left[\int_y^{d/2} t_1 y \, dy + \int_{d/2}^{D/2} By \, dy\right]$$

$$= \frac{Q}{2I}\left[\left(\frac{d}{2}\right)^2 - y^2\right] + \frac{QB}{8It_1}(D^2 - d^2) \tag{6.33}$$

This expression gives a parabolic distribution superimposed on a constant value as shown in Fig. 6.22.

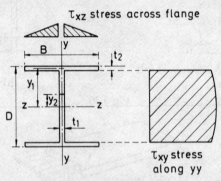

Fig 6.22

The maximum value occurs at the neutral axis, where $y = 0$, and is

$$\tau_{xy} = \frac{Q}{8I}\left[d^2 + \frac{B}{t_1}(D^2 - d^2)\right] \tag{6.34}$$

In that part of the flange directly above and below the web the vertical shear stress might be expressed as

$$\tau_{xy} = \frac{Q}{BI}\int_{y'}^{D/2} By \, dy = \frac{Q}{2I}\left[\left(\frac{D}{2}\right)^2 - y'^2\right]$$

However, in those parts of the flange on each side of the web the top and bottom surfaces are "free" from load, and therefore the longitudinal and complementary vertical shear stresses must be zero. Thus the distribution is parabolic as for a rectangular section as shown in Fig. 6.22. It is evident that in the flanges the vertical shear stress and its complementary component contribute little to balancing the longitudinal variation in bending stress. However, this may be achieved, as illustrated in Fig. 6.23, by means of a shear force Q_{zz} lying in an xy-plane which cuts off a segment of the flange. A complementary shear force Q_{zz} then occurs in a *yz*-plane. As in the analysis of Section 6.12, the net end load is equal to the shear force:

$$P_x = Q_{zx} = \tau_{zx}t_2 \, dx$$

Also

$$P_x = \int_A d\sigma_x \, dA$$

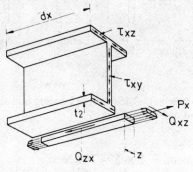

Fig 6.23

Therefore

$$\tau_{zx}t_2 \, dx = \int_A d\sigma_x \, dA$$

$$= \frac{dM}{I} \int_A y \, dA$$

$$\tau_{zx} = \frac{Q}{t_2 I} A\bar{y} \tag{6.35}$$

where Q is the *vertical* shear force on the section and $\bar{y}$ is still measured from the neutral axis to the centroid of the area.

$$A\bar{y} = zt_2 \frac{D - t_2}{2}$$

From eqn. (6.35) and the complementary shear-stress condition,

$$\tau_{zx} = \tau_{xz} = Qz \frac{D - t_2}{2I} \tag{6.36}$$

This is a linear distribution of shear stress in the *z*-direction, being zero at the outer edges of the flanges and a maximum at the joint with the web. The distribution is shown in Fig. 6.22, in which the maximum value is

$$\tau_{xz} = \frac{Q(D - t_2)(B - t_1)}{4I} \tag{6.37}$$

Example 6.10.   The vertical steel column of 5 m height and rolled I-section shown in Fig. 6.24 is built in at the lower end and subjected to a transverse

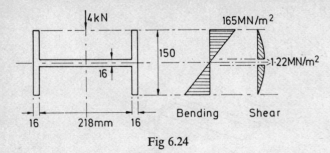

Fig 6.24

force of 4 kN at the free end. Calculate the bending and shear stress distributions at the fixed-end cross-section.

At the base

Bending moment $= 4 \times 5 = 20$ kN-m

Shear force $= 4$ kN

Second moment of area $= \dfrac{32 \times 150^3}{12} + \dfrac{218 \times 16^3}{12}$

$$= 9 \cdot 075 \times 10^6 \, \text{mm}^4$$

$y_{max} = \pm 0 \cdot 075$ m

Therefore

$$\sigma_{max} = \pm \frac{20 \times 10^3 \times 0 \cdot 075}{9 \cdot 075 \times 10^{-6}} = \pm 165 \, \text{MN/m}^2$$

In the flanges the shear stress in the y-direction is given by

$$\tau_{xy} = \frac{4000}{2 \times 0 \cdot 016 \times 9 \cdot 075 \times 10^{-6}} \int_y^{75} 2 \times 0 \cdot 016 y \, dy$$

$$= \frac{440 \times 10^6}{2} (0 \cdot 075^2 - y^2)$$

This is a parabolic distribution varying from zero at the outer surfaces to

$$\tau_{xy} = 220 \times 10^6 (0 \cdot 075^2 - 0 \cdot 008^2) = 1 \cdot 22 \, \text{MN/m}^2$$

at the section where the flanges join the web.

In the web itself, since the width, which appears in the denominator for shear stress above, is 250 mm, it is evident that the shear stress in the web can be neglected compared with that in the flanges.

A consideration of the geometry of the section indicates that shear stresses of the τ_{xz} type are also insignificant.

The distributions of bending and shear stresses are shown in Fig. 6.24.

6.14. ASYMMETRICAL OR SKEW BENDING

The previous analysis has been concerned with bending about an axis of symmetry. However, many occasions arise in practice where bending will occur either of a section which does not have any axes of symmetry or of a symmetrical section about an asymmetrical axis. In order to express the conditions of equilibrium in asymmetrical bending a knowledge of first and second moments of area about arbitrary axes through the centroid of the section is required. It is therefore necessary to study certain properties of areas before embarking on the analysis of stress distribution in asymmetrical bending. The reader is here referred to Appendix 2.

Asymmetrical pure bending of a beam is shown in Fig. 6.25 for positive

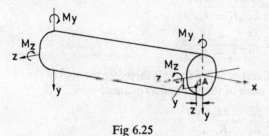

Fig 6.25

external moments M_y and M_z applied about an arbitrary set of centroidal axes. Considering first bending in the xy-plane only, the equilibrium equations are

$$\int_A \sigma_x \, dA = 0 \quad \text{(neutral surface through the centroid)}$$

$$\int_A \sigma_x y \, dA = M_z \quad \text{and} \quad \int_A \sigma_x z \, dA = M_y$$

But $\sigma_x = yE/R_y$, where R_y is the radius of curvature in the xy-plane; hence

$$M_z = \frac{E}{R_y} \int_A y^2 \, dA = \frac{EI_z}{R_y}$$

and

$$M_y = \frac{E}{R_y} \int_A yz \, dA = \frac{EI_{yz}}{R_y}$$

I_{yz} is called the *product moment of area* (see Appendix 2).

For bending in the xz-plane only a procedure similar to the above gives

$$M_y = \frac{E}{R_z} \int z^2 \, dA = \frac{EI_y}{R_z}$$

$$M_z = \frac{E}{R_z} \int zy \, dA = \frac{EI_{yz}}{R_z}$$

For simultaneous bending in the xy- and xz-planes the above relationships may be superimposed to give

$$\left. \begin{aligned} M_y &= \frac{EI_y}{R_z} + \frac{EI_{yz}}{R_y} \\[2mm] M_z &= \frac{EI_z}{R_y} + \frac{EI_{yz}}{R_z} \end{aligned} \right\} \tag{6.38}$$

The radii of curvature are obtained from the above equations as

$$\frac{1}{R_y} = \frac{M_z I_y - M_y I_{yz}}{E(I_y I_z - I_{yz}^2)}$$

$$\frac{1}{R_z} = \frac{M_y I_z - M_z I_{yz}}{E(I_y I_z - I_{yz}^2)}$$

The resultant bending stress is therefore the sum of the components for bending in each of the xy- and xz-planes:

$$\begin{aligned} \sigma_x &= \frac{yE}{R_y} + \frac{zE}{R_z} \\[2mm] &= \frac{y(M_z I_y - M_y I_{yz}) + z(M_y I_z - M_z I_{yz})}{I_y I_z - I_{yz}^2} \end{aligned} \tag{6.39}$$

If either M_z or M_y is in the opposite sense, i.e. negative to that shown in Fig. 6.25, then the appropriate signs must be changed in eqn. (6.39) and elsewhere.

The neutral surface, where $\sigma_x = 0$, is defined by the plane

$$y(M_z I_y - M_y I_{yz}) + z(M_y I_z - M_z I_{yz}) = 0 \tag{6.40}$$

For principal axes, $I_{yz} = 0$, and

$$\sigma_x = \frac{M_z y}{I_z} + \frac{M_y z}{I_y} \tag{6.41}$$

If there is only a single external applied moment M about an axis ss inclined at θ to one of the principal axes, as in Fig. 6.26, then the moment vector M can be resolved into components M_y and M_z about the y- and z-axes, so that

$$M_y = M \sin \theta \quad \text{and} \quad M_z = M \cos \theta$$

and substituting in eqn. (6.41),

$$\sigma_x = \frac{Mz \sin \theta}{I_y} + \frac{My \cos \theta}{I_z} \qquad (6.42)$$

The neutral plane will no longer be perpendicular to the plane of bending, as in the symmetrical problem, but can be determined by putting $\sigma_x = 0$ above; then

$$\frac{z \sin \theta}{I_y} + \frac{y \cos \theta}{I_z} = 0$$

and

$$\frac{y}{z} = -\frac{I_z}{I_y} \tan \theta = -\tan \phi$$

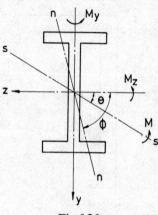

Fig 6.26

where ϕ is the inclination of nn, the neutral surface, to the z-axis.

The neutral surface is perpendicular to the plane of bending if either $I_z = I_y$ or $\theta = 0$.

Example 6.11. The angle section shown in Fig. 6.27 is subjected to a bending moment of 2 kN-m about the z-axis. Determine the bending stress distribution.

Considering the first moments of area about vertical and horizontal edges respectively,

$$\bar{z} = \frac{(90 \times 10 \times 5) + (60 \times 10 \times 30)}{(90 + 60)10} = 15 \, \text{mm}$$

$$\bar{y} = \frac{(90 \times 10 \times 55) + (60 \times 10 \times 5)}{(90 + 60)10} = 35 \, \text{mm}$$

which gives the co-ordinates of the centroid.

$$I_z = \frac{10 \times 90^3}{12} + (10 \times 90 \times 20^2) + \frac{60 \times 10^3}{12}$$
$$+ (60 \times 10 \times 30^2) = 151\cdot2 \times 10^4 \, \text{mm}^4$$
$$I_y = \frac{90 \times 10^3}{12} + (90 \times 10 \times 10^2) + \frac{10 \times 60^3}{12}$$
$$+ (10 \times 60 \times 15^2) = 41\cdot3 \times 10^4 \, \text{mm}^4$$
$$I_{zy} = [90 \times 10 \times (-10) \times (-20)] + [60 \times 10 \times 15 \times 30]$$
$$= 45 \times 10^4 \, \text{mm}^4$$

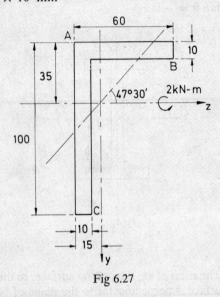

Fig 6.27

The position of the neutral axis may now be found using eqn. (6.40), from which, with $M_y = 0$,

$$y(2000 \times 41\cdot3 \times 10^{-8}) + z(-2000 \times 45 \times 10^{-8}) = 0$$

$$\frac{y}{z} = 1\cdot09 \quad \text{and} \quad \theta = 47^\circ \, 30'$$

The maximum tensile stress will occur at A, and using eqn. (6.39), in which M_y will be zero,

$$\sigma_A = \frac{2000[(0\cdot035 \times 41\cdot3 \times 10^{-8}) - (-0\cdot015 \times 45 \times 10^{-8})]}{[(41\cdot3 \times 151\cdot2) - 45^2]10^{-16}}$$

$$= 101 \, \text{MN/m}^2$$

At B,

$$\sigma_B = \frac{2000[(0\cdot025 \times 41\cdot3 \times 10^{-8}) - (0\cdot045 \times 45 \times 10^{-8})]}{[(41\cdot3 \times 151\cdot2) - 45^2]10^{-16}}$$

$$= -47\,\text{MN/m}^2$$

The maximum compressive stress occurs at C, where

$$\sigma_C = \frac{2000[(-0\cdot065 \times 41\cdot3 \times 10^{-8}) - (-0\cdot005 \times 45 \times 10^{-8})]}{[(41\cdot3 \times 151\cdot2) - 45^2]10^{-16}}$$

$$= -116\cdot5\,\text{MN/m}^2$$

The distribution is illustrated in Fig. 6.28.

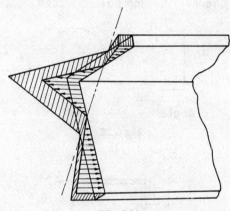

Fig 6.28

6.15. SHEAR STRESS IN THIN-WALLED OPEN SECTIONS AND SHEAR CENTRE

There are a number of beam sections widely used, particularly in aircraft construction, in which the thickness of material is small compared with the overall geometry and there is only one or no axis of symmetry. These members are termed *thin-walled open sections*, and some common shapes are shown in Fig. 6.29.

The arguments applied to the shear stress distribution in the flanges of the I-section (Section 6.13) may also be applied in the above cases; however, there is one important difference owing to the lack of symmetry in the latter.

If the external applied forces, which set up bending moments and shear forces, act through the centroid of the section, then in addition to bending, twisting of the beam will generally occur. To avoid twisting, and cause only bending, it is necessary for the forces to act through a particular point, which may not coincide with the centroid. The position of this point is a function only of the geometry of the beam section; it is termed the *shear centre*.

Before deriving a general theory for the bending of open-walled sections a simple example will be studied and the existence of a shear centre

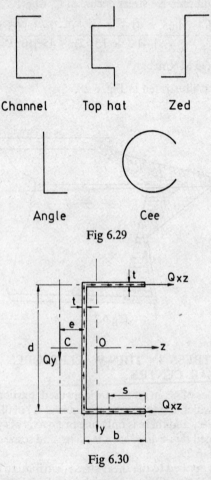

Channel   Top hat   Zed

Angle   Cee

Fig 6.29

Fig 6.30

established. Referring to Fig. 6.30, in which the channel section is loaded by a vertical force Q_y, the y- and z-axes are principal axes and hence $I_{yz} = 0$. For the shear stress in the flanges the analysis is similar to that in Section 6.13, eqn. (6.36), for the I-section.

$$\tau_{xz} = \frac{Q_y}{tI_z} \int_A y \, dA$$

and for a length z of the flange,

$$\tau_{xz} = \frac{Q_y}{tI_z} \int_0^z yt\, dz = \frac{Q_y}{tI_z} \frac{tzd}{2} \tag{6.43}$$

The shear stress varies linearly with z from zero at the left to a maximum at the centre line of the web:

$$\tau_{xz\, max} = \frac{Q_y bd}{2I_z}$$

The average shear stress is $Q_y bd/4I_z$, and therefore the horizontal shear force in the top and bottom flange is

$$Q_{xz} = \frac{Q_y b^2 t d}{4I_z}$$

The couple about the x-axis of these shear forces is

$$dQ_{xz} = \frac{Q_y b^2 d^2 t}{4I_z}$$

Let the vertical force Q_y act through a point C, the shear centre, at a distance e from the middle of the web. Then twisting of the section is avoided if the moment eQ_y balances the moment dQ_{xz} due to the horizontal shear forces, so that, for equilibrium,

$$eQ_y = \frac{Q_y b^2 d^2 t}{4I_z}$$

or

$$e = \frac{b^2 d^2 t}{4I_z} \tag{6.44}$$

which locates the position of the shear centre. The vertical shear stress τ_{xy} in the web may be found in the same way as was that for the I-section.

6.16. GENERAL CASE OF BENDING OF A THIN-WALLED OPEN SECTION

The analysis of the channel section was relatively simple, since there was one axis of symmetry about which bending was made to occur, and had much in common with the analysis of shear stresses in the I-section. The more general case is that of a geometrically asymmetrical open section subjected to bending which is not about a principal axis.

The method will be illustrated for a cantilever of arbitrarily curved open section, as shown in Fig. 6.31, subjected to a transverse force, W, through the centroid of the section at the free end. The resulting components of applied force and shear force in the co-ordinate directions will be W_z, W_y and Q_z, Q_y respectively acting through the centroid. The force components will give rise to bending moments M_y and M_z, which will vary along the length of the member.

It will be assumed that shear stress through the wall thickness is negligible and that it is the longitudinal and peripheral shear stresses

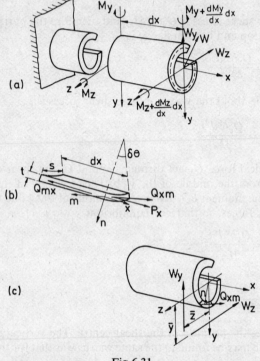

Fig 6.31

which are of importance. The equilibrium condition in the x direction for the strip, Fig. 6.31(b), cut from the section subjected to bending in the xy-plane alone, due to M_z, is

$$Q_{mx} = P_x = \int_{A'} d\sigma_x \, dA$$

where $d\sigma_x$ is the increase in bending stress along the strip and Q_{mx} is the shear force on the edge of the strip. The orthogonal axes m and n are tangential and radial to the wall at the particular strip under examination. Now, from eqn. (6.39), putting $M_y = 0$,

$$d\sigma_x = dM_z \frac{yI_y - zI_{yz}}{I_yI_z - I_{yz}{}^2}$$

Therefore

$$Q_{mx} = \frac{dM_z}{I_yI_z - I_{yz}{}^2} \int_{A'} (yI_y - zI_{yz}) \, dA$$

or

$$\tau_{mx} = \tau_{xm} = \frac{Q_{mx}}{t\,dx} = \frac{dM_z}{dx} \frac{1}{t(I_yI_z - I_{yz}{}^2)} \left[I_y \int_{A'} y \, dA - I_{yz} \int_{A'} z \, dA \right]$$

$$\tau_{xm} = \frac{Q_y}{t(I_yI_z - I_{yz}{}^2)} \left[I_y \int_{A'} y \, dA - I_{yz} \int_{A'} z \, dA \right] \tag{6.45}$$

For bending in the xz-plane alone, due to M_y,

$$\tau_{xm}{}' = \frac{Q_z}{t(I_yI_z - I_{yz}{}^2)} \left[I_z \int_{A'} z \, dA - I_{yz} \int_{A'} y \, dA \right] \tag{6.46}$$

The integrals in the above expressions are the first moments of the area A' about the y- and z-axes. The resultant value of the shear stress at a particular point is given by superposition of the above equations.

The shear stress τ_{xm} will give rise to a torque about the x-axis and hence twisting of the beam. In order that there shall be bending only and no twist, it is therefore necessary to arrange that W_y and W_z act through the shear centre having co-ordinates $\bar{y}$ and $\bar{z}$ as in Fig. 6.31(c). These co-ordinates can be found by considering torsional equilibrium of the section so that

$$W_y\bar{z} = \oint n\tau_{xm}t\rho \, d\theta \tag{6.47}$$

and

$$W_z\bar{y} = \oint n\tau_{xm}{}'t\rho \, d\theta \tag{6.48}$$

where ρ is the radius of curvature of the element and n is the perpendicular distance of τ_{xm} or $\tau_{xm}{}'$ from the x-axis. (τ_{xm} and $\tau_{xm}{}'$ are given by eqns. (6.45) and (6.46)). Since W_y and W_z equal Q_y and Q_z in the above equations and therefore cancel, it will be seen that the position of the shear centre is only a function of the geometry of the cross-section.

Example 6.12. A thin-walled tube of circular cross-section has inner and outer diameters of 50 and 70 mm respectively. If it is slit longitudinally on one side, at what position must a vertical force of 8 kN be applied so that there is only bending and no twisting of the section? Calculate the maximum shear stress in the section.

Referring to Fig. 6.32, since the slit is narrow the I for the section may

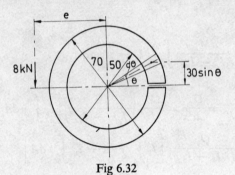

Fig 6.32

be taken as $(\pi/64)(70^4 - 50^4)$.

$$I = 843 \times 10^3 \, \text{mm}^4$$

For the element shown,

$$d\sigma = \frac{dM \times 0.03 \sin \theta}{843 \times 10^{-9}}$$

Net end load $= d\sigma \, tr \, d\theta$

$$= \frac{dM \times 0.03 \sin \theta}{843 \times 10^{-9}} 0.01 \times 0.03 \, d\theta$$

Total end load $= \dfrac{9 \times 10^{-6}}{843 \times 10^{-9}} dM \displaystyle\int_0^\theta \sin \theta \, d\theta$

For equilibrium with the horizontal shear stress,

$$\tau \times 0.01 dx = 10.67 dM(1 - \cos \theta)$$
$$\tau = 1067 Q(1 - \cos \theta)$$

The maximum shear stress occurs when $\theta = \pi/2$ and $3\pi/2$, so that

$$\hat{\tau} = 1067 \times 8000 = 8.54 \, \text{MN/m}^2$$

Torque set up by shear stress distribution

$$= \int_0^{2\pi} \tau \times 0.01 \times 0.03^2 \, d\theta$$

$$= 1067 \times 8000 \times 0.01 \times 0.03^2 \int_0^{2\pi} (1 - \cos \theta) \, d\theta$$

Let the shear centre be at a distance e from the centre of the tube. Then equilibrium of torques gives

$$8000e = 1067 \times 8000 \times 9 \times 10^{-6} \times 2\pi$$
$$e = 60 \cdot 5 \text{ mm}$$

6.17. BENDING OF INITIALLY CURVED BARS

The theory of bending developed so far has been related to initially straight bars and beams. The analysis will now be extended to include members which are initially curved.

The geometry of curved bars has an important bearing on the bending stress distribution. If the depth of the cross-section is small compared with the radius of curvature, then the stress distribution is linear as for straight beams. On the other hand, if the depth of section is of the same order as the radius of curvature, then a non-linear stress distribution occurs during bending.

Similar assumptions are made for curved beams as for straight beams,

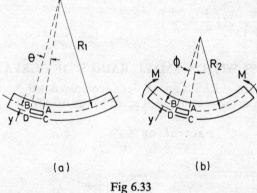

(a)   (b)

Fig 6.33

plane cross-sections remaining plane, etc., although a few of the assumptions are not strictly accurate for the case of a bar with a small radius of curvature.

Consider the curved bar shown unloaded in Fig. 6.33(a) and subjected to pure bending M (Fig. 6.33(b)) with initial and final radii of the neutral axis R_1 and R_2 respectively. The strain in a small element CD at a distance y from the neutral axis is derived as for the straight beam and is

$$\varepsilon_{CD} = \frac{(R_2 + y)\phi - (R_1 + y)\theta}{(R_1 + y)\theta}$$

but for an element AB at the neutral surface there is no change in length, so that $R_1\theta = R_2\theta$. Therefore

$$\varepsilon_{CD} = \frac{y(\phi - \theta)}{(R_1 + y)\theta}$$

Making the substitution $\phi = R_1\theta/R_2$ gives

$$\varepsilon_{CD} = \frac{y[(R_1/R_2) - 1]}{R_1 + y} = \frac{y(R_1 - R_2)}{R_2(R_1 + y)} \tag{6.49}$$

For the slender beam, y can be neglected compared with R_1 and

$$\varepsilon = y\left(\frac{1}{R_2} - \frac{1}{R_1}\right) \tag{6.50}$$

For R_1 infinite, i.e. a straight beam, the expression reduces to that found previously.

By using the same concept as for the straight beam it can be shown that for no end load the centroidal axis and the neutral axis coincide, and for equilibrium of moments

$$\frac{M}{I} = \frac{\sigma}{y} = E\left(\frac{1}{R_2} - \frac{1}{R_1}\right) \tag{6.51}$$

6.18. BEAMS WITH A SMALL RADIUS OF CURVATURE

For the beam in which y is not negligible compared with R_1, the strain at distance y from the neutral axis is given by eqn. (6.49). This is no longer

neutral axis

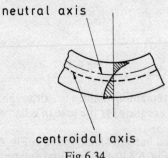

centroidal axis

Fig 6.34

a linear distribution of strain across the section as for the slender beam, and hence *the distribution of stress is non-linear*, as indicated in Fig. 6.34 and *the centroidal and neutral axes no longer coincide* as will now be shown.

Assuming that there is no applied end load,

$$\int_A \sigma \, dA = 0$$

and from eqn. (6.49),

$$\sigma = E\varepsilon = \frac{Ey(R_1 - R_2)}{R_2(R_1 + y)} \tag{6.52}$$

Therefore

$$\frac{E(R_1 - R_2)}{R_2} \int_A \frac{y}{R_1 + y} \, dA = 0$$

since the integral must be zero, and this is not the first moment of area about the centroid; therefore the centroidal and neutral axes do not coincide.

For equilibrium of internal and external moments,

$$M = \int_A \sigma y \, dA$$

$$M = \frac{E(R_1 - R_2)}{R_2} \int_A \frac{y^2}{R_1 + y} \, dA \tag{6.53}$$

The integral term may be re-expressed as

$$\int_A \frac{y^2}{R_1 + y} \, dA = \int_A y \, dA - R_1 \int_A \frac{y}{R_1 + y} \, dA$$

$$= \int_A y \, dA$$

since the second integral is zero. Now let the distance between the centroidal and neutral axes be n, then, from Fig. 6.35, $y = y' + n$; hence

$$\int_A y \, dA = \int_A y' \, dA + \int_A n \, dA = nA$$

since $\int_A y' \, dA$ is zero, being the first moment of area about the centroid.
Thus

$$\int_A \frac{y^2}{R_1 + y} \, dA = nA$$

Substituting into eqn. (6.53),

$$\frac{M}{nA} = \frac{E(R_1 - R_2)}{R_2}$$

and further substituting from eqn. (6.52),

$$\frac{M}{nA} = \frac{\sigma}{y}(R_1 + y)$$

or

$$\frac{\sigma}{y} = \frac{M}{nA(R_1 + y)} \qquad (6.54)$$

which is in a form similar to the bending stress relationship for slender beams, the second moment of area term, I, being replaced by $n(R_1 + y)A$.

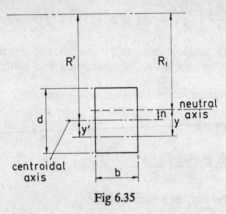

Fig 6.35

In order to determine the magnitude of bending stress it is first necessary to find the values of n and R_1 for the particular shape of cross-section. From the condition that

$$\int_A \frac{y}{R_1 + y}\, dA = 0$$

it follows that

$$\int_A \frac{y' + n}{R' + y'}\, dA = 0$$

or

$$\int_A \left(\frac{R' + y'}{R' + y'} - \frac{R'}{R' + y'} + \frac{n}{R' + y'} \right) dA = 0$$

$$\int_A dA - \int_A \frac{R'}{R' + y'}\, dA + \int_A \frac{n}{R' + y'}\, dA = 0$$

Hence

$$n = R' - \frac{A}{\displaystyle\int_A \frac{dA}{R' + y'}} \qquad (6.55)$$

and

$$R_1 = R' - n \qquad (6.56)$$

Rectangular Section

Width $= b$. Depth $= d$. From eqn. (6.55) and Fig. 6.35,

$$n = R' - \cfrac{bd}{\displaystyle\int_{-d/2}^{+d/2} \cfrac{b}{R' + y'}\, dy'}$$

$$= R' - \cfrac{d}{\log_e\left(\cfrac{R' + d/2}{R' - d/2}\right)} \qquad (6.57)$$

and

$$R_1 = \cfrac{d}{\log_e\left(\cfrac{R' + d/2}{R' - d/2}\right)}$$

To obtain an accurate value for n it is essential to have a very accurate value for the $\log_e$ term. This can best be achieved by using a series representation thus

$$\log_e\left(\frac{R' + d/2}{R' - d/2}\right) = \frac{d}{R'}\left[1 + \tfrac{1}{3}\left(\frac{d}{2R'}\right)^2 + \tfrac{1}{5}\left(\frac{d}{2R'}\right)^4 + \ldots\right] \qquad (6.58)$$

Circular Section of Radius r

Referring to Fig. 6.36,

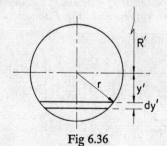

Fig 6.36

$$\int_A \frac{dA}{R' + y'} = 2\int_{-r}^{+r} \frac{\sqrt{(r^2 - y'^2)}}{R' + y'}\, dy$$

$$= 2\pi[R' - \sqrt{(R'^2 - r^2)}]$$

and

$$n = R' - \frac{r^2}{2[R' - \sqrt{(R'^2 - r^2)}]}$$

$$= R' - \frac{r^2}{2} \frac{R' + \sqrt{(R'^2 - r^2)}}{R'^2 - R'^2 + r^2}$$

$$= \frac{R'}{2} - \tfrac{1}{2}\sqrt{(R'^2 - r^2)}$$

$$= \frac{R'}{2}\left[\tfrac{1}{2}\left(\frac{r}{R'}\right)^2 + \tfrac{1}{8}\left(\frac{r}{R'}\right)^4 + \ldots\right] \tag{6.59}$$

Trapezoidal Section as in Fig. 6.37

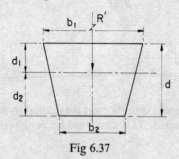

Fig 6.37

$$n = R' - \frac{\tfrac{1}{2}(b_1 + b_2)d}{\left[b_2 + \dfrac{b_1 - b_2}{d}(R' + d_2)\right]\log_e\left(\dfrac{R' + d_2}{R' - d_1}\right) - (b_1 - b_2)}$$

Example 6.13.   Find the maximum tensile and compressive stresses at the section AA of the curved bar shown in Fig. 6.38.

From eqns. (6.57) and (6.58),

$$n = 100\left[1 - \frac{1}{1 + \tfrac{1}{3}\left(\dfrac{50}{200}\right)^2 + \tfrac{1}{5}\left(\dfrac{50}{200}\right)^4}\right]$$

$$= 100\left(1 - \frac{1}{1\cdot0216}\right) = 2\cdot11\,\text{mm}$$

$$R_1 = 100 - 2\cdot11 = 97\cdot89\,\text{mm}$$

$$M = 5000 \times 0\cdot175 = 875\,\text{N-m}$$

$$A = 1250\,\text{mm}^2$$

$$y_{c\,max} = 27\cdot11 \qquad \text{and} \qquad y_{t\,max} = 22\cdot89$$

Substituting in eqn. (6.54),

$$\sigma_{c\,max} = -\frac{875 \times 0.0271}{0.00211 \times 1250 \times 10^{-6} \times 0.125} = -71.9\,\text{MN/m}^2$$

$$\sigma_{t\,max} = +\frac{875 \times 0.0229}{0.00211 \times 1250 \times 10^{-6} \times 0.075} = +101.3\,\text{MN/m}^2$$

Direct stress on section $= \dfrac{5000}{1250 \times 10^{-6}} = +4\,\text{MN/m}^2$

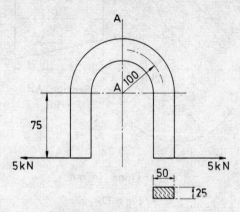

Dimensions in mm

Fig 6.38

The resultant stresses on AA are

$$\sigma_c = -67.9\,\text{MN/m}^2 \qquad \sigma_t = +105.3\,\text{MN/m}^2$$

For an initially straight beam having the same cross-section subjected to the same bending moment the maximum bending stresses would be

$$\sigma = \pm 84.2\,\text{MN/m}^2$$

Example 6.14. A crane hook as illustrated in Fig. 6.39(*a*) is designed to carry a maximum force of 12 kN. Calculate the maximum tensile and compressive stresses set up on the cross-section AB shown at (*b*).

The position of the centroid is given by

$$d_1 = \frac{[48 + (2 \times 24)]}{72}\frac{54}{3}$$

$$= 24\,\text{mm}$$

Hence

$$d_2 = 54 - 24 = 30\,\text{mm}$$

Using eqn. (6.60) to find n the distance between the centroid and the neutral axis,

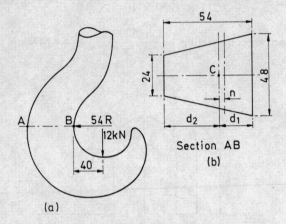

Section AB
(b)

(a)

Dimensions in mm

Fig 6.39

$$n = 78 - \frac{\frac{1}{2}(48 + 24)54}{[24 + \{(48 - 24)/54\}(78 + 30)] \log_e \left(\dfrac{78 + 30}{78 - 24}\right) - (48 - 24)}$$

$$= 78 - \frac{36 \times 54}{72 \log_e 2 - 24}$$

$$= 78 - 75 = 3\,\text{mm}$$

and

$$R_1 = 78 - 3 = 75\,\text{mm}$$
$$A = 36 \times 54 = 1944 \times 10^{-6}\,\text{m}^2$$
$$M = -12000 \times 0{\cdot}064 = -768\,\text{N-m (negative since curvature is reduced)}$$
$$\hat{y}_1 = -24 + 3 = -21\,\text{mm}$$
$$\hat{y}_2 = +30 + 3 = +33\,\text{mm}$$

Direct stress on section AB $= \dfrac{12000}{0{\cdot}001944} = +6{\cdot}17\,\text{MN/m}^2$

Bending stress at B $= \dfrac{-768 \times -0.021}{0.003 \times 0.001944(0.075 - 0.021)}$

$\qquad\qquad\qquad = +51.1 \, \text{MN/m}^2$

Total stress at B $= +57.27 \, \text{MN/m}^2$

Bending stress at A $= \dfrac{-768 \times 0.033}{0.003 \times 0.001944(0.075 + 0.033)}$

$\qquad\qquad\qquad = -40.1 \, \text{MN/m}^2$

Total stress at A $= -33.93 \, \text{MN/m}^2$

Chapter 7
Bending: Slope and Deflection

7.1. In Chapter 6 the stresses during bending were investigated. In this chapter the problem will be approached from an equally important direction, namely with regard to stiffness. The total deflection of a beam is due to a very large extent to the deflection caused by bending, and to a very much smaller extent to the deflection caused by shear. In practice it is usual to put a limit on the allowable deflection, in addition to the stresses. It is important that we should be able to calculate the deflection of a beam of given section, since for given conditions of span and load, it would be possible to adopt a section which would be quite strong enough but would give an excessive deflection.

Various methods are available for determining the slope and deflection of a beam due to elastic bending, and examples of the use of each method will be found in this chapter.

7.2. EQUATION OF THE DEFLECTION CURVE OF THE NEUTRAL AXIS

In Fig. 7.1, θ is the angle which the tangent to the curve at C makes with the x-axis, and $(\theta - d\theta)$ that which the tangent at D makes with the same axis. The normals to the curve at C and D meet at O. The point O is the centre of curvature and R is the radius of curvature of the small portion CD of the deflection curve of the neutral axis.

Numerically $ds = R\,d\theta$ and $1/R = d\theta/ds$. Using the sign convention of Section 3.3, it will be seen that a positive increase in ds is accompanied by a negative value of $d\theta$. Thus, when signs are taken into account, the above equation becomes

$$\frac{1}{R} = -\frac{d\theta}{ds} \tag{7.1}$$

Deflections of the neutral axis are denoted by the symbol v, measured positive downwards, and are relatively small giving a flat form of deflection curve; therefore no error is introduced in assuming that $ds \approx dx$, that $\theta \approx \tan\theta = dv/dx$, and hence that $d\theta/ds = d^2v/dx^2$. Therefore

$$\frac{1}{R} = -\frac{d^2v}{dx^2} \tag{7.2}$$

In Section 6.5 it was shown that, when elastic bending occurs,

$$\frac{1}{R} = \frac{M}{EI}$$

Therefore

$$\frac{d^2v}{dx^2} = -\frac{M}{EI} \tag{7.3}$$

This is the differential equation of the deflection curve.

The exact relationship between the radius of curvature and geometry of the deformed beam is also obtained from $1/R = -d\theta/ds$ by simple calculus, and is given by

$$\frac{1}{R} = \pm \frac{d^2v/dx^2}{[1 + (dv/dx)^2]^{3/2}}$$

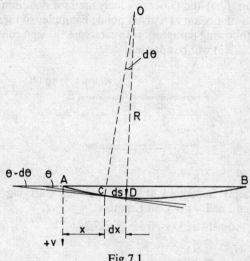

Fig 7.1

For small deflections, this again may be reduced to the form $1/R = \pm d^2v/dx^2$, since $(dv/dx)^2$ is small compared with unity. Therefore

$$M = \pm EI\frac{d^2v}{dx^2}$$

The correct sign is chosen by observing that in the beam shown M is positive and the rate of change of slope, d^2v/dx^2, is negative; hence

$$M = -EI\frac{d^2v}{dx^2}$$

or

$$EI\frac{d^2v}{dx^2} = -M \tag{7.4}$$

7.3. DOUBLE INTEGRATION METHOD

The first integration of eqn. (7.3) gives the slope of the beam at a distance x along its length when the origin is taken at A. Therefore

$$\frac{dv}{dx} = \int \frac{-M}{EI}\,dx + C \tag{7.5}$$

The second integration gives the deflection of the beam at the above point, or

$$v = \int \frac{dv}{dx}\,dx = \iint \frac{-M}{EI}\,dx\,dx + Cx + C_1 \tag{7.6}$$

C and C_1, the constants of integration, can be evaluated from the known conditions of slope and deflection at certain points, usually at the supports. Eqns. (7.5) and (7.6) are widely used for determining the slope and deflection of a beam at a given point. Examples of their use will be found in the following paragraphs. In each case the sign convention used to obtain eqn. (7.3) will be adopted.

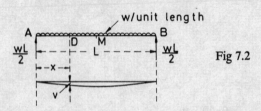

Fig 7.2

(a) **Beam Simply Supported at Each End and Carrying a Uniformly Distributed Load w per Unit Length**

The bending moment at D is, from Fig. 7.2,

$$M_D = \frac{wL}{2}\,x - \frac{wx^2}{2}$$

and

$$EI\frac{d^2v}{dx^2} = -\frac{wL}{2}x + \frac{wx^2}{2} \tag{7.7}$$

and

$$EI\frac{dv}{dx} = -\frac{wL}{4}x^2 + \frac{wx^3}{6} + C$$

From symmetry $dv/dx = 0$, at $x = \frac{1}{2}L$; therefore $C = wL^3/24$. Hence

$$\frac{dv}{dx} = -\frac{w}{2EI}\left[\frac{Lx^2}{2} - \frac{x^3}{3} - \frac{L^3}{12}\right] \tag{7.8}$$

The slope at the ends of the beam is given by $wL^3/24EI$ at A, where $x = 0$, and $-wL^3/24EI$ at B, where $x = L$.

$$v = -\frac{w}{2EI}\left[\frac{Lx^3}{6} - \frac{x^4}{12} - \frac{L^3x}{12}\right] + C_1$$

At $x = 0$, $v = 0$; therefore $C_1 = 0$ and

$$v = -\frac{w}{12EI}\left[Lx^3 - \frac{x^4}{2} - \frac{L^3x}{2}\right] \tag{7.9}$$

The maximum deflection occurs at mid-span, M, where $x = \frac{1}{2}L$. Therefore

$$v_{max} = -\frac{w}{12EI}\left[\frac{L^4}{8} - \frac{L^4}{32} - \frac{L^4}{4}\right] = \frac{5}{384}\frac{wL^4}{EI} \tag{7.10}$$

(b) Simply Supported Beam with Uniform Bending Moment

The beam shown in Fig. 7.3 is acted on by the terminal couples $\bar{M}$. The

Fig 7.3

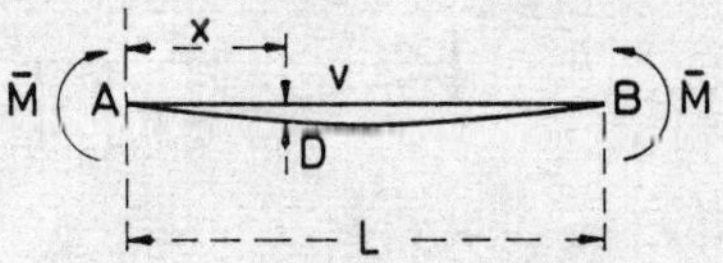

bending moment at D is $M_D = \bar{M}$, and with the previous sign convention,

$$EI\frac{d^2v}{dx^2} = -\bar{M} \tag{7.11}$$

Therefore

$$EI\frac{dv}{dx} = -\bar{M}x + C$$

and

$$EIv = -\frac{\bar{M}x^2}{2} + Cx + C_1$$

At $x = 0$, $v = 0$; therefore $C_1 = 0$; and at $x = L$, $v = 0$; therefore $C = \frac{1}{2}\bar{M}L$, and the slope dv/dx is given by

$$\frac{dv}{dx} = \frac{1}{EI}\left[-\bar{M}x + \frac{\bar{M}L}{2}\right] \tag{7.12}$$

from which the slope at the ends is $+ML/2EI$ at A, $-ML/2EI$ at B and at mid-span is zero. The deflection at any point is given by

$$v = \frac{1}{EI}\left[-\frac{\bar{M}x^2}{2} + \frac{\bar{M}L}{2}x\right] \tag{7.13}$$

The maximum deflection occurs at mid-span, and hence

$$v_{max} = \frac{\bar{M}L^2}{8EI} \tag{7.14}$$

(c) Cantilever with Uniformly Distributed Load w per Unit Length

The bending moment at D in Fig. 7.4 is

$$M_D = R_A x - M_A - \frac{wx^2}{2}$$

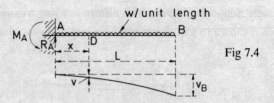

Fig 7.4

where $R_A = wL$ and $M_A = \frac{1}{2}wL^2$, the fixing moment at the support; see Section 1.16. Therefore

$$EI\frac{d^2v}{dx^2} = -M = -wLx + \frac{wL^2}{2} + \frac{wx^2}{2} \tag{7.15}$$

$$EI\frac{dv}{dx} = -\frac{wL}{2}x^2 + \frac{wL^2}{2}x + \frac{wx^3}{6} + C$$

At $x = 0$, $dv/dx = 0$; therefore $C = 0$. Hence

$$\frac{dv}{dx} = \frac{1}{EI}\left[-\frac{wL}{2}x^2 + \frac{wL^2}{2}x + \frac{wx^3}{6}\right] \tag{7.16}$$

At the free end, $x = L$. Therefore

$$\text{Slope at free end} = \frac{wL^3}{6EI} \tag{7.17}$$

$$v = \frac{1}{EI}\left[-\frac{wL}{6}x^3 + \frac{wL^2}{4}x^2 + \frac{wx^4}{24}\right] + C_1$$

and at $x = 0$, $v = 0$; therefore $C_1 = 0$. Hence

$$v = \frac{1}{EI}\left[-\frac{wL}{6}x^3 + \frac{wL^2}{4}x^2 + \frac{wx^4}{24}\right] \qquad (7.18)$$

The deflection at the free end B, where $w = L$, is given by

$$v_B = \frac{1}{EI}\left[-\frac{wL^4}{6} + \frac{wL^4}{4} + \frac{wL^4}{24}\right] = \frac{1}{8}\frac{wL^4}{EI} \qquad (7.19)$$

(d) Cantilever of Span L and Carrying a Concentrated Load at Distance l from the Fixed End

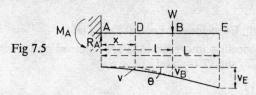

Fig 7.5

The bending moment at D in Fig. 7.5 is given by

$$M_D = -M_A + R_A x$$

where $M_A = Wl$, the fixing moment at the support, and $R_A = W$. Therefore

$$M = -Wl + Wx$$

$$EI\frac{d^2v}{dx^2} = -M = W(l - x) \qquad (7.20)$$

and

$$EI\frac{dv}{dx} = W\left(lx - \frac{x^2}{2}\right) + C$$

At $x = 0$, $dv/dx = 0$; therefore $C = 0$. Hence

$$\frac{dv}{dx} = \frac{W}{EI}\left(lx - \frac{x^2}{2}\right) \qquad (7.21)$$

Slope θ (where $x = l$) $= \dfrac{Wl^2}{2EI}$ \qquad (7.22)

$$v = \frac{W}{EI}\left(\frac{lx^2}{2} - \frac{x^3}{6}\right) + C_1$$

and at $x = 0$, $v = 0$; therefore $C_1 = 0$, which gives

$$v = \frac{W}{2EI}\left(lx^2 - \frac{x^3}{3}\right) \qquad (7.23)$$

For the deflection under the load we substitute $x = l$ in eqn. (7.23) and

$$v_B = \frac{1}{3}\frac{Wl^3}{EI} \qquad (7.24)$$

Deflection at free end E = deflection at B + slope at B × $(L - l)$
Therefore

$$v_E = \frac{1}{3}\frac{Wl^3}{EI} + \frac{Wl^2}{2EI}(L - l)$$

$$= \frac{Wl^2}{2EI}\left(L - \frac{l}{3}\right) \qquad (7.25)$$

(e) Cantilever with Concentrated Load at Free End

The slope and deflection under the load are obtained by substituting L for l in eqns. (7.22) and (7.24), and

$$v_{max} = \frac{1}{3}\frac{WL^3}{EI} \qquad (7.26)$$

7.4. DISCONTINUOUS LOADING: MACAULAY'S METHOD

It was seen in Section 1.8, when considering the bending moment distribution for a beam with discontinuous loading, that a bending moment expression has to be written for each part of the beam. This means that in deriving slope and deflection a double integration would have to be performed on each bending moment expression and two constants would result for each section of the beam. A further example of discontinuous loading is shown in Fig. 7.6(a); in this case there would be

(a)　　　(b)　　　Fig 7.6

three bending moment equations and thus six constants of integration. There are apparently only two boundary conditions, those of zero deflection at each end. However, at the points of discontinuity, B and C, both slope and deflection must be continuous from one section to the next, so that

$$At\ B \qquad \left(\frac{dv}{dx}\right)_{AB} = \left(\frac{dv}{dx}\right)_{BC} \qquad and \qquad v_{AB} = v_{BC}$$

$$At\ C \qquad \left(\frac{dv}{dx}\right)_{BC} = \left(\frac{dv}{dx}\right)_{CD} \qquad and \qquad v_{BC} = v_{CD}$$

The above four plus the two end conditions enable the six constants of integration to be determined. The derivation of the deflection curve by the above approach is rather tedious; it is therefore an advantage to use the mathematical technique termed a *step function*, commonly known as Macaulay's method* when applied to beam solutions. This approach requires only one expression to be written down to cover the bending moment conditions for the whole length of beam, and hence, on integration, only two unknown constants have to be determined.

Briefly, the step function is a function of x of the form $f_n(x) = [x - a]^n$ such that for $x < a$, $f_n(x) = 0$ and for $x > a$, $f_n(x) = (x - a)^n$. Note

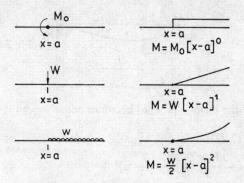

Fig 7.7

the change in the form of brackets used: the square brackets are particularly chosen to indicate the use of a step function, the curved brackets representing normal mathematical procedure. The important features when using the step function in analysis are that, if on substitution of a value for x the quantity inside the square brackets becomes negative, it is omitted from further analysis. Square bracket terms must be integrated in such a way as to preserve the identity of the bracket, i.e.

$$\int [x - a]^2 \, d(x - a) = \tfrac{1}{3}[x - a]^3$$

Finally, for mathematical continuity, distributed loading, as in Fig. 7.6.(a), must be arranged to continue to $x = l$ whether starting from $x = 0$ or $x = a$. This may be effected by the superposition of loadings which cancel each other in the required portions of the beam as shown in Fig. 7.6(b).

The three common step functions for bending moment are shown in Fig. 7.7.

* W. H. Macaulay, "Note on the deflection of beams", *Messenger of Mathematics*, **48**, pp. 129–130 (1919).

(a) Simply Supported Beam with Concentrated Load

Taking moments about one end, the reactions at the supports in Fig. 7.8 are

$$R_1 = \frac{W(L-a)}{L} \quad \text{and} \quad R_2 = \frac{Wa}{L}$$

When $x > a$, the bending moment at D is $M = R_1 x - W[x-a]$. Hence

$$EI\frac{d^2v}{dx^2} = -M = -R_1 x + W[x-a] \tag{7.27}$$

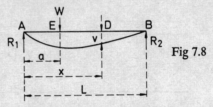

Fig 7.8

When $x < a$, the term $W[x-a]$ becomes negative and is inapplicable, and $M = R_1 x$.

From eqn. (7.27),

$$EI\frac{dv}{dx} = -\frac{R_1 x^2}{2} + \frac{W}{2}[x-a]^2 + C \tag{7.28}$$

and

$$EIv = -\frac{R_1 x^3}{6} + \frac{W}{6}[x-a]^3 + Cx + C_1 \tag{7.29}$$

If we omit the term inside the brackets on the right-hand side of eqns. (7.28) and (7.29) when $x < a$, the equations are then of the correct form for the portion AE of the beam, and since the second term on the right-hand side of these equations vanishes for a value of $x = a$, then when these equations are used for the whole beam, both dv/dx and v will be continuous at the point E.

When $x = 0$, $v = 0$, and since the term inside the brackets is omitted, $C_1 = 0$. For $x = L$, $v = 0$; therefore

$$0 = -\frac{R_1 L^3}{6} + \frac{W}{6}(L-a)^3 + CL$$

and

$$\begin{aligned}
C &= \frac{R_1 L^2}{6} - \frac{W}{6L}(L-a)^3 \\
&= \frac{W(L-a)L}{6} - \frac{W}{6L}(L-a)^3 \\
&= \frac{Wa}{6L}(L-a)(2L-a)
\end{aligned} \tag{7.30}$$

Substituting the values of C and R_1 in eqn. (7.29) and rearranging,

$$v = \frac{Wx}{6EI}\frac{L-a}{L}(2aL - a^2 - x^2) + \frac{W}{6EI}[x-a]^3 \qquad (7.31)$$

This equation gives the deflection at any point along the beam if the last term on the right-hand side is rejected when it becomes negative, i.e. for $x < a$. For the particular case when $x = a$, the deflection under the load is given by

$$v_E = \frac{Wa^2(L-a)^2}{3EIL} \qquad (7.32)$$

If W is placed at mid-span so that $a = \tfrac{1}{2}L$, this gives the deflection under the load,

$$v = \frac{WL^3}{48EI} \qquad (7.33)$$

(b) Simply Supported Beam with Distributed Load not reaching the Right-hand Support

As explained in Section 7.4, for mathematical continuity the loading must be continued to the right-hand end, and to maintain the same equilibrium upward loading must be inserted from D to B as shown in Fig. 7.9.

Fig 7.9

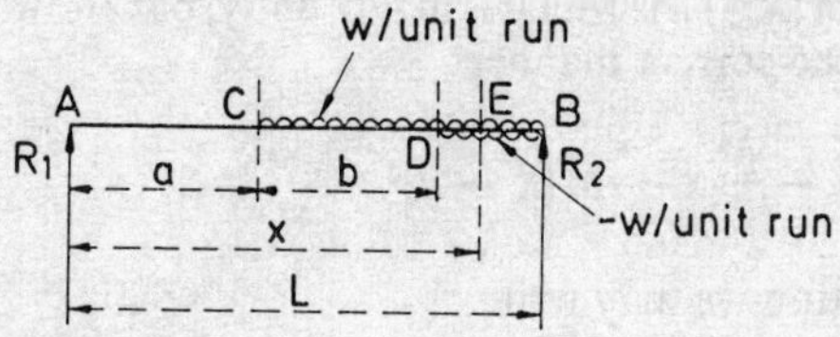

At point E between D and B,

$$M = R_1x - \frac{w}{2}[x-a]^2 + \frac{w}{2}[x-(a+b)]^2 \qquad (7.34)$$

the second and third terms being rejected when $x < a$ and the third term when $x < (a+b)$.

$$EI\frac{d^2v}{dx^2} = -R_1x + \frac{w}{2}[x-a]^2 - \frac{w}{2}[x-(a+b)]^2$$

and

$$EI\frac{dv}{dx} = -\frac{R_1x^2}{2} + \frac{w}{6}[x-a]^3 - \frac{w}{6}[x-(a+b)]^3 + C \qquad (7.35)$$

$$EIv = -\frac{R_1x^3}{6} + \frac{w}{24}[x-a]^4 - \frac{w}{24}[x-(a+b)]^4 + Cx + C_1 \qquad (7.36)$$

The values of C and C_1 are found from the conditions of $v = 0$ when $x = 0$ and $x = L$, the terms inside the brackets being rejected when negative.

(c) Simply Supported Beam with an Applied Couple

From moment equilibrium in Fig. 7.10,

$$R_1 = R_2 = \frac{\bar{M}}{L}$$

The bending moment at D, where $x > a$, is

$$M_D = R_1 x - \bar{M} = \frac{\bar{M}}{L}x - \bar{M}$$

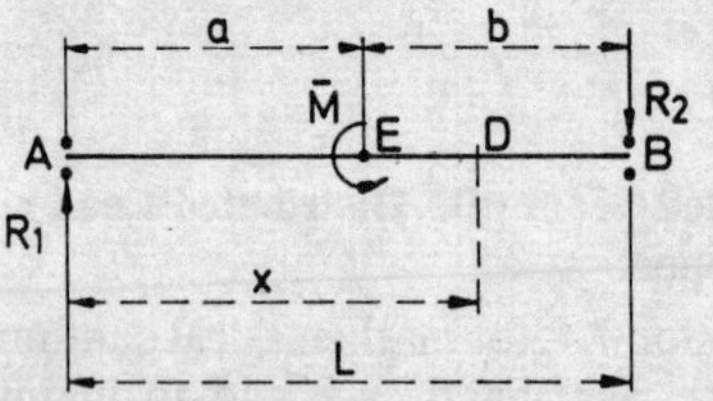

Fig 7.10

A convenient way of dealing with a couple by Macaulay's method is to introduce a term $[x - a]^0$ which is in fact unity, but allows for subsequent integration in the correct manner:

$$EI\frac{d^2v}{dx^2} = -\frac{\bar{M}}{L}x + \bar{M}[x - a]^0$$

and on integration we may write

$$EI\frac{dv}{dx} = -\frac{\bar{M}x^2}{2L} + \bar{M}[x - a]^1 + C$$

where the second term is integrated with respect to $(x - a)$. Therefore

$$EIv = -\frac{\bar{M}x^3}{6L} + \frac{\bar{M}}{2}[x - a]^2 + Cx + C_1 \tag{7.37}$$

When $x < a$, the bracketed term on the right-hand side of the equation becomes negative and is rejected.

At $x = 0$, $v = 0$; therefore $C_1 = 0$: and at $x = L$, $v = 0$; therefore

$$0 = -\frac{\bar{M}L^2}{6} + \frac{\bar{M}}{2}[L - a]^2 + CL$$

and

$$C = \frac{\bar{M}}{6L}\{-2L^2 + 6aL - 3a^2\} \tag{7.38}$$

Hence

$$v = \frac{1}{EI}\left\{ -\frac{\bar{M}}{6L}x^3 + \frac{\bar{M}}{2}[x-a]^2 - \frac{\bar{M}}{6L}(2L^2 - 6aL + 3a^2)x \right\} \text{(7.39)}$$

At E, where $x = a$, the deflection is given by

$$EIv_E = -\frac{\bar{M}}{6L}a^3 - \frac{\bar{M}}{6L}[2L^2 - 6aL + 3a^2]a$$

or, putting $L = (a + b)$,

$$EIv_E = -\frac{\bar{M}}{6L}[a^3 - a^3 - 2ab(a - b)]$$

$$v_E = \frac{\bar{M}}{3EI}\frac{(a-b)ab}{L} \tag{7.40}$$

The following rules in applying Macaulay's method are reiterated:

1. Write down the bending moment expression for a point close to the right-hand end by taking moments to the left of the point.
2. Where there are discontinuities of loading on the beam (point loads, etc.), integrate such expressions as $(x - a)$ in the form $\frac{1}{2}[x - a]^2$ the brackets [] being used to denote that such a term is rejected when on substitution of a value for x the portion inside the brackets becomes negative.
3. Uniformly distributed loads must be made to extend to the right-hand end of the beam, and a negative load introduced for equilibrium, as in Fig. 7.4.
4. Applied couples acting on the beam should be expressed in the form $\bar{M}[x - a]^0$ and integrated as in item 2 above.

Example 7.1. A simply supported beam is subjected to the loading shown in Fig. 7.11. Calculate the deflection at a section 1·8 m from the left-hand end. $E = 70\,\text{GN/m}^2$, $I = 832\,\text{cm}^4$.

To satisfy the Macaulay conditions, the distributed load must be extended to B and an equivalent negative load inserted to restore the correct resultant load distribution. Then

$$M = R_A x - 3[x - 1·5]^0 - 16\frac{[x - 1·5]^2}{2} - 20[x - 2·4]$$

$$+ 16\frac{[x - 2·4]^2}{2}$$

$M = 0$ when $x = 3$; therefore $R_A = 10\,\text{kN}$.

$$EI\frac{dv}{dx} = -\frac{10x^2}{2} + 3[x - 1\cdot5] + \frac{16[x - 1\cdot5]^3}{6} + \frac{20[x - 2\cdot4]^2}{2}$$

$$+ \frac{16[x - 2\cdot4]^3}{6} + C$$

$$EIv = -\frac{10x^3}{6} + \frac{3[x - 1\cdot5]^2}{2} + \frac{16[x - 1\cdot5]^4}{24} + \frac{20[x - 2\cdot4]^3}{6}$$

$$- \frac{16[x - 2\cdot4]^4}{24} + Cx + C_1$$

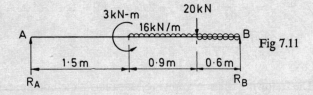

Fig 7.11

When $x = 0$, $v = 0$; therefore $C_1 = 0$: and when $x = 3$, $v = 0$; therefore

$$0 = -\frac{10(3)^3}{6} + \frac{3(1\cdot5)^2}{2} + \frac{16(1\cdot5)^4}{24} + \frac{20(0\cdot6)^3}{6} - \frac{16(0\cdot6)^4}{24} + 3C$$

from which $C = 12\cdot74$. When $x = 1\cdot8$ the third and fourth bracketed terms are omitted, and

$$EIv = -\frac{10 \times 1\cdot8^3}{6} + \frac{3 \times 0\cdot3^2}{2} + \frac{16 \times 0\cdot3^4}{24} + (12\cdot74 \times 1\cdot8)$$

$$= 14\cdot31 \,\text{kN-m}^3$$

$$v = \frac{14\cdot31 \times 10^3 \times 10^3}{70 \times 10^9 \times 832 \times 10^{-8}} = 24\cdot6\,\text{mm}$$

Example 7.2. Calculate the position and magnitude of the maximum deflection for the beam shown in Fig. 7.12. $EI = 1000\,\text{kN-m}^2$.

$$EI\frac{d^2v}{dx^2} = -4x + 5[x - 2] + \frac{2}{2}[x - 4]^2$$

$$EI\frac{dv}{dx} = -\frac{4x^2}{2} + \frac{5}{2}[x - 2]^2 + \frac{2}{6}[x - 4]^3 + A$$

$$EIv = -\frac{4x^3}{6} + \frac{5}{6}[x - 2]^3 + \frac{2}{24}[x - 4]^4 + Ax + B$$

Boundary Conditions

$$\frac{dv}{dx} = 0 \qquad x = 8$$

$$v = 0 \qquad x = 8$$

$$0 = -(2 \times 64) + (5 \times 18) + \frac{64}{3} + A$$

$$A = +16 \cdot 7$$

$$0 = -\left(4 \times \frac{8^3}{6}\right) + (5 \times 36) + \frac{16^2}{12} + (16 \cdot 7 \times 8) + B$$

$$B = -7$$

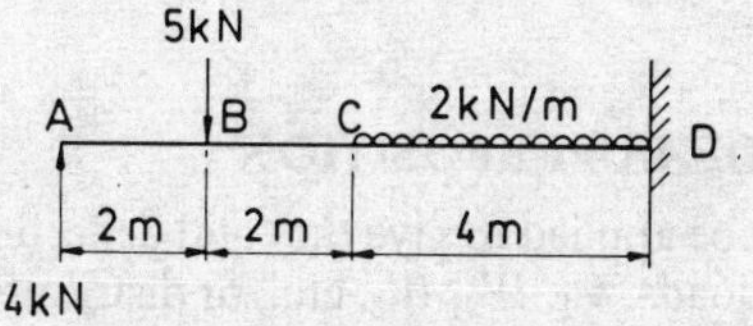

Fig 7.12

At the left end the deflection is obtained when $x = 0$; therefore

$$EIv = -7 \,\text{kN-m}^3$$

$$v = -\frac{7 \times 10^3}{1000} = -7 \,\text{mm}$$

This may not be the maximum deflection and we must check elsewhere in the span. However, it is not sufficient merely to equate dv/dx to zero since bracketed terms would then be included which might not be appropriate, depending on where v_{max} occurred. The best way is to make a sensible guess as to the section where the maximum deflection is likely to occur and to determine the slope at each end of that section. The slopes will be of opposite sign if the guess was correct. If not, then an adjacent section must be treated in the same way. For example, assuming $dv/dx = 0$ occurs between B and C, then

At B $\qquad EI\dfrac{dv}{dx} = -\left(4 \times \dfrac{2^2}{2}\right) + 16 \cdot 7 = +8 \cdot 7$

At C $\qquad EI\dfrac{dv}{dx} = -\left(4 \times \dfrac{4^2}{2}\right) + \left(\dfrac{5}{2} \times 2^2\right) + 16 \cdot 7 = -5 \cdot 3$

We may therefore now write

$$\frac{dv}{dx} = 0 = -\frac{4x^2}{2} + \frac{5}{2}(x - 2)^2 + 16 \cdot 7$$

Thus

$$x^2 - 20x + 53{\cdot}4 = 0$$

from which $x = 3{\cdot}17\,\text{m}$.

The deflection at this point is given by

$$EIv = -\left(\frac{4 \times 3{\cdot}17^3}{6}\right) + \left(\frac{5}{6} \times 1{\cdot}17^3\right) + (16{\cdot}7 \times 3{\cdot}17) - 7$$

$$= +26{\cdot}04\,\text{kN-m}^3$$

$$v = +\frac{26{\cdot}04 \times 10^3}{1000} = +26{\cdot}04\,\text{mm}$$

Hence the maximum deflection occurs at $3{\cdot}17\,\text{m}$ from A and is downwards.

7.5. METHOD OF SUPERPOSITION

This principle can be applied to give the total deflection of a beam which carries individual loads W_1, W_2, W_3, etc., or distributed loads w_1, w_2, w_3, etc. Let the bending moments at a section of the beam caused by each load when acting separately on the beam be M_1, M_2, M_3, etc., and the corresponding deflections be v_1, v_2, v_3, etc. Then the total bending moment is

$$M = M_1 + M_2 + M_3 + \ldots \tag{7.41}$$

But

$$M = -EI\frac{d^2v}{dx^2}$$

Therefore

$$v = -\frac{1}{EI}\iint M\,dx\,dx$$

$$= -\frac{1}{EI}\iint (M_1 + M_2 + M_3 + \ldots)\,dx\,dx$$

$$= -\frac{1}{EI}\iint M_1\,dx\,dx + \iint M_2\,dx\,dx + \iint M_3\,dx\,dx + \ldots$$

$$= v_1 + v_2 + v_3 + \ldots \tag{7.42}$$

Thus the deflection at a section of a beam subjected to complex loading can be obtained by the summation of the deflections caused at that section by the individual components of the loading.

Example 7.3. Use the principle of superposition to determine the deflections at the ends and centre of the beam shown in Fig. 7.13. $EI = 500\,\text{kN-m}^2$.

Splitting the problem into the three components shown in Figs. 7.14(*a*), (*b*) and (*c*), the respective deflections at the centre are

(*a*) $\delta_1 = \dfrac{Wl^3}{48EI} = +\dfrac{2000 \times 6^3 \times 10^3}{48 \times 500 \times 10^3} = +18\,\text{mm}$

(*b*) $\delta_2 = \dfrac{5wl^4}{384EI} = +\dfrac{5 \times 1000 \times 6^4 \times 10^3}{384 \times 500 \times 10^3} = +33\cdot8\,\text{mm}$

(*c*) This may be treated as a beam subjected to couples at B and C of 8 kN-m magnitude:

$$\delta_3 = \dfrac{Ml^2}{8EI} = -\dfrac{8000 \times 6^2 \times 10^3}{8 \times 500 \times 10^3} = -72\,\text{mm}$$

Resultant deflection $= +18 + 33\cdot8 - 72 = -20\cdot2\,\text{mm}$

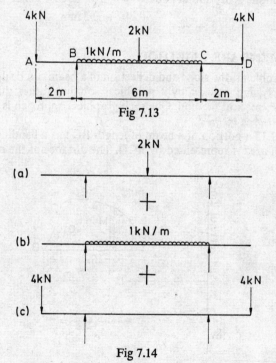

Fig 7.13

Fig 7.14

To find the deflection at A or D it is necessary to know the slope in each case at B or C. Then

(*a*) $\delta_{1A} = \theta_{1B}l_{AB} = \dfrac{Wl_{BC}^2}{16EI}l_{AB} = -\dfrac{2000 \times 6^2 \times 2 \times 10^3}{16 \times 500 \times 10^3}$

$= -18\,\text{mm}$

(b) $\delta_{2A} = \theta_{2B}l_{AB} = \dfrac{wl_{BC}^3}{24EI}l_{AB} = -\dfrac{1000 \times 6^3 \times 2 \times 10^3}{24 \times 500 \times 10^3}$

$= -36\,\text{mm}$

(c) $\delta_{3A} = \theta_{3B}l_{AB} + \dfrac{W_A l_{AB}^3}{3EI}$

$= \dfrac{Ml_{BC}l_{AB}}{2EI} + \dfrac{W_A l_{AB}^3}{3EI}$

$= \dfrac{10^3}{500 \times 10^3}\left[\dfrac{8000 \times 6 \times 2}{2} + \dfrac{4000 \times 8}{3}\right]$

$= 117\cdot3\,\text{mm}$

Resultant deflection at A or D $= -18 - 36 + 117\cdot3$
$= +63\cdot3\,\text{mm}$

7.6. MOMENT-AREA METHOD

In some problems, the slope and deflection of a beam can be determined more simply and quicker by a graphical solution rather than by the above mathematical method. One such graphical approach is known as the *moment-area method*.

In Fig. 7.15 a portion of a beam of length BD has a bending moment diagram of area A represented by BCD. The distance of the centroid G

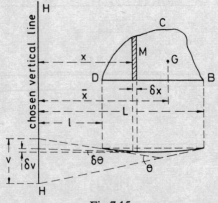

Fig 7.15

of the diagram from any chosen vertical line HH is $\bar{x}$. An exaggerated view of the deflected beam is shown below the bending moment diagram.

Consider a small piece of the beam of length δx over which the bending moment may be assumed to be constant and equal to M. The change of slope over the small piece δx is given by $\delta\theta$, where $\delta\theta$ is the angle included between tangents drawn at each extremity of δx. Let R be the radius of curvature of the small length δx when deflected; then since $\delta\theta$ is very small we have

$$\delta\theta = \frac{\delta x}{R}$$

But

$$\frac{M}{EI} = \frac{1}{R}$$

Hence

$$\delta\theta = \frac{1}{EI} M \, \delta x$$

Therefore

$$\theta = \int_l^L \frac{1}{EI} M \, dx = \frac{1}{EI}\int_l^L M \, dx$$

when the section of the beam is constant, and

$$\theta = \frac{1}{E}\int_l^L \frac{M}{I} \, dx$$

when the section of the beam varies, but $\displaystyle\int_l^{l_r} M \, dx$ is the area A of the bending moment diagram BCD over the whole piece of length BD. Therefore

$$\theta = \frac{A}{EI} \tag{7.43}$$

θ being the change of slope over the whole piece of the beam of length BD. Thus we have the important relationship between the bending moment diagram, over any portion of a loaded beam, and the change of slope over the same portion, that the change of slope is equal to the area of the bending moment diagram divided by EI. Also

$$\delta_v \approx x \, \delta\theta = x\frac{M}{EI}\delta x$$

Therefore

$$v = \frac{1}{EI}\int_l^L Mx \, dx$$

when EI is constant. But

$$\int_l^L Mx \, dx = A\bar{x}$$

where A is the area of the bending moment diagram and $\bar{x}$ is the distance of the centre of the area from the datum line; therefore

$$v = \frac{A\bar{x}}{EI} \qquad (7.44)$$

Thus the intercept, on any chosen line, between the tangents drawn to the ends of any portion of a loaded beam is equal to the product of the area of the bending moment diagram, over that portion of the beam, and the distance of the centre of area of this diagram from the chosen line, divided by EI.

If the vertical line is properly chosen, then v will represent the deflection of the beam. In the case where the bending moment diagram is not all of the same sign, care must be taken in finding the value of $A\bar{x}$. Examples of this will be given later.

Since the area of the bending moment diagram is proportional to the product of W and L^2, where W is the total load on the beam, and x is proportional to L, the deflection will be given by

$$v = k\frac{WL^3}{EI} \qquad (7.45)$$

In Fig. 7.16, ACB is the deflected form of a loaded beam, greatly exaggerated. θ_A and θ_B represent the slopes at A and B respectively, the

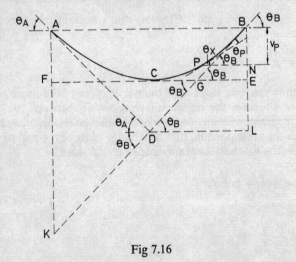

Fig 7.16

tangents to the curve at these points being AD and BD. At C, the point of maximum deflection, draw FE tangentially to the curve and intersecting BD at G, and a perpendicular to AB from B at E. At A draw a perpendicular to AB, to cut the tangent at C in F and meet BD produced in K. Let

> A = Area of bending moment diagram above AB
> $\bar{x}$ = Distance of the centroid of the bending moment diagram from the end A of the beam

Then

$$v = \text{AK} = \text{BE} + \text{FK}$$

and since θ_B is a small angle,

$$
\begin{aligned}
v &= \text{GE} \cdot \theta_B + \text{FG} \cdot \theta_B \\
&= \theta_B(\text{GE} + \text{FG}) \\
&= \theta_B \cdot \text{AB} \\
&= l\theta_B
\end{aligned}
$$

where $\text{AB} = l$. But

$$v = \frac{A\bar{x}}{EI}$$

from eqn. (7.44). Therefore

$$l\theta_B = \frac{A\bar{x}}{EI}$$

and

$$\theta_B = \frac{A\bar{x}}{EIl} \tag{7.46}$$

Similarly

$$\theta_A = \frac{A(l - \bar{x})}{EIl} \tag{7.47}$$

Also

$$\theta = \theta_A + \theta_B \tag{7.48}$$

and the slope of the beam at any point P is

$$\theta_X = \theta_B - \theta_P \tag{7.49}$$

where θ_P is the change of slope over PB. Therefore

$$v_P = \frac{A_P\bar{x}_P}{EI} + \text{PN} \cdot \theta_X \tag{7.50}$$

where A_P = Area of the bending moment diagram above PN
    $\bar{x}_P$ = Distance of centroid of bending moment diagram above PN from BL
    PN = Horizontal distance of P from BL

Substituting in eqn. (7.50) the value of θ_x from eqn. (7.49), we have

$$v_P = \frac{A_P \bar{x}_P}{EI} + \text{PN}(\theta_B - \theta_P)$$

$$= \frac{A_P \bar{x}_P}{EI} + \text{PN}\frac{A\bar{x}}{EIl} - \frac{A_P \cdot \text{PN}}{EI} \tag{7.51}$$

This equation enables us to find the deflection at any point along the span, but care must be taken when the slope of the beam changes sign.

7.7. EXAMPLES ILLUSTRATING THE APPLICATION OF $v = A\bar{x}/EI$

(a) Beam Simply Supported at Each End, with Concentrated Load W at Mid-span

The moment diagram and the deflected form of the beam are shown in Fig. 7.17. If the vertical line is chosen so that it passes through a support,

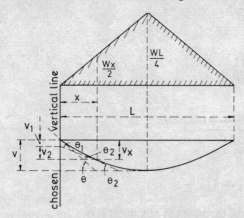

Fig 7.17

then, considering one-half of the span, the intercept between the tangents is v. Also

$$A = \frac{WL^2}{16} \quad \text{and} \quad \bar{x} = \frac{2}{3}\frac{L}{2} = \frac{L}{3}$$

Therefore

$$v = \frac{WL^2}{16} \frac{L}{3} \frac{1}{EI}$$

$$= \frac{1}{48} \frac{WL^3}{EI} \tag{7.52}$$

At a point distant x from the left-hand support, the deflection is given by

$$v_x = v_1 + v_2 = v_1 + x\theta_2$$

since θ_2 is very small.

$$v_1 = \frac{W}{2} \frac{x^2}{2} \frac{2}{3} \frac{x}{EI} = \frac{Wx^3}{6EI}$$

$$\theta_2 = \frac{WL}{4} \frac{L}{2} \frac{L}{2} \frac{1}{EIL} - \frac{Wx}{2} \frac{x}{2EI}$$

$$= \frac{W}{4EI} \left(\frac{L^2}{4} - x^2 \right)$$

$$v_2 = \frac{W}{4EI} \left(\frac{L^2}{4} - x^2 \right) x$$

$$v_x = v_1 + v_2 = \frac{Wx^3}{6EI} + \frac{W}{4EI} \left(\frac{L^2x}{4} - x^3 \right)$$

$$= \frac{Wx}{4EI} \left(\frac{L^2}{4} - \frac{x^2}{3} \right) \tag{7.53}$$

At mid-span, where $x = \frac{1}{2}L$, eqn. (7.53) reduces to the value obtained in eqn. (7.52).

(b) Beam Simply Supported at Each End, with Load w per Unit Length Uniformly Distributed over the Span

Referring to Fig. 7.18, and again considering one-half of the span,

$$A = \frac{2}{3} \frac{wL^2}{8} \frac{L}{2} = \frac{wL^3}{24} \quad \text{and} \quad \bar{x} = \frac{5}{8} \frac{L}{2} = \frac{5}{16}L$$

Therefore

$$v = \frac{5}{384} \frac{wL^4}{EI} \tag{7.54}$$

(c) Cantilever carrying w per Unit Length over the Span

The area of the bending moment diagram, Fig. 7.19, is given by

$$A = \frac{1}{3}\frac{wL^2}{2}L = \frac{wL^3}{6} \quad \text{and} \quad \bar{x} = \frac{3}{4}L$$

$$v = \frac{wL^3}{6}\frac{3}{4}L\frac{1}{EI} = \frac{1}{8}\frac{wL^4}{EI} \tag{7.55}$$

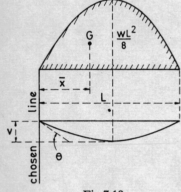

Fig 7.18   Fig 7.19

(d) Cantilever Carrying w per Unit Length from Fixed End for Distance l of the Span

In Fig. 7.20, total deflection is

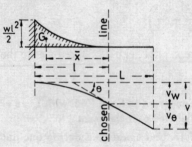

Fig 7.20

$$v = v_\theta + v_w$$

Taking the chosen line at the end of the uniform load, we have, from (c) above,

$$v_w = \frac{1}{8}\frac{wl^4}{EI} \quad \text{and} \quad \theta = \frac{wl^3}{6EI}$$

and

$$v_\theta = (L - l)\theta = \frac{wl^3}{6EI}(L - l)$$

Therefore

$$v = \frac{wl}{EI}\left(\frac{Ll^2}{6} - \frac{l^3}{6} + \frac{l^3}{8}\right)$$

$$= \frac{wl}{EI}\left(\frac{Ll^2}{6} - \frac{l^3}{24}\right) \tag{7.56}$$

(e) Beam with Overhanging End to which a Couple is Applied

The moment diagram and the deflected form of the beam are shown in

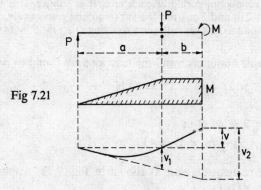

Fig 7.21

Fig. 7.21. The deflection at the free end is given by

$$v = v_2 - v_1\frac{a + b}{a}$$

and v_1 is obtained by considering the portion of span a:

$$v_1 = \frac{Ma}{2}\frac{a}{3}\frac{1}{EI}$$

$$= \frac{Ma^2}{6EI} \tag{7.57}$$

Also v_2 is found by considering the whole of the moment diagram:

$$v_2 = \left[\frac{Mb^2}{2} + \frac{Ma}{2}\left(\frac{a}{3} + b\right)\right]\frac{1}{EI}$$

$$= \left[\frac{Mb^2}{2} + \frac{Ma^2}{6} + \frac{Mab}{2}\right]\frac{1}{EI} \tag{7.58}$$

Therefore

$$v = \frac{1}{EI}\left[\frac{Mb^2}{2} + \frac{Ma^2}{6} + \frac{Mab}{2} - \frac{Ma^2}{6} - \frac{Mab}{6}\right]$$

$$= \frac{1}{EI}\left[\frac{Mb^2}{2} + \frac{Mab}{3}\right]$$

$$= \frac{Mb}{6EI}[2a + 3b] \tag{7.59}$$

where $M = Pa$.

Example 7.4. A rectangular section timber beam 4 m long, 75 mm wide and 100 mm deep is simply supported at each end. It is subjected to loads of 2 kN and 4 kN at 2 m and 3 m from the left-hand end respectively. Use the moment-area method to find the deflection resulting at 2·5 m. $E = 14 \, \mathrm{GN/m^2}$.

Taking moments about each end the reactions are found to be

$$R_A = \tfrac{1}{4}[(2 \times 2) + (4 \times 1)] = 2 \, \mathrm{kN}$$
$$R_B = 4 \, \mathrm{kN}$$
$$M_C = 2 \times 2 = 4 \, \mathrm{kN\text{-}m}$$
$$M_D = (2 \times 3) - (2 \times 1) = 4 \, \mathrm{kN\text{-}m}$$

The bending moment diagram is shown in Fig. 7.22. Applying the relationship $v = A\bar{x}/EI$,

$$v_1 = \frac{1}{EI}\left[\left(\frac{4 \times 2}{2} \times \frac{4}{3}\right) + (4 \times 1 \times 2{\cdot}5) + \left(\frac{4 \times 1}{2} \times 3\tfrac{1}{3}\right)\right]$$

$$= \frac{22}{EI} = \frac{22\,000 \times 12 \times 10^3}{14 \times 10^9 \times 75 \times 100^3 \times 10^{-12}} = 250 \, \mathrm{mm}$$

$$v_2 = \frac{1{\cdot}5}{4} v_1 = 93{\cdot}7 \, \mathrm{mm}$$

$$v_3 = \frac{1}{EI}\left[\left(4 \times \frac{1}{2} \times \frac{1}{4}\right) + \left(\frac{4 \times 1}{2} \times \frac{5}{6}\right)\right]$$

$$= \frac{2{\cdot}17}{EI} = 24{\cdot}7 \, \mathrm{mm}$$

Therefore

Deflection at 2·5 m from left end $= v_2 - v_3 = 69 \, \mathrm{mm}$

7.8. BEAMS WITH IRREGULAR LOADING

The deflection of an irregularly loaded beam can be found if both the bending moment M and term $\int M\,dx$ can be integrated with respect to x. It has been shown that

$$\frac{d^2v}{dx^2} = -\frac{M}{EI}$$

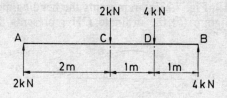

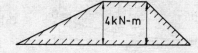

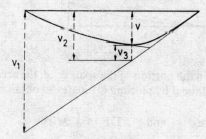

Fig 7.22

which can be written in the alternative form

$$\frac{d}{dx}\frac{dv}{dx} = -\frac{M}{EI}$$

Thus

$$\frac{dv}{dx} = -\frac{1}{EI}\int_0^x M\,dx + C \tag{7.60}$$

where I has a constant value, and

$$v = -\frac{1}{EI}\int_0^x\int_0^x M\,dx\,dx + Cx + C_1 \tag{7.61}$$

Thus, if M and $\int M\,dx$ can be integrated with respect to x, the value of v can be calculated since the end conditions fix the values of C and C_1.

If the analytical process cannot be carried out, a graphical method may be utilized as shown by the following example.

7.9. SIMPLY SUPPORTED BEAM WITH IRREGULAR LOADING

The curve AMB, Fig. 7.23(a), represents the bending moment diagram on AB. In diagram (b) the ordinate GH represents the area of the

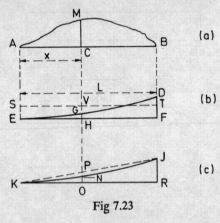

Fig 7.23

moment diagram on the portion of the span x, i.e. the area AMC. Hence the curve ED is obtained by plotting ordinates so obtained. Therefore

$$\text{GH} = \int_0^x M\,dx \qquad \text{and} \qquad \text{DF} = \int_0^L M\,dx$$

In diagram (c), the ordinate ON represents the area EGH in curve (b), or

$$\text{ON} = \int_0^x \text{GH}\,dx = \int_0^x \int_0^x M\,dx\,dx$$

and

$$\text{JR} = \int_0^L \int_0^L M\,dx\,dx$$

At $x = 0$, the deflection is $v = 0$; hence from eqn. (7.61), $C_1 = 0$. At B, $x = L$, $v = 0$, and the ordinate JR in (c) represents $\int_0^L \int_0^L M\,dx\,dx$ Therefore, from eqn. (7.61),

$$0 = -\frac{1}{EI}\text{JR} + CL$$

or

$$C = \frac{JR}{EIL} \tag{7.62}$$

Join KJ. Then

$$\frac{OP}{x} = \frac{JR}{L}$$

Therefore

$$OP = \frac{x}{L} JR = CEIx$$

$$PN = OP - ON = CEIx - \int_0^x \int_0^x M\,dx\,dx$$

$$= EI\left[Cx - \frac{1}{EI}\int_0^x \int_0^x M\,dx\,dx\right]$$

$$= EIv$$

from eqn. (7.61). Thus the ordinates at a given point, measured between the straight line KPJ and the curve KNJ, represent EI times the deflection at the point.

If the straight line ST be drawn at a height equal to $JR/L = CEI$, then

$$VG = VH - GH$$

$$= CEI - \int_0^x M\,dx$$

$$= EI\left[C - \frac{1}{EI}\int_0^x M\,dx\right]$$

$$= EI\frac{dv}{dx}$$

from eqn. (7.60). Thus the ordinates, measured at a given point between the straight line ST and the curve EGD, represent EI times the slope at the point. The slope changes sign where ST intersects EGD.

Example 7.5.   A horizontal beam, simply supported at each end, carries a load which increases, at a uniform rate, from zero at one end. Determine the position of, and the value of, the maximum deflection.

Let w be the intensity of loading at unit distance from A. The intensity at C, Fig. 7.24, is xw and that at B is Lw.

$$R_1 = \frac{L^2 w}{6}$$

Bending moment at $C = R_1 x - \dfrac{x^3 w}{6}$

$$= \frac{L^2 wx}{6} - \frac{x^3 w}{6} \tag{7.63}$$

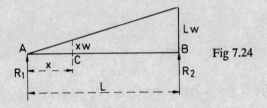

Fig 7.24

$$EI\frac{d^2v}{dx^2} = -\frac{L^2 wx}{6} + \frac{x^3 w}{6} \tag{7.64}$$

$$EI\frac{dv}{dx} = -\frac{L^2 w}{6}\frac{x^2}{2} + \frac{x^4 w}{24} + C$$

and

$$EIv = -\frac{L^2 w}{36}x^3 + \frac{x^5 w}{120} + Cx + C_1$$

At $x = 0$, $v = 0$; therefore $C_1 = 0$: and at $x = L$, $v = 0$; therefore

$$0 = -\frac{L^5 w}{36} + \frac{L^5 w}{120} + CL$$

$$C = \frac{7}{360}L^4 w \tag{7.65}$$

and

$$EI\frac{dv}{dx} = -\frac{L^2 w}{12}x^2 + \frac{x^4 w}{24} + \frac{7}{360}L^4 w \tag{7.66}$$

$$= \frac{1}{360}[-30L^2 wx^2 + 15x^4 w + 7L^4 w]$$

At maximum deflection, $dv/dx = 0$. Therefore

$$x^2 = \frac{30L^2 \pm \sqrt{(900L^4 - 420L^4)}}{30} = \frac{30L^2 - 21 \cdot 9L^2}{30}$$

$$= 0 \cdot 27L^2$$

so that

$$x = 0.52L$$

$$EIv = -\frac{L^2w}{36}x^3 + \frac{w}{120}x^5 + \frac{7}{360}L^4wx \quad (7.67)$$

$$= \frac{w}{360}[-10L^2x^3 + 3x^5 + 7L^4x]$$

$$v_{max} = \frac{w}{360}[-10L^2(0.52L)^3 + 3(0.52L)^5 + 7L^4(0.52L)]\frac{1}{EI}$$

$$= \frac{w}{360EI}[-1.4L^5 + 0.114L^5 + 3.64L^5]$$

$$= \frac{2.354L^5w}{360EI} \quad (7.68)$$

Example 7.6. A beam 4 m long is simply supported at its ends and carries a varying distributed load over the whole span. The equation to the loading curve is $w = ax^2 + bx + c$, where w is the load intensity in kN/m run, at a distance x along the beam, measured from an origin at the left-hand support, and a, b and c are constants. The load intensity is zero at each end of the beam and reaches a maximum value of 100 kN/m at the centre of the span. Calculate the slope of the beam at each support and the deflection at the centre.
$E = 208 \text{ GN/m}^2$; $I = 405 \times 10^{-6} \text{ m}^4$.

The loading conditions are such that at $x = 0$, $w = 0$; therefore $c = 0$. Also, at $x = 4$, $w = 0$; therefore

$$0 = 16a + 4b \quad \text{and} \quad b = -4a$$

At $x = 2$, $w = 100$; therefore

$$100 = 4a + 2b \quad \text{and} \quad 100 = 4a - 8a$$

i.e.

$$a = -25 \quad \text{and} \quad b = 100$$

Therefore

Loading distribution, $w = -25x^2 + 100x \text{ kN/m}$.

Total load on beam $= \displaystyle\int_0^4 w \, dx = \int_0^4 (-25x^2 + 100x) \, dx = 267 \text{ kN}$

Therefore the support reactions are each 133.5 kN.
The shear force distribution is given by

$$Q = -\int w \, dx = -\int (-25x^2 + 100x) \, dx = +25\frac{x^3}{3} - \frac{100x^2}{2} + A$$

At $x = 0$, $Q = R_1 = 133.5$; therefore $A = +133.5$.

$$\text{Bending moment, } M = \int Q\, dx = \int \left(+ \frac{25x^3}{3} - \frac{100x^2}{2} + 133.5 \right) dx$$

$$= + \frac{25x^4}{12} - \frac{100x^3}{6} + 133.5x + B$$

At $x = 0$, $M = 0$; therefore $B = 0$.

$$\text{Slope} = - \frac{1}{EI} \int M\, dx = - \frac{1}{EI} \int \left(+ \frac{25x^4}{12} - \frac{100x^3}{6} + 133.5x \right) dx$$

$$= \frac{1}{EI} \left[- \frac{25x^5}{60} + \frac{100x^4}{24} - \frac{133.5x^2}{2} + C \right]$$

At $x = 2$, $\theta = 0$; therefore

$$C = \frac{1}{208 \times 10^9 \times 405 \times 10^{-6}}$$

$$\left[\frac{25 \times 2^5}{60} - \frac{100 \times 2^4}{24} + \left(\frac{133.5}{2} \times 2^2 \right) \right]$$

$$= \frac{213 \times 10^3}{208 \times 405 \times 10^3} = 0.00253 \, \text{rad}$$

When $x = 0$ and $4\,\text{m}$, $\theta = \pm C = \pm 0.002\,53\,\text{rad}$.

The deflection is given by

$$v = \frac{1}{EI} \int \theta\, dx = \frac{1}{EI} \int \left(- \frac{25x^5}{60} + \frac{100x^4}{24} - \frac{133.5x^2}{2} + 213 \right) dx$$

$$= \frac{1}{EI} \left[- \frac{25x^6}{360} + \frac{100x^5}{120} - \frac{133.5x^3}{6} + 213x + D \right]$$

At $x = 0$, $v = 0$; therefore $D = 0$. At mid-span,

$$v = \frac{10^3}{208 \times 10^9 \times 405 \times 10^{-6}}$$

$$\left[- \frac{25 \times 2^6}{360} + \frac{100 \times 2^5}{120} - \frac{133.5 \times 2^3}{6} + (213 \times 2) \right]$$

$$= \frac{270.3 \times 10^3}{208 \times 405} = 3.2\,\text{mm}$$

7.10. DEFLECTION OF BEAMS OF UNIFORM STRENGTH

The bending equation gives $\sigma = My/I$. The condition, therefore, of uniform strength is that My/I shall be constant; that is, the section modulus shall be proportional to the bending moment. The value of I/y may be varied, in the case of rectangular beams, by altering the depth or altering the breadth.

(a) Constant depth

Taking the case of a beam supported at each end and considering one-half of the moment diagram, the maximum deflection is obtained from

$$v = \int_0^{L/2} \frac{M}{EI} x \, dx$$

Let M_1 be the bending moment and I_1 the second moment of area at mid-span, and M and I the bending moment and second moment of area at a point distant x from one end. Since the depth is constant,

$$\frac{M_1}{I_1} = \frac{M}{I}$$

Therefore

$$v = \frac{M_1}{EI_1} \int_0^{L/2} x \, dx \qquad (7.69)$$

$$= \frac{M_1 L^2}{8EI_1} \qquad (7.70)$$

For an isolated load at mid-span, $M_1 = \tfrac{1}{4}WL$, and therefore

$$v = \frac{WL^3}{32EI_1} \qquad (7.71)$$

For a uniformly distributed load, $M_1 = \tfrac{1}{8}wL^2$ and

$$v = \frac{wL^4}{64EI_1} \qquad (7.72)$$

The deflection of a cantilever may be obtained by substituting L for $\tfrac{1}{2}L$ in eqn. (7.69), when

$$v = \frac{M_1 L^2}{2EI_1}$$

I_1 being the second moment of area at the constraint and M_1 the bending moment at the same point. For a cantilever with a uniform load, $M_1 = \tfrac{1}{2}wL^2$ and

$$v = \frac{wL^4}{4EI_1} \qquad (7.73)$$

For a cantilever with a concentrated load at the free end, $M_1 = WL$ and

$$v = \frac{WL^3}{2EI_1} \tag{7.74}$$

(b) Constant breadth

$$\sigma = \frac{M_1 y_1}{I_1} = \frac{My}{I} \quad \text{or} \quad \frac{M_1}{\frac{1}{6}bD_1{}^2} = \frac{M}{\frac{1}{6}bd^2}$$

Therefore

$$\frac{D_1}{d} = \sqrt{\frac{M_1}{M}}$$

and

$$\frac{M}{I} = \frac{M_1}{I_1}\frac{D_1}{d} = \frac{M_1}{I_1}\sqrt{\frac{M_1}{M}}$$

where D_1 is the depth at mid-span and d the depth at a point distant x from the support. In the case of a beam resting on supports, then, as before, considering one-half of the span,

$$v = \int_0^{L/2} \frac{M}{EI} x \, dx$$

$$= \frac{M_1{}^{3/2}}{EI_1} \int_0^{L/2} M^{-1/2} x \, dx \tag{7.75}$$

For a concentrated load at mid-span, $M = \frac{1}{2}Wx$ and $M_1 = \frac{1}{4}WL$. Therefore

$$v = \frac{WL^3}{24EI_1} \tag{7.76}$$

Other cases may be solved by substituting the value of M in eqn. (7.75) and integrating.

Table 7.1

Principal Slope and Deflection for Beams with Basic Loading

Loading	Slope	Deflection
Cantilever A fixed, length l, moment M at free end B	$-\dfrac{Ml}{EI}$ at B	$-\dfrac{Ml^2}{2EI}$ at B
Simply supported A–B, length l, moment M at centre C ($l/2$, $l/2$)	$-\dfrac{Ml}{12EI}$ at C $\quad +\dfrac{Ml}{24EI}$ at A,B	0 at C
Simply supported A–C–B, equal moments M at each end	$\pm\dfrac{Ml}{2EI}$ at A,B	$+\dfrac{Ml^2}{8EI}$ at C
Cantilever A fixed, load W at free end B	$+\dfrac{Wl^2}{2EI}$ at B	$+\dfrac{Wl^3}{3EI}$ at B
Simply supported A–B, load W at centre C ($l/2$, $l/2$)	$\pm\dfrac{Wl^2}{16EI}$ at A,B	$+\dfrac{Wl^3}{48EI}$ at C
Fixed both ends A,B, load W at centre C	0 at A,B,C	$+\dfrac{Wl^3}{192EI}$ at C
Cantilever A fixed, w/unit length to B	$+\dfrac{wl^3}{6EI}$ at B	$+\dfrac{wl^4}{8EI}$ at B
Simply supported A–B, w/unit length, centre C	$\pm\dfrac{wl^3}{24EI}$ at A,B	$+\dfrac{5wl^4}{384EI}$ at C
Fixed both ends A,B, w/unit length, centre C	0 at A,B,C	$+\dfrac{wl^4}{384EI}$ at C

Chapter 8
Elastic Strain Energy

8.1. When a piece of material is deformed in simple tension, compression, bending or torsion, etc., within its elastic range, work is done by the applied loading. On removal of the loading the material returns to its undeformed state due to the release of stored energy. This is termed *elastic strain energy* and has the same magnitude as the external work done. The term *resilience* is sometimes used to define the strain energy stored per unit volume.

In the first part of this chapter expressions are developed for strain energy stored under various modes of deformation. In the remainder certain energy theorems are derived which are useful in the calculation of deflections.

8.2. STRAIN ENERGY UNDER DIRECT STRESS

Consider the load-extension diagram in Fig. 8.1: the external work done during a small increment of extension dx, is $w\,dx$ and the total area under

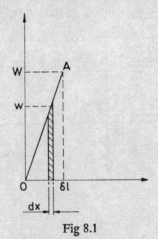

Fig 8.1

the curve up to the proportional limit at point A is $\frac{1}{2}W\,\delta l$. Dividing by the area a and length l of the specimen,

$$\frac{1}{2}\frac{W}{a}\frac{\delta l}{l}$$

180

which is the stored strain energy, denoted by U; hence

$$U = \frac{1}{2}\frac{W}{a}\frac{\delta l}{l} = \frac{1}{2}\sigma\varepsilon \quad \text{per unit volume}$$

Since $\varepsilon = \sigma/E$,

$$U = \frac{1}{2}\frac{\sigma^2}{E} \text{ per unit volume} \tag{8.1}$$

8.3. MAXIMUM STRESS DUE TO A SUDDENLY APPLIED LOAD

In the previous section the load was applied gradually so that the work done was the *average* load times the distance moved at the point where the load was applied. Now suppose that a bar fixed at the top with a flange at the lower end, as in Fig. 8.2(a) has a load W suddenly released

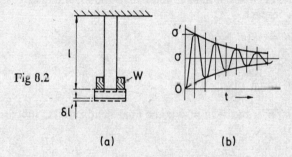

Fig 8.2

(a)   (b)

on to the flange. Let the momentary maximum extension, strain and stress in the bar be $\delta l'$, ε' and σ' respectively. In addition, if the masses of the bar and flange are small compared with the load W, then a reasonable approximation to the behaviour is made by neglecting the effect of the former. Because the *full* load moves through the extension $\delta l'$,

Work done $= W\delta l'$

The strain energy stored per unit volume momentarily is $\frac{1}{2}\sigma'\varepsilon'$, so that

$$U = \frac{1}{2}\sigma'\varepsilon'al = \frac{1}{2}\sigma'a\,\delta l'$$

Since the work done is equal to the strain energy,

$$W\,\delta l' = \frac{1}{2}\sigma'a\,\delta l'$$

Thus $\sigma' = 2W/a$; but $W/a = \sigma$, the stress due to a gradually applied load, so that

$$\sigma' = 2\sigma \tag{8.2}$$

or the momentary maximum stress due to a suddenly applied load is twice the stress for a gradually applied load. The bar will subsequently oscillate about the statical equilibrium position while the stresses and deformations rapidly die away, as shown in Fig. 8.2(b), to the value obtained for a gradually applied load. However, the momentary stress intensification by a factor of 2 might have serious consequences on a component.

8.4. MAXIMUM STRESS DUE TO IMPACT

An extension of the above problem is the case where the load W is dropped on to the flange from a height h, causing a momentary extension of the bar $\delta l'$. The total potential energy is $W(h + \delta l')$, and the momentarily stored strain energy is

$$U' = \tfrac{1}{2}\sigma' a \, \delta l'$$

then neglecting the mass of the bar and flange and assuming no losses of energy during impact,

$$\tfrac{1}{2}\sigma' a \, \delta l' = W(h + \delta l')$$

or

$$\tfrac{1}{2}\sigma' \, \delta l' = \frac{Wh}{a} + \frac{W}{a}\delta l' \tag{8.3}$$

Now $\delta l' = (\sigma'/E)l$, and $W/a = \sigma$ is the final steady stress; substituting into eqn. (8.3),

$$\sigma'^2 - 2\sigma\sigma' - 2\sigma\frac{Eh}{l} = 0$$

Therefore

$$\sigma' = \sigma + \left(\sigma^2 + 2\sigma\frac{Eh}{l}\right)^{1/2} \tag{8.4}$$

It will be seen from this equation that if $h = 0$ then $\sigma' = 2\sigma$, which is the result obtained in the previous section.

The true situation for the stress during impact of one body on another is more complicated than is indicated in this approximate analysis. In practice the deformation and stress imposed on the bar at the point of impact take time to propagate along the length of the bar. The *stress wave*, as it is called, on reaching the fixed end of the bar will be reflected towards the point of initiation and thus a complex state of stress will arise.

Example 8.1. A weight of 100 N falls freely through $\frac{1}{2}$ m and then impacts axially on to the end of a bar of 18 mm diameter and $1\frac{1}{2}$ m length. Find the maximum stress and strain induced in the bar. $E = 208\,\text{GN/m}^2$.

The final steady stress will be

$$\sigma = \frac{100 \times 4}{\pi \times 0.018^2} = 0.393 \times 10^6 \, \text{N/m}^2$$

From eqn. (8.4),

$$\sigma_{max} = (0.393 \times 10^6)$$

$$+ \sqrt{\left[(0.393 \times 10^6)^2 + \frac{2 \times 0.393 \times 10^6 \times 208 \times 10^9 \times \frac{1}{2}}{1.5} \right]}$$

$$= (0.393 \times 10^6) + (2.34 \times 10^8) = 234 \, \text{MN/m}^2$$

From eqn. (8.3),

$$\delta l_{max} = \frac{Wh/a}{\frac{1}{2}\sigma_{max} - W/a}$$

$$= 0.393 \times 10^6 \times \frac{1}{2}/(117 - 0.393) \times 10^6 = 1.68 \, \text{mm}$$

$$\text{Strain} = \frac{1.68}{1500} = 0.001\,12$$

Example 8.2. The lower part of a child's pogo stick is illustrated in Fig. 8.3 when the child and stick are just about to descend to the ground and the

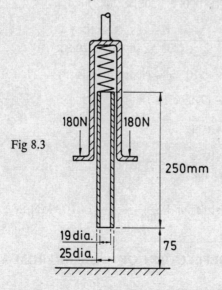

Fig 8.3

180N   180N

250mm

19 dia.
25 dia.   75

spring is underformed. Determine the momentary maximum stress in the steel compression tube on impact with the ground, and compare this with the final steady stress. It may be assumed that the outer sleeve and supports are rigid and that the ground does not deform. The weights of the various parts may be neglected. $E = 208\,\text{GN/m}^2$; spring stiffness $= 18\,\text{kN/m}$.

Let δ be the compression of the spring, and x the compression of the tube. If the force in the spring on impact is momentarily F then the strain energy stored in the spring and tube is

$$\frac{F\delta}{2} + \frac{\sigma^2}{2E} \times \text{volume}$$

Tube volume $= 250\dfrac{\pi}{4}(25^2 - 19^2) \times 10^{-9} = 51\cdot8 \times 10^{-6}\,\text{m}$

Potential energy lost on impact $= 2 \times 180(0\cdot075 + \delta + x)$

and

$$\delta = \frac{F}{18\,000}\,\text{m} \qquad x = \frac{Fl}{AE} = \frac{F \times 250 \times 10^{-3}}{208 \times 10^{-6} \times 208 \times 10^9}\,\text{m}$$

Equating potential and strain energies,

$$360\left(0\cdot075 + \frac{F}{18\,000} + \frac{250F \times 10^{-6}}{208 \times 208}\right)$$

$$= \frac{F^2}{36\,000} + \frac{F^2 \times 51\cdot8 \times 10^{-6}}{2 \times (208 \times 10^{-6})^2 \times 208 \times 10^9}$$

$$(F^2 \times 0\cdot0278 \times 10^{-3}) - (F \times 20 \times 10^{-3}) - 27 = 0$$

$$F^2 - 720F - (970 \times 10^3) = 0$$

from which $F = 1410\,\text{N}$.

$$\sigma_{max} = \frac{1410}{208 \times 10^{-6}} = 6\cdot78\,\text{MN/m}^2$$

Final steady stress, $\sigma = \dfrac{360}{208 \times 10^{-6}} = 1\cdot73\,\text{MN/m}^2$

8.5. BENDING DEFLECTION OF A BEAM FROM AN IMPACT

Let a load W strike a beam of span L simply supported at its ends, at midspan. If h is the distance fallen by W and δ is the deflection produced, then the work done is $W(h + \delta)$.

If W_1 is the equivalent static load applied at mid-span to produce the deflection δ then the work done by W_1 is given by $\frac{1}{2}W_1\delta$; therefore

$$\tfrac{1}{2}W_1\delta = W(h + \delta)$$

But the central deflection is given by

$$\delta = \frac{W_1 L^3}{48EI}$$

Thus

$$\frac{48EI\delta^2}{2L^3} = W(h + \delta)$$

$$\delta^2 - \frac{WL^3\delta}{24EI} - \frac{WL^3 h}{24EI} = 0$$

$$\delta = \frac{WL^3}{48EI} + \tfrac{1}{2}\sqrt{\left[\left(\frac{WL^3}{24EI}\right)^2 + \frac{WL^3 h}{6EI}\right]} \qquad (8.5)$$

8.6. STRAIN ENERGY IN PURE SHEAR

If a piece of material is subject to pure shear then the strain energy stored per unit volume is represented by the area under the shear-stress/shear-

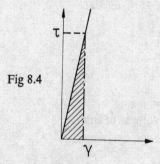

Fig 8.4

strain curve shown in Fig. 8.4.

Hence $U = \frac{1}{2}\tau\gamma$ per unit volume, and since $\gamma = \tau/G$,

$$U = \frac{\tau^2}{2G} \quad \text{per unit volume} \qquad (8.6)$$

This expression only applies if the shear stress is uniform over the element of material.

8.7. STRAIN ENERGY IN TORSION*

(a) Solid Circular Shaft

If a solid shaft of radius r and length L is subjected to a torque which increases gradually from zero to a value T, and θ is the corresponding angle of twist, then the energy stored in the shaft is

$$U = \tfrac{1}{2}T\theta = \tfrac{1}{2}\frac{T^2 L}{JG} \tag{8.7}$$

$$= \tfrac{1}{2}\frac{\tau^2}{r^2}\frac{JL}{G} = \frac{\tau^2}{2G}\frac{\pi r^2 L}{2}$$

when r is the outer radius, $\tau = \hat{\tau}$, and

$$\text{Total strain energy} = \frac{\tau^2}{4G} \times \text{volume of shaft} \tag{8.8}$$

(b) Hollow Circular Shaft

In the case of a hollow shaft of radii R and r the strain energy is

$$U = \tfrac{1}{2}T\theta = \tfrac{1}{2}\frac{T^2 L}{JG}$$

At the outer radius R, $\tau = \hat{\tau}$; therefore

$$U = \tfrac{1}{2}\frac{\left(\dfrac{\pi\hat{\tau}}{2}\dfrac{R^4 - r^4}{2}\right)^2 L}{\dfrac{\pi(R^4 - r^4)\,G}{2}}$$

$$= \frac{\hat{\tau}^2}{4G}\frac{R^2 + r^2}{R^2} \times \text{volume of shaft} \tag{8.9}$$

8.8. STRAIN ENERGY IN BENDING

The change in slope $\delta\theta$ between two cross-sections δs apart was shown in Chapter 7 to be related to bending moment by

$$\frac{\delta\theta}{\delta s} = \frac{M}{EI}$$

* See Chapter 9 for derivation of the torsion relationships used here.

The work done by the moments acting on the two sections is $\frac{1}{2}M\,\delta\theta$; therefore

Stored energy, $\delta U = \frac{1}{2}M\dfrac{M\,\delta s}{EI}$

Total strain energy, $U = \displaystyle\int \frac{M^2}{2EI}ds$ (8.10)

between required limits of length along the beam.

(a) For a Freely Supported Beam with Central Load W

Since the bending moment relationship is discontinuous over the length of the beam, it is necessary to split the integral of eqn. (8.10) into two parts. Thus for $0 < x < \frac{1}{2}L$, $M = \frac{1}{2}Wx$, and

Strain energy up to the load $= \displaystyle\int_0^{L/2} \frac{W^2x^2}{8EI}\,dx$ (8.11)

When $\frac{1}{2}L < x < L$, $M = \frac{1}{2}Wx - W(x - \frac{1}{2}L) = \frac{1}{2}W(L - x)$, and

Strain energy stored in second portion of beam $= \displaystyle\int_{L/2}^{L} \frac{W^2}{8EI}(L - x)^2\,dx$

Therefore the total strain energy is

$$U = \int_0^{L/2} \frac{W^2x^2}{8EI}\,dx + \int_{L/2}^{L} \frac{W^2(L - x)^2}{8EI}\,dx$$

$$= \frac{W^2L^3}{192EI} + \frac{W^2L^3}{192EI}$$

$$= \frac{W^2L^3}{96EI}$$ (8.12)

Also

$$M_{max} = \frac{WL}{4} \quad \text{or} \quad W = \frac{4M_{max}}{L} = \frac{4\sigma_{max}I}{Ly}$$

Therefore, substituting this expression in eqn. (8.12), the total strain energy in terms of the maximum bending stress is

$$U = \frac{1}{6}\frac{IL\sigma^2_{max}}{y^2E}$$ (8.13)

In this particular problem, since the load is at mid-span, the total strain energy could have been obtained by doubling the first integral (see eqn. (8.11)). The total strain energy could also have been obtained by considering the deflection of the beam under the load; hence

Work done $= \frac{1}{2}Wv_{max}$

Therefore

Strain energy, $U = \frac{1}{2}W\dfrac{WL^3}{48EI} = \dfrac{W^2L^3}{96EI}$

(b) **For a Uniformly Loaded Cantilever of Length L whose Load is w per Unit Length**

$$M = \frac{wx^2}{2}$$

Therefore eqn. (8.10) becomes

$$U = \frac{1}{2IE}\int_0^L \frac{w^2x^4}{4}\,dx$$

$$= \frac{1}{2IE}\frac{w^2L^5}{20}$$

$$= \frac{w^2L^5}{40IE}$$

so that the total strain energy is

$$U = \frac{F^2L^3}{40IE} \tag{8.14}$$

where $wL = F$, the total load on the beam.
   Since $M_{max} = \frac{1}{2}wL^2 = \frac{1}{2}FL$,

$$F = \frac{2}{L}\sigma_{max}\frac{I}{y}$$

Substituting for F in eqn. (8.14),

Total strain energy, $U = \dfrac{1}{10}\dfrac{IL}{y^2E}\sigma^2_{max}$ \hfill (8.15)

(c) **Beam in Pure Bending**

If M is constant, eqn. (8.10) becomes $\dfrac{M^2}{2IE}\displaystyle\int_0^L dx$. Therefore

Total strain energy, $U = \dfrac{M^2L}{2IE}$ \hfill (8.16)

which in terms of stress is

$$U = \tfrac{1}{2}\frac{IL}{y^2 E}\,\sigma^2{}_{max}$$

8.9. SHEAR DEFLECTION OF BEAMS

A deflection other than that due to bending moment occurs in beams owing to the shearing forces on transverse sections. This deflection may be found approximately from strain energy principles and by making use of the equation for shear stress at a point in the transverse section of a beam, which is, in itself, based on the assumption that pure bending occurs.

(a) Cantilever with Load at Free End

Assume the section to be rectangular, of breadth b and depth d, and the total length of the beam to be L. If v_s is the deflection, due to shear, at the free end, then

Work done by load $= \tfrac{1}{2}Wv_s$

The shear stress at a distance y from the neutral axis is

$$\tau = \frac{6W}{bd^3}\left(\frac{d^2}{4} - y^2\right)$$

where the shear force $Q = W$. Also, if dy is the height of the strip in the direction of the depth of the beam, and we consider a small portion of the beam of length dx and section $b\,dy$, we have, by eqn. (8.6),

Strain energy in strip $= \tfrac{1}{2}\dfrac{\tau^2}{G}dx\,b\,dy$

or

Strain energy $= \dfrac{1}{2G}dx\dfrac{36W^2}{b^2 d^6}\left(\dfrac{d^4}{16} - \dfrac{d^2 y^2}{2} + y^4\right)b\,dy$

Therefore the total strain energy for the piece of beam of length dx is

$$U = \frac{18W^2\,dx}{bd^6 G}\int_{-d/2}^{+d/2}\left(\frac{d^4}{16} - \frac{d^2 y^2}{2} + y^4\right)dy$$

$$= \frac{18W^2\,dx}{bd^6 G}\left[\frac{d^4 y}{16} - \frac{d^2 y^3}{6} + \frac{y^5}{5}\right]_{-d/2}^{d/2}$$

$$= \frac{3}{5}\frac{W^2\,dx}{bdG} \tag{8.17}$$

The strain energy for the whole beam of length L is

$$U_t = \frac{3}{5}\frac{W^2}{bdG}\int_0^L dx$$

$$= \frac{3}{5}\frac{W^2 L}{bdG} \tag{8.18}$$

Equating the strain energy to the work done by W,

$$\frac{1}{2}Wv_s = \frac{3}{5}\frac{W^2 L}{bdG}$$

Therefore

$$v_s = \frac{6}{5}\frac{WL}{bdG} \tag{8.19}$$

Thus the total deflection at the free end due to bending and shear is

$$v = v_b + v_s$$

$$= \frac{1}{3}\frac{WL^3}{EI} + \frac{6}{5}\frac{WL}{bdG} \tag{8.20}$$

(b) Horizontal Beam with Isolated Load at Mid-Span

If we treat each half of the span as a cantilever, L in eqn. (8.19) becomes $\frac{1}{2}L$, W becomes $\frac{1}{2}W$, and

$$v_s = \frac{3}{10}\frac{WL}{bdG} \tag{8.21}$$

Therefore

$$\text{Total deflection} = \frac{1}{48}\frac{WL^3}{EI} + \frac{3}{10}\frac{WL}{bdG} \tag{8.22}$$

(c) Cantilever with Uniformly Distributed Load

In (a) the shearing force Q is constant along the beam, but for a uniformly loaded cantilever at distance x from the fixed end, the shearing force is $w(L - x)$; hence eqn. (8.17) becomes

$$U = \frac{3}{5}\frac{w^2(L - x)^2}{bdG}dx$$

The shearing force acting on a piece of length dx is $w(L - x)$, and the external work done by this force is $\frac{1}{2}w(L - x)\,dv_s$, where dv_s is the deflection of the piece due to shear. Equating the external work done to the strain energy,

$$\frac{1}{2} w(L - x) \, dv_s = \frac{3}{5} \frac{w^2(L - x)^2}{bdG} \, dx$$

$$dv_s = \frac{6}{5} \frac{w(L - x)}{bdG} \, dx$$

$$v_s = \frac{6}{5} \frac{w}{bdG} \int_0^L (L - x) \, dx$$

Therefore

$$v_s = \frac{3}{5} \frac{wL^2}{bdG} \tag{8.23}$$

$$\text{Total deflection} = \frac{1}{8} \frac{wL^4}{EI} + \frac{3}{5} \frac{wL^2}{bdG} \tag{8.24}$$

(d) Horizontal Beam with Uniformly Distributed Load

Treating each half of the span as a cantilever, L becomes $\frac{1}{2}L$ in eqn. (8.23). Therefore

$$v_s = \frac{3}{20} \frac{wL^2}{bdG} \tag{8.25}$$

$$\text{Total deflection} = \frac{5}{384} \frac{wL^4}{EI} + \frac{3}{20} \frac{wL^2}{bdG} \tag{8.26}$$

For the majority of beams, where the span L is large compared to the cross-section of the beam, it is seen that the deflection due to shear is negligible in comparison with that due to bending.

Example 8.3. A beam of 3 m length is simply supported at each end and is subjected to a couple of 9 kN-m at a point B, 2 m from the left end as shown in Fig. 8.5. Determine the slope at B. $EI = 30$ kN-m².

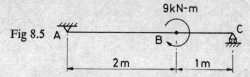

Fig 8.5

The reactions at A and C are $\bar{M}/L = 9000/3 = 3$ kN

When $0 < x < a$ $M = \dfrac{\bar{M}x}{L}$

When $a < x < L$ $M = \dfrac{\bar{M}}{L}(x - L)$

The strain energy stored is

$$U = \frac{1}{2EI} \int_0^2 \left(\frac{\bar{M}x}{L} \right)^2 dx + \frac{1}{2EI} \int_2^3 \left[\frac{\bar{M}}{L} (x - L) \right]^2 dx$$

$$= \frac{1}{60 \times 10^3} \left(\frac{9000}{3} \right)^2 \left\{ \left[\frac{x^3}{3} \right]_0^2 + \left[\frac{(x-3)^3}{3} \right]_2^3 \right\}$$

$$= \frac{3}{60 \times 10^3} \left(\frac{9000}{3} \right)^2$$

The work done at B is

$$\tfrac{1}{2} \bar{M} \theta = \frac{9000}{2} \theta$$

Therefore

$$\frac{9000}{2} \theta = \frac{3}{60 \times 10^3} \left(\frac{9000}{3} \right)^2$$

and

$$\theta = 0 \cdot 1 \, \text{rad}$$

8.10. STRAIN AND COMPLEMENTARY ENERGY SOLUTIONS FOR DEFLECTIONS

Energy functions may be very usefully employed in the determination of deflections of frameworks, beams, shells, etc. The methods depend on the principle of virtual work, which was introduced in Chapter 4 and will now again be briefly reviewed.

If a body is subjected to the system of forces shown in Fig. 8.6 and is

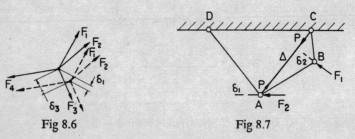

Fig 8.6   Fig 8.7

then given an arbitrary virtual displacement for which the corresponding displacements in the directions of the forces are δ_1, δ_2, δ_3 and δ_4, then static equilibrium exists for the system of forces if

$$F_1 \delta_1 + F_2 \delta_2 + F_3 \delta_3 + F_4 \delta_4 = 0 \qquad (8.27)$$

That is to say, if a body, subjected to a system of forces, is given any virtual displacement, the net work done by the forces must be zero for static equilibrium of the system.

Let the resultant force of the above system be P and assume a virtual displacement Δ in the direction of P; then for static equilibrium $P\Delta$ must be zero. Since Δ need not be zero, it follows that P must be. Thus the resultant of a system of forces in equilibrium is zero.

We can now make use of the above principle in deriving energy solutions for deflections. Consider the simple framework in Fig. 8.7 acted upon by forces F_1 and F_2. Let the displacements at the joints in the direction of the forces be δ_1 and δ_2. The internal reactions and deformations of the members of the frame are P_1, P_2, etc., and Δ_1, Δ_2, etc., respectively. Then, from the principle of virtual work,

$$F_1\delta_1 + F_2\delta_2 = P_1\Delta_1 + P_2\Delta_2 + \ldots P_n\Delta_n$$
$$= \Sigma P\Delta \qquad (8.28)$$

for static equilibrium. Now let us suppose that there is a small change in displacement at joint A of an amount $\delta\delta_1$, δ_2 remaining constant, which results in changes $\delta\Delta_1$, $\delta\Delta_2$, etc., in the members for compatibility. Then

$$F_1(\delta_1 + \delta\delta_1) + F_2\delta_2 = P_1(\Delta_1 + \delta\Delta_1) + P_2(\Delta_2 + \delta\Delta_2)$$
$$+ \ldots P_n(\Delta_n + \delta\Delta_n)$$
$$= \sum_n P(\Delta + \delta\Delta) \qquad (8.29)$$

Subtracting eqn. (8.28) from eqn. (8.29),

$$F_1\,\delta\delta_1 = \sum_n P\,\delta\Delta \qquad (8.30)$$

But $P\,\delta\Delta$ is the increment of strain energy stored in a member of the system due to the increments of deformation $\delta\Delta$ caused by the change in displacement $\delta\delta_1$. Therefore for the system

$$\sum_n P\,\delta\Delta = \delta U$$

or

$$F_1\delta\delta_1 = \delta U \qquad (8.31)$$

Thus, for an infinitely small change in displacement,

$$F_1 = \frac{\partial U}{\partial \delta_1} \qquad (8.32)$$

By a similar argument we have that

$$F_2 = \frac{\partial U}{\partial \delta_2}$$

Thus the external force on a member is given by the partial derivative of the strain energy with respect to the displacement at the point of application of and in the direction of the force.

We now return to the original proposition, and instead of changing the displacement δ_1, we change the force F_1 by an amount δF_1, keeping F_2 constant; then there will be a reaction in the system causing changes $\delta P_1, \delta P_2$, etc., in the internal forces in the members. Now, by the principle of virtual work, we have

$$(F_1 + \delta F_1)\delta_1 + F_2\delta_2 = (P_1 + \delta P_1)\Delta_1 + (P_2 + \delta P_2)\Delta_2$$
$$+ \ldots (P_n + \delta P_n)\Delta_n$$
$$= \sum_n (P + \delta P)\Delta \qquad (8.33)$$

Subtracting eqn. (8.28) from eqn. (8.33),

$$\delta F_1 \delta_1 = \sum_n \delta P\, \Delta \qquad (8.34)$$

Now considering Fig. 8.8 (*a*) or (*b*), the shaded area $P\,\delta\Delta$ is the increment of strain energy δU below the load-deformation curve used in

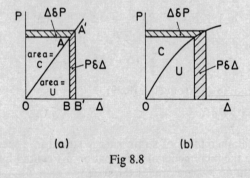

(a)   (b)

Fig 8.8

eqn. (8.30). The shaded area above the load-deformation curve represents $\Delta\,\delta P$ in eqn. (8.34); this is termed the *complementary energy* and is denoted by C; thus

$$\sum_n \delta P\, \Delta = \delta C$$

and therefore

$$\delta F_1 \delta_1 = \delta C \qquad (8.35)$$

For an infinitely small change in the force,

$$\delta_1 = \frac{\partial C}{\partial F_1} \qquad (8.36)$$

Thus the deflection at a point on a member in the direction of a force applied at that point is given by the partial derivative of the complementary energy with respect to the external force at the point.

The above energy theorems provide a most useful method of attack on many structural analysis problems.

A further point of interest is illustrated in the load-deformation characteristics of Figs. 8.8 (*a*) and (*b*). The former illustrates linear elasticity for a member or system of members, while the latter represents non-linear elasticity, which can occur in certain frameworks and materials. In both cases the sum of the strain energy and complementary energy is given by the force times the deformation, i.e.

$$U + C = P\Delta \tag{8.37}$$

but in the particular case of linear elasticity,

$$\delta U = \delta C = P\,\delta\Delta = \Delta\,\delta P$$

and

$$U = C = \tfrac{1}{2}P\Delta \tag{8.38}$$

Because of this last relationship we can express displacements in terms of strain energy instead of complementary energy for linear elastic systems. One of the earliest theorems of this form was due to Castigliano (1875) in which it was stated that the partial derivative of the strain energy with respect to a force gives the displacement corresponding to that force, or

$$\frac{\partial U}{\partial P} = \Delta \tag{8.39}$$

This relationship may be proved in the following way. Consider a force P applied to a body giving a displacement Δ. Then the work done or the stored strain energy is equal to OAB which equals $\tfrac{1}{2}P\Delta$ in Fig. 8.8(*a*). If an additional force δP is applied giving an additional deformation $\delta\Delta$, then the extra strain energy is

$$\text{BAA'B'} = P\,\delta\Delta + \tfrac{1}{2}\delta P\,\delta\Delta = \delta U$$

(or $\delta U/\delta\Delta = P$, neglecting second-order products). Therefore

$$\text{Total energy, OA'B'} = \tfrac{1}{2}P\Delta + P\,\delta\Delta + \tfrac{1}{2}\delta P\,\delta\Delta$$

If both forces had acted simultaneously, the stored strain energy would have been OA'B' $= \tfrac{1}{2}(P + \delta P)(\Delta + \delta\Delta)$. Since work done is independent of the order of application of the forces, we have

$$\tfrac{1}{2}P\Delta + P\,\delta\Delta + \tfrac{1}{2}\delta P\,\delta\Delta = \tfrac{1}{2}(P + \delta P)(\Delta + \delta\Delta)$$

On simplifying, and neglecting small products, we find that

$$P\,\delta\Delta = \Delta\,\delta P \tag{8.40}$$

(i.e. $\delta U = \delta C$ for linear elasticity). Thus, substituting above,

$$\delta U = \Delta\,\delta P + \tfrac{1}{2}\delta P\,\delta\Delta$$

Therefore

$$\frac{\delta U}{\delta P} = \Delta$$

neglecting the second-order term on the right. Hence

$$\frac{\partial U}{\partial P} = \Delta$$

which proves *Castigliano's hypothesis*.

In the case of a bar under simple tension,

$$U = \frac{P^2 L}{2AE}$$

and

$$\frac{\partial U}{\partial P} = \frac{PL}{AE} = \Delta, \text{ the extension of the bar}$$

and in torsion

$$U = \frac{T^2 L}{2GJ}$$

or

$$\frac{\partial U}{\partial T} = \frac{TL}{GJ} = \theta, \text{ the angle of twist}$$

8.11. BENDING DEFLECTION OF BEAMS

The complementary energy function can be used very conveniently to solve for beam deflections, since

$$\frac{\partial C}{\partial F} = \delta$$

or, using the notation for beams,

$$\frac{\partial C}{\partial W} = v$$

where W is a concentrated load whose displacement (beam deflection) is v.

The complementary energy in bending of a small length of beam δx

is shown in Fig. 8.9(*a*), which is the moment-slope relationship. The shaded area is

$$\delta C = \theta \, \delta M$$

or

$$C = \int_0^M \theta \, dM \tag{8.41}$$

In the most frequently used case,

$$\theta = \frac{M}{EI} \delta x$$

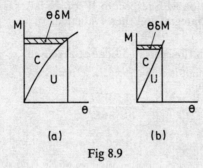

(a)　　　　(b)

Fig 8.9

for a linear elastic beam, Fig. 8.9(*b*). Therefore

$$C = \int_0^M \frac{M}{EI} dx \, dM = \frac{M^2}{2EI} dx \tag{8.42}$$

For the whole length of beam, therefore,

$$C = \int_0^L \frac{M^2}{2EI} dx \tag{8.43}$$

This result could also have been arrived at by the fact that $U = C$ in a linear elastic system, and it has already been shown, Section 8.8, that the strain energy is

$$U = \int_0^L \frac{M^2}{2EI} dx$$

It is also apparent that

$$\frac{\partial C}{\partial W} = \frac{\partial U}{\partial W} = v$$

and the latter part of the expression shows that Castigliano's analysis can be applied to beam deflection.

We can solve for the deflection in one of two ways, either

$$v = \frac{\partial U}{\partial W} = \frac{\partial}{\partial W}\left[\int_0^L \frac{[Wf(x)]^2}{2EI}\,dx\right] \tag{8.44}$$

where $Wf(x) = M$, or we can write

$$v = \frac{\partial U}{\partial W} = \int_0^L \frac{M}{EI}\frac{\partial M}{\partial W}\,dx \tag{8.45}$$

It is merely a question of whether the bending moment expression in terms of W and x is substituted and the integral evaluated, followed by partial differentiation with respect to W, or the latter is carried out first, substituted in the integral and then evaluated.

(a) Cantilever with Concentrated Load at Free End

The bending moment is $M = Wx$; therefore

$$v = \frac{\partial}{\partial W}\int_0^L \frac{W^2 x^2}{2EI}\,dx = \frac{\partial}{\partial W}\frac{W^2 L^3}{6EI}$$

$$= \frac{WL^3}{3EI} \tag{8.46}$$

Alternatively, $\partial M/\partial W = x$; therefore

$$v = \int_0^L \frac{Wx}{EI}x\,dx = \int_0^L \frac{Wx^2}{EI}\,dx$$

$$= \frac{WL^3}{3EI}$$

(b) Simply Supported Beam with Uniformly Distributed Load w over the Span

The bending moment in this case is

$$M = \frac{wL}{2}x - \frac{wx^2}{2}$$

and $\partial/\partial W$ of the above is zero, indicating no deflection, which is obviously not true. To get round this difficulty, we introduce an imaginary concentrated load W at some point in the span, let us say mid-span for simplicity. Then, for $0 < x < L/2$,

$$M = \frac{W}{2}x + \frac{wL}{2}x - \frac{wx^2}{2} \tag{8.47}$$

and $\partial M/\partial W = \frac{1}{2}x$; therefore

$$v = 2\int_0^{L/2} \frac{1}{EI}\left(\frac{W}{2}x + \frac{wL}{2}x - \frac{wx^2}{2}\right)\frac{x}{2}\,dx$$

$$= \frac{WL^3}{48EI} + \frac{5wL^4}{384EI}$$

Putting $W = 0$, we obtain

$$v_{max} = \frac{5}{384}\frac{wL^4}{EI} \tag{8.48}$$

If we require the deflection due to the point load only, we put $w = 0$; then

$$v = \frac{WL^3}{48EI} \tag{8.49}$$

Example 8.4. A freely supported beam (Fig. 8.10) carries a concentrated load at a distance a from the left-hand support and at a distance b from the other support. Determine the deflection of the beam underneath the load.

Fig 8.10

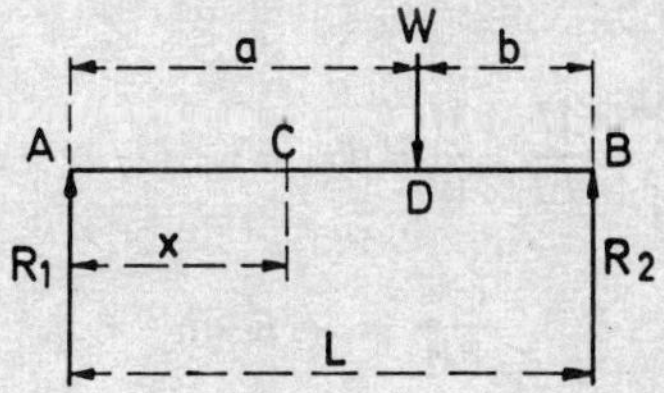

$R_1 = Wb/L$ and the bending moment at C is $(Wb/L)x$. For the portion of beam AD, we have the complementary energy

$$C = \int_0^a \frac{(Wb/L)^2 x^2\,dx}{2EI} = \int_0^a \frac{W^2 b^2}{L^2 2EI}x^2\,dx = \frac{W^2 b^2 a^3}{6EIL^2} \tag{8.50}$$

Similarly, we may write that C for the portion of beam DB is $W^2 a^2 b^3/6EIL^2$; therefore

Total value of C for the beam
$$= \frac{W^2 b^2 a^3}{6EIL^2} + \frac{W^2 a^2 b^3}{6EIL^2} = \frac{W^2 a^2 b^2}{6EIL^2}(a + b)$$

$$= \frac{W^2 a^2 (L - a)^2}{6EIL} \tag{8.51}$$

and

Deflection underneath load $= \dfrac{dC}{dW} = \dfrac{Wa^2(L - a)^2}{3EIL}$ $\tag{8.52}$

Example 8.5. Determine the vertical and horizontal displacements of the end of the curved member shown in Fig. 8.11.

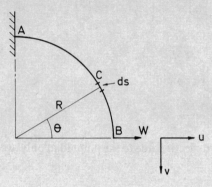

Fig. 8.11

Considering first the displacement u in the direction of the load,

$$U = \int_0^{\pi/2} \frac{M^2}{2EI} R\, d\theta$$

$$u = \frac{\partial U}{\partial W} = \int_0^{\pi/2} \frac{M}{EI} \frac{\partial M}{\partial W} R\, d\theta$$

$$M = -WR \sin\theta \qquad \frac{\partial M}{\partial W} = -R \sin\theta$$

$$u = \int_0^{\pi/2} -\frac{WR \sin\theta}{EI}(-R \sin\theta)R\, d\theta$$

$$= \int_0^{\pi/2} \frac{WR^3}{EI} \sin^2\theta\, d\theta$$

$$= \frac{\pi}{4} \frac{WR^3}{EI} \tag{8.53}$$

To find the vertical displacement v, an imaginary vertical force W_0 is applied. The bending moment on C will be

$$M_c = -WR \sin\theta + W_0 R(1 - \cos\theta)$$

and

$$\frac{\partial M_c}{\partial W_0} = R(1 - \cos\theta)$$

Putting $W_0 = 0$ in the expression for M_c,

$$v = \frac{\partial U}{\partial W_0} = \int_0^{\pi/2} -\frac{WR}{EI} \sin \theta \, R(1 - \cos \theta) \, R \, d\theta$$

$$= \int_0^{\pi/2} -\frac{WR^3}{EI} \sin \theta (1 - \cos \theta) \, d\theta$$

from which

$$v = -\frac{WR^3}{2EI} \tag{8.54}$$

Example 8.6. Determine the horizontal deflection of the member shown in Fig. 8.12.

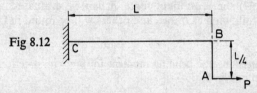

Fig 8.12

The strain-energy function is scalar; therefore the separate strain-energy quantities for the two parts of the member can be added before proceeding to use the Castigliano theorem.

From A to B

$$M = +Px \quad \text{and} \quad \frac{\partial M}{\partial P} = + x$$

From B to C

$$M = +\frac{PL}{4} \quad \text{and} \quad \frac{\partial M}{\partial P} = +\frac{L}{4}$$

$$\Delta = \frac{\partial U}{\partial P} = \frac{1}{EI} \int_0^{L/4} Px \, x \, dx + \frac{1}{EI} \int_0^L \frac{PL}{4} \frac{L}{4} \, dx$$

$$= \frac{1}{EI} \left[\frac{Px^3}{3} \right]_0^{L/4} + \frac{1}{EI} \left[\frac{PL^2 x}{16} \right]_0^L$$

$$= \frac{13PL^3}{192EI} \tag{8.55}$$

8.12. CASTIGLIANO'S SECOND THEOREM

This theorem is of value in finding redundant forces in members and frames. Let the force in a redundant member be R and also let it have an initial lack of fit λ. Then the theorem states that, if the total strain energy for the structure is partially differentiated with respect to the load in a redundant member, then any initial lack of fit of the member is obtained thus:

$$\frac{\partial U}{\partial R} = -\lambda \qquad (8.56)$$

If there is no initial lack of fit,

$$\frac{\partial U}{\partial R} = 0 \qquad (8.57)$$

which is a condition for a minimum value of the strain energy. This relationship also expresses what is termed the *principle of least work*.

From eqn. (8.57) the redundant force R can be evaluated. This is illustrated in the following example, and others will be found in Chapter 10.

Example 8.7. Determine the bending moment for any cross-section of the slender ring shown in Fig. 8.13(a).

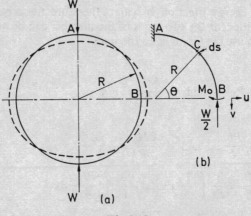

Fig 8.13

In view of the symmetry of the ring, only one quadrant need be considered as shown in Fig. 8.13(b). Cutting the ring at any section the bending moment is given by

$$M = M_0 - \frac{WR}{2}(1 - \cos\theta) \qquad (8.58)$$

but M_0 is unknown and is a redundancy.

The strain energy is given by

$$U = \int_0^{\pi/2} \frac{1}{2EI}\left[M_0 - \frac{WR}{2}(1 - \cos\theta)\right]^2 R\,d\theta \qquad (8.59)$$

and for $\partial U/\partial M_0 = 0$,

$$\frac{1}{EI}\int_0^{\pi/2}\left[M_0 - \frac{WR}{2}(1 - \cos\theta)\right]R\,d\theta = 0$$

from which

$$M_0 = WR\left(\tfrac{1}{2} - \frac{1}{\pi}\right) \qquad (8.60)$$

and

$$M = WR\left(\tfrac{1}{2}\cos\theta - \frac{1}{\pi}\right) \qquad (8.61)$$

8.13. THE RECIPROCAL THEOREM

In a linear structural system (Fig. 8.14), the deflection at point 1 due to

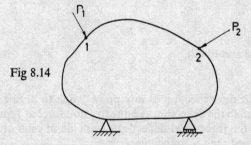

Fig 8.14

forces P_1 at point 1 and P_2 at point 2 is, from the principle of super-position,

$$\Delta_1 = \Delta_{11} + \Delta_{12}$$

and the deflection at point 2 is

$$\Delta_2 = \Delta_{22} + \Delta_{21}$$

The deflections may be expressed in terms of *flexibility coefficients*, which are the displacements per unit force, as follows

$$\Delta_1 = f_{11}P_1 + f_{12}P_2 \qquad (8.62)$$
$$\Delta_2 = f_{21}P_1 + f_{22}P_2 \qquad (8.63)$$

If the strain energy of the system due to the application of these forces is U then we may write

$$\Delta_1 = \frac{\partial U}{\partial P_1} \tag{8.64}$$

and

$$\Delta_2 = \frac{\partial U}{\partial P_2} \tag{8.65}$$

Partially differentiating eqns. (8.62) and (8.64) with respect to P_2,

$$\frac{\partial \Delta_1}{\partial P_2} = f_{12}$$

and

$$\frac{\partial^2 U}{\partial P_1 \partial P_2} = f_{12} \tag{8.66}$$

Similarly, from eqns. (8.63) and (8.65),

$$\frac{\partial \Delta_2}{\partial P_1} = f_{21}$$

and

$$\frac{\partial^2 U}{\partial P_2 \partial P_1} = f_{21} \tag{8.67}$$

From eqns. (8.66) and (8.67),

$$f_{21} = f_{12} \tag{8.68}$$

This shows that the deflection at any point 1 due to a unit force at any point 2 is equal to the deflection at 2 due to a unit force at 1, providing the directions of the forces and deflections coincide in each of the two cases. This is termed the *reciprocal theorem*.

Chapter 9

Theory of Torsion

9.1. One of the common engineering modes of deformation is that of torsion, in which a solid or tubular member is subjected to torque about its longitudinal axis resulting in twisting deformation. A theory is required in order to estimate shear stress distribution and angular twist for solid and hollow shafts of circular cross-section and thin-walled closed and open sections of non-circular section. Engineering examples of the above are obtained in shafts transmitting power in machinery and transport, structural members in aeroplanes, springs, etc.

9.2. TORSION OF A THIN-WALLED CYLINDER

The thin cylinder of mean radius r, thickness t and length L, shown in Fig. 9.1, is subjected to an axial torque T, which causes a uniform shear

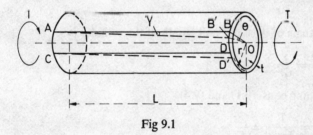

Fig 9.1

stress of intensity τ to act in a circumferential direction in the wall of the cylinder. This is a problem which is statically determinate, and the shear stress can be found simply by considering equilibrium as was shown in Chapter 2.

Equilibrium

The area of material on which the shear stress acts is $2\pi r t$; hence

$$\text{Torque, } T = 2\pi r t \times \tau \times r = 2\pi r^2 t \tau$$

or

$$\tau = \frac{T}{2\pi r^2 t} \tag{9.1}$$

Geometry of Deformation

If AC is a small length of the circumference of the cylinder, then the element ABDC is in a state of pure shear. AB and CD, being initially parallel to the axis of the cylinder, will assume the positions AB′ and CD′. The shear strain is given by γ, and if θ is the angle through which the right-hand end of the cylinder moves in relation to the left-hand end, then $r\theta = \gamma L$, or

$$\theta = \frac{\gamma L}{r} \tag{9.2}$$

Stress–Strain Relationship

Remembering that a shear is accompanied by a complementary shear, it will be seen that the intensity of shear stress both in the cylinder wall and parallel to the longitudinal axis is given by τ, and assuming the material to be elastic,

$$\gamma = \frac{\tau}{G} \tag{9.3}$$

Therefore

$$\theta = \frac{\tau}{G}\frac{L}{r}$$

or the angle of twist per unit length of the cylinder is

$$\frac{\theta}{L} = \frac{\tau}{Gr} \tag{9.4}$$

Combining eqns. (9.1) and (9.4),

$$\frac{\tau}{r} = \frac{T}{2\pi r^3 t} = \frac{G\theta}{L} \tag{9.5}$$

9.3. TORSION OF A SOLID CIRCULAR SHAFT

In the case of the thin-walled cylinder, in the previous section the shear stress was assumed to be constant throughout the thin wall, but in the case of a solid cylinder the shear stress may vary over the cross-section. In order to evaluate the torsional stresses in the cylindrical shaft, it is assumed that:

1. The shaft is straight and of uniform cross-section over its length.
2. The torque is constant along the length of the shaft.
3. Cross-sections which are plane before twisting remain plane during twisting.
4. Radial lines remain radial during twisting.

5. Induced stresses do not exceed the limit of proportionality of the material.

9.4. RELATION BETWEEN STRESS, STRAIN AND ANGLE OF TWIST

A cylindrical rod is said to be subject to pure torsion when the torsion is caused by a couple, applied so that the axis of the couple coincides with the axis of the rod. The state of stress, at any point in the cross-section of the rod, is one of pure shear, and the strain is such that one cross-section of the rod rotates relative to another.

Considering the cylindrical rod of length L and radius r, shown in Fig. 9.2, a couple of magnitude T is applied to one end, and the other

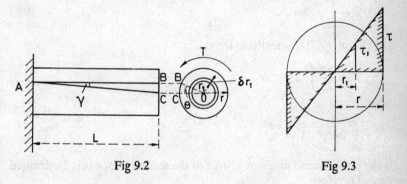

Fig 9.2   Fig 9.3

end of the rod is held, or constrained, by a balancing couple of equal magnitude.

Geometry of Deformation

This is of exactly the same form as for the thin-walled tube considered in Section 9.2; thus

$$\frac{\theta}{L} = \frac{\gamma}{r} \tag{9.6}$$

Stress–Strain Relationship

Substituting the shear-stress/shear-strain relationship $\gamma = \tau/G$ in eqn. (9.6),

$$\frac{G\theta}{L} = \frac{\tau}{r} = \text{constant} \tag{9.7}$$

Thus the shear stress in the cross-section is distributed linearly in proportion to the distance from the centre of the shaft as shown in Fig. 9.3.

9.5. RELATION BETWEEN TORQUE AND SHEAR STRESS

Consider an annulus of radius r_1 and thickness δr_1 on which the shear stress is τ_1.

Equilibrium

Force per unit length $= \tau_1 \delta r_1$

Torque per unit length of ring about shaft axis $= \tau_1 r_1 \delta r_1$
Resisting torque on whole ring $= \tau_1 r_1 \delta r_1 \times 2\pi r_1$

Resisting torque for whole cross-section $= \displaystyle\int_0^r \tau_1 2\pi r_1{}^2 \, dr$

This is equal to the applied torque; therefore

$$T = \int_0^r \tau_1 2\pi r_1{}^2 \, dr \tag{9.8}$$

Using eqn. (9.7) to substitute for τ_1,

$$T = \frac{G\theta}{L} \int_0^r 2\pi r_1{}^3 \, dr_1 \tag{9.9}$$

Now

$$\int_0^r 2\pi r_1{}^3 \, dr_1 = \frac{\pi r^4}{2}$$

is the polar second moment of area of the section (Appendix 2), denoted by J. Therefore

$$T = \frac{G\theta}{L} J$$

or

$$\frac{T}{J} = \frac{G\theta}{L} \tag{9.10}$$

and from eqn. (9.7),

$$\frac{T}{J} = \frac{G\theta}{L} = \frac{\tau}{r} \tag{9.11}$$

The quantities concerned and their units are

$T =$ Torque, N-m
$J =$ Polar second moment of area, m⁴ or mm⁴
$\tau =$ Shear stress, N/m² at radius r, m
$G =$ Shear modulus, MN/m²
$\theta =$ Angle of twist, rad, over length L, m

Example 9.1. Calculate the size of shaft which will transmit $40\,\mathrm{kW}$ at $2\,\mathrm{rev/sec}$. The shear stress is to be limited to $50\,\mathrm{MN/m^2}$ and the twist of the shaft is not to exceed $1°$ for each $2\,\mathrm{m}$ length of shaft. The modulus of shear, G, is $77\,\mathrm{GN/m^2}$.

$$T = \frac{40\,000}{2\pi \times 2} = 3185\,\mathrm{N\text{-}m}$$

From eqn. (9.11),

$$\tau = \frac{T}{\dfrac{\pi r^3}{2}}$$

$$r^3 = \frac{2 \times 3185 \times 10^9}{\pi \times 50 \times 10^6} = 40\cdot6 \times 10^3\,\mathrm{mm^3}$$

$$r = 34\cdot4\,\mathrm{mm}\ \text{on a stress basis}$$

Considering the twist criterion,

$$r^4 = \frac{2TL}{\pi G\theta}$$

$$= \frac{2 \times 3185 \times 2 \times 57\cdot3 \times 10^{12}}{\pi \times 77 \times 10^9 \times 1}$$

$$= 302 \times 10^4\,\mathrm{mm^4}$$

$$r = 41\cdot7\,\mathrm{mm}\ \text{on a twist basis}$$

Therefore shaft diameter $= 83\cdot4\,\mathrm{mm}$.

9.6. HOLLOW CIRCULAR SHAFTS

The above analysis for the solid shaft is similarly applicable to the hollow shaft. Thus the torsion relationship equation (9.11) also expresses the conditions of equilibrium and compatibility for a hollow circular shaft. However, the radial boundaries are now $r = r_1$ and $r = r_2$, the outer and inner radii respectively, and thus the polar second moment of area is

$$J = \int_{r_2}^{r_1} 2\pi r^3 \, dr$$

$$= \frac{\pi}{2}\left(r_1{}^4 - r_2{}^4\right)$$

The shear stress varies linearly from Tr_2/J at the bore to Tr_1/J at the outer surface as shown in Fig. 9.4.

The hollow shaft is more efficient in its use of stressed material than the solid shaft because the core of a solid shaft has relatively low stresses as compared with the outer layers. However, hollow shafts are not used widely in practice owing to the cost of machining, unless saving of weight is at a premium or it is necessary to pass services down the centre of the shaft.

Example 9.2. Compare the strength of a hollow shaft with that of a solid shaft of the same material, weight and length.

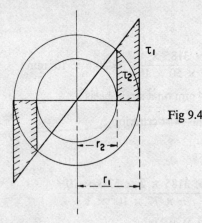

Fig 9.4

Let r_1 and r_2 be the outer and inner radii of the hollow shaft and τ the maximum shear stress; also let r be the radius of the solid shaft. Then

$$T_{hollow} = \frac{\tau\pi}{2}\frac{r_1{}^4 - r_2{}^4}{r_1}$$

and

$$T_{solid} = \frac{\tau\pi}{2}r^3$$

since the maximum stress is the same for each shaft.

$$\frac{T_{hollow}}{T_{solid}} = \frac{r_1{}^4 - r_2{}^4}{r_1 r^3} \tag{9.12}$$

Since $r^2 = r_1{}^2 - r_2{}^2$ for shafts of the same weight,

$$\frac{T_{hollow}}{T_{solid}} = \frac{r_1{}^2 + r_2{}^2}{r_1 r}$$

$$= \frac{r_1}{r}\left(1 + \frac{1}{n^2}\right) \tag{9.13}$$

where $r_2 = r_1/n$; but

$$r^2 = r_1{}^2 - \left(\frac{r_1}{n}\right)^2 \qquad \text{or} \qquad \left(\frac{r_1}{r}\right) = \frac{n}{\sqrt{(n^2-1)}}$$

Therefore

$$\frac{T_{hollow}}{T_{solid}} = \frac{n^2+1}{n\sqrt{(n^2-1)}} \tag{9.14}$$

It is common practice to take $n = 2$, which gives

$$\frac{T_{hollow}}{T_{solid}} = \frac{5}{2\sqrt{3}} = 1{\cdot}44$$

Thus the hollow shaft can carry 44% greater torque than the solid shaft for the same weight, etc.

9.7. NON-UNIFORM AND COMPOSITE SHAFTS

In certain shaft arrangements, the complete shaft is continuous but not of uniform diameter, or the arrangement may be such that one shaft is hollow with another shaft arranged coaxially. In each case it is necessary to investigate the conditions of both the twisting moment and angle of twist in each part of the system in order to obtain a sufficient number of equations for the solution.

(a) Continuous Shaft having Two Different Diameters

In this case the total twisting moment T is transmitted by each portion of the shaft; thus

$$T = T_1 = T_2 \tag{9.15}$$

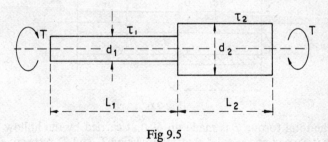

Fig 9.5

and the total deformation θ is that due to L_1 and L_2 in Fig. 9.5; therefore

$$\theta = \theta_1 + \theta_2 \tag{9.16}$$

and from eqn. (9.11),

$$\tau_1 \tfrac{1}{2}(\pi r_1{}^3) = \tau_2 \tfrac{1}{2}(\pi r_2{}^3)$$

since $(\tau_1/r_1)J_1 = (\tau_2/r_2)J_2 = T$; or

$$\frac{\tau_1}{\tau_2} = \frac{r_2{}^3}{r_1{}^3} = \left(\frac{d_2}{d_1}\right)^3 \tag{9.17}$$

i.e. the ratio of the stresses is inversely proportional to the ratio of the diameters cubed. Now,

$$\theta_1 = \frac{\tau_1}{r_1}\frac{L_1}{G} \quad \text{and} \quad \theta_2 = \frac{\tau_2}{r_2}\frac{L_2}{G}$$

Therefore

$$\theta = \frac{\tau_1}{r_1}\frac{L_1}{G} + \frac{\tau_2}{r_2}\frac{L_2}{G}$$

$$= \frac{\tau_1 L_1}{r_1 G} + \frac{\tau_1 r_1{}^3}{r_2{}^3}\frac{L_2}{r_2 G}$$

$$= \frac{\tau_1}{G}\left[\frac{L_1}{r_1} + \frac{L_2}{r_2}\left(\frac{r_1}{r_2}\right)^3\right] \tag{9.18}$$

and the ratio of the angle of twist per unit length is given by

$$\frac{\theta_1/L_1}{\theta_2/L_2} = \frac{T/J_1 G}{T/J_2 G} = \frac{J_2}{J_1} = \left(\frac{d_2}{d_1}\right)^4 \tag{9.19}$$

(b) Concentric Shafts

In Fig. 9.6 the shafts have a common axis and are joined at the ends so

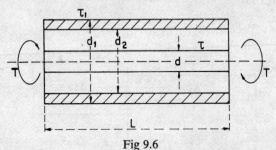

Fig 9.6

that the total torque T is made up of that carried by the hollow shaft and that carried by the solid shaft, these being T_1 and T_2 respectively:

$$T = T_1 + T_2$$

$$= \tau_1 \frac{\pi}{2} \frac{r_1{}^4 - r_2{}^4}{r_1} + \tau \frac{\pi}{2} r^3 \tag{9.20}$$

Both shafts twist through the same angle θ since their ends are rigidly connected; hence

$$\theta = \theta_1 = \theta_2$$

or

$$\frac{T_1 L}{G J_1} = \frac{T_2 L}{G J_2}$$

when both shafts are of the same material. Therefore

$$\frac{T_1}{T_2} = \frac{J_1}{J_2} = \frac{r_1^4 - r_2^4}{r^4} \tag{9.21}$$

Substituting for T_1 and T_2 in terms of the maximum shear stresses τ_1 and τ,

$$\frac{\tau_1 \dfrac{\pi}{2} \dfrac{r_1^4 - r_2^4}{r_1}}{\tau \dfrac{\pi}{2} r^3} = \frac{r_1^4 - r_2^4}{r^4}$$

Hence

$$\frac{\tau_1}{\tau} = \frac{r_1}{r} \tag{9.22}$$

Example 9.3. A solid alloy shaft of 50 mm diameter is to be coupled in series with a hollow steel shaft of the same external diameter, Fig. 9.7. Find

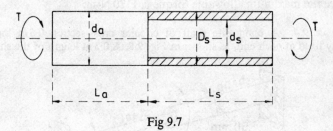

Fig 9.7

the internal diameter of the steel shaft if the angle of twist per unit length is to be 75% of that of the alloy shaft.

What is the maximum torque that can be transmitted if the limiting shear stresses in the alloy and the steel are 50 MN/m² and 75 MN/m² respectively?

$$G_{steel} = 2 \cdot 2 \, G_{alloy}$$

Equilibrium

$$T_{alloy} = T_{steel} = T \tag{9.23}$$

Deformation

$$\frac{\theta_s}{L_s} = 0.75\frac{\theta_a}{L_a} \qquad (9.24)$$

Hence

$$\frac{T_s}{J_sG_s} = 0.75\frac{T_a}{J_aG_a}$$

Using eqn. (9.23) and $G_s = 2.2G_a$,

$$J_a = 2.2 \times 0.75J_s$$

$$\frac{\pi d_a{}^4}{32} = 2.2 \times 0.75\frac{\pi}{32}(D_s{}^4 - d_s{}^4)$$

$$50^4 = (2.2 \times 0.75 \times 50^4) - (2.2 \times 0.75 \times d_s{}^4)$$

$$d_s{}^4 = \frac{0.65 \times 50^4}{2.2 \times 0.75} \qquad d_s = 39.6\,\text{mm}$$

The torque that can be carried by the alloy is

$$T = \frac{\pi d^3}{16}\tau = \frac{\pi \times 50^3}{16 \times 10^9} \times 50 \times 10^6 = 1227\,\text{N-m}$$

The torque that can be carried by the steel is

$$T = \frac{\pi}{16}\frac{50^4 - 39.6^4}{50 \times 10^9} \times 75 \times 10^6 = 1120\,\text{N-m}$$

Hence the maximum allowable torque is 1120 N-m.

Example 9.4. A composite shaft of circular cross-section 0.5 m long is rigidly fixed at each end, as shown in Fig. 9.8. A 0.3 m length of the shaft is

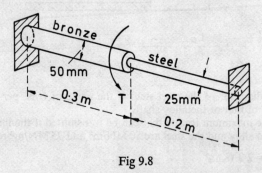

Fig 9.8

50 mm in diameter and is made of bronze to which is joined the remaining 0.2 m length of 25 mm diameter made of steel. If the limiting shear stress in

the steel is $55 \, \text{MN/m}^2$ determine the maximum torque that can be applied at the joint. What is then the maximum shear stress in the bronze?

$$G_{steel} = 82 \, \text{GN/m}^2 \qquad G_{bronze} = 41 \, \text{GN/m}^2$$

The torque that can be carried by the steel is

$$T_s = \frac{2 \times 55 \times 10^6}{0 \cdot 025} \times \frac{\pi \times 0 \cdot 025^4}{32} = 169 \, \text{N-m}$$

The corresponding angle of twist is

$$\theta_s = \frac{169}{\dfrac{\pi}{32} \times 0 \cdot 025^4} \times \frac{0 \cdot 2}{82 \times 10^9} = 0 \cdot 0108 \, \text{rad}$$

Similarly $\theta_b = 0 \cdot 0108 \, \text{rad}$. Therefore

$$T_b = \frac{41 \times 10^9 \times 0 \cdot 0108}{0 \cdot 3} \times \frac{\pi \times 0 \cdot 05^4}{32} = 904 \, \text{N-m}$$

The total torque that can be applied at the joint is

$$T_b + T_s = 904 + 169 = 1073 \, \text{N-m}$$

The maximum shear stress in the bronze is

$$\tau_b = \frac{41 \times 10^9 \times 0 \cdot 0108}{0 \cdot 3} \times \frac{25}{10^3} = 36 \cdot 8 \, \text{MN/m}^2$$

9.8. TORSION OF A TAPER SHAFT

Suppose a twisting moment T is applied to the taper shaft of length L, shown in Fig. 9.9. The resisting moment of all cross-sections of the shaft must be equal to T. Let

$\tau_1 = $ Maximum shear stress at end of radius r_1
$\tau_2 = $ Maximum shear stress at end of radius r_2
$\tau = $ Maximum shear stress at cross-section of radius r

Then

$$\frac{\pi \tau_1 r_1^3}{2} = \frac{\pi \tau_2 r_2^3}{2} = \frac{\pi \tau r^3}{2}$$

or

$$\tau_1 r_1^3 = \tau_2 r_2^3 = \tau r^3 \qquad\qquad (9.25)$$

Let δx be a small length of the shaft at distance x from the larger end and whose radius we may assume to be r; if $\delta\theta$ is the angle of twist of the small length δx, then

$$\delta\theta = \frac{T}{JG}\,\delta x$$

$$= \frac{2T}{G\pi}\frac{\delta x}{r^4} \tag{9.26}$$

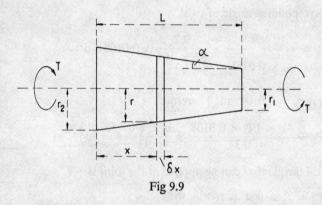

Fig 9.9

But

$$r = r_2 - x\tan\alpha$$

$$= r_2 - x\frac{r_2 - r_1}{L}$$

$$= r_2 - ax \tag{9.27}$$

where $a = (r_2 - r_1)/L$. Substituting eqn. (9.27) in eqn. (9.26),

$$\delta\theta = \frac{2T}{G\pi}(r_2 - ax)^{-4}\,\delta x$$

or, in the limit,

$$d\theta = \frac{2T}{G\pi}(r_2 - ax)^{-4}\,dx$$

The total angle of twist θ for the length L is given by

$$\theta = \int_0^L d\theta = \int_0^L \frac{2T}{G\pi}(r_2 - ax)^{-4}\,dx$$

$$= \frac{2T}{G\pi}\int_0^L (r_2 - ax)^{-4}\,dx$$

$$= \frac{2T}{G\pi} \frac{1}{3a} \left[(r_2 - ax)^{-3} \right]_0^L$$

$$= \frac{2T}{G\pi} \frac{L}{3(r_2 - r_1)} \left[\left\{ r_2 - \frac{(r_2 - r_1)L}{L} \right\}^{-3} - r_2^{-3} \right]$$

$$= \frac{2T}{G\pi} \frac{L}{3(r_2 - r_1)} \left[\frac{1}{r_1^3} - \frac{1}{r_2^3} \right]$$

Therefore

$$\theta = \frac{2TL}{G\pi} \frac{r_1^2 + r_1 r_2 + r_2^2}{3r_1^3 r_2^3} \qquad (9.28)$$

In the special case when $r_1 = r_2$, i.e. the shaft is parallel,

$$\theta = \frac{2T}{G\pi} \frac{L}{r_1^4} = \frac{TL}{GJ} \qquad (9.29)$$

which is the result already obtained in Section 9.5.

9.9. TORSION OF A THIN TUBE OF NON-CIRCULAR SECTION

The torsion of a solid shaft of non-circular section involves a very complex problem; but in the case of a thin hollow shaft, or tube, a simple theory can be developed even if the tube thickness is not constant.

The thin-walled tube shown in Fig. 9.10 is assumed to be of constant

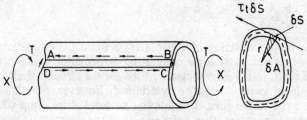

Fig 9.10

cross-section throughout its length. The wall thickness is variable but at any point is taken to be t. Assume the applied torque T to act about the longitudinal axis XX and to introduce shearing stresses over the end of the tube, these stresses have a direction parallel to that of a tangent to the tube at a given point. A shearing stress of magnitude τ at any point in the circumference has a complementary shear stress of the same magnitude acting in a longitudinal direction.

Consider the small portion ABCD of the tube and assume that the shearing stress τ is constant throughout the wall thickness t. The shearing force along the thin edge AB is τt per unit length, and for longitudinal equilibrium of ABCD, this force must be equal to that on the thin edge CD.

Now, since ABCD was an arbitrary choice, it follows that τt is constant for all parts of the tube. The value $\tau t = q$ is called the *shear flow* and is an internal shearing force per unit length of the circumference of the section of the thin tube.

The force acting in a tangential direction on an element of the perimeter of length δs is $\tau t \, \delta s$, and if r is the mean radius at the point, then the moment of this force about the axis XX is $\tau t r \, \delta s$ and the total torque on the cross-section of the tube is

$$T = \oint \tau t r \, ds \tag{9.30}$$

The integration extends over the whole circumference. Since $\tau t = q$ is constant, we may write

$$T = \tau t \oint r \, ds = q \oint r \, ds \tag{9.31}$$

Now, $r \, ds$ is twice the shaded area shown, and $\oint r \, ds$ for the whole circumference is therefore equal to $2A$, where A is the area enclosed by the centre-line of the wall of the tube; therefore

$$T = 2Aq \tag{9.32}$$

Thus at any point the shearing stress is given by

$$\tau = \frac{q}{t} = \frac{T}{2At} \tag{9.33}$$

Owing to the variation of shear stress around the circumference of the tube it is not possible to predict the deformations, and hence the angle of twist, in the simple manner used for the circular shaft or tube. The angle of twist, θ, can be determined, however, from the strain energy stored in the tube. Considering an axial strip along which the shear stress is constant, then from eqn. (8.6) the shear strain energy per unit volume is

$$U_s = \frac{\tau^2}{2G}$$

If the strip is of length l, thickness t and width ds, then the energy stored in the strip is

$$U_s = \frac{\tau^2}{2G} \times lt \, ds$$

Therefore the energy stored in the complete tube is

$$U_s = \oint \frac{\tau^2}{2G} lt \, ds$$

Substituting for τ from eqn. (9.33),

$$U_s = \oint \frac{T^2}{8A^2t^2G} lt \, ds$$

$$= \frac{T^2l}{8A^2G} \oint \frac{ds}{t} \tag{9.34}$$

But the stored energy is equal to the work, $\frac{1}{2}T\theta$, since in the elastic range the torque is proportional to the angle of twist, θ; therefore

$$\tfrac{1}{2}T\theta = \frac{T^2l}{8A^2G} \oint \frac{ds}{t}$$

so that

$$\theta = \frac{Tl}{4A^2G} \oint \frac{ds}{t} \tag{9.35}$$

If the tube is of constant thickness t around the circumference, s, then

$$\theta = \frac{Tls}{4A^2Gt}$$

or, substituting for T from eqn. (9.33),

$$\theta = \frac{\tau ls}{2AG} \tag{9.36}$$

Example 9.5. The light-alloy stabilizing strut of a high-wing monoplane is 2 m long and has the cross-section shown in Fig. 9.11. Determine the torque that can be sustained and the angle of twist if the maximum shear stress is limited to 28 MN/m². $G = 27\,\text{GN/m}^2$.

The area enclosed by the median line of the wall thickness is $A = (\pi \times 25^2) + (50 \times 50) = 4460\,\text{mm}^2$. The total length of the median line is

$$S = (2 \times 50) + (\pi \times 50) = 257\,\text{mm}$$

The allowable torque is obtained using the minimum wall thickness:

$$T = 2At\tau$$

$$= 2 \times \frac{4460}{10^6} \times \frac{2}{10^3} \times 28 \times 10^6 = 500\,\text{N-m}$$

The angle of twist is obtained from eqn. (9.35):

$$\theta = \frac{Tl}{5A^2G} \oint \frac{ds}{t}$$

Therefore

$$\theta = \frac{500 \times 2}{4 \times 4460^1 \times 10^{-12} \times 27 \times 10^9} \left[\frac{100}{2} + \frac{50\pi}{3} \right]$$

$$= 0 \cdot 0476 \, \text{rad} = 2 \cdot 73°$$

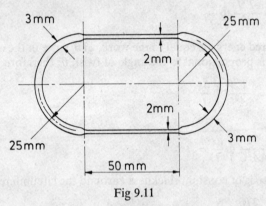

Fig 9.11

9.10. TORSION OF A THIN RECTANGULAR STRIP

An approximate solution for the torsion of a prismatic rectangular strip, whose thickness is small compared with the width, Fig. 9.12 (a), can be obtained by considering the strip to be built up of a series of thin-walled concentric tubes which all twist by the same amount. One of these tubes is shown at (b), and the area enclosed by the median line is

$$A = (b - 2h)2h + \pi h^2$$

If b is large compared with h then

$$A \approx 2bh$$

If the tube is subjected to a torque δT then, from eqn. (9.33), the shear stress is

$$\tau = \frac{\delta T}{4bh \, \delta h} \tag{9.37}$$

As the tube becomes infinitely thin,

$$\frac{dT}{dh} = 4bh\tau$$

and the torque carried by the strip is

$$T = \int_0^{t/2} 4bh\tau \, dh \qquad (9.38)$$

From eqn. (9.36),

$$\tau = \frac{G\theta}{l} \frac{2A}{S}$$

$$= \frac{G\theta}{l} \frac{4bh}{2b} \qquad (9.39)$$

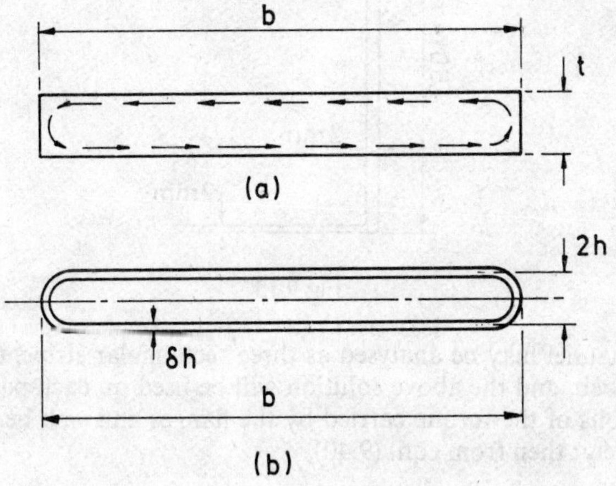

(a)

(b)

Fig. 9.12

Substituting in eqn. (9.38) for τ,

$$T = \int_0^{t/2} \frac{G\theta}{l} 8bh^2 \, dh$$

$$= \frac{1}{3} bt^3 \frac{G\theta}{l} \qquad (9.40)$$

The quantity $bt^3/3$ is termed the *torsion constant*, but it is *not* the polar second moment of area for the section.

Eqn. (9.39) shows that the shear stress parallel to the long edge of the cross-section is proportional to the distance h from the central axis. The maximum shear stress occurs at the outer surface and is

$$\tau_{max} = \frac{tG\theta}{l} \qquad (9.41)$$

Example 9.6.   Determine the angle of twist per unit length and the maximum shear stress in the aluminium channel section shown in Fig. 9.13 when subjected to a pure torque of 20 N-m. Shear modulus = 27 GN/m².

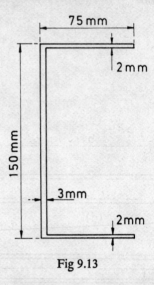

Fig 9.13

The channel may be analysed as three rectangular strips, the flanges and the web, and the above solution will be used on each part. Let the proportions of the torque carried by the flanges and web be T_1 and T_2 respectively; then from eqn. (9.40),

$$T_1 = \frac{1}{3} \times \frac{75}{10^3} \times \left(\frac{2}{10^3}\right)^3 \times 27 \times 10^9 \frac{\theta}{l}$$

$$= 5.4 \frac{\theta}{l}$$

$$T_2 = \frac{1}{3} \times \frac{150}{10^3} \times \left(\frac{3}{10^3}\right)^3 \times 27 \times 10^9 \frac{\theta}{l}$$

$$= 36.5 \frac{\theta}{l}$$

But $2T_1 + T_2 = 20$; therefore

$$47.3 \frac{\theta}{l} = 20$$

and

$$\frac{\theta}{l} = 0.422 \, \text{rad/m} = 24.2°/\text{m}$$

The maximum shear stress in the flanges is

$$\tau_{max} = \frac{2}{10^3} \times 27 \times 10^9 \times 0.422 = 22.8\,\text{MN/m}^2$$

and in the web is

$$\tau_{max} = \frac{3}{10^3} \times 27 \times 10^9 \times 0.422 = 34.2\,\text{MN/m}^2$$

9.11. EFFECT OF WARPING DURING TORSION

In the previous section and the above example dealing with a thin-walled open section subjected to torsion, the pure torque was supposed to be applied to each end of the member in such a way that there was no axial restraint. Owing to the variation in transverse shear stress, for example in the flanges of the above channel, there is also a variation in longitudinal complementary shear stress which results in axial movement of one flange with respect to the other. Therefore, cross-sections which were initially plane do not remain so during torsion and there is warping of any cross-section. If one or more sections of a member are constrained in some manner to remain plane during torsion then warping is restrained.

Resisting torque in the section is supplied in two ways, by simple torsion and also by torque set up through the restraint of warping. Thus an applied torque will cause a smaller angle of twist than when the section is free to warp, and torsional stiffness may be considerably increased if warping is restrained. Non-circular closed and solid section members also exhibit warping, but the effect is much smaller in comparison with the open section. A detailed treatment of warping behaviour is beyond the scope of this book.

9.12. TORSION OF SOLID RECTANGULAR AND SQUARE CROSS-SECTIONS

The solution in Section 9.10 is only applicable for a rectangular section in which the longer side is much greater than the other. Unfortunately the solution for torsion of general rectangular and square sections, although of some engineering interest, is very complex and beyond the scope of this text. (The reader who is interested in the full solution should consult a text such as *Theory of Elasticity* by Timoshenko and Goodier.) The result in terms of torque, maximum shear stress at the centre of the long side and angle of twist may be expressed as

$$T = \alpha bh^2 \tau_{max} = \beta bh^3 G \frac{\theta}{l} \tag{9.42}$$

where b is the longer and h the shorter side, and α, β are factors dependent on the geometry as given in Table 9.1.

Table 9.1

b/h	1	1·5	2	2·5	3	4	6	10	∞
α	0·208	0·231	0·246	0·256	0·267	0·282	0·299	0·312	0·333
β	0·141	0·196	0·229	0·249	0·263	0·281	0·299	0·312	0·333

9.13. HELICAL SPRINGS

Springs are directly concerned with theories of torsion, bending and strain energy and form an excellent example of their application. There are comparatively few machines which do not incorporate a spring to assist in their operation. The principal function of a spring is to absorb energy, store it for a long or short period, and then return it to the surrounding material. Two extremes of operation are found in a watch and on an engine valve. In the former case energy is stored for a long period and in the latter case the process is very rapid. The energy stored by a spring up to a certain limiting stress is termed the *resilience*. The force required to produce a unit deformation is called the *stiffness* of a spring.

The geometry of a helical spring is shown in Fig. 9.14. The centre-line of the wire forming the spring is a helix on a cylindrical surface such that the helix angle is α. Helical springs are designed and manufactured in two categories: open coiled and close coiled. In the former the helix angle α is very small and the coils almost touch each other. In the second case the helix angle is larger and the coils are spaced farther apart.

The most common form of loading on a spring is a force W acting along the central axis. Since this force acts at a distance R, the coil radius, from the axis of the wire, there will be torque and bending moment set up about the mutually perpendicular axes $0x$, $0y$ and $0z$ on the cross-section at 0 in Fig. 9.14.

$M_z = WR \cos \alpha$ (causing torsion on the wire) $= T$
$M_y = WR \sin \alpha$ (causing bending about the y-axis and a change in R)
$M_x = 0$

In addition, cross-sections of the wire are subjected to a transverse shear force $W \cos \alpha$ and an axial force $W \sin \alpha$. The stresses due to these forces are considerably smaller than those due to torsion and bending and are generally neglected.

Open-coiled Spring

The work done in deflecting the spring by the axial load W is $\frac{1}{2}W\delta$, and the stored energies due to torsion and bending are

$$\frac{1}{2}T\theta \quad \text{and} \quad \frac{1}{2}M\phi$$

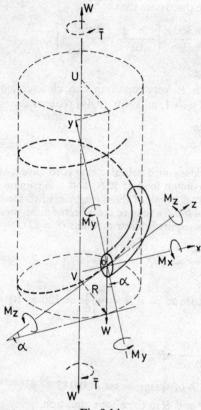

Fig 9.14

Therefore

$$\frac{1}{2}W\delta = \frac{1}{2}T\theta + \frac{1}{2}M\phi$$
$$= \frac{1}{2}(WR\cos\alpha)\theta + \frac{1}{2}(WR\sin\alpha)\phi$$

Hence

$$\delta = R\left[\cos\alpha\,\frac{Tl}{JG} + \sin\alpha\,\frac{Ml}{EI}\right]$$

$$= WR^2l\left[\frac{\cos^2\alpha}{JG} + \frac{\sin^2\alpha}{EI}\right] \tag{9.43}$$

The length of wire in the spring is $2\pi nR \sec \alpha$, where n is the number of complete coils.

$$J = \frac{\pi d^4}{32} \quad \text{and} \quad I = \frac{\pi d^4}{64}$$

where d is the wire diameter; thus

$$\delta = \frac{64WR^3 n \sec \alpha}{d^4} \left[\frac{\cos^2 \alpha}{G} + \frac{2 \sin^2 \alpha}{E} \right] \tag{9.44}$$

Close-coiled Spring

The helix angle α is very small in the close-coiled spring so that $\sin \alpha \to 0$ and $\cos \alpha \to 1$, and eqn. (9.44) reduces to

$$\delta = \frac{64WR^3 n}{Gd^4} \tag{9.45}$$

Example 9.7. A close-coiled helical spring is to have a stiffness of 1 kN/m compression, a maximum load of 50 N, and a maximum shearing stress of 120 MN/m². The solid length of the spring, i.e. when the coils are touching, is to be 45 mm. Find the diameter of the wire, the mean radius of the coils and the number of coils required. Shear modulus $G = 82$ GN/m².

$$\text{Stiffness} = \frac{W}{\delta} = 1000 = \frac{Gd^4}{64R^3 n}$$

$$\text{Maximum torque} = WR = \frac{\tau \pi d^3}{16} = 120 \times 10^6 \frac{\pi d^3}{16} = 50R$$

Hence

$$R = \frac{120 \times 10^6}{800} \pi d^3$$

Closed length of spring $= nd = 0.045$

Substituting for n and R in the above equation,

$$1000 = \frac{Gd^4 \times d}{64 \times (15 \times 10^4 \pi d^3)^3 \times 0.045}$$

$$d^4 = \frac{82 \times 10^9}{64 \times (15\pi)^3 \times 10^{12} \times 45} = 273 \times 10^{-12}$$

Thus $d = 0.004\,06$ m $= 4.06$ mm; $R = 31.6$ mm; and $n = 11$ coils.

Axial Torque on Open-coiled Spring

The only other type of loading on a spring which is of interest is that where the spring is subjected to a torque about the central axis. The resolved components of the axial torque $\bar{T}$ about any cross-section are $\bar{T}\sin\alpha$ about the axis of the wire and $\bar{T}\cos\alpha$ changing the curvature of the coils. If one end of the spring moves round the longitudinal axis an amount ψ relative to the other end due to the torque $\bar{T}$ then the work done is $\frac{1}{2}\bar{T}\psi$ and the stored energies are $\frac{1}{2}\bar{T}\theta\sin\alpha$ and $\frac{1}{2}\bar{T}\phi\cos\alpha$, so that

$$\psi = \theta\sin\alpha + \phi\cos\alpha \qquad (9.46)$$

$$= \frac{Tl}{JG}\sin\alpha + \frac{Ml}{EI}\cos\alpha$$

$$= \bar{T}\pi nR\sec\alpha\left[\frac{\sin^2\alpha}{JG} + \frac{\cos^2\alpha}{EI}\right]$$

$$= \frac{128\bar{T}nR\sec\alpha}{d^4}\left[\frac{\sin^2\alpha}{2G} + \frac{\cos^2\alpha}{E}\right] \qquad (9.47)$$

Axial Torque on Close-coiled Spring

Putting $\sec\alpha = \cos\alpha = 1$ and $\sin\alpha = 0$ in eqn. (9.47) gives

$$\psi = \frac{128\bar{T}Rn}{Ed^4} \qquad (9.48)$$

Example 9.8   Compare the stiffness of a close-coiled spring with that of an open-coiled spring of helix angle 30°. The two springs are made of the same steel and have the same coil radius, number of coils and wire diameter, and are subjected to axial loading. $E = 2\cdot5G$.

From eqn. (9.45),

$$k_{cc} = \frac{W}{\delta} = \frac{Gd^4}{64R^3n}$$

From eqn. (9.44),

$$k_{oc} = \frac{W}{\delta} = \frac{d^4}{64R^3n\sec\alpha\left[\dfrac{\cos^2\alpha}{G} + \dfrac{2\sin^2\alpha}{E}\right]}$$

$$\frac{k_{cc}}{k_{oc}} = G\sec\alpha\left[\frac{\cos^2\alpha}{G} + \frac{2\sin^2\alpha}{E}\right]$$

$$= \sec\alpha\left[\cos^2\alpha + \frac{2\sin^2\alpha}{2\cdot5}\right]$$

$$= \frac{2}{\sqrt{3}}\left[\frac{3}{4} + \frac{1}{5}\right] = 1\cdot1$$

Chapter 10
Statically Indeterminate Beams and Frames

10.1. In Chapter 1 the forces in pin-jointed frames and the reactions at beam supports could be determined simply from suitable equilibrium equations. The present chapter develops a number of solutions to beam and frame situations where, owing to redundancies of reactions or members respectively, equilibrium is not a sufficient condition for solution. These cases are defined as statically indeterminate, and account has to be taken of the geometry of deformation in order to obtain values of forces, moments, etc. The first part of the chapter deals with built-in (fixed) and multiply supported beams and the remainder considers redundant plane frames.

10.2. STATICALLY INDETERMINATE BEAMS

Since there are only two equilibrium equations ($\Sigma M = 0$ and $\Sigma F = 0$) applicable to transverse loading of beams then only two unknown reactions can be determined. Thus, a beam which is supported in such a way as to produce three or more reaction forces or moments is statically indeterminate. Some typical examples are shown in Fig. 10.1.

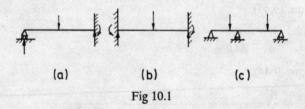

(a)   (b)   (c)

Fig 10.1

Three of the principal methods which are used for analysis are (a) double integration, (b) superposition, (c) moment-area. The applications of these methods, which have been discussed previously, will now be illustrated for a number of beam problems.

10.3. DOUBLE INTEGRATION METHOD

This method was first developed in Chapter 7. To reiterate briefly, the curvature is expressed in terms of bending moment at any point along the beam, this equation is then integrated twice, and the constants of integration are found from known boundary conditions of slope and deflection at the supports. Whereas, however, in the case of the statically determinate beam the reactions could be found *prior* to the above procedure, this is not possible for the indeterminate beam and the reactions must be carried through the analysis as unknown quantities. Since there are *always* enough boundary conditions to determine all the unknowns in the equations, the reactions can be evaluated together with the constants of integration.

(a) Beam fixed Horizontally at Each End carrying Uniformly Distributed Load

The problem is illustrated in Fig. 10.2, and it is evident that, owing to symmetry, $M_1 = M_2$ and $R_1 = R_2 = wL/2$. From eqn. (7.4),

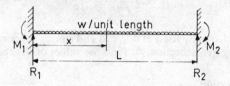

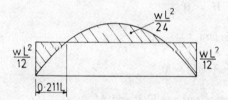

Fig 10.2

$$EI\frac{d^2v}{dx^2} = -M = -\frac{wLx}{2} + M_1 + \frac{wx^2}{2} \qquad (10.1)$$

$$EI\frac{dv}{dx} = -\frac{wLx^2}{4} + M_1x + \frac{wx^3}{6} + A \qquad (10.2)$$

$$EIv = -\frac{wLx^3}{12} + \frac{M_1x^2}{2} + \frac{wx^4}{24} + Ax + B \qquad (10.3)$$

The boundary conditions are that when $x = 0$, $dv/dx = 0$ and $v = 0$ from which, respectively, $A = 0$ and $B = 0$; and when $x = L$, $dv/dx = 0$ and $v = 0$.

Either of these conditions can now be used to solve for M_1, which is found to be $M_1 = M_2 = wL^2/12$.

At mid-span

$$M = \frac{wL^2}{24}$$

The deflection is a maximum at mid-span and is

$$v_{max} = \frac{wL^4}{384EI}$$

The bending moment diagram, Fig. 10.2, shows two points of contra-flexure (where BM changes sign), occurring at $x = 0\cdot211L$ and $0\cdot789L$.

Example 10.1. A cantilever, Fig. 10.3, carrying a concentrated load W at

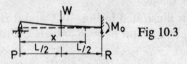

Fig 10.3

mid-span, is horizontally fixed at one end and simply supported at the other end giving a negative deflection, Δ. Find the values of the reactions.

Vertical Equilibrium

$$P + R - W = 0 \tag{10.4}$$

Moment Equilibrium

$$PL - \frac{WL}{2} + M_0 = 0 \tag{10.5}$$

The unknown reactions are P, R and M_0.

It is now convenient to apply the step-function or Macaulay method to deal with the discontinuous loading as in Chapter 7, so that

$$EI\frac{d^2v}{dx^2} = -Px + W\left[x - \frac{L}{2}\right] \tag{10.6}$$

$$EI\frac{dv}{dx} = -\frac{Px^2}{2} + \frac{W}{2}\left[x - \frac{L}{2}\right]^2 + A \tag{10.7}$$

$$EIv = -\frac{Px^3}{6} + \frac{W}{6}\left[x - \frac{L}{2}\right]^3 + Ax + B \tag{10.8}$$

Boundary Conditions

(i) When $x = 0$, $v = -\Delta$. (ii) When $x = L$, $v = 0$ and $dv/dx = 0$.
   For (i), from eqn. (10.8),

$$B = -EI\Delta$$

(The term in square brackets has to be omitted.)

For (ii) from eqn. (10.7),

$$0 = -\frac{PL^2}{2} + \frac{WL^2}{8} + A \tag{10.9}$$

and from eqn. (10.8),

$$0 = -\frac{PL^3}{6} + \frac{WL^3}{48} + AL - EI\Delta \tag{10.10}$$

Solving eqns. (10.9) and (10.10),

$$P = \frac{5}{16}W + \frac{3EI\Delta}{L^3} \tag{10.11}$$

and

$$A = +\frac{WL^2}{32} + \frac{3EI\Delta}{2L}$$

Substituting for P in eqns. (10.4) and (10.5),

$$R = \frac{11}{16}W - \frac{3EI\Delta}{L^3}$$

and

$$M_0 = \frac{3}{16}WL - \frac{3EI\Delta}{L^2}$$

Any required S.F., B.M., slope or deflection can now be determined.

(b) Beam fixed Horizontally at Each End carrying a Concentrated Load

There are four unknown reactions illustrated in Fig. 10.4.

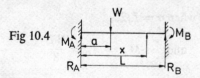

Fig 10.4

Vertical Equilibrium

$$R_A + R_B - W = 0 \tag{10.12}$$

Moment Equilibrium

$$R_A L - W(L - a) - M_A + M_B = 0 \tag{10.13}$$

Curvature Relation

$$EI\frac{d^2v}{dx^2} = -R_A x + W[x - a] + M_A \tag{10.14}$$

$$EI\frac{dv}{dx} = -\frac{R_A x^2}{2} + \frac{W}{2}[x - a]^2 + M_A x + A \tag{10.15}$$

$$EIv = -\frac{R_A x^3}{6} + \frac{W}{6}[x - a]^3 + M_A \frac{x^2}{2} + Ax + B \tag{10.16}$$

Boundary Conditions

(i) When $x = 0$, $v = 0$ and $dv/dx = 0$. (ii) When $x = L$, $v = 0$ and $dv/dx = 0$.

From (i)

$$A = 0 \quad \text{and} \quad B = 0$$

From (ii) and solving for R_A and M_A,

$$R_A = \frac{W}{L^3}(L - a)^2(L + 2a)$$

$$M_A = \frac{Wa}{L^2}(L - a)^2$$

Using eqn. (10.13),

$$M_B = \frac{Wa^2}{L^2}(L - a)$$

From eqn. (10.16) the deflection under the load is

$$v = \frac{Wa^3(L - a)^3}{3EIL^3} \tag{10.17}$$

For the particular case when $a = L/2$,

$$R_A = R_B = \frac{W}{2} \quad \text{and} \quad M_A = M_B = \frac{WL}{8}$$

$$v_{max} = \frac{WL^3}{192EI} \tag{10.18}$$

(c) Continuous Beam on Multiple Simple Supports

A fairly general example is illustrated in Fig. 10.5, in which Δ_B and Δ_C are known displacements.

Vertical Equilibrium

$$R_A + R_B + R_C + R_D - W - wL = 0 \tag{10.19}$$

Moment Equilibrium

$$3R_AL + 2R_BL + R_CL - \frac{5}{2}WL - \frac{wL^2}{2} = 0 \tag{10.20}$$

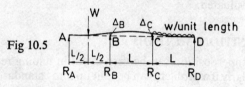

Fig 10.5

Curvature Relation

$$EI\frac{d^2v}{dx^2} = -R_Ax + W\left[x - \frac{L}{2}\right] - R_B[x - L] - R_C[x - 2L]$$

$$+ \frac{w}{2}[x - 2L]^2 \tag{10.21}$$

$$EI\frac{dv}{dx} = -\frac{R_Ax^2}{2} + \frac{W}{2}\left[x - \frac{L}{2}\right]^2 - \frac{R_B}{2}[x - L]^2 - \frac{R_C}{2}[x - 2L]^2$$

$$+ \frac{w}{6}[x - 2L]^3 + A \tag{10.22}$$

$$EIv = -\frac{R_Ax^3}{6} + \frac{W}{6}\left[x - \frac{L}{2}\right]^3 - \frac{R_B}{6}[x - L]^3 - \frac{R_C}{6}[x - 2L]^3$$

$$+ \frac{w}{24}[x - 2L]^4 + Ax + B \tag{10.23}$$

Boundary Conditions
(i) $x = 0, v = 0$, (ii) $x = L, v = -\Delta_B$, (iii) $x = 2L, v = -\Delta_C$, (iv) $x = 3L, v = 0$.

From (i)

$$B = 0$$

From (ii)

$$A = -\frac{WL^2}{48} + \frac{R_AL^2}{6} - \frac{EI\Delta_B}{L}$$

From (iii)

$$EI\Delta_C = R_AL^3 - \frac{25WL^3}{48} + \frac{R_BL^3}{6} + 2EI\Delta_B$$

From (iv)

$$0 = -4R_A L^3 + \frac{122WL^3}{48} - \frac{8R_B L^3}{6} - \frac{R_C L^3}{6} + \frac{wL^4}{24} - 3EI\Delta_B$$

Although perhaps somewhat laborious, these latter two equations together with eqns. (10.19) and (10.20) can be solved to give values for R_A, R_B, R_C and R_D. From this stage any required aspect of this beam problem can be evaluated.

10.4. SUPERPOSITION METHOD

The principle of superposition can be very useful in finding redundant reactions, particularly if a problem can be split up into "standard" cases. (See table at end of Chapter 7.)

(a) Beam carrying a Uniformly Distributed Load, Fixed at One End and Simply Supported at the Same Level at the Other

This problem is illustrated in Fig. 10.6 and can be represented by superposition of the two parts shown in Figs. 10.7 (a) and (b). If there was no

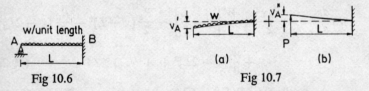

Fig 10.6   Fig 10.7

support at A there would be a downward deflection due to the distributed load. If there was no distributed load and the reaction, P, at the support was considered as a force which could cause an upward deflection of the beam, then the necessary boundary condition is that the sum of these deflections must be zero.

Due to loading, w,

$$v_A' = +\frac{wL^4}{8EI}$$

Due to reaction force, P,

$$v_A'' = -\frac{PL^3}{3EI}$$

But

$$v_A' + v_A'' = 0$$

Therefore

$$+\frac{wL^4}{8EI} - \frac{PL^3}{3EI} = 0 \tag{10.24}$$

Hence

$$P = \frac{3}{8}wL$$

From vertical and moment equilibrium,

$$R = \frac{5}{8}wL \quad \text{and} \quad M_0 = \frac{wL^2}{8}$$

(b) Alternative Superposition for (a) above

The case of Fig. 10.6 can be split up in the manner shown in Figs. 10.8 (a) and (b). The superposition must now satisfy the boundary condition of zero slope at B. Thus

$$\theta_B' + \theta_B'' = 0$$

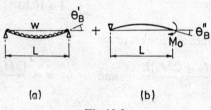

(a)   (b)

Fig 10.8

Now,

$$\theta_B' = -\frac{wL^3}{24EI}$$

and

$$\theta_B'' = +\frac{M_0L}{3EI}$$

Therefore

$$-\frac{wL^3}{24EI} + \frac{M_0L}{3EI} = 0 \qquad (10.25)$$

$$M_0 = \frac{wL^2}{8}$$

Knowing M_0 we can find P and R from the two equilibrium equations.

(c) Beam Fixed Horizontally at Each End

Before considering any particular form of applied loading, we will examine the effect of the fixing moments alone. Taking the general case of, say, $M_B > M_A$ as illustrated in Fig. 10.9, this may itself be put into the two parts in Figs. 10.10 (a) and (b), giving slopes at each end.

$$\theta_A = -\frac{M_A L}{2EI} \quad \text{and} \quad \theta_B = +\frac{M_A L}{2EI}$$

$$\theta_A = -\frac{(M_B - M_A)L}{6EI} \quad \text{and} \quad \theta_B = +\frac{(M_B - M_A)L}{3EI}$$

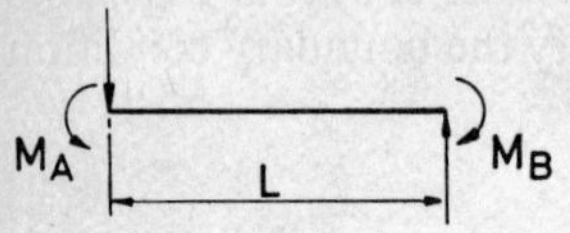

Fig 10.9

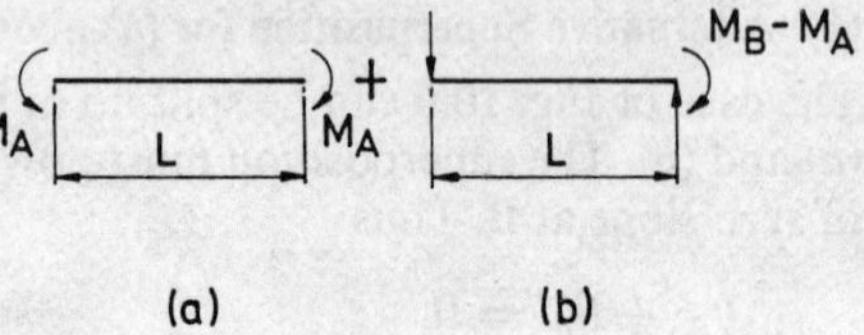

(a)   (b)

Fig 10.10

The resultant slopes are therefore

$$\theta_A = -\frac{(M_B + 2M_A)L}{6EI} \quad \text{and} \quad \theta_B = +\frac{(2M_B + M_A)L}{6EI}$$

$$(10.26)$$

(d) Fixed Horizontal Beam carrying Uniformly Distributed Load

This problem may be represented as in Fig. 10.11. The slopes at the ends for the simply supported part are $\pm wL^3/24EI$. Using the condition of zero slope at each end and the result from (c) above,

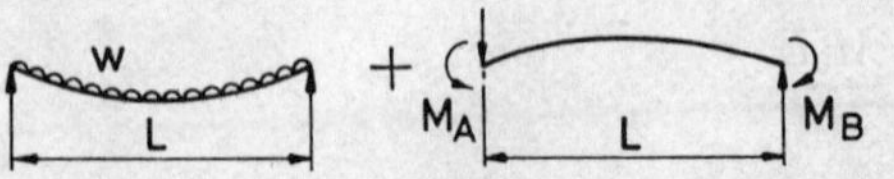

Fig 10.11

$$+\frac{wL^3}{24EI} - \frac{(M_B + 2M_A)L}{6EI} = 0 \tag{10.27}$$

and

$$-\frac{wL^3}{24EI} + \frac{(2M_B + M_A)L}{6EI} = 0 \tag{10.28}$$

from which

$$M_A = M_B = \frac{wL^2}{12}$$

(e) Fixed Beam carrying Uniformly Distributed Load with Ends not at Same Level

Let v be the difference in level between the ends of the beam, Fig. 10.12. A point of contraflexure occurs at mid-span, owing to symmetry, and

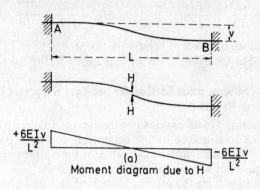

(a)
Moment diagram due to H

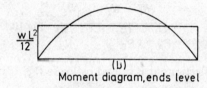

(b)
Moment diagram, ends level

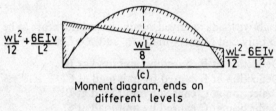

(c)
Moment diagram, ends on different levels

Fig 10.12

each half of the beam might be taken as a cantilever, the free end of which is caused to deflect $\frac{1}{2}v$ by a load H at the free end. Then

$$\frac{v}{2} = \frac{1}{3}\frac{H(\frac{1}{2}L)^3}{EI} \qquad \text{so that} \qquad H = \frac{12EIv}{L^3}$$

The bending moment diagram on each half, due to H, is triangular, the maximum ordinate being

$$M_H = \pm \frac{12EIv}{L^3}\frac{L}{2} = \pm \frac{6EIv}{L^2}$$

The fixing bending moments due to the distributed load w will be altered, owing to the differences in level of A and B, by the amount M_H, and hence this quantity must be added to, or subtracted from, M_A and M_B, depending on which side is the higher. When A is above B,

$$M_A = \frac{wL^2}{12} + \frac{6EIv}{L^2} \quad \text{and} \quad M_B = \frac{wL^2}{12} - \frac{6EIv}{L^2} \quad (10.29)$$

The combined diagram of bending moment, due to H, to the distributed load, and to the fixing of the ends, is shown in Fig. 10.12(c).

(f) Wilson's Superposition Method for Multiply Supported or Continuous Beams

In this method the reactions at the supports are found by equating the upward deflections caused by the supporting forces to the downward deflections which the loads would cause at the various supports, if in each case the beam were supported at the ends only.

In Fig. 10.13 assume the beam is supported at A and D only. Let

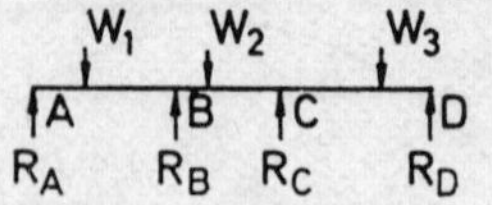

Fig 10.13

v_B = Sum of deflections at B caused by loads W_1, W_2 and W_3 separately

v_C = Sum of deflections at C caused by loads W_1, W_2 and W_3 separately

With the above loads removed, again assume the beam to be supported at A and D only and to carry loads R_B and R_C. Let

$\delta_1{}^B$ = Deflection at B caused by unit load at B
$\delta_1{}^C$ = Deflection at B caused by unit load at C
$\delta_2{}^B$ = Deflection at C caused by unit load at B
$\delta_2{}^C$ = Deflection at C caused by unit load at C

$$v_B = R_B\delta_1{}^B + R_C\delta_1{}^C$$
$$v_C = R_B\delta_2{}^B + R_C\delta_2{}^C$$
$$v_D = R_B\delta_3{}^B + R_C\delta_3{}^C$$

From these equations R_B and R_C can be determined. R_A and R_D can then be found by taking moments about A and D respectively.

10.5. MOMENT-AREA METHOD

For a beam of uniform section, the change in slope is given by $\theta = A/EI$. In the case of a horizontally fixed beam, $\theta = 0$ for the complete span, and since the product EI is not zero, it follows that A, the resultant area of the moment diagram for the beam, must be zero. This enables the deflection and slope at any point on the beam to be evaluated, and various examples of its use will follow.

(a) Fixed Beam with Irregular Loading

Referring to Fig. 10.14, let A_1 be the area of the free bending diagram

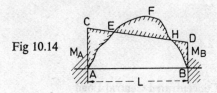

Fig 10.14

AEFHB, and let $\bar{x}_1$ be the distance of its centroid from A; also let A_2 be the area of ACDB and $\bar{x}_2$ the distance of its centroid from A. Then

$$A_1 = A_2 = \frac{(M_A + M_B)L}{2}$$

or

$$M_A + M_B = \frac{2A_1}{L} \qquad (10.30)$$

$$\frac{A_1\bar{x}_1 + (-A_2\bar{x}_2)}{EI} = 0$$

since each end of the beam remains horizontal. Therefore

$$A_1\bar{x}_1 = A_2\bar{x}_2$$

Since $A_1 = A_2$

$$\bar{x}_1 = \bar{x}_2$$

Taking moments about A,

$$A_2\bar{x}_2 = \frac{M_B L^2}{2} + \frac{M_A - M_B}{2} L \frac{L}{3}$$

$$= \frac{M_B L^2}{3} + \frac{M_A L^2}{6} \qquad (10.31)$$

or

$$A_1\bar{x}_1 = \frac{M_B L^2}{3} + \frac{M_A L^2}{6}$$

Hence

$$2M_B + M_A = \frac{6A_1\bar{x}_1}{L^2} \tag{10.32}$$

From eqns. (10.30) and (10.32),

$$M_A = \frac{4A_1}{L} - \frac{6A_1\bar{x}_1}{L^2} \tag{10.33}$$

and

$$M_B = \frac{6A_1\bar{x}_1}{L^2} - \frac{2A_1}{L} \tag{10.34}$$

(b) Fixed Beam with Concentrated Central Load

By combining the positive and negative moment diagram, Fig. 10.15,

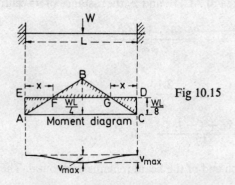

Fig 10.15

the resultant diagram AEFBGDC is obtained. Since area ABC + EACD = 0,

$$EAL = \frac{WL}{4}\frac{L}{2}$$

Fixing moment $= EA = \dfrac{WL}{8}$

and the points of contraflexure will be at G and F, distant x from each end, where $x = \frac{1}{4}L$.

The value of M at mid-span is equal to that of M at each end, i.e. $\frac{1}{8}WL$. Considering half the span, and taking a chosen line through one of the constraints, we have the deflection at mid-span given by

$$v = \frac{A\bar{x}}{EI}$$

Therefore

$$v_{max} = \frac{1}{EI}\left[\left(\frac{WL}{4} \times \frac{L}{4} \times \frac{2}{3}\frac{L}{2}\right) - \left(\frac{WL}{8} \times \frac{L}{2} \times \frac{L}{4}\right)\right]$$

$$= \frac{WL^3}{EI}\left[\frac{1}{48} - \frac{1}{64}\right]$$

$$= \frac{1}{192}\frac{WL^3}{EI} \tag{10.35}$$

(c) Continuous Beams

A beam resting on more than two supports is said to be *continuous*. Such a beam is represented by Fig. 10.16. Changes of curvature occur in each span, due to negative bending moments at the supports. In the case represented, the supports are assumed to be at different levels, being distant v_0, v_1 and v_2 from a horizontal line AB. Suppose the loading to

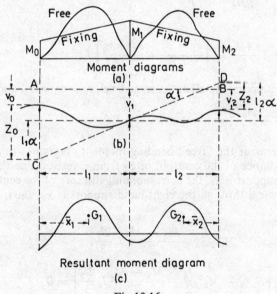

Resultant moment diagram
(c)

Fig 10.16

be such that the "free" and "fixing" bending moment diagrams are as shown at (*a*), the resultant diagram being shown at (*c*). The area of the resultant diagram on the span l_1 is A_1, and the distance of its centroid from a vertical line through the left-hand support is $\bar{x}_1$; also the area of the resultant diagram on the span l_2 is A_2, and $\bar{x}_2$ is the distance of its centroid from a vertical line through the right-hand support.

Draw CD, a common tangent at the point of contact of the central support, and let α be its inclination to the horizontal; agreeing to take intercepts, between tangents to the deflected beam, on a vertical line as positive when measured downwards, and vice versa as negative, we have

$$z_0 = \frac{A_1 \bar{x}_1}{EI} = v_1 - v_0 + l_1 \alpha$$

$$- z_2 = \frac{A_2 \bar{x}_2}{EI} = -l_2 \alpha + (v_1 - v_2)$$

and

$$\frac{A_1 \bar{x}_1}{EIl_1} - \frac{v_1 - v_0}{l_1} = \alpha = -\frac{A_2 \bar{x}_2}{EIl_2} + \frac{v_1 - v_2}{l_2}$$

Therefore

$$\frac{A_1 \bar{x}_1}{EIl_1} + \frac{A_2 \bar{x}_2}{EIl_2} = \frac{v_1 - v_0}{l_1} + \frac{v_1 - v_2}{l_2}$$

and

$$\frac{A_1 \bar{x}_1}{l_1} + \frac{A_2 \bar{x}_2}{l_2} = \left[\frac{v_1 - v_0}{l_1} + \frac{v_1 - v_2}{l_2} \right] EI \tag{10.36}$$

When the supports are all at the same level, $v_0 = v_1 = v_2$, and

$$\frac{A_1 \bar{x}_1}{l_1} + \frac{A_2 \bar{x}_2}{l_2} = 0 \tag{10.37}$$

If the areas of the "free" bending moment diagrams are S_1 and S_2 and the distance of the centroid of S_1 from a vertical line through the left-hand support is x_1, the corresponding distance of the centroid of S_2 from a vertical through the right-hand support is x_2; then, from eqn. (10.37),

$$\frac{1}{l_1} \left[S_1 x_1 - M_0 l_1 \frac{l_1}{2} - \frac{M_1 - M_0}{2} l_1 \frac{2l_1}{3} \right]$$

$$+ \frac{1}{l_2} \left[S_2 x_2 - M_2 l_2 \frac{l_2}{2} - \frac{M_1 - M_2}{2} l_2 \frac{2l_2}{3} \right] = 0$$

or

$$\frac{S_1 x_1}{l_1} + \frac{S_2 x_2}{l_2} - \frac{M_0 l_1}{6} - \frac{M_1}{3} (l_1 + l_2) - \frac{M_2 l_2}{6} = 0$$

and

$$M_0 l_1 + 2M_1 (l_1 + l_2) + M_2 l_2 = 6 \left[\frac{S_1 x_1}{l_1} + \frac{S_2 x_2}{l_2} \right] \tag{10.38}$$

This is Clapeyron's *theorem of three moments*, and with its aid, by taking the spans in pairs, sufficient equations are obtained to solve for all the "fixing" bending moments.

Next we turn to the calculation of the support reactions. Consider the spans l_1 and l_2 of the continuous beam, Fig. 10.17, and let the reactions

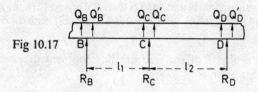

Fig 10.17

at B, C and D be R_B, R_C and R_D respectively. If the shearing forces on each side of the support C are Q_C and Q'_C, then $R_C = Q_C + Q'_C$.

Let W_1 be the total load on l_1 and $\bar{x}_B$ the horizontal distance of its centroid from B, and let W_2 be the total load on l_2, the horizontal distance of its centroid from D being $\bar{x}_D$.

Considering the span BC and taking moments about B,

$$M_B - M_C + Q_C l_1 = W_1 \bar{x}_B$$

or

$$Q_C = \frac{M_C}{l_1} \quad \frac{M_B}{l_1} + \frac{W_1 \bar{x}_B}{l_1} \tag{10.39}$$

Considering span CD and taking moments about D,

$$M_D - M_C + Q'_C l_2 = W_2 \bar{x}_D$$

or

$$Q'_C = \frac{M_C - M_D}{l_2} + \frac{W_2 \bar{x}_D}{l_2} \tag{10.40}$$

Therefore

$$R_C = Q_C + Q'_C$$

$$= \frac{M_C - M_B}{l_1} + \frac{M_C - M_D}{l_2} + \frac{W_1 \bar{x}_B}{l_1} + \frac{W_2 \bar{x}_D}{l_2} \tag{10.41}$$

If the loadings on l_1 and l_2 are w_1 and w_2 per unit run, then

$$\frac{W_1 \bar{x}_B}{l_1} = \frac{w_1 l_1}{2} \quad \text{and} \quad \frac{W_2 \bar{x}_D}{l_2} = \frac{w_2 l_2}{2}$$

and

$$R_C = \frac{M_C - M_B}{l_1} + \frac{M_C - M_D}{l_2} + \frac{w_1 l_1}{2} + \frac{w_2 l_2}{2} \tag{10.42}$$

The reaction and shearing forces at the other supports are found by a similar method.

Example 10.2.   Draw the bending moment and shearing force diagrams for a continuous beam which is supported at three points at the same level, but free at its extremities. The spans are 15·2 m and 10·6 m; the 15·2 m span supports two loads of values 8900 N and 4450 N distant 6 m and 12 m respectively from a free end, and the 10·6 m span is loaded uniformly with 1459 N per metre run.

The maximum ordinate of the free bending moment diagram (Fig. 10.18) for the 10·6 m span is

$$\frac{wL^2}{8} = \frac{1459 \times 10·6 \times 10·6}{8} = 20\,500 \text{ N-m}$$

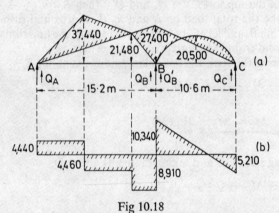

Fig 10.18

In the free bending moment diagram for the 15·2 m span,

Bending moment at 8900 N load $= 6240 \times 6$
$= 37\,440$ N-m

Bending moment at 4450 N load $= (6240 \times 12) - (8900 \times 6)$
$= 21\,480$ N-m

Referring to Fig. 10.18, $M_A = 0$ and $M_C = 0$, since the ends are free. Then eqn. (10.38),

$$M_A l_1 + 2M_B(l_1 + l_2) + M_C l_2 = 6\left[\frac{S_1 x_1}{l_1} + \frac{S_2 x_2}{l_2}\right]$$

becomes

$$2M_B(l_1 + l_2) = 6\left[\frac{S_1 x_1}{l_1} + \frac{S_2 x_2}{l_2}\right]$$

$$2M_B(15\cdot2 + 10\cdot6) = 6\left[\frac{1}{15\cdot2}\left\{\left(\frac{37\,400 \times 6\cdot08}{2} \times \frac{12\cdot16}{3}\right)\right.\right.$$

$$\left.+(21\,480 \times 6\cdot18 \times 9\cdot12)\right\}$$

$$+\frac{1}{15\cdot2}\left\{\left(\frac{15\,960 \times 6\cdot08}{2} \times 8\cdot11\right) + \left(\frac{21\,480 \times 3\cdot04}{2} \times 13\cdot27\right)\right\}$$

$$\left.+\frac{1}{10\cdot6}\left(\frac{20\,500 \times 10\cdot6 \times 2}{3} \times 5\cdot3\right)\right]$$

$$M_B = 27\,400 \text{ N-m}$$

Taking moments about B for span AB,

$$Q_A = 6240 - \frac{27\,400}{15\cdot2} = 6240 - 1800 = 4440 \text{ N}$$

The shear force at the 8900 N load is

$$Q_A - 8900 = 4440 - 8900 = -4460 \text{ N}$$

$$Q_B = Q_A - 8900 - 4450 = -8910 \text{ N}$$

From eqn. (10.40),

$$Q_B' = \frac{15\,500 \times \frac{1}{2}(10\cdot6)}{10\cdot6} + \frac{27\,400 - 0}{10\cdot6} = 7750 + 2590$$

$$= 10\,340 \text{ N}$$

$$Q_C = -\left[7790 + \frac{0 - 27\,400}{10\cdot6}\right] = -5210 \text{ N}$$

The shearing force diagram is shown in Fig. 10.18.

Example 10.3. A cantilever is built into a wall and propped level when un-loaded. Find where the prop must be placed in order that it shall support one-half of an evenly distributed load that may be placed on the cantilever. (*Lond. Univ.*)

Let the constraint be A, the prop B and the free end C. Then, imagining a similar span to the left of A, we can get the value of M_A.

$$M'_B = M_B = \frac{wl^2}{2}$$

where $l = $ BC and w is the evenly distributed load. Then, if AB $= l_1$, AC $= l_1 + l = L$.

In eqn. (10.38),

$$S_1 = S_2 = \frac{wl_1^2}{8} \times \frac{2}{3} l_1$$

and

$$x_1 = x_2 = \frac{l_1}{2}$$

$$M_B l_1 + 2M_A(l_1 + l_1) + M'_B l_1 = 6\left[\frac{1}{l_1}\left(\frac{wl_1^2}{8} \times \frac{2}{3} l_1 \times \frac{l_1}{2}\right)\right] 2$$

$$wl^2 l_1 + 4M_A l_1 = \frac{wl_1^3}{2}$$

$$wl^2 + 4M_A = \frac{wl_1^2}{2}$$

Therefore

$$4M_A = \frac{wl_1^2}{2} - wl^2$$

$$M_A = \left[\frac{wl_1^2}{2} - wl^2\right]\frac{1}{4}$$

From eqn. (10.42),

$$R_B = \frac{wl_1}{2} + \frac{M_B - M_A}{l_1} + \frac{wl}{2} + \frac{M_B - M_C}{l} = \frac{w(l_1 + l)}{2}$$

Therefore

$$\frac{wl_1}{2} + \frac{wl^2}{2l_1} - \frac{wl_1}{8} + \frac{wl^2}{4l_1} + \frac{wl}{2} + \frac{wl}{2} - 0 = \frac{w(l_1 + l)}{2}$$

$$l_1 + \frac{l^2}{l_1} - \frac{l_1}{4} + \frac{l^2}{2l_1} + l + l = l_1 + l$$

$$l_1^2 - 4ll_1 - 6l^2 = 0$$

Therefore

$$l_1 = 5\cdot16l = 0\cdot840L$$

10.6. ADVANTAGES AND DISADVANTAGES OF FIXED AND CONTINUOUS BEAMS

At first sight, it would appear to be a decided advantage to use a fixed beam in preference to a simply-supported beam, since the fixed beam is much stiffer and stronger, also the maximum bending moment often occurs at the fixing; and thus the strengthening of the beam to take this bending moment does not increase its weight to any great extent, as would occur in a simply-supported beam. Referring to Section 10.4(e), it will be observed that, if the ends are not at the same level, then the bending moment is greatly increased; hence any settling of the supports of a fixed beam would cause a great alteration in the stresses. In a fixed beam also the slope at each end must not change, and any variation in this will cause alteration in the estimated stresses.

These two factors, which make the actual stresses uncertain, are the main reasons for not generally using fixed beams. If a fixed beam is hinged at the points of contraflexure, then each end portion forms a cantilever, and the centre portion a simply supported beam. This method obviates the above disadvantages and is used in cantilever bridges.

The maximum bending moment on a continuous beam is less than that which occurs if the spans are bridged by simply-supported beams; also the maximum bending moment occurs at the supports. It would appear that a decided advantage is gained by the use of a continuous beam, since lighter material will be required to resist the bending moment, and also, since the heavy portion of the beam, required to resist the bending moment, is near the supports, the weight of the girder will not materially affect the stresses.

It must be remembered, however, that the bending moment is calculated on the assumption that the supports are all at the same level. Any deviation from this condition seriously affects the bending moment, and consequently the estimated stresses. It is owing to the difficulty of arranging all the supports at the same level, and keeping them there, that continuous beams are not more generally used.

When a continuous beam consists of a number of spans, two points of contraflexure usually occur between each pair of supports. The bending moment being zero at these points, it is possible to hinge the girder here. The end portions then act as cantilevers and the central portion as a simply-supported beam; hence changes in level of the supports do not affect the bending moment. This is the principle underlying the multi-span bridge, an excellent example being the Forth Bridge.

10.7. ANALYSIS OF STATICALLY INDETERMINATE PLANE FRAMES

This topic is one of the main areas of study in the theory of structures, and it is almost presumptuous to attempt an introduction in the space available here. However, it is felt that a rensonable insight into the basic concepts can be got across now, and the reader who needs a greater depth of study may then proceed to the texts specializing in structural analysis.

There are three basic methods of analysis, which use principles that have been established in earlier chapters, namely (*a*) virtual work, (*b*) compatibility (or flexibility) method due to Maxwell, (*c*) equilibrium (or stiffness) method due to Navier.

10.8. VIRTUAL WORK METHOD

This is a relatively simple extension of the virtual work applications in Chapter 8. Consider the plane pin-jointed frame in Fig. 10.19, in which

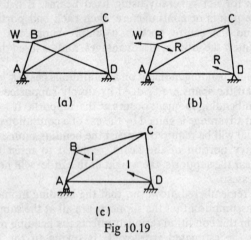

Fig 10.19

any one member may be regarded as redundant. Let the force in one member, say BD, be tensile of magnitude R. If this member is now removed and equal and opposite forces of magnitude R are applied at B and D, we still have the same status as in the original frame. However, the forces in all the other members can now be calculated in terms of the applied external load, W, and the force R. The deformations, Δ, of the members can then be found using the forces above in terms of W and R. The next step is to replace the forces R by unit loads and to remove all external loading. We now determine the forces, P, set up in the members by the unit loads and write the virtual work equation to express the real displacement δ_D of joint D in terms of the hypothetical forces, P, and the real deformations, Δ, of the members as follows:

$$1 \times \delta_D = \sum_m P\Delta \tag{10.43}$$

Now in the actual frame, since BD is subjected to tension R, then BD will *extend* an amount, say Δ_{BD}, which is in the opposite sense to δ_D. Therefore the compatibility condition is

$$\Delta_{BD} = -\delta_D$$

Thus eqn. (10.43) becomes

$$\Delta_{BD} + \sum_m P\Delta = 0 \qquad (10.44)$$

and since all the quantities are in terms of W and R, the value of the latter redundant force can be found.

If there is an initial lack of fit, that is a member is not of an exact length when unloaded to fit in with the remaining members, then internal forces will be set up in the frame. Denoting a positive lack of fit by λ, the member being longer than it should be, then eqn. (10.44) becomes

$$(\lambda + \Delta_{BD}) + \sum_m P\Delta = 0 \qquad (10.45)$$

Example 10.4. Determine the forces in the members of the plane frame shown in Fig. 10.20. The area and modulus are the same for all members.

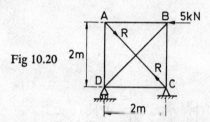

Fig 10.20

This type of problem is most conveniently solved in tabular form (overleaf). Member AC will be considered as the redundancy.

Firstly, the forces are found by, say, resolution at joints due to the 5 kN load and the redundancy, R. These values are given in column (2) of the table. The next column (3) gives the deformations of the members. Now the 5 kN load and R are removed, and unit loads are placed at A and C from which forces in the members are found as shown in column (4). The virtual work is expressed in column (5) as the product of columns (3) and (4). Eqn. (10.44) may now be expressed as the summation of the terms in column (5) and R evaluated.

$$\left(R\frac{2\sqrt{2}}{EA} \times 1 \right) + [-10(2 + \sqrt{2}) + 4R + 2\sqrt{2}R]\frac{1}{EA} = 0$$

from which $R = 3.53$ kN.

Substitution back into column (2) gives the values of all other member forces.

10.9. COMPATIBILITY METHOD

This method uses the principle that displacements and rotations at joints must be compatible with the deformation of the members. Forces

(1) Member	(2) Force due to 5 kN and R	(3) Deformations due to 5 kN and R	(4) Forces due to unit loads at A and C	(5) (3) × (4)
AB	$-\dfrac{R}{\sqrt{2}}$	$-\dfrac{R}{\sqrt{2}}\dfrac{2}{EA}$	$-\dfrac{1}{\sqrt{2}}$	$+\dfrac{R}{EA}$
BC	$5-\dfrac{R}{\sqrt{2}}$	$\left(5-\dfrac{R}{\sqrt{2}}\right)\dfrac{2}{EA}$	$-\dfrac{1}{\sqrt{2}}$	$-\left(5-\dfrac{R}{\sqrt{2}}\right)\dfrac{\sqrt{2}}{EA}$
CD	$5-\dfrac{R}{\sqrt{2}}$	$\left(5-\dfrac{R}{\sqrt{2}}\right)\dfrac{2}{EA}$	$-\dfrac{1}{\sqrt{2}}$	$-\left(5-\dfrac{R}{\sqrt{2}}\right)\dfrac{\sqrt{2}}{EA}$
AD	$-\dfrac{R}{\sqrt{2}}$	$-\dfrac{R}{\sqrt{2}}\dfrac{2}{EA}$	$-\dfrac{1}{\sqrt{2}}$	$+\dfrac{R}{EA}$
AC	R	$R\dfrac{2\sqrt{2}}{EA}$	$+1$	$R\dfrac{2\sqrt{2}}{EA}\times 1$
BD	$-5\sqrt{2}+R$	$-(5\sqrt{2}-R)\dfrac{2\sqrt{2}}{EA}$	$+1$	$-(5\sqrt{2}-R)\dfrac{2\sqrt{2}}{EA}$
				$\Sigma(5)$

and moments must be such as to satisfy the foregoing condition and thus a solution may be obtained.

Example 10.5. Determine the forces in the members of the plane frame illustrated in Fig. 10.21(*a*).

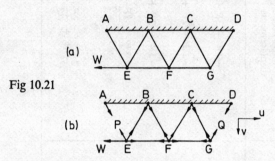

Fig 10.21

Since there are two redundancies in this framework the first step is to choose two members to remove, say AE and DG, and replace by the unknown forces, P and Q respectively, acting at the joints. The frame now appears as in Fig. 10.21(*b*). The next step is to determine the horizontal and vertical displacements of the joints E, F, G for the three situations, when W is acting alone, then P only and Q alone. Each of these cases is, of course, statically determinate, so forces and thus displacements are readily found using, say, a Williot diagram. The results are tabulated overleaf. The right-hand column gives the components of displacement due to the superposition of the W, P and Q cases.

The next stage is to ensure compatibility at joints E and G. Member AE will extend by an amount $PL/$AE, and DG, similarly, by $QL/$AE. These extensions can be expressed in terms of the joint displacements thus:

$$\frac{PL}{AE} = u_E \cos 60° + v_E \sin 60° = \frac{u_E}{2} + v_E \frac{\sqrt{3}}{2} \qquad (10.46)$$

$$\frac{QL}{AE} = -u_G \cos 60° + v_G \sin 60° = -\frac{u_G}{2} + v_G \frac{\sqrt{3}}{2} \qquad (10.47)$$

Substituting the appropriate values from the above table into eqns. (10.46) and (10.47) gives

$$\frac{2PL}{AE} = (-3W - 3P + 2Q)\frac{L}{AE} + (-3W - 5P + 2Q)\frac{L}{AE}$$

$$4Q - 10P = 6W \qquad (10.48)$$

Displacement	Due to W	Due to P	Due to Q	Superposed displacements $\times \dfrac{L}{AE}$
u_E	$-3\dfrac{WL}{AE}$	$-3\dfrac{PL}{AE}$	$+2\dfrac{QL}{AE}$	$-3W - 3P + 2Q$
u_F	$-2\dfrac{WL}{AE}$	$-2\dfrac{PL}{AE}$	$+2\dfrac{QL}{AE}$	$-2W - 2P + 2Q$
u_G	$-2\dfrac{WL}{AE}$	$-2\dfrac{PL}{AE}$	$+3\dfrac{QL}{AE}$	$-2W - 2P + 3Q$
v_E	$-\dfrac{3}{\sqrt{3}}\dfrac{WL}{AE}$	$-\dfrac{5}{\sqrt{3}}\dfrac{PL}{AE}$	$+\dfrac{2}{\sqrt{3}}\dfrac{QL}{AE}$	$-\dfrac{3W}{\sqrt{3}} - \dfrac{5P}{\sqrt{3}} + \dfrac{2Q}{\sqrt{3}}$
v_F	0	0	0	0
v_G	$+\dfrac{2}{\sqrt{3}}\dfrac{WL}{AE}$	$+\dfrac{2}{\sqrt{3}}\dfrac{PL}{AE}$	$-\dfrac{5}{\sqrt{3}}\dfrac{QL}{AE}$	$\dfrac{2}{\sqrt{3}}W + \dfrac{2P}{\sqrt{3}} - \dfrac{5Q}{\sqrt{3}}$

and

$$\frac{2QL}{AE} = -(-2W - 2P + 3Q)\frac{L}{AE} + (2W + 2P - 5Q)\frac{L}{AE}$$

$$10Q - 4P = 4W \qquad\qquad (10.49)$$

Solving eqns. (10.48) and (10.49) gives

$$P = -\frac{11}{21}W \quad \text{and} \quad Q = \frac{4}{21}W$$

The forces in members AE and DG having been found, the remainder can be determined by simple statics. The complete solution is

AE	BE	BF	CF	CG	DG	EF	FG
$-\dfrac{11}{21}W$	$\dfrac{11}{21}W$	$-\dfrac{2}{7}W$	$\dfrac{2}{7}W$	$-\dfrac{4}{21}W$	$\dfrac{4}{21}W$	$\dfrac{10}{21}W$	$\dfrac{4}{21}W$

A somewhat shorter and simpler solution to the above, but by exactly the same method, would be obtained by taking EF and FG as the redundant members.

10.10. EQUILIBRIUM METHOD

In this method the forces in the members are expressed in terms of the components of joint displacement using the modulus, area and length of each member. The equilibrium equations for each joint are first expressed in terms of the forces in the members meeting at this joint. They are then transposed into displacement equations using the above force-displacement relations. After solving the equations for displacements, these are back-substituted to give values for the forces.

Example 10.6.   Calculate the forces in the members of the plane frame illustrated in Fig. 10.22.

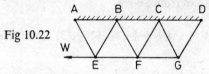

Fig 10.22

Joints A, B, C and D are position fixed and joints E, F and G have displacement components u_E, u_F, u_G horizontally and v_E, v_F, v_G vertically.

The extensions of a member such as AE can be expressed in terms of the displacement components of E as

$$u_E \sin 60° + v_E \cos 60° = u_E\frac{\sqrt{3}}{2} + \frac{v_E}{2}$$

from which the force in the member can be expressed as

$$F_{AE} = \frac{AE}{L}\left(\frac{u_E}{2} + \frac{v_E}{2}\sqrt{3}\right)$$

The forces in all the other members, by a similar approach, are

$$F_{BE} = \frac{AE}{L}\left(-\frac{u_E}{2} + \frac{v_E}{2}\sqrt{3}\right)$$

$$F_{BF} = \frac{AE}{L}\left(\frac{u_F}{2} + \frac{v_F}{2}\sqrt{3}\right)$$

$$F_{CF} = \frac{AE}{L}\left(-\frac{u_F}{2} + \frac{v_F}{2}\sqrt{3}\right)$$

$$F_{CG} = \frac{AE}{L}\left(\frac{u_G}{2} + \frac{v_G}{2}\sqrt{3}\right)$$

$$F_{DG} = \frac{AE}{L}\left(-\frac{u_G}{2} + \frac{v_G}{2}\sqrt{3}\right)$$

$$F_{EF} = \frac{AE}{L}(u_F - u_E)$$

$$F_{FG} = \frac{AE}{L}(u_G - u_F)$$

The above equations satisfy the geometry of deformation or compatibility condition. Next we must consider the equilibrium condition at each joint.

Joint E

Horizontal $-W - F_{AE}\cos 60° + F_{EB}\cos 60° + F_{EF} = 0$

Substituting for the forces from above,

$$-\frac{2WL}{AE} - (u_E + v_E\sqrt{3})\tfrac{1}{2} + (-u_E + v_E\sqrt{3})\tfrac{1}{2} + 2(u_F - u_E) = 0$$

$$-3u_E + 2u_F - \frac{2WL}{AE} = 0 \qquad\qquad (10.50)$$

Vertical $-F_{AE}\sin 60° - F_{BE}\sin 60° = 0$

$$u_F + v_E\sqrt{3} - u_E + v_E\sqrt{3} = 0$$

Hence

$$v_E = 0 \qquad\qquad (10.51)$$

Joint F

Horizontal $-F_{EF} + F_{FG} - F_{BF} \cos 60° + F_{CF} \cos 60° = 0$

$$2u_E - 5u_F + 2u_G = 0 \tag{10.52}$$

Vertical $-F_{BF} \sin 60° - F_{CF} \sin 60° = 0$

$$u_F + v_F - u_F + v_F = 0$$

Hence

$$v_F = 0 \tag{10.53}$$

Joint G

Horizontal $-F_{FG} - F_{CG} \cos 60° + F_{DG} \cos 60° = 0$

$$-3u_G + 2u_F = 0 \tag{10.54}$$

Vertical $-F_{CG} \sin 60° - F_{DG} \sin 60° = 0$

from which

$$v_G = 0 \tag{10.55}$$

Eqns. (10.50), (10.52) and (10.54) can now be solved to give

$$u_E = -\frac{22}{21}\frac{WL}{AE}$$

$$u_F = -\frac{4}{7}\frac{WL}{AE}$$

$$u_G = -\frac{8}{21}\frac{WL}{AE}$$

These values are substituted back to find the above forces. For example,

$$F_{FG} = \frac{AE}{L}\left(-\frac{8}{21}\frac{WL}{AE} + \frac{12}{21}\frac{WL}{AE}\right) = \frac{4}{21}W$$

which is the same result as obtained by the compatibility method.

Chapter 11
Buckling Instability

11.1. INTRODUCTION

In previous chapters a fundamental condition in all the problems was the equilibrium of internal and external forces. Now, if the system of forces is disturbed owing to a small displacement of a body, two principal situations are possible: either the body will return to its original configuration owing to restoring forces during displacement, or the body will accelerate farther away from its original state owing to displacing forces. The former situation is termed *stable equilibrium* and the latter is termed *unstable equilibrium*.

Consider the simple case in Fig. 11.1(*a*) of a vertical bar pinned at the

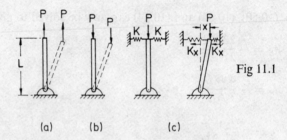

Fig 11.1

(a)   (b)   (c)

lower end and carrying an axial tensile force at the upper end. If there is a slight displacement from the vertical, the force will tend to restore the bar to its original position. In Fig. 11.1(*b*), however, the same bar subjected to a compressive load when displaced slightly from the vertical will accelerate towards a horizontal position, illustrating unstable equilibrium. A slightly more sophisticated case is shown in Fig. 11.1(*c*), where the bar is assisted in remaining vertical by the action of the horizontal springs. When the bar is displaced by an amount x in either direction, there are a displacing moment Px and a restoring moment $2KxL$, where K is the stiffness of a spring; hence we have

$$Px < 2KxL \rightarrow \text{stable}$$
$$Px > 2KxL \rightarrow \text{unstable}$$

The critical condition is when

$$Px = 2KxL \qquad \text{or} \qquad P = 2KL$$

and P is termed the *critical load*, being the borderline between stable and unstable equilibrium.

The analysis of stress and strain in previous chapters was aimed at the efficient and safe working of a component, i.e. the avoidance of failure. The instability of structural members subjected to compressive loading may also be regarded as a mode of failure, even though the stresses are elastic, owing to excessive deformation and distortion of the structure.

Columns or struts subjected to compression can be classified in three ways:

1. Those of very short length.
2. Those not longer than about thirty diameters.
3. Those longer than about thirty diameters.

Each class can be subdivided into (a) those having axial loading and (b) those in which the loading is eccentric.

Columns under class (1) have been dealt with in earlier chapters. This chapter will be devoted to the investigation of columns in classes (2) and (3), axial and eccentric loading being considered in each case. It will be assumed that the column is of uniform cross-section.

11.2. EULER'S THEORY OF BUCKLING FOR LONG COLUMNS

(a) Pinned Ends

The column shown in Fig. 11.2(a) carries an axial load P. Its ends are considered as pin-jointed, in that they are perfectly free to change their slope, but all other movement is prevented.

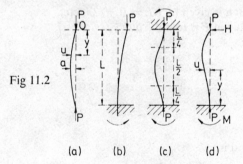

Fig 11.2

(a)   (b)   (c)   (d)

Suppose P to be the critical or buckling load of the column, i.e. the load is such that if the column receives slight lateral displacement, as shown, then the elastic forces tending to straighten the column will just not be able to balance the effect of P. Let O be the top end of the column; then the bending moment at a point on the column distant y from O is given by

$$M = Pu$$

using the same sign convention as for beams; but

$$EI\frac{d^2u}{dy^2} = -M$$

where I is the *least* second moment of area. Therefore

$$\frac{d^2u}{dy^2} = -\frac{P}{EI}u = -K^2u$$

where $K^2 = P/EI$, or

$$\frac{d^2u}{dy^2} + K^2u = 0 \qquad (11.1)$$

The solution of this equation is given by

$$u = A\cos Ky + B\sin Ky \qquad (11.2)$$

At $y = 0$, $u = 0$; therefore $A = 0$. Hence

$$u = B\sin Ky \qquad (11.3)$$

At $y = L$, $u = 0$; therefore

$$0 = B\sin KL$$

and B cannot be equal to zero, otherwise no deflection could occur; hence

$$\sin KL = 0$$

the solution of which is given by $KL = 0, \pi, 2\pi$, etc.
  The smallest value of P is found from

$$KL = \pi \quad \text{or} \quad K^2 = \frac{\pi^2}{L^2}$$

Therefore

$$\frac{P}{EI} = \frac{\pi^2}{L^2}$$

and the critical load is

$$P_c = \frac{\pi^2 EI}{L^2} \qquad (11.4)$$

The maximum deflection occurs when $y = \frac{1}{2}L$, i.e. $\sin \frac{1}{2}KL = 1$; therefore $B = u_{max}$.
  The buckling loads for cases (b) and (c) can be found in a similar manner, but a solution is obtained more rapidly by considering the "equivalent length" of a column having pinned ends.

(b) One End Fixed, the Other End Free

This is shown in Fig. 11.2(b) and would be equivalent to a column having pinned ends of length $2L$. Substituting this in eqn. (11.4), the buckling load is given by

$$P_c = \frac{\pi^2 EI}{(2L)^2}$$

$$= \frac{\pi^2 EI}{4L^2} \tag{11.5}$$

The critical load for this column is thus only one-fourth of that for a similar column having pinned ends.

(c) Both Ends Fixed

This is represented by Fig. 11.2(c), and an examination of the diagram shows that the equivalent length of a column having pinned ends is given by $\frac{1}{2}L$. Substituting this value in eqn. (11.4), the buckling load for this case is given by

$$P_c = \frac{\pi^2 EI}{(\frac{1}{2}L)^2}$$

$$= \frac{4\pi^2 EI}{L^2} \tag{11.6}$$

This column can carry a load four times greater than that represented by case (a), and sixteen times greater than that represented by case (b). In practice, cases of absolute fixing are rare, the great majority of columns being either pinned or partially restrained.

(d) One End Fixed, the Other End Free to Rotate

This is shown in Fig. 11.2(d). The bending moment, M, is introduced by the "fixing" which involves the use of the horizontal force H for equilibrium. At a distance y from the fixed end, let u be the amount of deviation from the perpendicular. The bending moment at this point is given by

$$M = Pu - H(L - y)$$

Therefore

$$\frac{d^2u}{dy^2} = \frac{1}{EI}[H(L - y) - Pu]$$

or

$$\frac{d^2u}{dy^2} + \frac{P}{EI}u = \frac{H}{EI}(L - y) \tag{11.7}$$

Let $\sqrt{(P/EI)} = K$. The solution of eqn. (11.7) is given by

$$u = A \cos Ky + B \sin Ky + \frac{H}{P}(L - y) \tag{11.8}$$

At $y = 0$, $u = 0$; therefore

$$A = -\frac{H}{P}L \tag{11.9}$$

Differentiating eqn. (11.8),

$$\frac{du}{dy} = -AK \sin Ky + BK \cos Ky - \frac{H}{P}$$

and at $y = 0$, $du/dy = 0$; therefore

$$B = \frac{H}{P}\frac{1}{K} \tag{11.10}$$

Eqn. (11.8) may now be written as

$$u = -\frac{H}{P}L \cos Ky + \frac{H}{P}\frac{1}{K}\sin Ky + \frac{H}{P}(L - y) \tag{11.11}$$

When $y = L$, $u = 0$; hence

$$0 = -\frac{H}{P}L \cos KL + \frac{H}{P}\frac{1}{K}\sin KL$$

or

$$\tan KL = KL \tag{11.12}$$

The smallest value of P, from eqn. (11.12) (other than $P = 0$), is given by

$$\sqrt{\frac{P}{EI}}L = 4{\cdot}49$$

or

$$\frac{P}{EI}L^2 = 4{\cdot}49^2 = 20 \approx 2\pi^2$$

Therefore

$$P_c = 2\frac{\pi^2 EI}{L^2} \tag{11.13}$$

From eqns. (11.4), (11.5), (11.6) and (11.13), it will be observed that, according to the Euler theory, the buckling load for a column carrying an axial load is given by

$$P_c = \beta\frac{\pi^2 EI}{L^2} \tag{11.14}$$

	Case (a)	Case (b)	Case (c)	Case (d)
Value of β	1	$\frac{1}{4}$	4	2

11.3. ECCENTRIC LOADING OF LONG COLUMNS

It is seldom in practice that a column or strut can be loaded exactly along its central axis as the Euler analysis implies. A general solution will now be developed for the case of a long column subjected to a load parallel to, but eccentric from, the central axis.

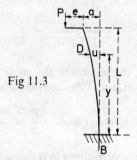

Fig 11.3

Consider the example illustrated in Fig. 11.3.

Bending moment at D $= -P(e + a - u)$

Therefore

$$EI\frac{d^2u}{dy^2} = P(e + a - u)$$

$$\frac{d^2u}{dy^2} = -\frac{P}{EI}(u - a - e) \qquad (11.15)$$

Let $\sqrt{(P/EI)} = K$. Then

$$u - a - e = A \cos Ky + B \sin Ky \qquad (11.16)$$

At $y = 0$, $du/dy = 0$; therefore $B = 0$ and $u - a - e = A \cos Ky$. At $y = 0$, $u = 0$; therefore $A = -a - e$, and

$$u - a - e = (-a - e) \cos Ky$$

Therefore

$$u = (a + e)[1 - \cos Ky] \qquad (11.17)$$

At $y = L$, $u = a$, and

$$a = (a + e)[1 - \cos KL]$$

Hence

$$a = e[\sec KL - 1] \qquad (11.18)$$

Bending moment at B $= M_B = P(e + a) = Pe \sec KL$

from eqn. (11.18); and

$$\text{Bending stress at B} = \pm \frac{M_B}{Z}$$

$$= \pm \frac{Pe \sec KL}{Z}$$

$$= \pm \frac{Pes \sec KL}{Ak^2} \tag{11.19}$$

where s is half the depth of the section in the plane of bending, k the radius of gyration of the section, and A the cross-sectional area. Therefore

$$\text{Maximum compressive stress at B} = -\frac{P}{A} - \frac{Pes \sec KL}{Ak^2} \tag{11.20}$$

If σ_c is the allowable compressive stress in the material, then

$$\sigma_c = -\frac{P}{A}\left[1 + \frac{es \sec KL}{k^2}\right] \tag{11.21}$$

Therefore

$$P = -\frac{\sigma_c A}{1 + \dfrac{es \sec KL}{k^2}} \tag{11.22}$$

In the case of a cast-iron column, if σ_t is the safe tensile stress for cast-iron, then

$$\sigma_t = \frac{P}{A}\left[\frac{es \sec KL}{k^2} - 1\right]$$

Therefore

$$P = \frac{\sigma_t A}{\dfrac{es \sec KL}{k^2} - 1} \tag{11.23}$$

For a column having "free ends", we may write $\frac{1}{2}L$ for L, in any of the above equations.

Thus

$$P = -\frac{\sigma_c A}{1 + \dfrac{es \sec K\dfrac{L}{2}}{k^2}} \tag{11.24}$$

and for a cast-iron column,

$$P = \frac{\sigma_t A}{\dfrac{es \sec K \dfrac{L}{2}}{k^2} - 1} \tag{11.25}$$

Since P appears on both sides of eqns. (11.22)–(11.25), a solution can only be obtained by trial and error. A good approximation may be found as follows: let

$$K\frac{L}{2} = \theta = \sqrt{\left(\frac{P}{EI}\right)}\frac{L}{2} = \frac{\pi}{2}\sqrt{\left(\frac{L^2 P}{\pi^2 EI}\right)} = \frac{\pi}{2}\sqrt{\left(\frac{P}{P_e}\right)}$$

When θ lies between 0 and $\frac{1}{2}\pi$, a close approximation for the value of $\sec \theta$ is given by

$$\sec \theta = \frac{1 + 0\cdot 26 \left(\dfrac{2\theta}{\pi}\right)^2}{1 - \left(\dfrac{2\theta}{\pi}\right)^2} = \frac{P_e + 0\cdot 26 P}{P_e - P}$$

where P_e is the Euler critical load for the strut when axially loaded Substituting this value of $\sec \theta$ in eqn. (11.24),

$$P = -\frac{\sigma_c A}{1 + \dfrac{es}{k^2}\dfrac{P_e + 0\cdot 26 P}{P_e - P}}$$

from which we obtain

$$P^2 \left(1 - 0\cdot 26\frac{es}{k^2}\right) - P\left[P_e\left(1 + \frac{es}{k^2}\right) + \sigma_c A\right] + \sigma_c A P_e = 0 \tag{11.26}$$

This quadratic equation can be readily solved in a given case. Equations of a similar nature may be obtained from eqns. (11.22), (11.23) and (11.25).

Example 11.1. A vertical steel tube having 75 mm external and 62 mm internal diameters is 3 m long, fixed at the lower end and completely un-restrained at the upper end. The tube is subjected to a vertical compressive load parallel to but eccentric by 6 mm from the central axis. Determine the limiting value of the load so that there is no tensile stress at the base of the tube.

If the column had been loaded along its axis what would be the value of the Euler buckling load?

$E = 208\,\mathrm{GN/m^2}$.

From eqn. (11.19) the surface bending stress at the base is

$$\sigma_b = \pm \frac{Pes \sec KL}{I}$$

Direct stress, $\sigma_d = -\dfrac{P}{A}$

$$A = \frac{\pi}{4}(0 \cdot 075^2 - 0 \cdot 062^2) = 0 \cdot 001\,384\,\text{m}^2$$

$$I = \frac{\pi}{64}(0 \cdot 075^4 - 0 \cdot 062^4) = 0 \cdot 815 \times 10^{-6}\,\text{m}^4$$

For zero tensile stress,

$$-\frac{P}{A} + \frac{Pes \sec KL}{I} = 0$$

$$\sec KL = \frac{I}{Aes} = \frac{0 \cdot 815 \times 10^{-6} \times 2}{0 \cdot 001\,38 \times 0 \cdot 006 \times 0 \cdot 075} = 2 \cdot 63$$

$$KL = 1 \cdot 1659$$

$$K = 0 \cdot 388$$

$$P = 0 \cdot 388^2 \times 208 \times 10^9 \times 0 \cdot 815 \times 10^{-6}$$

$$= 25 \cdot 6\,\text{kN}$$

For the axially loaded column,

$$P_c = \frac{\pi^2 EI}{4L^2} = \frac{\pi^2 \times 208 \times 10^9 \times 0 \cdot 815 \times 10^{-6}}{4 \times 9}$$

$$= 46 \cdot 5\,\text{kN}$$

Example 11.2. The column shown in Fig. 11.4 has pinned ends. The upper part, of length $(1 - n)L$, which is slender and of constant stiffness EI, is fixed to the lower part, of length nL, which is rigid. Show that at instability, $\tan k(1 - n)L = -knL$. What is the critical load if the upper and lower parts are of equal length?

In this particular problem the differential equation (11.1) and its solution (11.2) are not suitable as there is only *one* known boundary condition of deflection to solve for the two constants of integration. There are, however, known conditions of bending moment together with required continuity at the joint of the upper and lower parts. We start, therefore, with the basic bending moment equation

$$M = Pu$$

Differentiating twice,

$$\frac{d^2M}{dy^2} = \frac{Pd^2u}{dy^2}$$

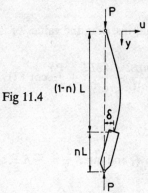

Fig 11.4

But

$$\frac{d^2u}{dy^2} = -\frac{M}{EI}$$

Therefore

$$\frac{d^2M}{dy^2} + \frac{P}{EI}M = 0 \qquad (11.27)$$

This is an alternative and very useful differential equation to express strut behaviour. The solution is

$$M = A \sin Ky + B \cos Ky \qquad (11.28)$$

where $K^2 = P/EI$. Using the bending-moment/curvature relation,

$$-EI\frac{d^2u}{dy^2} = A \sin Ky + B \cos Ky \qquad (11.29)$$

Integration gives

$$EI\frac{du}{dy} = +\frac{A}{K}\cos Ky - \frac{B}{K}\sin Ky + C \qquad (11.30)$$

$$EIu = -\frac{A}{K^2}\sin Ky - \frac{B}{K^2}\cos Ky + Cy + D \qquad (11.31)$$

Boundary Conditions

At $y = 0$, $M = 0$; therefore $B = 0$. At $y = (1 - n)L$, $M = -P\delta$, and

$$A = -\frac{P\delta}{\sin K(1 - n)L}$$

At $y = 0$, $u = 0$; hence $D = 0$. At $y = (1 - n)L$, $du/dy = -\delta/nL$ from eqn. (11.30), so that

$$C = -\frac{EI\delta}{nL} - \frac{P\delta}{K} \cot K(1 - n)L$$

At $y = (1 - n)L$, $u = \delta$, and inserting the values of the constants into eqn. (11.31),

$$EI\delta = \frac{P\delta \sin K(1 - n)L}{K^2 \sin K(1 - n)L} - \left[\frac{\delta EI}{nL} + \frac{P\delta}{K} \cot K(1 - n)L\right](1 - n)L$$

which on simplifying gives

$$\tan K(1 - n)L = - KnL \qquad (11.32)$$

If $(1 - n)L = nL$, then $n = \frac{1}{2}$, and $\tan K\dfrac{L}{2} = -K\dfrac{L}{2}$, so that

$$\frac{KL}{2} = 2 \cdot 03$$

and

$$P_c = \left(\frac{4 \cdot 06}{L}\right)^2 EI = 1 \cdot 67 \frac{\pi^2 EI}{L^2}$$

11.4. BEHAVIOUR OF IDEAL AND REAL STRUTS

The load-deflection behaviour for an ideal Euler strut is illustrated in Fig. 11.5. The applied load, P, increases from zero, with no deflection

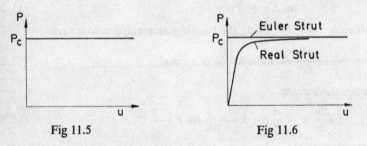

Fig 11.5   Fig 11.6

occurring until the critical load, P_c, is reached, at which point deflection commences and continues at constant load. Obviously the strut can only accommodate a certain deflection up to its elastic limit; thereafter yielding and "failure" would occur by plasticity.

In the case of a real strut, which incorporates some deficiency such as eccentricity of loading, deflection will occur from the moment when load is applied as shown in Fig. 11.6. The curve becomes asymptotic to the Euler load at large deflection. Again, this situation will probably not be attained owing to yielding.

In order to appreciate the significance of stress during buckling behaviour we consider the Euler equation (11.4):

$$P_c = \frac{\pi^2 EI}{L^2} = \frac{\pi^2 EAk^2}{L^2}$$

where A is the cross-sectional area and k is the minimum radius of gyration of the section. Therefore

$$\sigma_c = \frac{P_c}{A} = \frac{\pi^2 E}{(L/k)^2} \qquad (11.33)$$

The ratio L/k is termed the *slenderness ratio*, and plotting this against σ_c gives a curve known as the *Euler hyperbola*, as shown in Fig. 11.7. If a

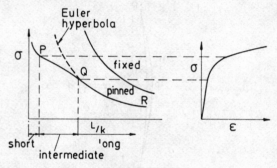

Fig 11.7

comparison is now made with the compression stress–strain curve for the material, one sees that only struts of large L/k in the range Q to R will buckle elastically.

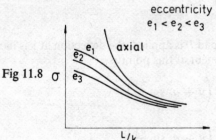

Fig 11.8

For intermediate values of L/k, from P to Q, instability will be accompanied by yielding. At a particular point on the stress–strain curve the stiffness is given by E_t, known as the *tangent modulus*. It is found that, in this elastic–plastic range, buckling loads can be predicted from the original Euler expression if the ordinary modulus E is replaced by the tangent modulus E_t. For short columns, instability does not occur and the problem is one of simple compression.

Fig. 11.7 also brings out the influence of end condition in relation to slenderness ratio. As soon as eccentricity is involved then the relationship between slenderness ratio and critical stress changes to that shown in Fig. 11.8.

11.5. STRUTS HAVING INITIAL CURVATURE

After eccentricity the next practical departure from the Euler idealization is that in some cases a column or strut may not be perfectly straight before loading. This will influence the onset of instability. The following

Fig 11.9

analysis was developed by Perry. The strut is illustrated in Fig. 11.9, in which the initial maximum deflection is a_0, the value of the deflection distant y from P is u_0, and

$$u_0 = a_0 \sin \frac{\pi y}{L}$$

When the buckling load P is applied, the deflection at y is increased by u and the bending moment at this point is

$$P(u + u_0) = P\left(u + a_0 \sin \frac{\pi y}{L}\right)$$

Hence

$$EI\frac{d^2u}{dy^2} = -P\left(u + a_0 \sin \frac{\pi y}{L}\right)$$

or

$$\frac{d^2u}{dy^2} = -\frac{P}{EI}\left(u + a_0 \sin \frac{\pi y}{L}\right)$$

Therefore

$$\frac{d^2u}{dy^2} + K^2 \left(u + a_0 \sin \frac{\pi y}{L} \right) = 0 \tag{11.34}$$

where $K^2 = P/EI$; hence

$$u = A \cos Ky + B \sin Ky + \frac{K^2 a_0 \sin \dfrac{\pi y}{L}}{\dfrac{\pi^2}{L^2} - K^2}$$

At $y = 0$ and $y = L$, $u = 0$; therefore $A = 0$ and $B \sin KL = 0$; and since K is not zero, it follows that $B = 0$; therefore

$$\begin{aligned}
u &= \frac{K^2 a_0 \sin \dfrac{\pi y}{L}}{\dfrac{\pi^2}{L^2} - K^2} \\[2em]
&= \frac{P a_0}{EI \left(\dfrac{\pi^2}{L^2} - K^2 \right)} \sin \frac{\pi y}{L} \\[2em]
&= \frac{P a_0}{P_e - P} \sin \frac{\pi y}{L} \tag{11.35}
\end{aligned}$$

where $P_e = \pi^2 EI/L^2$, or

$$u = \frac{u_0}{(P_e/P) - 1} \tag{11.36}$$

Thus the effect of the end thrust P is to increase the initial maximum deflection a_0 by the multiplying factor $[(P_e/P) - 1]^{-1}$.

From eqn. (11.36) it will be observed that, as the value of P approaches that of P_e, the value of u increases, tending to become infinite. At $y = \frac{1}{2}L$, the increased deflection is given by

$$u' = \frac{a_0}{(P_e/P) - 1} \tag{11.37}$$

Plotting values of P and u, the curve shown in Fig. 11.10 is obtained. Breakdown of the strut would occur before P reached the theoretical value P_e.

Eqn. (11.37) may be written in the form

$$(P_e/P)u' - u' = a_0 \tag{11.38}$$

which shows that there is a linear connection between u' and u'/P, Fig. 11.11, the intercept on the axis of u' being equal to $-a_0$.

Let $\sigma = P/A$ and $\sigma_e = P_e/A$, where A is the cross-sectional area of the strut. Hence, from eqn. (11.35),

$$u = \frac{\sigma}{\sigma_e - \sigma}\, a_0 \sin\frac{\pi y}{L}$$

and the total deflection at y is given by

$$u + u_0 = \frac{\sigma}{\sigma_e - \sigma}\, a_0 \sin\frac{\pi y}{L} + a_0 \sin\frac{\pi y}{L}$$

$$= \frac{\sigma_e}{\sigma_e - \sigma}\, a_0 \sin\frac{\pi y}{L}$$

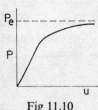

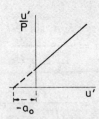

Fig 11.10   Fig 11.11

The maximum deflection occurs at 0, where $y = \tfrac{1}{2}L$; therefore

$$u_{max} = \frac{\sigma_e}{\sigma_e - \sigma}\, a_0 \tag{11.39}$$

Maximum bending moment $= Pa_0 \dfrac{\sigma_e}{\sigma_e - \sigma}$

and the maximum compressive stress is given by

$$\sigma_c = -\frac{Pa_0 \left(\dfrac{\sigma_e}{\sigma_e - \sigma}\right) c}{I} - \frac{P}{A} \tag{11.40}$$

where c is the distance from the neutral axis to the point of maximum compressive stress. If k is the least radius of gyration of the section, then, substituting $P/A = \sigma$,

$$\sigma_c = -\sigma\left(\frac{\eta\sigma_e}{\sigma_e - \sigma} + 1\right) \tag{11.41}$$

where $\eta = a_0 c/k^2$. Taking σ_c equal to the yield stress in compression, σ_Y, we get a quadratic equation

$$\sigma^2 - \sigma[\sigma_Y + \sigma_e(\eta + 1)] + \sigma_Y\sigma_e = 0$$

from which

$$\sigma_c = \frac{\sigma_Y + \sigma_e(\eta + 1)}{2} - \sqrt{\left[\left\{\frac{\sigma_Y + (\eta + 1)\sigma_e}{2}\right\}^2 - \sigma_Y\sigma_e\right]} \quad (11.42)$$

This equation appears to give good results for pin-jointed struts, where an average value of η is taken as $0\cdot001L/k$, and at its lower limit, $0\cdot003L/k$. British Standard 449:1948 recommends the following:

If K is a load factor, taken as 2 for steel, then the average stress on the sectional area of the strut shall not exceed the value of σ_c obtained from eqn. (11.36):

$$K\sigma_e = \frac{\sigma_Y + (\eta + 1)\sigma_e}{2} - \sqrt{\left[\left\{\frac{\sigma_Y + (\eta + 1)\sigma_e}{2}\right\}^2 - \sigma_Y\sigma_e\right]} (11.43)$$

The reader is recommended to consult the relevant British Standard for further information relating to the use of this formula for struts of different materials and having different end connections.

In the case of brittle materials strut failure would probably be due to reaching a limiting tensile stress. An equation of the same form as (11.36) can be written in terms of the tension criterion.

11.6. EMPIRICAL FORMULAE

The Euler formula for a column or strut is only applicable when the length is great in comparison with the cross-sectional dimensions, and, in the case of a circular column, a rough rule may be taken to be that the length must be greater than thirty diameters, the limit for any particular case being reached when the buckling stress is equal to the yield stress. The Euler theory neglects the direct stress in the column, and, for long columns, this is negligible in comparison with the buckling stress. Many cases are met, however, where both stresses are of importance. Various empirical formulae have been produced to meet cases of this kind, the following being representative of these:

(i) Rankine–Gordon

$$P = \frac{\sigma A}{1 + a(L/k)^2} \quad (11.44)$$

where P = Buckling load (kN or MN)
      σ = Intensity of stress at yield point in compression (MN/m²) (see Chapter 19)
      A = Area of cross-section (m²)
      L = Length of column (m)
      k = Least radius of gyration of cross-section (m)
      a = A constant depending on the end conditions and on the material

The safe load is obtained by dividing P by a suitable factor of safety.

Material	σ (MN/m^2)	Values of a for cases in Fig. 11.2			
		(a)	(b)	(c)	(d)
Mild steel	324	$\dfrac{4}{30\,000}$	$\dfrac{16}{30\,000}$	$\dfrac{1}{30\,000}$	$\dfrac{16}{9 \times 30\,000}$
Wrought iron	247	$\dfrac{4}{36\,000}$	$\dfrac{16}{36\,000}$	$\dfrac{1}{36\,000}$	$\dfrac{16}{9 \times 36\,000}$
Cast iron	560	$\dfrac{4}{6400}$	$\dfrac{16}{6400}$	$\dfrac{1}{6400}$	$\dfrac{16}{9 \times 6400}$
Timber	34·7	$\dfrac{4}{3000}$	$\dfrac{16}{3000}$	$\dfrac{1}{3000}$	$\dfrac{16}{9 \times 3000}$

The Rankine–Gordon formula may be written in the form

$$P[1 + a(L/k)^2] = \sigma A$$

or

$$\sigma = \frac{P}{A} + \frac{aL^2P}{Ak^2} = \frac{P}{A} + \frac{aL^2P}{I} \tag{11.45}$$

The formula thus holds for very short struts, where buckling stress is negligible, and gives $P = \sigma A$. For long struts, where the direct stress is negligible,

$$P = \frac{\sigma I}{aL^2}$$

which is of the same form as the Euler formula, where

$$\frac{\sigma}{a} = \beta\pi^2 E$$

(ii) Straight Line

Formulae of this type are commonly used in America, and give results which are good enough for a rough approximation of the load. They are usually given in the form

$$P = \sigma A[1 - c(L/k)] \tag{11.46}$$

where P = Safe load (kN or MN)
σ = Safe compressive stress of a small length of the material (MN/m^2)

A, L and k have the previous meaning

c = A constant depending on the material and on the method of constraint

$c = 0.005$ is a common value for mild steel
$c = 0.008$ is a common value for cast iron

(iii) Parabolic

This formula, which is intended to agree with the Euler formula for long columns, is

$$P = \sigma A[1 - c(L/k)^2] \tag{11.47}$$

where c is a constant; the other symbols have the same meanings as in (ii) above. With pin ends and $L/k < 150$, then for mild steel c may be taken as $0.000\,023$.

(iv) Fidler

Fidler, in his *Bridge Construction*, investigated the column problem and obtained the following formula:

$$P = \frac{A}{c}[(\sigma + H) - \sqrt{\{(\sigma + H)^2 - 2c\sigma H\}}] \tag{11.48}$$

where P = Crippling or breaking load (kN or MN)
A = Cross-sectional area (m²)
σ = Ultimate compressive stress in material (MN/m²)
H = Euler buckling stress = $\pi^2 Ek^2/L^2$
c = A constant of average value 1.2

(v) Eccentric Loading

The above empirical formulae can be used quite simply where there is eccentricity of the applied loading, as follows.

Let the permissible eccentric load be P'. Then the maximum fibre stress due to combined direct and bending stresses is

$$\sigma_c = -\frac{P'}{A} - \frac{P'ec}{Ak_b^2}$$

where e = Eccentricity of load from central axis
c = Distance of surface from neutral axis
A = Cross-sectional area
k_b = Radius of gyration in the plane of bending

If the Rankine–Gordon relation between axial load, P, and compressive stress, σ_c, is used, then from eqn. (11.45),

$$\sigma_c = -\frac{P}{A} - \frac{aL^2P}{Ak_l^2}$$

in which k_l is the *least* radius of gyration.

Equating the values of σ_c,

$$-\frac{P}{A} - \frac{aL^2P}{Ak_l^2} = -\frac{P'}{A} - \frac{P'ec}{Ak_b^2}$$

Thus

$$P' = \frac{\left(1 + \dfrac{aL^2}{k_l^2}\right)P}{1 + \dfrac{ec}{k_b^2}} \qquad (11.49)$$

Example 11.3. A piece of timber of rectangular section 100 × 50 mm and 1 m in length can be regarded as having pinned ends. It is subjected to a compressive load acting on one centre-line and eccentric from the other as shown in Fig. 11.12. Find the value of eccentricity which will result in an

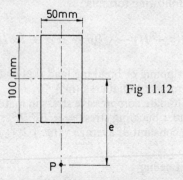

Fig 11.12

equal likelihood of reaching the limiting compressive stress for buckling in either of the principal planes. Use the Rankine–Gordon formula, with a constant of 1.33×10^{-3}. If the compressive stress is limited to 35 MN/m² calculate the allowable compressive load.

For buckling in either direction of the principal axes, P' must equal P in eqn. (11.49). Therefore

$$1 + \frac{ec}{k_b^2} = 1 + \frac{aL^2}{k_l^2}$$

$$e = \frac{aL^2k_b^2}{ck_l^2}$$

$$I_l = \frac{0.1 \times 0.05^3}{12} \qquad I_b = \frac{0.05 \times 0.1^3}{12} \qquad A = 0.05 \times 0.1$$

$$k_i{}^2 = \frac{0.05^2}{12} \qquad k_b{}^2 = \frac{0.1^2}{12}$$

$$e = \frac{1.33 \times 10^{-3} \times 1 \times 0.1^2}{0.05 \times 0.05^2} = 106 \, \text{mm}$$

From eqn. (11.44),

$$P = \frac{35 \times 10^6 \times 0.05 \times 0.1}{1 + \left[1.33 \times 10^{-3} \left(\dfrac{12}{0.05^2} \right) \right]} = \frac{175 \times 10^3}{7.4} \, \text{N}$$

$$= 23.7 \, \text{kN}$$

11.7. "CRINKLING" INSTABILITY OF TUBES

A tube, whose thickness is small compared with its diameter, may "crinkle" or form into folds when under compressive load. These folds may appear either just before or just after the direct stress is equal in magnitude to the yield stress of the material. Southwell in his investigation obtained the formula

$$\sigma = E \frac{t}{R} \sqrt{\frac{1}{3(1 - v^2)}} \qquad (11.50)$$

where σ = Stress causing yielding (MN/m²)
$\quad t$ = Thickness of column (mm)
$\quad R$ = Mean radius of tube (mm)
$\quad E$ = Young's modulus for the material (GN/m²)
$\quad v$ = Poisson's ratio for the material

Very few tubes used in practice are liable to fail in this way; for, taking a mild steel whose yield stress is 324 MN/m², E = 195 GN/m² and v = 0.25, then t/R must be less than 1/370 in order that "crinkling" may occur.

11.8. MEMBERS SUBJECTED TO AXIAL AND TRANSVERSE LOADING

Since transverse loading of a slender member causes bending, the effect on instability under axial compression is similar to that of initial curvature of the strut, and consequently a more rapid rate of deflection occurs. The following two examples will illustrate this type of problem.

(i) Pin-ended Strut carrying a Concentrated Lateral Load at Mid-span

With reference to Fig. 11.13 for $0 < y < L/2$,

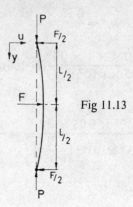

Fig 11.13

$$EI\frac{d^2u}{dy^2} = -Pu - \frac{F}{2}y$$

$$\frac{d^2u}{dy^2} + K^2u = -\frac{K^2F}{2P}\,y \tag{11.51}$$

The standard solution for this type of equation is

$$u = A\sin Ky + B\cos Ky - \frac{Fy}{2P} \tag{11.52}$$

$$\frac{du}{dy} = AK\cos Ky - BK\sin Ky - \frac{F}{2P} \tag{11.53}$$

Boundary Conditions

At $y = 0$, $u = 0$; therefore $B = 0$. At $y = L/2$, $du/dy = 0$.

$$A = \frac{F}{2PK}\sec K\frac{L}{2}$$

Substituting in eqn. (11.52),

$$u = \frac{F}{2PK}\sec K\frac{L}{2}\sin Ky - \frac{F}{2P}\,y$$

At $y = L/2$, $u = u_{max}$:

$$u_{max} = \frac{F}{2PK}\tan K\frac{L}{2} - \frac{FL}{4P} \tag{11.54}$$

Now,

$$u_{max} \to \infty \quad \text{when} \quad K\frac{L}{2} \to \frac{n\pi}{2}$$

Hence

$$P_c = \frac{n^2\pi^2}{L^2}EI$$

for which the lowest value is

$$P_c = \frac{\pi^2 EI}{L^2}$$

which is the Euler load for a simple strut without transverse loading. Although theoretically the load-deflection relationship would become asymptotic to the Euler load, in fact "failure" is governed by yielding at a lower load.

$$M = +Pu + \frac{F}{2}y$$

$$= \frac{F}{2K}\sec K\frac{L}{2}\sin Ky$$

and

$$M_{max} = \frac{F}{2K}\tan K\frac{L}{2} \tag{11.55}$$

(ii) Pin-ended Strut carrying a Uniformly Distributed Lateral Load w per Unit Length

The bending moment at an arbitrary section along the beam in Fig. 11.14 is

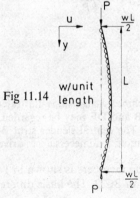

Fig 11.14 w/unit length

$$M = Pu + \frac{wL}{2}y - \frac{wy^2}{2}$$

Hence

$$EI\frac{d^2u}{dy^2} = -Pu - \frac{wL}{2}y + \frac{wy^2}{2}$$

$$\frac{d^2u}{dy^2} + K^2u = \frac{wK^2y^2}{2P} - \frac{wLK^2y}{2P} \tag{11.56}$$

The standard solution for this type of equation is

$$u = A \sin Ky + B \cos Ky + \frac{wy^2}{2P} - \frac{wL}{2P}y - \frac{w}{PK^2} \tag{11.57}$$

At $y = 0$, $u = 0$; hence $B = w/PK^2$; and at $y = L$, $u = 0$.

$$0 = A \sin KL + \frac{w}{PK^2} \cos KL - \frac{w}{PK^2}$$

$$A = \frac{w}{PK^2}(\text{cosec } KL - \cot KL)$$

$$= \frac{w}{PK^2} \tan \frac{KL}{2}$$

The deflection equation (11.57) becomes

$$u = \frac{w}{PK^2}\left(\tan K\frac{L}{2} \sin Ky + \cos Ky\right) + \frac{wy^2}{2P} - \frac{wLy}{2P} - \frac{w}{PK^2}$$

At $y = L/2$, $u = u_{max}$:

$$u_{max} = \frac{w}{PK^2}\left(\sec K\frac{L}{2} - 1\right) - \frac{wL^2}{8P} \tag{11.58}$$

from which

$$M_{max} = \frac{w}{K^2}\left(\sec K\frac{L}{2} - 1\right) \tag{11.59}$$

Example 11.4.   A part of a machine mechanism is illustrated in Fig. 11.15(*a*.) The oscillating end portions AB and DE may be regarded as rigid, but can pivot freely at A, B, D and E. The central slender strut BCD has a simple restraint at C. What is the maximum load necessary to drive the mechanism?

The deflected position of the members is shown in Fig. 11.15(*b*) with the appropriate forces acting on BCD. The basic differential equation is

$$EI\frac{d^2u}{dy^2} = P(\delta - u) + \frac{P\delta}{L}y$$

Putting $P/EI = K^2$,

$$\frac{d^2u}{dy^2} + K^2u = K^2\delta\left(\frac{y}{L} + 1\right)$$

Differentiating,

$$\frac{d^3u}{dy^3} + K^2\frac{du}{dy} = K^2\frac{\delta}{L}$$

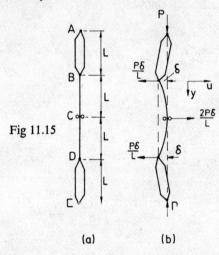

Fig 11.15

(a)   (b)

The solution of this equation is

$$u = A + B\sin Ky + C\cos Ky + \frac{\delta}{L}y$$

Boundary Conditions

(i) $y = 0$, $M = 0$; (ii) $y = L$, $du/dy = 0$; (iii) $y = 0$, $u = \delta$; (iv) $y = L$, $u = 0$.

From (iii) and (iv),

$$\delta = A + C \quad \text{and} \quad 0 = A + B\sin KL + C\cos KL + \delta$$

From (ii),

$$\frac{du}{dy} = BK\cos Ky - CK\sin Ky + \frac{\delta}{L}$$

$$0 = BK\cos KL - CK\sin KL + \frac{\delta}{L}$$

From (i),

$$\frac{d^2u}{dy^2} = -BK^2 \sin Ky - CK^2 \cos Ky$$

$$0 = -CK^2$$

Hence $C = 0$, $A = \delta$, and

$$B = -\frac{\delta}{KL \cos KL} = -\frac{2\delta}{\sin KL}$$

From this last equation,

$$\sin KL = 2KL \cos KL$$

or

$$\tan KL = 2KL$$

and

$$KL = 1\cdot166$$

So that

$$P_c = \left(\frac{1\cdot166}{L}\right)^2 EI = \frac{1\cdot36EI}{L^2}$$

Chapter 12

Stress and Strain Transformations and Relationships

12.1. The preceding chapters have been concerned with "one-dimensional" problems of stress and strain, i.e. in any particular example, consideration has only been given to one type of stress acting in one direction. However, the majority of engineering components and structures are subjected to loading conditions, or are of such a shape that, at any point in the material, a complex state of stress and strain exists involving direct (tension, compression) and shear components in various directions. A simple example of this is a shaft which transmits power through a pulley and belt drive. An element of material in the shaft would be subjected to direct stress and shear stress due to bending action and additional shear stress from the torque required to transmit power. It is not necessarily sufficient to be able to determine the individual values of these stresses in order to select a suitable material from which to make the shaft because, as will be seen later, on certain planes within the element more severe conditions of stress and strain exist.

12.2. SYMBOLS AND SIGNS

Since conditions will be studied in which several different stresses occur simultaneously, it is essential to be consistent in the use of distinctive symbols, and a sign convention must be established and adhered to. A subscript notation will be used as follows:

Direct stress $\quad \sigma_x, \sigma_y, \sigma_z \ldots$

where the subscript denotes the direction of the stress.

Shear stress $\quad \tau_{xy}, \tau_{yx}, \tau_{yz}, \tau_{zy}, \tau_{xz}, \tau_{zx} \ldots$

where the first subscript denotes the direction of the normal to the plane on which the shear stress acts, and the second subscript the direction of the shear stress.

As in the previous work, tensile stress will be taken as positive and compressive stress negative. A shear stress is defined as positive when the direction of the arrow and the normal to the plane are both in the positive sense or both in the negative sense in relation to the co-ordinate axes. If the directions of the shear stress and the normal to the plane are opposed in sign, then the shear stress is negative. Pairs of complementary

281

shear stress components are therefore either both positive or both negative.

The angle between an inclined plane and a co-ordinate axis is positive when measured in the anticlockwise sense from the co-ordinate axis.

A general three-dimensional stress system is shown in Fig. 12.1.

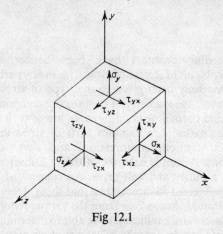

Fig 12.1

Plane Stress

If there are no normal and shear stresses on the two planes perpendicular to the z-direction (this also implies that the complementary shear stresses τ_{yz} and τ_{zz} are zero), then this system is known as *plane stress*. It is a situation found, or approximately so, in a number of important engineering problems. The analysis of complex stresses which follows is only concerned with plane stress conditions.

12.3. STRESSES ON A PLANE INCLINED TO THE DIRECTION OF LOADING

In preceding chapters, the analysis has dealt with a stress set up on a plane perpendicular (direct stress) or parallel (shear stress) to the direction of loading. However, if a piece of material is cut along a plane inclined at some angle θ to the direction of loading, then, in order to maintain equilibrium, a system of forces would have to be applied to the plane. This implies that there must be a stress system acting on that plane.

In Fig. 12.2 a bar is shown subjected to an axial tensile load W. The area of cross-section normal to the axis of the bar is denoted by A. If the bar is cut along the plane BC, inclined at an angle θ to the y-direction, then the load W applied to this portion of the bar can be reacted by component forces N normal and S tangential to the plane BC. Thus for equilibrium,

$$N = W \cos \theta \quad \text{and} \quad S = W \sin \theta$$

The area of the plane BC will be sec θ times the area of the bar normal to its axis, or $A \sec \theta$.

Denoting the stress normal to the plane BC by σ_n and the tangential stress by τ_s, then

$$\sigma_n = \frac{N}{A \sec \theta} = \frac{W \cos \theta}{A \sec \theta} = \frac{W}{A} \cos^2 \theta$$

and

$$\tau_s = \frac{S}{A \sec \theta} = \frac{W \sin \theta}{A \sec \theta} = \frac{W}{A} \sin \theta \cos \theta$$

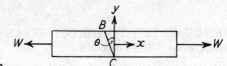

Fig 12.2

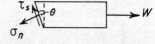

σ_n is the direct stress, in this case tensile, and τ_s is the shearing stress on the plane. W/A is the direct stress normal to the axis of the bar and is equal to σ_x. Therefore

$$\sigma_n = \sigma_x \cos^2 \theta \tag{12.1}$$

and

$$\tau_s = \sigma_x \sin \theta \cos \theta = \tfrac{1}{2}\sigma_x \sin 2\theta \tag{12.2}$$

Note that, in eqn. (12.1), when $\theta = 0$, σ_n is a maximum and equal to σ_x, and for $\theta = 90°$, σ_n is zero, indicating that there is no transverse stress in the bar.

Again, in eqn. (12.2), τ_s will be a maximum when $\sin 2\theta$ is a maximum, i.e. when $2\theta = 90°$ and $270°$, or $\theta = 45°$ and $135°$; the value of τ_s is then $\tfrac{1}{2}\sigma_x$ on the planes prescribed by $\theta = 45°$ and $135°$. This result is borne out in practice, and for materials whose shear strength is less than half the tensile strength, direct tensile or compressive loading results in failure along planes of maximum shear stress.

Example 12.1.   A cylindrical bar of 20 mm diameter is subjected to an axial tensile load of 10 kN. Determine the normal and shear stresses on planes inclined at 60° to the axis of the bar.

$$\sigma_x = \frac{10\,000 \times 10^6}{\frac{\pi}{4} \times 400} = 31 \cdot 8 \, \text{MN/m}^2$$

In one case $\theta = 30°$; hence

$$\sigma_n = 31 \cdot 8 \cos^2 30° = 24 \cdot 6 \, \text{MN/m}^2$$

and

$$\tau_s = \tfrac{1}{2} \times 31 \cdot 8 \sin 60° = 13 \cdot 78 \, \text{MN/m}^2$$

For the other plane $\theta = 60°$; therefore

$$\sigma_n = 31 \cdot 8 \cos^2 60° = 7 \cdot 95 \, \text{MN/m}^2$$

and

$$\tau_s = \tfrac{1}{2} \times 31 \cdot 8 \sin 120° = 13 \cdot 78 \, \text{MN/m}^2$$

12.4. ELEMENT SUBJECTED TO DIRECT STRESS IN TWO PERPENDICULAR DIRECTIONS

The rectangular element of material of unit thickness, Fig. 12.3, is subjected to tensile stresses in the x- and y- directions as shown.

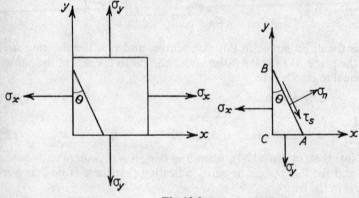

Fig 12.3

Considering a corner cut off the element by the plane AB inclined at θ to the y-axis, then normal and tangential stresses will act on the plane as in the diagram. For equilibrium of ABC, the forces on AB, BC and CA must be in balance. As the element is of constant unit thickness, the areas of the faces are proportional to the lengths of the sides of the triangle. Resolving forces normal to the plane AB,

$$\sigma_n \, \text{AB} - \sigma_x \, \text{BC} \cos \theta - \sigma_y \, \text{AC} \sin \theta = 0 \qquad (12.3)$$

Dividing through by AB,

$$\sigma_n - \sigma_x \frac{\text{BC}}{\text{AB}} \cos \theta - \sigma_y \frac{\text{AC}}{\text{AB}} \sin \theta = 0$$

Therefore

$$\sigma_n = \sigma_x \cos^2 \theta + \sigma_y \sin^2 \theta = \tfrac{1}{2}(\sigma_x + \sigma_y) + \tfrac{1}{2}(\sigma_x - \sigma_y) \cos 2\theta$$
(12.4)

Resolving forces parallel to AB,

$$\tau_s \,\text{AB} - \sigma_x \,\text{BC} \sin \theta + \sigma_y \,\text{AC} \cos \theta = 0$$
(12.5)

Dividing by AB,

$$\tau_s - \sigma_x \frac{\text{BC}}{\text{AB}} \sin \theta + \sigma_y \frac{\text{AC}}{\text{AB}} \cos \theta = 0$$

$$\tau_s = \sigma_x \cos \theta \sin \theta - \sigma_y \sin \theta \cos \theta = \tfrac{1}{2}(\sigma_x - \sigma_y) \sin 2\theta$$
(12.6)

If σ_y is made zero, eqns. (12.4) and (12.6) reduce to those obtained in Section 12.3 for the normal and shear stresses on a plane inclined to the direction of loading.

12.5. ELEMENT SUBJECTED TO SHEARING STRESSES ONLY

The rectangular element of the previous section is now considered with shearing stresses on the faces instead of direct stresses. The plane AB inclined at θ to the y-axis, Fig. 12.4, is subjected to normal and tangential

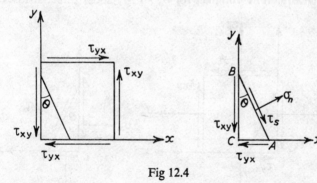

Fig 12.4

stresses as previously. Consider the equilibrium of the triangular portion ABC. Resolving forces normal to the plane AB,

$$\sigma_n \,\text{AB} - \tau_{xy} \,\text{BC} \sin \theta - \tau_{yx} \,\text{AC} \cos \theta = 0$$
(12.7)

Dividing through by AB as before,

$$\sigma_n - \tau_{xy} \frac{\text{BC}}{\text{AB}} \sin \theta - \tau_{yx} \frac{\text{AC}}{\text{AB}} \cos \theta = 0$$

$$\sigma_n = \tau_{xy} \cos \theta \sin \theta + \tau_{yx} \sin \theta \cos \theta$$
(12.8)

But, from a consideration of complementary shear stresses,

$$\tau_{xy} = \tau_{yx} \tag{12.9}$$

Therefore

$$\sigma_n = 2\tau_{xy} \sin \theta \cos \theta = \tau_{xy} \sin 2\theta \tag{12.10}$$

Resolving forces parallel to the plane AB,

$$\tau_s \text{AB} + \tau_{xy} \text{BC} \cos \theta - \tau_{yx} \text{AC} \sin \theta = 0 \tag{12.11}$$

Dividing by AB,

$$\tau_s + \tau_{xy} \frac{\text{BC}}{\text{AB}} \cos \theta - \tau_{yx} \frac{\text{AC}}{\text{AB}} \sin \theta = 0$$

Therefore

$$\tau_s = \tau_{yx} \sin^2 \theta - \tau_{xy} \cos^2 \theta = -\tau_{xy} \cos 2\theta \tag{12.12}$$

12.6. ELEMENT SUBJECTED TO GENERAL TWO-DIMENSIONAL STRESS SYSTEM

The general two-dimensional stress system shown in Fig. 12.5 may be obtained by a summation of the conditions of stress in Figs. 12.3 and 12.4. Hence the equations obtained for σ_n and τ_s under direct stresses and

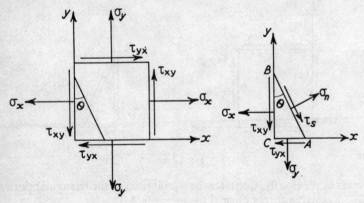

Fig 12.5

shearing stresses separately may be added together to give values for the normal and shear stress on the inclined plane AB, Fig. 12.5, in the general stress system. Therefore

$$\sigma_n = \tfrac{1}{2}(\sigma_x + \sigma_y) + \tfrac{1}{2}(\sigma_x - \sigma_y) \cos 2\theta + \tau_{xy} \sin 2\theta \tag{12.13}$$

$$\tau_s = \tfrac{1}{2}(\sigma_x - \sigma_y) \sin 2\theta - \tau_{xy} \cos 2\theta \tag{12.14}$$

The validity of this method of superposition may be checked by considering the equilibrium of the triangular element of Fig. 12.5 and resolving forces perpendicular and parallel to the plane AB as was done in the previous sections.

It must be remembered that the signs in eqns. (12.13) and (12.14) are dependent on the directions chosen for the arrows (indicating stress) on the element in Fig. 12.5. If for any reason the directions of the stresses are different, then the appropriate signs in eqns. (12.13) and (12.14) must be changed.

Resultant Stress

The normal and shear stresses acting on AB can be combined to give a resultant stress. In this case, since σ_n and τ_s are applied on the same area of the element (AB × unit thickness) and the resultant will also act on this area, it is not necessary to consider forces. Hence the resultant stress is

$$\sigma_r = \sqrt{(\sigma_n{}^2 + \tau_s{}^2)} \tag{12.15}$$

from the vectorial addition.

The angle α between the normal to the plane and the resultant stress is given by

$$\tan \alpha = \frac{\tau_s}{\sigma_n} \tag{12.16}$$

Example 12.2.   A tubular shaft is subjected to combined bending and torsion. An element on the surface experiences a tensile bending stress of $30\,MN/m^2$

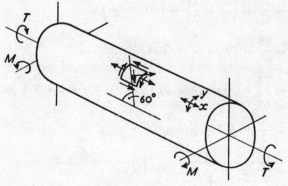

Fig 12.6

and shearing stresses of $40\,MN/m^2$ as shown in Fig. 12.6. Calculate the normal, shear and resultant stresses on the plane shown, inclined at 60° to a generator through the element.

Firstly, it will be observed that the stresses on the element are not all acting in the same sense as in Fig. 12.5. The shear stresses are reversed in direction and σ_y is zero. The angle of 60° is also not the angle corresponding to θ.

The conditions, therefore, are $\sigma_x = +30 \text{ MN/m}^2$, $\sigma_y = 0$, $\tau_{yx} = -40 \text{ MN/m}^2$, the negative sign allowing for the opposite sense of shear

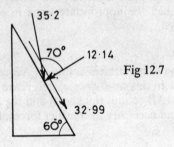

Fig 12.7

stress, and $\theta = 90° - 60° = 30°$. Substituting in eqns. (12.13) and (12.14),

$$\sigma_n = \left[\frac{1}{2} \times 30\right] + \left[\frac{1}{2} \times 30 \cos (2 \times 30°)\right]$$
$$+ [(-40) \times \sin (2 \times 30°)]$$
$$= \frac{30}{2} + \frac{30}{2} \cos 60° - 40 \sin 60°$$
$$= \frac{30}{2} + \left(\frac{30}{2} \times \frac{1}{2}\right) - \left(40 \times \frac{\sqrt{3}}{2}\right)$$
$$= \frac{90}{4} - 20\sqrt{3} = -12 \cdot 14 \text{ MN/m}^2$$

$$\tau_s = \left[\frac{1}{2} \times 30 \sin (2 \times 30°)\right] - [(-40) \times \cos (2 \times 30°)]$$
$$= \frac{30}{2} \sin 60° + 40 \cos 60°$$
$$= \frac{30}{2}\frac{\sqrt{3}}{2} + \frac{40}{2} = 32 \cdot 99 \text{ MN/m}^2$$

$$\sigma_r = \sqrt{(\sigma_n^2 + \tau_s^2)}$$
$$= \sqrt{(-12\cdot14^2 + 32\cdot99^2)}$$
$$= 35\cdot2 \text{ MN/m}^2$$

The direction of the resultant stress is given by

$$\tan \alpha = \frac{32 \cdot 99}{12 \cdot 14} = -2 \cdot 715 \qquad \text{so that} \qquad \alpha = 70°$$

The stress system on the inclined plane is therefore as shown in Fig. 12.7.

12.7. MOHR'S STRESS CIRCLE

Considering eqns. (12.13) and (12.14) once more and rewriting,

$$\sigma_n - \tfrac{1}{2}(\sigma_x + \sigma_y) = \tfrac{1}{2}(\sigma_x - \sigma_y)\cos 2\theta + \tau_{xy}\sin 2\theta$$
$$\tau_s = \tfrac{1}{2}(\sigma_x - \sigma_y)\sin 2\theta - \tau_{xy}\cos 2\theta$$

Squaring both sides and adding the equations,

$$[\sigma_n - \tfrac{1}{2}(\sigma_x + \sigma_y)]^2 + \tau_s{}^2 = \tfrac{1}{4}(\sigma_x - \sigma_y)^2 + \tau_{xy}{}^2 \qquad (12.17)$$

This is the equation of a circle of radius

$$\sqrt{[\tfrac{1}{4}(\sigma_x - \sigma_y)^2 + \tau_{xy}{}^2]}$$

and whose centre has the co-ordinates

$$[\tfrac{1}{2}(\sigma_x + \sigma_y), 0]$$

The circle represents all possible states of normal and shear stress on any plane through a stressed point in a material, and was developed by the German engineer Otto Mohr. The element of Fig. 12.5 and the corresponding Mohr diagram are shown in Fig. 12.8.

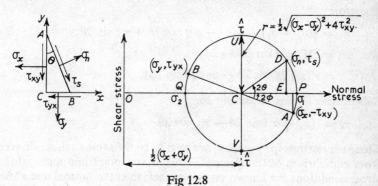

Fig 12.8

The sign convention used on the circle will be, for normal stress, positive to the right and negative to the left of the origin. Shear stresses which might be described as trying to cause a clockwise rotation of an element are plotted above the abscissa axis, i.e. "positive", and shear stresses appearing as anticlockwise rotation are plotted below the axis, i.e. "negative".

It is important to remember that shear stress plotted, say, below the σ-axis, although being regarded as negative in the circle construction, may be either positive or negative on the physical element according to the shear stress convention previously defined. Likewise, positive shear stresses on the circle may be either positive or negative on the element.

The diagram is constructed as follows: using co-ordinate axes of normal stress and shear stress as shown, point A $(\sigma_x, -\tau_{xy})$ is plotted representing the direct and shear stress (negative) acting on the plane AC of the element. Assuming in this case $\sigma_y < \sigma_x$, the point B (σ_y, τ_{xy}) is plotted to represent the stresses on the plane BC of the element. The normal stress axis bisects the line joining AB at C, and with centre C and radius AC a circle is drawn.

An angle equal to twice that in the element, i.e. 2θ, is set off from AC in the anticlockwise direction (the same sense as in the element), and the line CD then cuts the circle at the point whose co-ordinates are (σ_n, τ_s). These are then the normal and shearing stresses on the plane AB in the element.

The validity of the diagram is demonstrated thus:

$$\sigma_n = \text{OC} + \text{CE} = \tfrac{1}{2}(\sigma_x + \sigma_y) + r\cos(2\theta - 2\phi)$$
$$= \tfrac{1}{2}(\sigma_x + \sigma_y) + r\cos 2\theta \cos 2\phi + r\sin 2\theta \sin 2\phi$$

But

$$\cos 2\phi = \frac{\tfrac{1}{2}(\sigma_x - \sigma_y)}{r} \quad \text{and} \quad \sin 2\phi = \frac{\tau_{xy}}{r}$$

Therefore

$$\sigma_n = \tfrac{1}{2}(\sigma_x + \sigma_y) + \tfrac{1}{2}(\sigma_x - \sigma_y)\cos 2\theta + \tau_{yx}\sin 2\theta$$
$$\tau_s = \text{DE} = r\sin(2\theta - 2\phi)$$
$$= r(\sin 2\theta \cos 2\phi - \cos 2\theta \sin 2\phi)$$

Substituting for $\cos 2\phi$ and $\sin 2\phi$,

$$\tau_s = \tfrac{1}{2}(\sigma_x - \sigma_y)\sin 2\theta - \tau_{xy}\cos 2\theta$$

These expressions for σ_n and τ_s are seen to be the same as those derived from equilibrium of the element. Thus, if at a point in a material the stress conditions are known on two planes, then the normal and shear stresses on any other plane through the point can be found using Mohr's circle.

Certain features of the diagram are worthy of note. The sides of the element AC and CB, which are 90° apart, are represented on the circle by AC and CB, 180° apart. A compressive direct stress would be plotted to the left of the shear-stress axis. The resultant stress on AB, $\sigma_r = \sqrt{(\sigma_n{}^2 + \tau_s{}^2)}$, is given by OD on the Mohr diagram. The maximum shearing stress in an element is given by the top and bottom points of the circle, i.e.

$$\tau_{s\,max} = \pm\sqrt{[\tfrac{1}{4}(\sigma_x - \sigma_y)^2 + \tau_{xy}^2]} \qquad (12.18)$$

and the corresponding normal stress is $\tfrac{1}{2}(\sigma_x + \sigma_y)$. The angle θ to the plane on which a maximum shear stress acts is obtained from the circle as

$$\tan 2\theta = \tan(90° + 2\phi) = -\cot 2\phi$$

Therefore

$$\tan 2\theta = -\frac{\sigma_x - \sigma_y}{2\tau_{xy}} \qquad (12.19)$$

The second plane of maximum shear stress is displaced by 90° from that above.

Example 12.3. Solve Example 12.2 using Mohr's circle. $\sigma_x = +30\,\mathrm{MNm/^2}$, $\sigma_y = 0$, $\tau_{yx} = -40\,\mathrm{MN/m^2}$.*

Construct a circle on the diameter $(+30, +40)$, $(0, -40)$ as in Fig.

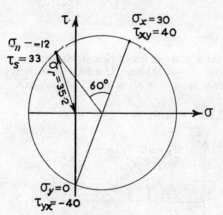

Fig 12.9

12.9, and set off the angle $2\theta = 60°$ in the anticlockwise direction, from which are obtained

$$\sigma_n = -12\cdot 1\,\mathrm{MN/m^2} \qquad \tau_s = 33\,\mathrm{MN/m^2} \qquad \sigma_r = 35\cdot 2\,\mathrm{MN/m^2}$$

Example 12.4. Solve the previous example for the bending stress compressive. $\sigma_x = -30\,\mathrm{MN/m^2}$, $\sigma_y = 0$, $\tau_{xy} = -40\,\mathrm{MN/m^2}$.

* In each of the examples the correct sense of the shear stress should be observed from the stress subscripts, the direction and plane, and the sign.

Construct on the diameter $(-30, +40)$, $(0, -40)$ setting off the angle from the σ_x-plane as before. It will be found from Fig. 12.10 that

$$\sigma_n = -57 \cdot 1 \, \text{MN/m}^2 \qquad \tau_s = 7 \, \text{MN/m}^2 \qquad \sigma_r = 57 \cdot 5 \, \text{MN/m}^2$$

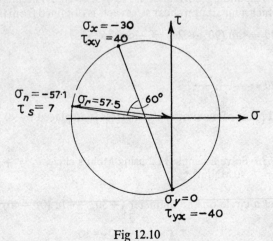

Fig 12.10

Example 12.5. At a point in a complex stress field $\sigma_x = 40 \, \text{MN/m}^2$, $\sigma_y = 80 \, \text{MN/m}^2$, and $\tau_{yx} = -20 \, \text{MN/m}^2$. Use Mohr's circle solution to find the normal and shear stresses on a plane at $45°$ to the y axis.

From Fig. 12.11,

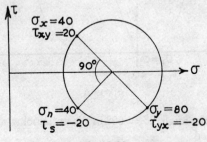

Fig 12.11

$$\sigma_n = 40 \, \text{MN/m}^2 \qquad \text{and} \qquad \tau_s = -20 \, \text{MN/m}^2$$

Example 12.6. Construct Mohr's circle for the following point stresses: $\sigma_x = 60 \, \text{MN/m}^2$, $\sigma_y = 10 \, \text{MN/m}^2$ and $\tau_{xy} = +20 \, \text{MN/m}^2$, and hence determine the stress components and planes on which the shear stress is a maximum.

From Fig. 12.12, the normal and maximum shear stress components are

$$\sigma_n = 35 \text{ MN/m}^2 \quad \text{and} \quad \tau_s = 32 \text{ MN/m}^2$$

acting on the planes $\theta_1 = 64 \cdot 5°$, $\theta_2 = 154 \cdot 5°$.

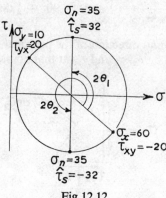

Fig 12.12

12.8. PRINCIPAL STRESSES AND PLANES

It has been shown that Mohr's circle represents all possible states of normal and shear stress at a point. From Fig. 12.8 it can be seen that there are two planes, QC and CP, 180° apart on the diagram and therefore 90° apart in the material, on which the shear stress τ_s is zero. These planes are termed *principal planes* and the normal stresses acting on them are termed *principal stresses*. The latter are denoted by σ_1 and σ_2 at **P** and **Q** respectively, and are the maximum and minimum values of normal stress that can be obtained at a point in a material. The values of the principal stresses can be found either from eqn. (13.17) by putting $\tau_s = 0$, or directly from the Mohr diagram; hence

$$\sigma_1 = \frac{\sigma_x + \sigma_y}{2} + \tfrac{1}{2}\sqrt{[(\sigma_x - \sigma_y)^2 + 4\tau_{xy}^2]} \tag{12.20}$$

$$\sigma_2 = \frac{\sigma_x + \sigma_y}{2} - \tfrac{1}{2}\sqrt{[(\sigma_x - \sigma_y)^2 + 4\tau_{xy}^2]} \tag{12.21}$$

where σ_1 is the maximum and σ_2 the minimum principal stress. The planes are specified by

$$2\theta = 2\phi \text{ and } 180° + 2\phi$$

or

$$\theta = \phi \text{ and } 90° + \phi$$

But

$$\tan 2\phi = \frac{\tau_{xy}}{\frac{1}{2}(\sigma_x - \sigma_y)} \qquad \left(\sin 2\phi = \frac{\tau_{xy}}{\sqrt{[\frac{1}{2}(\sigma_x - \sigma_y)^2 + \tau_{xy}^2]}}\right)$$

Therefore

$$\theta = \tfrac{1}{2}\tan^{-1}\frac{2\tau_{xy}}{\sigma_x - \sigma_y} \quad \text{and} \quad 90° + \tfrac{1}{2}\tan^{-1}\frac{2\tau_{xy}}{\sigma_x - \sigma_y} \quad (12.22)$$

Thus the magnitude and direction of the principal stresses at any point in a material depend on σ_x, σ_y and τ_{xy} at that point, Fig. 12.13.

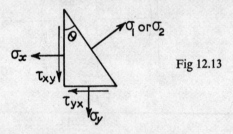

Fig 12.13

Maximum Shear Stress in Terms of Principal Stresses

It was shown earlier, eqn. (12.18), that the maximum shearing stress at a point is given by

$$\tau_{s\,max} = \pm\sqrt{[(\sigma_x - \sigma_y)^2 + 4\tau_{xy}^2]}$$

If expressions (12.20) and (12.21) for σ_1 and σ_2 are subtracted, then

$$\sigma_1 = \sigma_2 = \pm\sqrt{[(\sigma_x - \sigma_y)^2 + 4\tau_{xy}^2]} \qquad (12.23)$$

and therefore

$$\tau_{s\,max} = \frac{\sigma_1 - \sigma_2}{2} \qquad (12.24)$$

It should be noted that principal stresses are considered a maximum or minimum mathematically, e.g. a compressive or negative stress is less than a positive stress, irrespective of numerical value.

In Mohr's circle the principal planes PC and OC are at 90° to those of maximum shear stress, UC and VC, and therefore in the material the angles between these two sets of planes become 45°, or the maximum shear-stress planes bisect the principal planes.

Example 12.7. Determine the principal stresses and maximum shear stress in Examples 12.3 and 12.4.

Firstly, considering the problem analytically we have

$$\sigma_1, \sigma_2 = \frac{\sigma_x + \sigma_y}{2} \pm \tfrac{1}{2}\sqrt{[(\sigma_x - \sigma_y)^2 + 4\tau_{xy}^2]}$$

(a) $\sigma_x = +30\,\text{MN/m}^2$ $\sigma_y = 0$ $\tau_{xy} = -40\,\text{MN/m}^2$

$$\sigma_1, \sigma_2 = \frac{30 + 0}{2} \pm \tfrac{1}{2}\sqrt{[(30 - 0)^2 + 4 \times 40^2]}$$

$$= 15 \pm \frac{10}{2}\sqrt{[9 + 64]}$$

$$= 15 \pm 42.7$$

$$\sigma_1 = +57.7\,\text{MN/m}^2 \qquad \sigma_2 = -27.7\,\text{MN/m}^2$$

Maximum shear stress $= \dfrac{\sigma_1 - \sigma_2}{2} = +\dfrac{57.7 + 27.7}{2}$

$$= 42.7\,\text{MN/m}^2$$

By Mohr's circle the above values can be obtained from Fig. 12.9.

(b) $\sigma_x = -30\,\text{MN/m}^2$ $\sigma_y = 0$ $\tau_{yx} = -40\,\text{MN/m}^2$

$$\sigma_1, \sigma_2 = \frac{-30 + 0}{2} \pm \tfrac{1}{2}\sqrt{[(-30 - 0)^2 + (4 \times 40^2)]}$$

$$= -15 \pm \frac{10}{2}\sqrt{[9 + 64]}$$

$$= -15 \pm 42.7$$

$$\sigma_1 = +27.7\,\text{MN/m}^2 \qquad \sigma_2 = -57.7\,\text{MN/m}^2$$

Maximum shear stress $= \dfrac{\sigma_1 - \sigma_2}{2} = \dfrac{+27.7 + 57.7}{2}$

$$= 42.7$$

Therefore

$$\tau_{s\,max} = 42.7\,\text{MN/m}^2$$

Fig. 12.10 gives the corresponding Mohr's circle values.

12.9. COMBINED BENDING AND TORSION

A good example of a two-dimensional state of stress is found in members subjected to combined bending and torsion. Consider the solid circular shaft shown in Fig. 12.14 acted on by bending moment M and torque T.

The maximum bending and shear stresses occur at the outer surface of the shaft and are given by

$$\sigma_x = \frac{32M}{\pi d^3} \quad \text{and} \quad \tau_{xz} = \frac{16T}{\pi d^3}$$

where d is the diameter of the shaft.

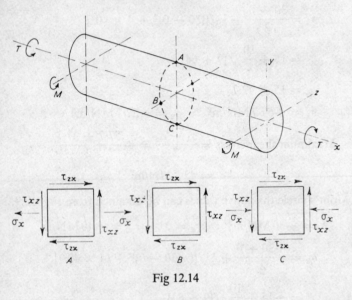

Fig 12.14

The stress components on an element of material at the surface points A, B and C are shown in Fig. 12.14. The shear stresses are the same in each; however, the bending stress is maximum tension at A, zero at B (the neutral plane) and maximum compression at C. The principal stresses at these three points are therefore

At A $\sigma_1, \sigma_2 = \frac{1}{2}\sigma_x \pm \frac{1}{2}\sqrt{(\sigma_x{}^2 + 4\tau_{xz}{}^2)}$

At B $\sigma_1 = -\sigma_2 = \tau_{xz}$

At C $\sigma_1, \sigma_2 = -\frac{1}{2}\sigma_x \pm \frac{1}{2}\sqrt{(\sigma_x{}^2 + 4\tau_{xz}{}^2)}$

At B, therefore, there is a state of pure shear.

The principal stresses can be expressed in terms of the bending moment and torque by substituting for σ_x and τ_{xz}:

At A $\sigma_1, \sigma_2 = \dfrac{16M}{\pi d^3} \pm \sqrt{\left[\left(\dfrac{16M}{\pi d^3}\right)^2 + \left(\dfrac{16T}{\pi d^3}\right)^2\right]}$

$$= \frac{16}{\pi d^3}[M \pm \sqrt{(M^2 + T^2)}]$$

At B $\qquad \sigma_1 = -\sigma_2 = \dfrac{16T}{\pi d^3}$

At C the values are the same as at A but are negative.

The inclinations of the principal planes to the z-axis are

$$\theta = \tfrac{1}{2}\tan^{-1}\frac{T}{M} \quad \text{and} \quad 90^\circ + \tfrac{1}{2}\tan^{-1}\frac{T}{M}$$

The maximum shear stress is given by

$$\tau = \frac{16}{\pi d^3}\sqrt{(M^2 + T^2)}$$

It is sometimes useful to express the above principal stresses in terms of a shaft subjected to bending only. If the equivalent bending moment is M_e, then

$$\frac{32M_e}{\pi d^3} = \frac{16}{\pi d^3}[M \pm \sqrt{(M^2 + T^2)}]$$

and

$$M_e = \tfrac{1}{2}[M \pm \sqrt{(M^2 + T^2)}]$$

If expressed in terms of an equivalent torque only,

$$T_e = M \pm \sqrt{(M^2 + T^2)}$$

Example 12.8. For a curved bar loaded as in Fig. 12.15 determine at what position the maximum principal stress has its greatest value. Calculate the

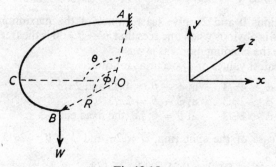

Fig 12.15

latter and also the maximum shear stress for a split ring of radius 250 mm, bar diameter 50 mm and loaded with 5 kN at one end perpendicular to the plane of the ring.

Maximum principal stress, $\sigma_1 = \dfrac{16}{\pi d^3}[M + \sqrt{(M^2 + T^2)}]$

where $M = WR \sin(\phi - \theta)$, and $T = WR[1 - \cos(\phi - \theta)]$. Therefore

$$\sigma_1 = \frac{16WR}{\pi d^3}\{\sin(\phi - \theta)$$
$$+ \sqrt{[\sin^2(\phi - \theta) + \{1 - \cos(\phi - \theta)\}^2]}\}$$

The greatest value for σ_1 occurs when the expression in brackets has its greatest value. Simplifying the expression under the square root, differentiating σ_1 with respect to the variable θ and equating to zero gives

$$\frac{d\sigma_1}{d\theta} = \frac{16WR}{\pi d^3}\left\{\cos(\phi - \theta) + \frac{\sqrt{2}}{2}\sin(\phi - \theta)[1 - \cos(\phi - \theta)]^{-1/2}\right\}$$
$$= 0$$

Rearranging and squaring the terms inside the brackets,

$$\cos^2(\phi - \theta) = \frac{\sin^2(\phi - \theta)}{2[1 - \cos(\phi - \theta)]}$$

from which

$$2\cos^3(\phi - \theta) - 3\cos^2(\phi - \theta) + 1 = 0$$

This may be factorized into

$$[2\cos(\phi - \theta) + 1][\cos(\phi - \theta) - 1]^2 = 0$$

The solutions of this equation are

$$\cos(\phi - \theta) = 1 \quad \text{or} \quad -\tfrac{1}{2}$$

or

$$\phi - \theta = 0 \quad \text{and} \quad 2\pi \quad \text{or} \quad 2\pi/3 \quad \text{and} \quad 4\pi/3$$

The solutions 0 and 2π give least values of the maximum principal stress; this is obvious when one sees that $\phi - \theta = 0$ is the free end of the bar where the bending moment is zero.

The greatest values of σ_1 occur as follows:

For $\phi > 4\pi/3$ at $\phi - \theta = 4\pi/3$ and $2\pi/3$
For $\phi > 2\pi/3$ at $\phi - \theta = 2\pi/3$
For $\phi < 2\pi/3$ at $\theta = 0$, i.e. the fixed end

In the case of the split ring, $\phi = 2\pi$ and $\phi - \theta = 2\pi/3$ and $4\pi/3$. Therefore

$$\sigma_1 = \frac{16 \times 5000 \times 0.25}{\pi \times 0.05^3}\left[\frac{\sqrt{3}}{2} + \sqrt{\left(\frac{3}{4} + \frac{4}{9}\right)}\right]$$

$$= \frac{160 \times 10^6}{\pi} \times \frac{3\sqrt{3}}{2} = 132.2\,\text{MN/m}^2$$

$$\sigma_2 = \frac{160 \times 10^6}{\pi} \times -\frac{\sqrt{3}}{2} = -44.07\,\text{MN/m}^2$$

Therefore the maximum shear stress is

$$\tau = \frac{132\cdot2 - (-44\cdot07)}{2} = 88\cdot14\,\text{MN/m}^2$$

12.10. STRESSES ON A PLANE INCLINED TO THE PRINCIPAL PLANES

Let it be assumed that the principal stresses at a point are known, and that we wish to find the normal and shear stresses on some other plane as

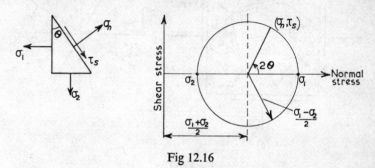

Fig 12.16

in Fig. 12.16. This is the same problem as in Section 12.4 and Fig. 12.3, where $\sigma_x = \sigma_1$, and $\sigma_y = \sigma_2$. Hence eqns. (12.4) and (12.6) become

$$\sigma_n = \frac{\sigma_1 + \sigma_2}{2} + \frac{\sigma_1 - \sigma_2}{2}\cos 2\theta \tag{12.25}$$

$$\tau_s = \frac{\sigma_1 - \sigma_2}{2}\sin 2\theta \tag{12.26}$$

Rewriting eqn. (12.25) as

$$\sigma_n - \frac{\sigma_1 + \sigma_2}{2} = \frac{\sigma_1 - \sigma_2}{2}\cos 2\theta \tag{12.27}$$

Squaring eqns. (12.26) and (12.27), and adding gives

$$\left(\sigma_n - \frac{\sigma_1 + \sigma_2}{2}\right)^2 + \tau_s^2 = \left(\frac{\sigma_1 - \sigma_2}{2}\right)^2 \tag{12.28}$$

which is the equation of a circle whose centre has co-ordinates $[\tfrac{1}{2}(\sigma_1 + \sigma_2), 0]$, and whose radius is $\tfrac{1}{2}(\sigma_1 - \sigma_2)$; in other words a Mohr's circle, Fig. 12.16. It will be seen that this is the same circle as in Fig. 12.8, but with the planes AC and BC coinciding with PC and QC, i.e. $2\phi = 0$.

12.11. GENERAL TWO-DIMENSIONAL STATE OF STRESS AT A POINT

In the preceding paragraphs the various stress conditions which exist in an element subjected to direct and shear stress have been derived, and these are summarized in Fig. 12.17.

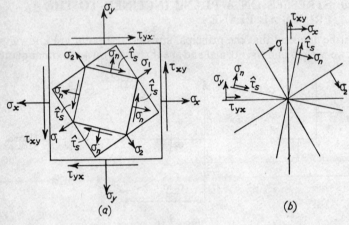

Fig 12.17

12.12. TWO-DIMENSIONAL STRAIN ANALYSIS

In the following analysis all strains are considered to be small in magnitude. *Direct strains* are defined as the ratio of change in length to original length in a particular direction, and a subscript notation similar to that for stresses will be adopted, ε_x and ε_y being the direct strains in the x- and y-directions respectively, positive for tension and negative for compression. *Shear strain* is defined as the change in angle between two planes initially at right angles, and the symbol and subscripts γ_{xy} will be used for shear referred to the x- and y-planes.

Plane strain is the term used to describe the strain system in which the normal strain in, say, the z-direction, along with the shear strains γ_{zx} and γ_{zy}, is zero. It should be noted that plane stress is not the stress system associated with plane strain. It will be evident from Section 12.16 that plane strain, i.e. $\varepsilon_z = 0$, is associated with a three-dimensional stress system and likewise plane stress is related to a three-dimensional strain system.

The stress system in Fig. 12.18(a) will give rise to a strain system combining direct and shear strains as shown in an exaggerated manner at (b). The object now is to determine the direct, ε_n, and shear, γ_s, strains for directions normal and tangential to a plane, inclined at θ to a co-ordinate direction, in terms of the direct, ε_x, ε_y, and shear, γ_{xy}, γ_{yx}, strains referred to the co-ordinate planes.

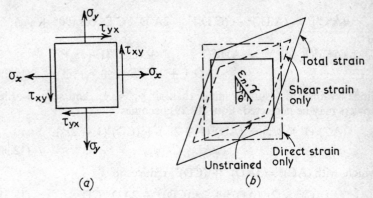

(a)   (b)

Fig 12.18

12.13. NORMAL STRAIN IN TERMS OF CO-ORDINATE STRAINS

Referring to Fig. 12.19 $(A'C' - AC)/AC$ gives the strain normal to

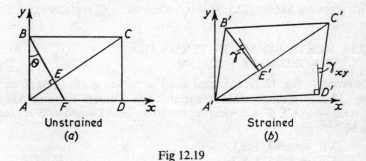

Unstrained   Strained
(a)   (b)

Fig 12.19

the plane FB related to the normal stress σ_n. Considering the triangles ACD and A'C'D' and with δx as the increase in length from AD to A'D', and δy the increase in length from CD to C'D', then

$$A'D' = AD + \delta x = AD \left(1 + \frac{\delta x}{AD}\right) = AD(1 + \varepsilon_x)$$

$$C'D' = CD + \delta y = CD \left(1 + \frac{\delta y}{CD}\right) = CD(1 + \varepsilon_y)$$

Similarly

$$A'C' = AC(1 + \varepsilon_n)$$

Now,

$$(A'C')^2 = (A'D')^2 + (C'D')^2 - 2A'D' \cdot C'D' \cos(90° + \gamma_{xy})$$

or

$$(AC)^2(1 + \varepsilon_n)^2 = (AD)^2(1 + \varepsilon_x)^2 + (CD)^2(1 + \varepsilon_y)^2$$
$$+ 2AD(1 + \varepsilon_x)CD(1 + \varepsilon_y)\sin\gamma_{xy} \quad (12.29)$$

Since strains are assumed small, then $\sin\gamma_{xy} \approx \gamma_{xy}$ and second-order powers may be neglected. Eqn. (12.29) becomes

$$(AC)^2(1 + 2\varepsilon_n) = (AD)^2(1 + 2\varepsilon_x) + (CD)^2(1 + 2\varepsilon_y)$$
$$+ 2AD \cdot CD\,\gamma_{xy} \quad (12.30)$$

which, with $(AC)^2 = (AD)^2 + (CD)^2$, reduces to

$$2\varepsilon_n(AC)^2 = 2\varepsilon_x(AD)^2 + 2\varepsilon_y(CD)^2 + 2AD \cdot CD\,\gamma_{xy} \quad (12.31)$$

Dividing through by $2(AC)^2$ and introducing $\sin\theta$ and $\cos\theta$,

$$\varepsilon_n = \varepsilon_x \cos^2\theta + \varepsilon_y \sin^2\theta + \gamma_{xy}\sin\theta\cos\theta \quad (12.32)$$

Therefore

$$\varepsilon_n = \frac{\varepsilon_x + \varepsilon_y}{2} + \frac{\varepsilon_x - \varepsilon_y}{2}\cos 2\theta + \tfrac{1}{2}\gamma_{xy}\sin 2\theta \quad (12.33)$$

This is similar to eqn. (12.13) for σ_n except that $\tfrac{1}{2}\gamma_{xy}$ replaces τ_{xy}.

12.14. SHEAR STRAIN IN TERMS OF CO-ORDINATE STRAINS

Referring to Fig. 12.19, the shear strain γ_s related to the shear stress τ_s, Fig. 12.5, is given by the change in angle between EB and AE, or $\angle AEB - \angle A'E'B'$. Considering the triangles AEB and A'E'B', then, as before

$$A'B' = AB(1 + \varepsilon_y)$$
$$A'E' = AE(1 + \varepsilon_n)$$
$$E'B' = EB(1 + \varepsilon_{n+90})$$

ε_{n+90} is the direct strain in a direction at 90° to ε_n. Now

$$(A'B')^2 = (A'E')^2 + (E'B')^2 - 2A'E' \cdot E'B' \cos(90° - \gamma_s) \quad (12.34)$$

or

$$(AB)^2(1 + \varepsilon_y)^2 = (AE)^2(1 + \varepsilon_n)^2 + (EB)^2(1 + \varepsilon_{n+90})^2$$
$$- 2AE(1 + \varepsilon_n)EB(1 + \varepsilon_{n+90})$$
$$\cos(90° - \gamma_s) \quad (12.35)$$

But $\cos (90° - \gamma_s) = \sin \gamma_s \approx \gamma_s$, and neglecting the second order of small quantities,

$$(AB)^2(1 + 2\varepsilon_y) = (AE)^2(1 + 2\varepsilon_n) + (EB)^2(1 + 2\varepsilon_{n+90})$$
$$-2AE \cdot EB \gamma_s \qquad (12.36)$$

But $(AB)^2 = (AE)^2 + (EB)^2$, and dividing by $2(AB)^2$ gives

$$\varepsilon_y = \varepsilon_n \sin^2 \theta + \varepsilon_{n+90} \cos^2 \theta - \gamma_s \sin \theta \cos \theta \qquad (12.37)$$

$$\tfrac{1}{2}\gamma_s \sin 2\theta = \frac{\varepsilon_{n+90} + \varepsilon_n}{2} + \frac{\varepsilon_{n+90} - \varepsilon_n}{2} \cos 2\theta - \varepsilon_y \qquad (12.38)$$

Now,

$$\varepsilon_{n+90} = \frac{\varepsilon_x + \varepsilon_y}{2} + \frac{\varepsilon_x - \varepsilon_y}{2} \cos 2\,(\theta + 90°) + \tfrac{1}{2}\gamma_{xy} \sin2\,(\theta + 90°)$$

$$= \frac{\varepsilon_x + \varepsilon_y}{2} - \frac{\varepsilon_x - \varepsilon_y}{2} \cos 2\theta - \tfrac{1}{2}\gamma_{xy} \sin 2\theta \qquad (12.39)$$

and

$$\varepsilon_n = \frac{\varepsilon_x + \varepsilon_y}{2} + \frac{\varepsilon_x - \varepsilon_y}{2} \cos 2\theta + \tfrac{1}{2}\gamma_{xy} \sin 2\theta \qquad (12.40)$$

Therefore

$$\frac{\varepsilon_{n+90} + \varepsilon_n}{2} = \frac{\varepsilon_x + \varepsilon_y}{2} \qquad (12.41)$$

and

$$\frac{\varepsilon_{n+90} - \varepsilon_n}{2} = - \frac{\varepsilon_x - \varepsilon_y}{2} \cos 2\theta - \tfrac{1}{2}\gamma_{xy} \sin 2\theta \qquad (12.42)$$

Substituting the above expressions in eqn. (12.38),

$$\tfrac{1}{2}\gamma_s \sin 2\theta = \frac{\varepsilon_x + \varepsilon_y}{2} - \frac{\varepsilon_x - \varepsilon_y}{2} \cos^2 2\theta - \tfrac{1}{2}\gamma_{xy} \sin 2\theta \cos 2\theta - \varepsilon_y$$

$$= \frac{\varepsilon_x - \varepsilon_y}{2} (1 - \cos^2 2\theta) - \tfrac{1}{2}\gamma_{xy} \sin 2\theta \cos 2\theta \qquad (12.43)$$

Dividing through by $\sin 2\theta$,

$$\tfrac{1}{2}\gamma_s = \frac{\varepsilon_x - \varepsilon_y}{2} \sin 2\theta - \tfrac{1}{2}\gamma_{xy} \cos 2\theta \qquad (12.44)$$

This is similar to eqn. (12.14) in terms of stresses but with $\tfrac{1}{2}\gamma_s$ and $\tfrac{1}{2}\gamma_{xy}$ in place of τ_s and τ_{xy}.

12.15. MOHR'S STRAIN CIRCLE

The graphical method of obtaining normal and shear stress on any plane by Mohr's circle can also be employed to determine normal and shear strains at a point. Considering eqns. (12.33) and (12.44) and rewriting, we have

$$\varepsilon_n - \frac{\varepsilon_x + \varepsilon_y}{2} = \frac{\varepsilon_x - \varepsilon_y}{2} \cos 2\theta + \tfrac{1}{2}\gamma_{xy} \sin 2\theta \qquad (12.45)$$

and

$$\tfrac{1}{2}\gamma_s = \frac{\varepsilon_x - \varepsilon_y}{2} \sin 2\theta - \tfrac{1}{2}\gamma_{xy} \cos 2\theta \qquad (12.46)$$

Squaring each and adding produces the expression

$$\left(\varepsilon_n - \frac{\varepsilon_x + \varepsilon_y}{2}\right)^2 + (\tfrac{1}{2}\gamma_s)^2 = \left(\frac{\varepsilon_x - \varepsilon_y}{2}\right)^2 + (\tfrac{1}{2}\gamma_{xy})^2 \qquad (12.47)$$

which is the equation of a circle of radius $\tfrac{1}{2}\sqrt{[(\varepsilon_x - \varepsilon_y)^2 + \gamma_{xy}{}^2]}$, and with centre at $[\tfrac{1}{2}(\varepsilon_x + \varepsilon_y), 0]$ relating ε_n and γ_s.

The Mohr's circle as shown in Fig. 12.20 is constructed in the same manner as for stresses. The correct position for plotting shear strain on the circle, i.e. above or below the ε-axis, may be found either by relating

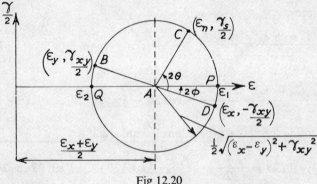

Fig 12.20

the deformation of the element to the corresponding shear stress system, or by the convention that a positive shear strain in the element corresponds to the sides of the deformed element having positive slope in relation to the co-ordinate axes. On co-ordinate axes of normal strain ε and semi-shear strain $\tfrac{1}{2}\gamma$, the points $(\varepsilon_x, -\tfrac{1}{2}\gamma_{xy})$ and $(\varepsilon_y, +\tfrac{1}{2}\gamma_{xy})$ are set up, and a circle is drawn with the line joining these two points as diameter. The normal and semi-shear strain $\varepsilon_n, \tfrac{1}{2}\gamma_s$ in a direction at θ to the x-direction is obtained from the intersection of a radius with the circle, at 2θ (anticlockwise) from AD.

12.16. PRINCIPAL STRAIN AND MAXIMUM SHEAR STRAIN

The maximum and minimum values of the normal strain at a point are given by P and Q in the diagram, whence

$$\varepsilon_1 = \frac{\varepsilon_x + \varepsilon_y}{2} + \tfrac{1}{2}\sqrt{[(\varepsilon_x - \varepsilon_y)^2 + \gamma_{xy}{}^2]} \quad \text{(maximum)} \qquad (12.48)$$

and

$$\varepsilon_2 = \frac{\varepsilon_x + \varepsilon_y}{2} - \tfrac{1}{2}\sqrt{[(\varepsilon_x - \varepsilon_y)^2 + \gamma_{xy}{}^2]} \quad \text{(minimum)} \qquad (12.49)$$

These are termed the *principal strains* and may be compared for similarity with the expressions for principal stresses. The former occur on mutually perpendicular planes making angles $2\theta = 2\phi$ and $180° + 2\phi$, or $\theta = \phi$ and $90° + \phi$ with the x-direction. From the diagram, when $2\theta = 2\phi$,

$$\theta = \tfrac{1}{2}\tan^{-1}\frac{\gamma_{xy}}{\varepsilon_x - \varepsilon_y} \quad \text{and} \quad 90° + \tfrac{1}{2}\tan^{-1}\frac{\gamma_{xy}}{\varepsilon_x - \varepsilon_y} \qquad (12.50)$$

Either by substituting for θ in eqn. (12.46) or by reference to the circle, it is found that the shear strain is zero for the planes of principal strain.

Since $\tau_{xy} = G\gamma_{xy}$ when the shear strain is zero, so also must be the shear stress. But it was previously shown that the shear stress was zero on the principal stress planes, and therefore the planes of principal stress and principal strain must coincide.

The maximum shear strain occurs at the top and bottom points of the circle when $\varepsilon_n = \tfrac{1}{2}(\varepsilon_x + \varepsilon_y)$ and $\tfrac{1}{2}\gamma_s = \tfrac{1}{2}\sqrt{[(\varepsilon_x - \varepsilon_y)^2 + \gamma_{xy}{}^2]}$; therefore

$$\hat{\gamma}_s = \sqrt{[(\varepsilon_x - \varepsilon_y)^2 + \gamma_{xy}{}^2]} \qquad (12.51)$$

Alternatively this may be written in terms of the principal strains as

$$\tfrac{1}{2}\hat{\gamma}_s = \frac{\varepsilon_1 - \varepsilon_2}{2}$$

or

$$\hat{\gamma}_s = \varepsilon_1 - \varepsilon_2 \qquad (12.52)$$

The Mohr's strain circle can equally well be drawn in terms of principal strains as was done in Fig. 12.16 for the principal stresses. The radius is $\tfrac{1}{2}(\varepsilon_1 - \varepsilon_2)$ and the centre has co-ordinates $[\tfrac{1}{2}(\varepsilon_1 + \varepsilon_2), 0]$. An example of the use of Mohr's strain circle is given at the end of Chapter 25 on resistance strain gauges.

Example 12.9. A marine propeller shaft is subjected to thrust and a clockwise torque, giving rise to a compressive strain of $0{\cdot}0003$ and a shear strain of $0{\cdot}0004$. Calculate the principal strains and the maximum shear strain at this point, and also construct the Mohr strain circle.

The torque is such that shear stress and shear strain will have the same sense as in the elements previously analysed.

$$\varepsilon_x = -0.0003 \qquad \varepsilon_y = 0 \qquad \text{and} \qquad \gamma_{yx} = 0.0004$$

The maximum and minimum principal strains are given as

$$\varepsilon_1, \varepsilon_2 = \frac{\varepsilon_x + \varepsilon_y}{2} \pm \tfrac{1}{2}\sqrt{[(\varepsilon_x - \varepsilon_y)^2 + \gamma_{xy}^2]}$$

$$= \frac{-0.0003}{2} \pm \tfrac{1}{2}\sqrt{[(-0.0003)^2 + (0.0004)^2]}$$

$$= -0.00015 \pm \tfrac{1}{2} \times 5 \times 10^{-4}$$

$$= -0.00015 \pm 0.00025$$

Therefore

$$\varepsilon_1 = +0.0001 \qquad \varepsilon_2 = -0.0004$$

$$\text{Maximum shear strain} = \varepsilon_1 - \varepsilon_2$$
$$= 0.0001 - (-0.0004) = 0.0005$$

The Mohr's circle solution is shown in Fig. 12.21.

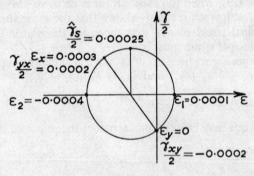

Fig 12.21

Example 12.10. Two strain gauges are fitted at $\pm 45°$ to the axis of a 75 mm diameter shaft. The shaft is rotating, and in addition to transmitting power, it is subjected to an unknown bending moment and a direct thrust. The readings of the gauges are recorded, and it is found that the maximum or minimum values for each gauge occur at 180° intervals of shaft rotation and are -0.0006 and $+0.0003$ for the two gauges at one instant and -0.0005 and $+0.0004$ for the same gauges 180° of rotation later. Determine the transmitted torque, the applied bending moments and the end thrust. Assume all the forces and moments are steady, i.e. do not vary during each rotation of the shaft. $E = 208 \text{ GN/m}^2$, $\nu = 0.29$, $G = 80 \text{ GN/m}^2$. The shaft and strain gauges are illustrated diagrammatically in Fig. 12.22.

The simplest starting point is to find the torque. This may be found from the shear stress which is in turn related to shear strain. From eqn. (12.44) we have

$$\gamma_{xy} = (\varepsilon_p - \varepsilon_q) \sin 2\theta - \gamma_{pq} \cos 2\theta$$
$$= (0.000\,3 + 0.000\,6) \sin 90 - \gamma_{pq} \cos 90°$$
$$= 0.000\,9$$
$$\tau_{xy} = 0.000\,9 \times 80\,\text{GN/m}^2 = 72\,\text{MN/m}^2$$

Hence the torque is

$$T = \frac{\pi}{16} \times 0.075^{3\bullet} \times 72 \times 10^6 = 5.97\,\text{kN-m}$$

Fig 12.22

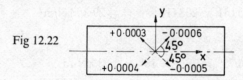

In order to find the bending moment and end thrust it is necessary to know the strains in the x-direction for the two specific rotational positions of the shaft. From eqn. (12.33) we have

Top $\qquad -0.000\,6 = \dfrac{\varepsilon_x{}^T + \varepsilon_y{}^T}{2} - 0.000\,45$

or

$$+0.000\,3 = \frac{\varepsilon_x{}^T + \varepsilon_y{}^T}{2} + 0.000\,45$$

Hence

$$\varepsilon_x{}^T + \varepsilon_y{}^T = -0.000\,3$$

Bottom $\qquad -0.000\,5 = \dfrac{\varepsilon_x{}^B + \varepsilon_y{}^B}{2} - 0.000\,45$

or

$$+0.000\,4 = \frac{\varepsilon_x{}^B + \varepsilon_y{}^B}{2} + 0.000\,45$$

Hence

$$\varepsilon_x{}^B + \varepsilon_y{}^B = -0.000\,1$$

To eliminate ε_y from the above equations we use the relationship in Sections 3.6:

$$\varepsilon_y = -\nu\varepsilon_x$$

Thus

$$\varepsilon_x^T = -\frac{0\cdot000\,3}{1-\nu} = -0\cdot000\,423$$

$$\varepsilon_x^B = -\frac{0\cdot000\,1}{1-\nu} = -0\cdot000\,141$$

These strains are the sum of the bending strain which reverses in sign for 180° rotation and the steady compressive strain (due to end thrust).

Bending strain $= \pm\frac{1}{2}(\varepsilon_x^T - \varepsilon_x^B) = \pm0\cdot000\,141$
Compressive strain $= -\frac{1}{2}(\varepsilon_x^T + \varepsilon_x^B) = -0\cdot000\,282$

Hence

$$\sigma_{xb} = \pm0\cdot000\,141 \times 208\,\text{GN/m}^2 = \pm29\cdot3\,\text{MN/m}^2$$

from which

$$M = \frac{\pi}{32} \times 0\cdot075^3 \times 29\cdot3 \times 10^6 = 1\cdot21\,\text{kN-m}$$

$$\sigma_{xc} = -0\cdot000\,282 \times 208\,\text{GN/m}^2 = -58\cdot6\,\text{MN/m}^2$$

and

$$P = \frac{\pi}{4} \times 0\cdot075^2 \times (-58\cdot6) \times 10^6 = -259\,\text{kN}$$

12.17. RELATIONSHIPS BETWEEN STRESS AND STRAIN

At the beginning of this chapter the stresses set up in a rectangular bar subjected to tensile forces at each end were analysed, Fig. 12.2. It is apparent that under the action of these forces an extension and hence a tensile strain will take place in the x-direction, that is ε_x. This deformation will also be accompanied by a narrowing of the cross-section, and this we term *lateral* or *transverse strain*. Note that this is occurring without any external forces or stresses in the y- and z- directions. (This concept was first introduced in Section 3.6.) This lateral strain is proportional to the direct strain and the constant of proportionality is known as *Poisson's ratio*, ν. Hence

Lateral strain $= -\nu \times$ Longitudinal strain (due to stress)

For most metals, ν is in the range 0·28 to 0·32. The principle stated above would apply similarly if the bar had been in compression, where a lateral expansion would have occurred.

If the case is now extended to a two-dimensional rectangular element under tensile direct stresses in two perpendicular directions x and y, the strain in the x-direction is composed of the direct strain caused by the stress σ_x and the lateral contraction in the x-direction due to the stress σ_y, Fig. 12.23.

The strain due to σ_x will be σ_x/E, and the lateral strain due to σ_y will

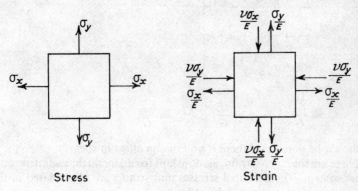

Stress   Strain

Fig 12.23

be $-v\sigma_y/E$, negative because it is a contraction. Hence the total strain in the x-direction will be

$$\varepsilon_x = \frac{\sigma_x}{E} - \frac{v\sigma_y}{E} \qquad (12.53)$$

If the element is three-dimensional with a tensile stress σ_z in the z-direction, then there will be an additional lateral contraction in the x-direction of an amount $-v\sigma_z/E$. The total strain will then become

$$\varepsilon_x = \frac{\sigma_x}{E} - \frac{v\sigma_y}{E} - \frac{v\sigma_z}{E} \qquad (12.54)$$

or

$$\varepsilon_x = \frac{\sigma_x}{E} - \frac{v}{E}(\sigma_y + \sigma_z) \qquad (12.55)$$

This then is the form of the general relationship between stress and strain in a two- and three-dimensional stress system. It has been derived for a positive system of stresses, and hence signs must be changed if negative (compressive) stresses are introduced.

Eqn. (12.55) has been obtained for the x-direction, but by rotation of the subscripts, similar expressions are obtained for the y- and z-directions, so that

$$
\left.
\begin{aligned}
\varepsilon_x &= \frac{\sigma_x}{E} - \frac{v}{E}(\sigma_y + \sigma_z) \\[2mm]
\varepsilon_y &= \frac{\sigma_y}{E} - \frac{v}{E}(\sigma_z + \sigma_x) \\[2mm]
\varepsilon_z &= \frac{\sigma_z}{E} - \frac{v}{E}(\sigma_x + \sigma_y)
\end{aligned}
\right\}
\tag{12.56}
$$

$$
\gamma_{xy} = \frac{\tau_{xy}}{G} \qquad \gamma_{yz} = \frac{\tau_{yz}}{G} \qquad \gamma_{zx} = \frac{\tau_{zx}}{G}
$$

It should be noted that there is no Poisson effect in shear.

These are the stress–strain relationships for a general three-dimensional stress system. The principal stresses and strains are also related in the same manner as in eqn. (12.56); thus

$$
\left.
\begin{aligned}
\varepsilon_1 &= \frac{\sigma_1}{E} - \frac{v}{E}(\sigma_2 + \sigma_3) \\[2mm]
\varepsilon_2 &= \frac{\sigma_2}{E} - \frac{v}{E}(\sigma_3 + \sigma_1) \\[2mm]
\varepsilon_3 &= \frac{\sigma_3}{E} - \frac{v}{E}(\sigma_1 + \sigma_2)
\end{aligned}
\right\}
\tag{12.57}
$$

In the two-dimensional system, σ_3 in the above equations is made zero.

Example 12.11. A block of steel is subjected to stresses of $-60\,\mathrm{MN/m^2}$ and $20\,\mathrm{MN/m^2}$ in the x- and y-directions, and strain is prevented in the z-direction. Determine the strain in the former directions and the stress in the latter. $E = 208\,\mathrm{GN/m^2}$; $v = 0\cdot3$.

The conditions are that $\sigma_x = -60$, $\sigma_y = +20$ and $\varepsilon_z = 0$. Using the relationship

$$
\varepsilon_z = \frac{\sigma_z}{E} - \frac{v}{E}(\sigma_x + \sigma_y)
$$

then

$$
0 = \frac{\sigma_z}{E} - \frac{0\cdot3}{E}(-60 + 20)
$$

and

$$
\sigma_z = -12\,\mathrm{MN/m^2}
$$

Also

$$\varepsilon_x = \frac{\sigma_x}{E} - \frac{\nu}{E}(\sigma_y + \sigma_z)$$

Therefore

$$\varepsilon_x = \frac{-60 \times 10^6}{208 \times 10^9} - \frac{0 \cdot 3}{208 \times 10^9}(20 - 12)10^6$$

$$= \frac{-60 - 2 \cdot 4}{208 \times 10^3} = -0 \cdot 000\,3$$

and

$$\varepsilon_y = \frac{\sigma_y}{E} - \frac{\nu}{E}(\sigma_z + \sigma_x)$$

Therefore

$$\varepsilon_y = \frac{20 \times 10^6}{208 \times 10^9} - \frac{0 \cdot 3}{208 \times 10^9}(-12 - 60) \times 10^6$$

$$= \frac{20 + 21 \cdot 6}{208 \times 10^3} = +0 \cdot 000\,2$$

12.18. RELATIONSHIPS BETWEEN THE ELASTIC CONSTANTS

During the analysis in this and preceding chapters, four constants of elasticity have been defined relating various conditions of stress and strain. These are Young's modulus, E, shear modulus, G, bulk modulus, K, and Poisson's ratio, ν. It will now be shown that these constants are not independent of one another.

If an element of material is subjected to hydrostatic tension or compression, then the ratio of the hydrostatic stress, σ, to the resulting volumetric strain, ε_v, is constant in the elastic range, and is termed the *bulk modulus, K*.

It was shown in eqn. (3.1) that volumetric strain is given by the sum of the three linear strains along the axes of the element. Hence

$$\varepsilon_v = \varepsilon_x + \varepsilon_y + \varepsilon_z$$

But

$$\varepsilon_x = \frac{\sigma_x}{E} - \frac{\nu}{E}(\sigma_y + \sigma_z)$$

Also for hydrostatic stress, $\sigma_x = \sigma_y = \sigma_z = \sigma$; therefore

$$\varepsilon_x = \frac{\sigma}{E}(1 - 2\nu)$$

Similarly

$$\varepsilon_y = \frac{\sigma}{E}(1 - 2\nu) \qquad (12.58)$$

and

$$\varepsilon_z = \frac{\sigma}{E}(1 - 2\nu)$$

Therefore, summing the above three strains, we have

$$\varepsilon_v = \frac{3\sigma}{E}(1 - 2\nu) \qquad (12.59)$$

or

$$E = 3\frac{\sigma}{\varepsilon_v}(1 - 2\nu)$$

But $\sigma/\varepsilon_v = K$, the bulk modulus; therefore

$$E = 3K(1 - 2\nu) \qquad (12.60)$$

The square element of unit thickness shown in Fig. 12.24 (a) is acted on by pure shearing stresses. This system is equivalent to the system of direct stresses on the element of Fig. 12.24(b), and from equilibrium $\sigma_{n_1} = \sigma_{n_2} = \tau_s$. The strain along the diagonal AB in terms of the stresses is given by

$$\varepsilon_{n_1} = \frac{\sigma_{n_1}}{E} - \frac{\nu}{E}(-\sigma_{n_2}) \qquad (12.61)$$

that is, the extension due to σ_{n_1} plus the lateral expansion in the direction AB due to the compression σ_{n_2}. But

$$\sigma_{n_1} = \sigma_{n_2} = \tau_s$$

Therefore

$$\varepsilon_{n_1} = \frac{\tau_s}{E}(1 + \nu) \qquad (12.62)$$

Since for pure shear ε_x and ε_y are zero, then from eqn. (12.33),

$$\varepsilon_{n_1} = \tfrac{1}{2}\gamma_s \sin 2\theta \qquad (12.63)$$

In this case $\theta = 45°$; therefore

$$\varepsilon_{n_1} = \tfrac{1}{2}\gamma_s \qquad (12.64)$$

Equating eqns. (12.64) and (12.62),

$$\tfrac{1}{2}\gamma_s = \frac{\tau_s}{E}(1 + \nu) \qquad (12.65)$$

or

$$E = \frac{2\tau_s}{\gamma_s}(1 + \nu) \tag{12.66}$$

But $\tau_s/\gamma_s = G$, the shear modulus; therefore

$$E = 2G(1 + \nu) \tag{12.76}$$

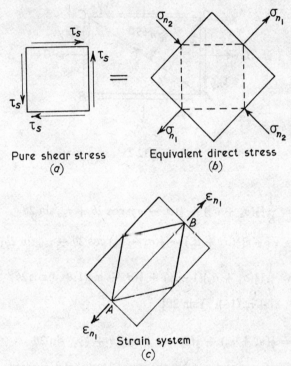

Pure shear stress
(a)

Equivalent direct stress
(b)

Strain system
(c)

Fig 12.24

The above solution may seem rather particularized by taking a square element, but it does enable a clearer physical picture to be obtained of the stresses and deformations. A more rigorous approach considering a general rectangular element, as in Fig. 12.25, now follows.

The normal stress in the direction AC is given by

$$\sigma_n = \tfrac{1}{2}(\sigma_x + \sigma_y) + \tfrac{1}{2}(\sigma_x - \sigma_y)\cos 2\theta + \tau_{xy}\sin 2\theta \tag{12.68}$$

and in a perpendicular direction,

$$\sigma_{n+90} = \tfrac{1}{2}(\sigma_x + \sigma_y) - \tfrac{1}{2}(\sigma_x - \sigma_y)\cos 2\theta - \tau_{xy}\sin 2\theta \tag{12.69}$$

But the strain along AC is

$$\varepsilon_n = \frac{\sigma_n}{E} - \frac{\nu\sigma_{n+90}}{E} \tag{12.70}$$

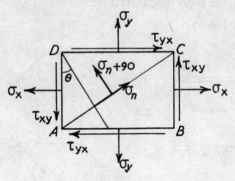

Fig 12.25

Therefore

$$\varepsilon_n = \frac{1}{E}\left[\tfrac{1}{2}(\sigma_x + \sigma_y) + \tfrac{1}{2}(\sigma_x - \sigma_y)\cos 2\theta + \tau_{xy}\sin 2\theta\right.$$

$$\left. -\nu\{\tfrac{1}{2}(\sigma_x + \sigma_y) - \tfrac{1}{2}(\sigma_x - \sigma_y)\cos 2\theta - \tau_{xy}\sin 2\theta\}\right] \tag{12.71}$$

$$\varepsilon_n = \frac{1}{E}\left[\tfrac{1}{2}(\sigma_x + \sigma_y)(1 - \nu) + \tfrac{1}{2}(\sigma_x - \sigma_y)(1 + \nu)\cos 2\theta\right.$$

$$\left. +\tau_{xy}(1 + \nu)\sin 2\theta\right] \tag{12.72}$$

Also

$$\varepsilon_n = \tfrac{1}{2}(\varepsilon_x + \varepsilon_y) + \tfrac{1}{2}(\varepsilon_x - \varepsilon_y)\cos 2\theta + \tfrac{1}{2}\gamma_{xy}\sin 2\theta \tag{12.73}$$

Substituting for ε_x and ε_y from the stress–strain relationships,

$$\varepsilon_n = \tfrac{1}{2}\left[\frac{1}{E}(\sigma_x + \sigma_y)(1 - \nu)\right] + \tfrac{1}{2}\left[\frac{1}{E}(\sigma_x - \sigma_y)(1 + \nu)\right]\cos 2\theta$$

$$+\tfrac{1}{2}\gamma_{xy}\sin 2\theta \tag{12.74}$$

Subtracting eqn. (12.74) from eqn. (12.72),

$$0 = \frac{\tau_{xy}}{E}(1 + \nu)\sin 2\theta - \tfrac{1}{2}\gamma_{xy}\sin 2\theta \tag{12.75}$$

Therefore

$$\frac{\tau_{xy}}{\gamma_{xy}} = \frac{E}{2(1 + \nu)} = G \tag{12.76}$$

or

$$E = 2G(1 + v) \qquad (12.77)$$

It follows from eqns. (12.60) and (12.77) that

$$3K(1 - 2v) = 2G(1 + v)$$

or

$$K = \frac{2G(1 + v)}{3(1 - 2v)} \qquad (12.78)$$

Thus if any two of the four constants are known, or can be measured, then the other two can be determined.

Chapter 13
Analysis of Variation of Stress and Strain

13.1. In the previous chapter the conditions of stress and strain at a point in a material were considered. It is now necessary to take the analysis a stage further by examining the variation of stress between adjacent points and deriving suitable expressions for this variation. As in the earlier work, relationships for stresses may be found by considering the equilibrium of a small element of material. The solution of these equations of equilibrium must satisfy the boundary conditions of the problem as defined by the applied forces. However, it is not possible to obtain the individual components of stress directly from the above equations, owing to the statically indeterminate nature of the problem. It is necessary, therefore, to consider the elastic deformations of the material such that, in a continuous strain field, the displacements are compatible with the stress distribution. These relationships are termed the *equations of compatibility*. From this point it is only required to have a relationship between stress and strain, e.g. Hooke's law, to obtain a complete solution of the stress components in a body.

The equations of equilibrium and compatibility are quite general and may be derived in terms of various co-ordinate systems. The mathematical solution of a problem may often be simplified if an appropriate set of co-ordinates is chosen. With this in mind, and suitable illustrative applications in the following pages, the various equations will be derived in two-dimensional Cartesian and cylindrical co-ordinates.*

13.2. EQUILIBRIUM EQUATIONS: PLANE STRESS CARTESIAN CO-ORDINATES

Consider the equilibrium of a small rectangular element of dimensions δx, δy, δz, Fig. 13.1. Owing to the variation of stress through the material, $\sigma_{x_{AB}}$ is a little different from $\sigma_{x_{CD}}$, and likewise for the other stresses σ_y and τ_{xy}. The variation that must occur over any particular face may be neglected, as it cancels out when the force equilibrium on opposite pairs of faces is considered. On this occasion body forces arising from gravity, inertia, etc., must be taken into account, and these are shown as X and Y per unit volume.

---

* For a detailed study of equilibrium, strain-displacement and compatibility relationships in various three-dimensional co-ordinate systems, see H. Ford, *Advanced Mechanics of Materials* (Longmans, 1963).

For equilibrium in the x-direction,

$$(\sigma_{x_{CD}} - \sigma_{x_{AB}})\, \delta_y\, \delta_z + (\tau_{yx_{BC}} - \tau_{yx_{AD}})\, \delta_x\, \delta_z + X\, \delta x\, \delta y\, \delta z = 0$$

$$(13.1)$$

Dividing by $\delta x\, \delta y\, \delta z$ gives

$$\frac{\sigma_{x_{CD}} - \sigma_{x_{AB}}}{\delta x} + \frac{\tau_{yx_{BC}} - \tau_{yx_{AD}}}{\delta y} + X = 0$$

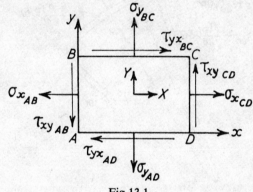

Fig 13.1

In the limit, as $\delta x \to 0$ and $\delta y \to 0$ and the element becomes smaller and smaller, the terms become partial differentials with respect to x and y, and thus

$$\frac{\partial \sigma_x}{\partial x} + \frac{\partial \tau_{yx}}{\partial y} + X = 0$$

$$(13.2)$$

Considering the y-direction,

$$(\sigma_{y_{BC}} - \sigma_{y_{AD}})\, \delta x\, \delta z + (\tau_{xy_{CD}} - \tau_{xy_{AB}})\, \delta y\, \delta z + Y\, \delta x\, \delta y\, \delta z = 0$$

$$(13.3)$$

Dividing by $\delta x\, \delta y\, \delta z$ and limiting leads to

$$\frac{\partial \sigma_y}{\partial y} + \frac{\partial \tau_{xy}}{\partial x} + Y = 0$$

$$(13.4)$$

It is often the case that the only body force is the weight of the component and that it can be neglected in comparison with the applied forces. Then

$$\frac{\partial \sigma_x}{\partial x} + \frac{\partial \tau_{yx}}{\partial y} = 0$$

$$(13.5)$$

$$\frac{\partial \sigma_y}{\partial y} + \frac{\partial \tau_{xy}}{\partial x} = 0$$

$$(13.6)$$

13.3. EQUILIBRIUM EQUATIONS: PLANE STRESS CYLINDRICAL CO-ORDINATES

As has been previously stated, there are certain cases such as cylinders, discs, curved bars, etc., in which it is rather more convenient to use r, θ, z co-ordinates. Consider the element ABCD, Fig. 13.2, which is bounded

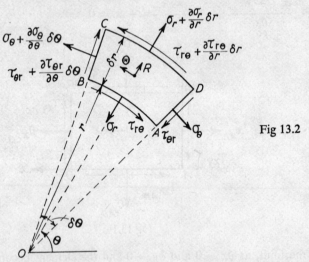

Fig 13.2

by radial lines OC and OD, subtending an angle $\delta\theta$ at the origin, and circular arcs AB and CD at radii r and $r + \delta r$ respectively. The element is of thickness δz.

In the preceding section and Fig. 13.1, the stress variation on the element was represented symbolically by the different letter subscripts in order that the first analysis could be written down simply. However, it is quite usual to show stress variation in mathematical symbols, and this has been adopted in Fig. 13.2 for this case. The body forces are shown as R radially and Θ tangentially per unit volume.

Considering equilibrium along the radial centre-line of the element, there will be, in addition to forces from the radial stresses, the resolved components of force from the hoop and shear stresses. Hence

$$\left(\sigma_r + \frac{\partial \sigma_r}{\partial r}\, \delta r\right)(r + \delta r)\,\delta\theta\,\delta z - \sigma_r r\,\delta\theta\,\delta z$$

$$- \left(\sigma_\theta + \frac{\partial \sigma_\theta}{\partial \theta}\, \delta\theta\right)\delta r\,\delta z \sin\frac{\delta\theta}{2} - \sigma_\theta\,\delta r\,\delta z \sin\frac{\delta\theta}{2}$$

$$+ \left(\tau_{\theta r} + \frac{\partial \tau_{\theta r}}{\partial \theta}\, \delta\theta\right)\delta r\,\delta z \cos\frac{\delta\theta}{2} - \tau_{\theta r}\,\delta r\,\delta z \cos\frac{\delta\theta}{2} + Rr\,\delta\theta\,\delta r\,\delta z$$

$$= 0 \tag{13.7}$$

As $\delta\theta \to 0$, $\sin \frac{1}{2}\delta\theta \to \frac{1}{2}\delta\theta$ and $\cos \frac{1}{2}\delta\theta \to 1$. Also, neglecting second- and higher-order terms and dividing by $r\,\delta r\,\delta\theta\,\delta z$, the equation reduces to

$$\frac{\sigma_r}{r} + \frac{\partial\sigma_r}{\partial r} - \frac{\sigma_\theta}{r} + \frac{1}{r}\frac{\partial\tau_{\theta r}}{\partial\theta} + R = 0$$

or

$$\frac{\partial\sigma_r}{\partial r} + \frac{1}{r}\frac{\partial\tau_{\theta r}}{\partial\theta} + \frac{\sigma_r - \sigma_\theta}{r} + R = 0 \tag{13.8}$$

Resolving in the tangential direction,

$$\left(\sigma_\theta + \frac{\partial\sigma_\theta}{\partial\theta}\,\delta\theta\right)\delta r\,\delta z\cos\frac{\delta\theta}{2} - \sigma_\theta\,\delta r\,\delta z\cos\frac{\delta\theta}{2}$$

$$+ \left(\tau_{\theta r} + \frac{\partial\tau_{\theta r}}{\partial\theta}\,\delta\theta\right)\delta r\,\delta z\sin\frac{\delta\theta}{2} + \tau_{\theta r}\,\delta r\,\delta z\sin\frac{\delta\theta}{2}$$

$$+ \left(\tau_{r\theta} + \frac{\partial\tau_{r\theta}}{\partial r}\,\delta r\right)(r + \delta r)\,\delta\theta\,\delta z - \tau_{r\theta}r\,\delta\theta\,\delta z$$

$$+ \Theta r\,\delta r\,\delta\theta\,\delta z = 0 \tag{13.9}$$

In the limit, as $\delta\theta \to 0$, neglecting the appropriate terms and dividing by $r\,\delta r\,\delta\theta\,\delta z$,

$$\frac{1}{r}\frac{\partial\sigma_\theta}{\partial\theta} + \frac{\partial\tau_{r\theta}}{\partial r} + 2\frac{\tau_{\theta r}}{r} + \Theta = 0 \tag{13.10}$$

Axial Symmetry

In certain cases, such as a ring, disc or cylinder, the body is symmetrical about a central axis, z, through O, Fig. 13.2. Then by symmetry the stress components depend on r only, and σ_θ at any particular radius is constant. Also, from the consideration of symmetry, the shear stress components, $\tau_{\theta r}$ must vanish. Eqn. (13.10) no longer exists and the equilibrium equation (13.8) becomes

$$\frac{d\sigma_r}{dr} + \frac{\sigma_r - \sigma_\theta}{r} + R = 0 \tag{13.11}$$

If the body force can be neglected, then

$$\frac{d\sigma_r}{dr} + \frac{\sigma_r - \sigma_\theta}{r} = 0 \tag{13.12}$$

13.4. STRAIN IN TERMS OF DISPLACEMENT: TWO-DIMENSIONAL CARTESIAN CO-ORDINATES

With reference to a set of fixed axes the movement of an elastic body consists of displacement and rotation of the body combined with strain in the material. Consider a continuous strain field and the displacement of elements OA, of length δx, and OB, of length δy, referred to the axes Ox and Oy, Fig. 13.3. The point O moves to O' having co-ordinates u and v, which are in general functions of x and y. The rate of change of u in the x-direction with respect to x will be $\partial u/\partial x$. Therefore, since OA is of length δx, the point A will move to A', where the displacement in the x-direction will be $u + (\partial u/\partial x)\delta x$. The change in length along this axis is thus $(\partial u/\partial x)\delta x$ and the strain is

$$\varepsilon_x = \frac{\dfrac{\partial u}{\partial x}\delta x}{\delta x}$$

$$= \frac{\partial u}{\partial x} \tag{13.13}$$

Strain in the y-direction is obtained from a consideration of the displacement of OB to O'B'. The rate of change of v in the y-direction with respect to y will be $\partial v/\partial y$, and therefore the point B'-will have been displaced from B in the y-direction an amount $v + (\partial v/\partial y)\delta y$. The strain occurring will thus be

$$\varepsilon_y = \frac{\dfrac{\partial v}{\partial y}\delta y}{\delta y}$$

$$= \frac{\partial v}{\partial y} \tag{13.14}$$

The shearing strain in the element AOB will be given by the change from the original right angle to the new angle A'O'B'. Hence

$$\gamma_{xy} = \angle \text{CO'A'} + \angle \text{DO'B'}$$

For small displacements,

$$\angle \text{CO'A'} \approx \frac{\text{CA'}}{\text{O'C}} \quad \text{and} \quad \text{DO'B'} \approx \frac{\text{DB'}}{\text{O'D}}$$

Now, CA' is the rate of change of v in the x-direction for an amount δx or $(\partial v/\partial x)\delta x$. Similarly DB' is the rate of change of u in the y-direction for a length δy giving $(\partial u/\partial y)\delta y$. Therefore

$$\angle \text{CO'A'} \approx \frac{\dfrac{\partial v}{\partial x}\delta x}{\delta x} \approx \frac{\partial v}{\partial x}$$

and

$$\angle \text{DO}'\text{B}' \approx \frac{\frac{\partial u}{\partial y}\delta y}{\delta y} \approx \frac{\partial u}{\partial y}$$

Thus, for very small displacements,

$$\gamma_{xy} = \frac{\partial v}{\partial x} + \frac{\partial u}{\partial y} \tag{13.15}$$

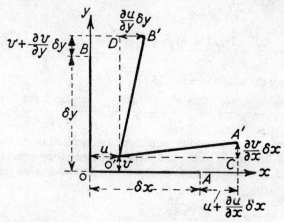

Fig 13.3

In a two-dimensional strain field, the strains in terms of displacements are therefore

$$\varepsilon_x = \frac{\partial u}{\partial x} \qquad \varepsilon_y = \frac{\partial v}{\partial y} \qquad \gamma_{xy} = \frac{\partial v}{\partial x} + \frac{\partial u}{\partial y} \tag{13.16}$$

13.5. COMPATIBILITY EQUATIONS

Because the above three strain components are in terms of the functions u and v, there must be a relationship between them. This may be obtained by differentiating ε_x twice with respect to y, ε_y twice with respect to x, and γ_{xy} with respect to both x and y. Hence

$$\frac{\partial^2 \varepsilon_x}{\partial y^2} = \frac{\partial^3 u}{\partial x\,\partial y^2} \qquad \frac{\partial^2 \varepsilon_y}{\partial x^2} = \frac{\partial^3 v}{\partial y\,\partial x^2} \qquad \frac{\partial^2 \gamma_{xy}}{\partial x\,\partial y} = \frac{\partial^2}{\partial x\,\partial y}\left(\frac{\partial v}{\partial x} + \frac{\partial u}{\partial y}\right)$$

Eliminating u and v between these three equations provides the relationship

$$\frac{\partial^2 \varepsilon_x}{\partial y^2} + \frac{\partial^2 \varepsilon_y}{\partial x^2} = \frac{\partial^2 \gamma_{xy}}{\partial x\,\partial y} \tag{13.17}$$

which is known as a *compatibility equation* in terms of strains. To obtain a compatibility relationship in terms of stresses, it is only necessary to substitute for the strains in eqn. (13.17), using the stress–strain relationships given in Section 12.16.

For the case of plane stress, $\sigma_z = 0$, the stress–strain relationships are

$$\varepsilon_x = \frac{\sigma_x}{E} - \frac{\nu\sigma_y}{E}$$

$$\varepsilon_y = \frac{\sigma_y}{E} - \frac{\nu\sigma_x}{E}$$

$$\gamma_{xy} = \frac{\tau_{xy}}{G} = \frac{2\tau_{xy}(1 + \nu)}{E}$$

Substituting the above in eqn. (13.17),

$$\frac{1}{E}\frac{\partial^2\sigma_x}{\partial y^2} - \frac{\nu}{E}\frac{\partial^2\sigma_y}{\partial y^2} + \frac{1}{E}\frac{\partial^2\sigma_y}{\partial x^2} - \frac{\nu}{E}\frac{\partial^2\sigma_x}{\partial x^2} = \frac{2(1 + \nu)}{E}\frac{\partial^2\tau_{xy}}{\partial x\,\partial y} \qquad (13.18)$$

Considering the equilibrium equations and neglecting the body force, if it is only the weight of the body, then differentiating eqn. (13.5) with respect to x and eqn. (13.6) with respect to y and adding we get

$$\frac{\partial^2\sigma_x}{\partial x^2} + \frac{\partial^2\sigma_y}{\partial y^2} = -2\frac{\partial^2\tau_{xy}}{\partial x\,\partial y} \qquad (13.19)$$

Eliminating τ_{xy} between eqns. (13.18) and (13.19),

$$\frac{1}{E}\frac{\partial^2\sigma_x}{\partial y^2} - \frac{\nu}{E}\frac{\partial^2\sigma_y}{\partial y^2} + \frac{1}{E}\frac{\partial^2\sigma_y}{\partial x^2} - \frac{\nu}{E}\frac{\partial^2\sigma_x}{\partial x^2} = -\frac{1 + \nu}{E}\frac{\partial^2\sigma_x}{\partial x^2} + \frac{\partial^2\sigma_y}{\partial y^2}$$

Simplifying,

$$\frac{\partial^2\sigma_x}{\partial x^2} + \frac{\partial^2\sigma_y}{\partial x^2} + \frac{\partial^2\sigma_x}{\partial y^2} + \frac{\partial^2\sigma_y}{\partial y^2} = 0$$

or

$$\left(\frac{\partial^2}{\partial x^2} + \frac{\partial^2}{\partial y^2}\right)(\sigma_x + \sigma_y) = 0 \qquad (13.20)$$

An analysis similar to that above can be used to show that the compatibility equation (13.20) also applies to the case of plane strain.

13.6. STRAIN IN TERMS OF DISPLACEMENT: TWO-DIMENSIONAL CYLINDRICAL CO-ORDINATES

When dealing with the analysis of elements having a circular geometry (curved bars, discs, etc.) it is often more convenient to consider strain and displacements in terms of cylindrical co-ordinates as was done for the equilibrium equations. Consider the element ABCD, Fig. 13.4, sub-tending an angle $\delta\theta$, with AB at radius r and CD at $r + \delta r$. This element is displaced at A'B'C'D' so that the radial and tangential movements to

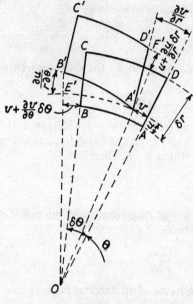

Fig 13.4

A' are u and v respectively. The displacement of the point D to D' in the r-direction will be $u + (\partial u/\partial r)\delta r$, where $\partial u/\partial r$ is the rate of change of u with respect to r. The change in length of AD is therefore $(\partial u/\partial r)\delta r$, and hence the strain in the radial direction is

$$\varepsilon_r = \frac{\dfrac{\partial u}{\partial r}\,\delta r}{\delta r}$$

$$= \frac{\partial u}{\partial r} \tag{13.21}$$

In the tangential direction there are two effects of displacement on strain. As B moves to B' there is the change of v with respect to θ, giving $(\partial v/\partial\theta)\delta\theta$ as the increase in length and hence a strain of

$$\frac{(\partial v/\partial\theta)\delta\theta}{r\,\delta\theta} \qquad \text{or} \qquad \frac{\partial v}{r\,\partial\theta}$$

There is also the tangential strain due to the element moving out to the new radius, $r + u$. This part is

$$\frac{(r + u)\delta\theta - r\,\delta\theta}{r\,\delta\theta} = \frac{u}{r}$$

Therefore, the total tangential strain is

$$\varepsilon_\theta = \frac{\partial v}{r\,\partial\theta} + \frac{u}{r} \tag{13.22}$$

The shear strain, $\gamma_{r\theta}$, is given by the difference between $\angle\mathrm{DAB}$ and $\angle\mathrm{D'A'B'}$, which is

$$\gamma_{r\theta} = \angle\mathrm{D'A'F'} + \angle\mathrm{B'A'E'}$$

Now, $\angle\mathrm{D'A'F'}$ is the difference between the tangential displacement of D to D', $(\partial v/\partial r)\delta r$, which as an angle is $\partial v/\partial r$, and the rigid body rotation about O, $\angle\mathrm{DOF'}$, which is v/r. Therefore

$$\angle\mathrm{D'A'F'} = \frac{\partial v}{\partial r} - \frac{v}{r}$$

$\angle\mathrm{B'A'E'}$ is in respect of displacement in the radial direction u and is therefore $\partial u/r\,\partial\theta$. Hence

$$\gamma_{r\theta} = \frac{\partial v}{\partial r} - \frac{v}{r} + \frac{\partial u}{r\,\partial\theta} \tag{13.23}$$

The three strains in terms of displacements are

$$\varepsilon_r = \frac{\partial u}{\partial r} \tag{13.24a}$$

$$\varepsilon_\theta = \frac{\partial v}{r\,\partial\theta} + \frac{u}{r} \tag{13.24b}$$

$$\gamma_{r\theta} = \frac{\partial v}{\partial r} + \frac{\partial u}{r\,\partial\theta} - \frac{v}{r} \tag{13.24c}$$

Axial Symmetry

For problems which are symmetrical about a z-axis through O, there will be no tangential displacement, v, and no shear strain, $\gamma_{r\theta}$, since $\tau_{r\theta}$ is zero. The above equations then reduce to

$$\varepsilon_r = \frac{\partial u}{\partial r} = \frac{du}{dr} \tag{13.25}$$

since σ_θ is constant at any one value of r; and

$$\varepsilon_\theta = \frac{u}{r} \tag{13.26}$$

13.7. COMPATIBILITY EQUATIONS: CYLINDRICAL CO-ORDINATES

As in the case of Cartesian co-ordinates there must be a relationship between the strains in cylindrical co-ordinates. This may be obtained by eliminating u and v between the strain-displacement equations (13.24).

Differentiating eqn. (13.24b) with respect to r and eqn. (13.24c) with respect to θ, dividing the latter by r, and subtracting gives

$$\frac{\partial \varepsilon_\theta}{\partial r} - \frac{1}{r}\frac{\partial \gamma_{r\theta}}{\partial \theta} = \frac{1}{r}\frac{\partial u}{\partial r} - \frac{u}{r^2} - \frac{1}{r^2}\frac{\partial^2 u}{\partial \theta^2}$$

Multiplying by r^2 and substituting for $\partial u/\partial r$, we have

$$r^2\frac{\partial \varepsilon_\theta}{\partial r} - r\frac{\partial \gamma_{r\theta}}{\partial \theta} = r\varepsilon_r - u - \frac{\partial^2 u}{\partial \theta^2} \tag{13.27}$$

Differentiating this equation with respect to r, and eqn. (13.24a) with respect to θ twice, and eliminating u gives

$$2r\frac{\partial \varepsilon_\theta}{\partial r} + r^2\frac{\partial^2 \varepsilon_\theta}{\partial r^2} - \frac{\partial \gamma_{r\theta}}{\partial \theta} - r\frac{\partial^2 \gamma_{r\theta}}{\partial \theta\, \partial r} = \varepsilon_r + r\frac{\partial \varepsilon_r}{\partial r} - \varepsilon_r - \frac{\partial^2 \varepsilon_r}{\partial \theta^2}$$

and simplifying,

$$\frac{\partial \gamma_{r\theta}}{\partial \theta} + r\frac{\partial^2 \gamma_{r\theta}}{\partial \theta\, \partial r} = \frac{\partial^2 \varepsilon_r}{\partial \theta^2} + r^2\frac{\partial^2 \varepsilon_\theta}{\partial r^2} + 2r\frac{\partial \varepsilon_\theta}{\partial r} - r\frac{\partial \varepsilon_r}{\partial r} \tag{13.28}$$

This is the compatibility equation in terms of strain. To obtain a similar relationship for stresses it is necessary to use the stress–strain relationships and equilibrium equations in cylindrical co-ordinates.

The stress–strain equations in two dimensions are

$$\varepsilon_r = \frac{\sigma_r}{E} - \frac{v\sigma_\theta}{E}$$

$$\varepsilon_\theta = \frac{\sigma_\theta}{E} - \frac{v\sigma_r}{E}$$

$$\gamma_{r\theta} = \frac{\tau_{r\theta}}{G} = \frac{2(1+v)}{E}\tau_{r\theta}$$

Substituting in eqn. (13.28),

$$\frac{2(1+v)}{E}\frac{\partial \tau_{r\theta}}{\partial \theta} + \frac{2(1+v)}{E}r\frac{\partial^2 \tau_{r\theta}}{\partial \theta\, \partial r} = \frac{\partial^2 \sigma_r}{E\,\partial \theta^2} - \frac{v\partial^2 \sigma_\theta}{E\,\partial \theta^2} + r^2\frac{\partial^2 \sigma_\theta}{E\,\partial r^2}$$

$$-r^2\frac{v\partial^2 \sigma_r}{E\,\partial r^2} + \frac{2r}{E}\frac{\partial \sigma_\theta}{\partial r} - \frac{2rv}{E}\frac{\partial \sigma_r}{\partial r} - r\frac{\partial \sigma_r}{E\,\partial r} + \frac{rv}{E}\frac{\partial \sigma_\theta}{\partial r} \tag{13.29}$$

Now the equilibrium equations (13.8) and (13.10) are used to eliminate $\tau_{r\theta}$. Multiply through eqn. (13.8) by r, and differentiate with respect to r; then

$$\frac{\partial \sigma_r}{\partial r} + r\frac{\partial^2 \sigma_r}{\partial r^2} + \frac{\partial^2 \tau_{r\theta}}{\partial r \, \partial\theta} + \frac{\partial \sigma_r}{\partial r} - \frac{\partial \sigma_\theta}{\partial r} = 0$$

Differentiating eqn. (13.10) with respect to θ,

$$\frac{1}{r}\frac{\partial^2 \sigma_\theta}{\partial \theta^2} + \frac{\partial^2 \tau_{r\theta}}{\partial r \, \partial\theta} + \frac{2}{r}\frac{\partial \tau_{r\theta}}{\partial \theta} = 0$$

Multiplying each of the above equations by r, adding and simplifying, we get

$$2\frac{\partial \tau_{r\theta}}{\partial \theta} + 2r\frac{\partial^2 \tau_{r\theta}}{\partial r \, \partial\theta} = -\frac{\partial^2 \sigma_\theta}{\partial \theta^2} - r^2\frac{\partial^2 \sigma_r}{\partial r^2} - 2r\frac{\partial \sigma_r}{\partial r} + r\frac{\partial \sigma_\theta}{\partial r} \qquad (13.30)$$

Substituting for $\tau_{r\theta}$ in eqn. (13.29) and simplifying,

$$\frac{\partial^2 \sigma_r}{\partial r^2} + \frac{\partial^2 \sigma_\theta}{\partial r^2} + \frac{1}{r}\frac{\partial \sigma_r}{\partial r} + \frac{1}{r}\frac{\partial \sigma_\theta}{\partial r} + \frac{1}{r^2}\frac{\partial^2 \sigma_r}{\partial \theta^2} + \frac{1}{r^2}\frac{\partial^2 \sigma_\theta}{\partial \theta^2} = 0$$

or

$$\left(\frac{\partial^2}{\partial r^2} + \frac{1}{r}\frac{\partial}{\partial r} + \frac{1}{r^2}\frac{\partial^2}{\partial \theta^2}\right)(\sigma_r + \sigma_\theta) = 0 \qquad (13.31)$$

This is the equation of compatibility in terms of stresses. For cases of axial symmetry, since stress and displacement are independent of θ, the equation becomes

$$\left(\frac{\partial^2}{\partial r^2} + \frac{1}{r}\frac{\partial}{\partial r}\right)(\sigma_r + \sigma_\theta) = 0 \qquad (13.32)$$

The complete analysis of stress distribution in a body may now be made using the equilibrium equations (13.8) and (13.10), the above compatibility equation, and the boundary conditions appropriate to the applied forces or displacements.

13.8. STRESS FUNCTIONS

It is beyond the scope of this text to discuss and make use of stress functions in detail. However, since the stress function method of solution plays an essential part in the mathematical theory of elasticity, it is perhaps of some value to mention it here.

In the last paragraph of the preceding section the principles of obtaining the complete solution for the stresses in a body were stated. A stress function is a mathematical function of the co-ordinate variables x and y, or r and θ. The stress components can be expressed in terms of such a function which, on substitution, will satisfy the equilibrium equations. The compatibility equations can also be written in terms of the stress function.

For a particular problem, the solution of the various partial differential equations is therefore a matter of choosing a suitable mathematical function by which to express the stress components. In Cartesian co-ordinates the stress components may be written as

$$\sigma_x = \frac{\partial^2 \phi}{\partial y^2} \qquad \sigma_y = \frac{\partial^2 \phi}{\partial x^2} \qquad \tau_{xy} = -\frac{\partial^2 \phi}{\partial x \, \partial y} \tag{13.33}$$

where ϕ is the stress function. It may be easily verified that the above three equations completely satisfy the equilibrium equations without body force, namely (13.5) and (13.6). The compatibility equation (13.20) may now be expressed in the form

$$\left(\frac{\partial^2}{\partial x^2} + \frac{\partial^2}{\partial y^2} \right)^2 \phi = 0 \tag{13.34}$$

In cylindrical co-ordinates the stress components in terms of a stress function are as follows

$$\left. \begin{aligned} \sigma_r &= \frac{1}{r} \frac{\partial \phi}{\partial r} + \frac{1}{r^2} \frac{\partial^2 \phi}{\partial \theta^2} \\[2mm] \sigma_\theta &= \frac{\partial^2 \phi}{\partial r^2} \\[2mm] \tau_{r\theta} &= \frac{1}{r^2} \frac{\partial \phi}{\partial \theta} - \frac{1}{r} \frac{\partial^2 \phi}{\partial r \, \partial \theta} \end{aligned} \right\} \tag{13.35}$$

It can also be shown that these expressions satisfy the equilibrium equations (13.8) and (13.10), without body forces R and Θ. The compatibility equation in polar co-ordinates is, from the above,

$$\frac{\partial^2 \sigma_r}{\partial r^2} + \frac{\partial^2 \sigma_\theta}{\partial r^2} + \frac{1}{r} \frac{\partial \sigma_r}{\partial r} + \frac{1}{r} \frac{\partial \sigma_\theta}{\partial r} + \frac{1}{r^2} \frac{\partial^2 \sigma_r}{\partial \theta^2} + \frac{1}{r^2} \frac{\partial^2 \sigma_\theta}{\partial \theta^2} = 0$$

Substituting for σ_r and σ_θ from above and simplifying, we obtain

$$\left(\frac{\partial^2}{\partial r^2} + \frac{1}{r} \frac{\partial}{\partial r} + \frac{1}{r^2} \frac{\partial^2}{\partial \theta^2} \right) \left(\frac{\partial^2 \phi}{\partial r^2} + \frac{1}{r} \frac{\partial \phi}{\partial r} + \frac{1}{r^2} \frac{\partial^2 \phi}{\partial \theta^2} \right) = 0 \tag{13.36}$$

as the compatibility equation expressed in terms of the stress function.

Chapter 14

Some Applications of the Equilibrium and Strain–Displacement Relationships

14.1. It has been explained in principle in the last chapter how the stress components may be determined in a body by use of the equilibrium and compatibility equations and the particular boundary conditions of the problem. However, in a majority of cases, the solution of the various equations is sufficiently complex to require the stress-function method of attack mentioned in the previous chapter.

There are a few problems in beams and axi-symmetrical bodies in which a rather more elementary analysis is possible using the equilibrium, strain–displacement and stress–strain relationships. As it is considered to be of importance to understand how to apply these principles, the present chapter is devoted to this end.

14.2. SHEARING STRESS IN A BEAM

The distribution of transverse shear stress in a beam in terms of the shear force and the geometry of the cross-section was obtained in Chapter 6, using simple bending theory.

Consider the beam in Fig. 14.1, which, for simplicity of solution, is shown simply supported and carrying a uniformly distributed load, w per unit length. The origin of Cartesian co-ordinates is taken on the neutral axis with x positive left to right and y positive downwards.

This is treated as a two-dimensional problem with no variation of stress through the thickness of the beam, and therefore only two equations of equilibrium are applicable. From Section 13.2,

$$\frac{\partial \sigma_x}{\partial x} + \frac{\partial \tau_{xy}}{\partial y} = 0 \tag{14.1}$$

$$\frac{\partial \sigma_y}{\partial y} + \frac{\partial \tau_{xy}}{\partial x} = 0 \tag{14.2}$$

If the exact solution for pure bending is employed, that is

$$\sigma_x = \frac{My}{I} \tag{14.3}$$

then eqn. (14.2) is not required, neither is any strain–displacement relationship. It should be remembered, however, that in the derivation of eqn. (14.3) assumptions were made as to geometry of deformation, and Hooke's law was employed. Substituting for σ_x in eqn. (14.1) gives

$$\frac{\partial\left(\dfrac{My}{I}\right)}{\partial x} + \frac{\partial \tau_{xy}}{\partial y} = 0$$

Therefore

$$\partial \tau_{xy} = -\frac{\partial M}{\partial x}\frac{y}{I}\,\partial y$$

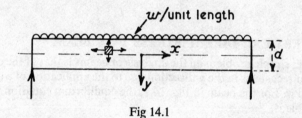

Fig 14.1

But

$$\frac{\partial M}{\partial x} = Q$$

the shear force on the section, so that

$$\partial \tau_{xy} = -\frac{Qy}{I}\,\partial y \tag{14.4}$$

Integrating gives

$$\tau_{xy} = -\frac{Q}{I}\int y\,dy \,\therefore\, C$$

$$= -\frac{Qy^2}{2I} + C \tag{14.5}$$

At the top and bottom free surface of the beam the shear stress must be zero; therefore

$$\tau_{xy} = 0 \quad \text{at} \quad y = \pm\frac{d}{2}$$

from which

$$C = \frac{Qd^2}{8I}$$

and

$$\tau_{xy} = -\frac{Qy^2}{2I} + \frac{Qd^2}{8I} \tag{14.6}$$

At the neutral axis $y = 0$ and the shear stress has its maximum value

$$\tau_{xy} = \frac{Qd^2}{8I} \tag{14.7}$$

This agrees with the value obtained in Chapter 6.

14.3. VERTICAL DIRECT STRESS IN A BEAM

A further simple problem on the analysis of beams is that of the distribution of direct stress in the y-direction due to the application of a distributed load w. For the beam in Fig. 14.1, the equilibrium equation which is applicable is

$$\frac{\partial \sigma_y}{\partial y} + \frac{\partial \tau_{xy}}{\partial x} = 0 \tag{14.8}$$

Now, the shear stress, τ_{xy}, was determined in the previous section, and substituting that value in eqn. (14.8) gives

$$\frac{\partial \sigma_y}{\partial y} + \frac{\partial}{\partial x}\left(-\frac{Qy^2}{2I} + \frac{Qd^2}{8I}\right) = 0$$

But

$$\frac{\partial Q}{\partial x} = -w$$

Therefore

$$\frac{\partial \sigma_y}{\partial y} + \frac{wy^2}{2I} - \frac{wd^2}{8I} = 0 \tag{14.9}$$

or

$$\sigma_y = -\int\left(\frac{wy^2}{2I} - \frac{wd^2}{8I}\right)dy + C$$

$$= -\frac{w}{I}\left(\frac{y^3}{6} - \frac{d^2y}{8}\right) + C \tag{14.10}$$

Using the boundary condition that at the upper surface $y = -\frac{1}{2}d$, the compressive stress is $\sigma_y = -w/b$, where b is the beam thickness; then

$$C = -\frac{w}{b} + \frac{w}{I}\left(-\frac{d^3}{48} + \frac{d^3}{16}\right)$$

$$= -\frac{w}{I}\frac{d^3}{12} + \frac{wd^3}{24I}$$

$$= -\frac{wd^3}{24I}$$

Therefore

$$\sigma_y = -\frac{w}{I}\left(\frac{y^3}{6} - \frac{d^2 y}{8}\right) - \frac{wd^3}{24I}$$

$$= -\frac{w}{I}\left(\frac{y^3}{6} - \frac{d^2 y}{8} + \frac{d^3}{24}\right) \tag{14.11}$$

A check on this solution may be made by considering the condition at the lower free surface. Here $y = +\frac{1}{2}d$; therefore

$$\sigma_y = -\frac{w}{I}\left(\frac{d^3}{48} - \frac{d^3}{16} + \frac{d^3}{24}\right)$$

giving $\sigma_y = 0$, which is correct.

14.4. STRESS DISTRIBUTION IN A THICK-WALLED CYLINDER

This problem is of considerable practical importance in pressure vessels and gun barrels. It is a further case which can be tackled without recourse to a stress-function solution, and is a convenient example in which to use the cylindrical co-ordinate system.

A section through a long hollow cylinder which is subjected to uniformly distributed internal and external pressure is shown in Fig. 14.2. The deformations produced are symmetrical about the axis of the cylinder, and the small element of material in the wall supports the stress system shown. This is the same as in Fig. 13.2 for the general plane stress cylindrical co-ordinate system, except that for axial symmetry $\tau_{r\theta} = 0$ and σ_θ is constant at any particular radius.

There are two methods of maintaining the pressure in the cylinder, either by end caps which are subjected to pressure or by pistons in each end of the cylinder. In the former case, axial stress and strain, σ_z and ε_z respectively, will be set up, and in the latter case $\sigma_z = 0$, but there is still an axial strain due to the Poisson effect of σ_r and σ_θ. From the symmetry of the system and for a long cylinder, we come to the conclusion that plane cross-sections remain plane when subjected to pressure, and therefore axial deformation, w, across the section is independent of r and $dw/dr = 0$.

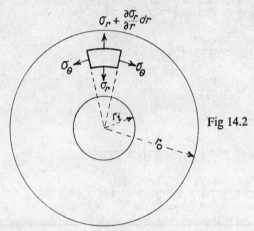

Fig 14.2

The equations of equilibrium for an element of material are

$$\frac{d\sigma_r}{dr} + \frac{\sigma_r - \sigma_\theta}{r} = 0 \quad \text{(eqn. (13.12))} \tag{14.12}$$

and

$$\frac{d\sigma_z}{dz} = 0 \tag{14.13}$$

The strain–displacement equations, from Section 13.6, are

$$\varepsilon_r = \frac{du}{dr} \tag{14.14}$$

$$\varepsilon_\theta = \frac{u}{r} \tag{14.15}$$

$$\varepsilon_z = \frac{dw}{dz} \tag{14.16}$$

The stress–strain relationships are

$$\varepsilon_r = \frac{\sigma_r}{E} - \frac{v}{E}(\sigma_\theta + \sigma_z) = \frac{du}{dr} \tag{14.17}$$

$$\varepsilon_\theta = \frac{\sigma_\theta}{E} - \frac{v}{E}(\sigma_z + \sigma_r) = \frac{u}{r} \tag{14.18}$$

$$\varepsilon_z = \frac{\sigma_z}{E} - \frac{v}{E}(\sigma_r + \sigma_\theta) = \frac{dw}{dz} \tag{14.19}$$

Differentiating eqn. (14.18) with respect to r gives

$$\frac{E}{r}\left(\frac{du}{dr} - \frac{u}{r}\right) = \frac{d\sigma_\theta}{dr} - v\frac{d\sigma_z}{dr} - v\frac{d\sigma_r}{dr}$$

Substituting for du/dr and u/r from eqns. (14.17) and (14.18) and simplifying,

$$\frac{1 + v}{r}(\sigma_r - \sigma_\theta) = \frac{d\sigma_\theta}{dr} - v\frac{d\sigma_r}{dr} - v\frac{d\sigma_z}{dr} \tag{14.20}$$

Now, since $\varepsilon_z = $ constant, from eqn. (14.19),

$$\frac{d\sigma_z}{dr} = v\frac{d\sigma_r}{dr} + v\frac{d\sigma_\theta}{dr} \tag{14.21}$$

and substituting for $d\sigma_z/dr$ from eqn. (14.21) and $(\sigma_r - \sigma_\theta)/r$ from eqn. (14.12) into eqn. (14.20) and simplifying gives

$$(1 - v^2)\left(\frac{d\sigma_\theta}{dr} + \frac{d\sigma_r}{dr}\right) = 0 \tag{14.22}$$

Therefore

$$\sigma_\theta + \sigma_r = \text{constant} = 2A \quad \text{(say)} \tag{14.23}$$

From eqns. (14.22) and (14.21),

$$\frac{d\sigma_z}{dr} = 0$$

Therefore σ_z is constant. Eliminating σ_θ between eqns. (14.23) and (14.12) gives

$$\frac{d\sigma_r}{dr} + \frac{2\sigma_r - 2A}{r} = 0 \tag{14.24}$$

from which

$$2Ar - 2r\sigma_r - r^2\frac{d\sigma_r}{dr} = 0$$

and

$$2Ar - \frac{d}{dr}(r^2\sigma_r) = 0$$

By integration,

$$Ar^2 - r^2\sigma_r = B$$

Hence

$$\sigma_r = A - \frac{B}{r^2} \tag{14.25}$$

and from eqn. (14.23),

$$\sigma_\theta = A + \frac{B}{r^2} \tag{14.26}$$

where A and B are constants which may be found using the boundary conditions. Before taking this step it is worth noting that the above expressions for σ_r and σ_θ could have been obtained directly from the equilibrium equation (13.12) and the compatibility equation (13.32) as follows:

$$\frac{\partial \sigma_r}{\partial r} + \frac{\sigma_r - \sigma_\theta}{r} = 0 \tag{14.27}$$

$$\left(\frac{\partial^2}{\partial r^2} + \frac{1}{r} \frac{\partial}{\partial r} \right) (\sigma_r + \sigma_\theta) = 0 \tag{14.28}$$

Multiplying out the latter gives

$$\frac{\partial^2 \sigma_r}{\partial r^2} + \frac{1}{r} \frac{\partial \sigma_r}{\partial r} + \frac{\partial^2 \sigma_\theta}{\partial r^2} + \frac{1}{r} \frac{\partial \sigma_\theta}{\partial r} = 0 \tag{14.29}$$

Now, from eqn. (14.27),

$$\sigma_\theta = r \frac{\partial \sigma_r}{\partial r} + \sigma_r$$

Therefore

$$\frac{\partial \sigma_\theta}{\partial r} = \frac{\partial \sigma_r}{\partial r} + r \frac{\partial^2 \sigma_r}{\partial r^2} + \frac{\partial \sigma_r}{\partial r}$$

and

$$\frac{\partial^2 \sigma_\theta}{\partial r^2} = 3 \frac{\partial^2 \sigma_r}{\partial r^2} + r \frac{\partial^3 \sigma_r}{\partial r^3}$$

Substituting these expressions for σ_θ in eqn. (14.29) and gathering terms together gives

$$r \frac{\partial^3 \sigma_r}{\partial r^3} + 5 \frac{\partial^2 \sigma_r}{\partial r^2} + \frac{3}{r} \frac{\partial \sigma_r}{\partial r} = 0 \tag{14.30}$$

which is the general equation for σ_r in an axially-symmetrical stress system with no body force. It can be verified by substitution that $\sigma_r = A - (B/r^2)$ is one particular solution of this equation. Substituting for σ_r in eqn. (14.27) yields

$$\sigma_\theta = A + \frac{B}{r^2}$$

Returning now to the determination of the constants A and B, the boundary conditions of the problem are: when $r = r_i$, $\sigma_r = -p_i$ (pressure being negative in sign); and when $r = r_o$, $\sigma_r = -p_o$,

$$-p_i = A - \frac{B}{r_i^2}$$

and

$$-p_o = A - \frac{B}{r_o^2}$$

from which, eliminating A, we get

$$B\left(\frac{1}{r_i^2} - \frac{1}{r_o^2}\right) = p_i - p_o$$

$$B = \frac{(p_i - p_o)r_i^2 r_o^2}{r_o^2 - r_i^2}$$

and

$$A = \frac{p_i r_i^2 - p_o r_o^2}{r_o^2 - r_i^2}$$

Therefore the radial and hoop stresses become

$$\sigma_r = \frac{p_i r_i^2 - p_o r_o^2}{r_o^2 - r_i^2} - \frac{(p_i - p_o)r_i^2 r_o^2}{r^2(r_o^2 - r_i^2)} \tag{14.31}$$

$$\sigma_\theta = \frac{p_i r_i^2 - p_o r_o^2}{r_o^2 - r_i^2} + \frac{(p_i - p_o)r_i^2 r_o^2}{r^2(r_o^2 - r_i^2)} \tag{14.32}$$

These equations were first derived by Lamé and Clapeyron in 1833.

An important special case of the above is when the external pressure is atmospheric only and can be neglected in relation to the internal pressure. Then with $p_o = 0$,

$$\sigma_r = \frac{p_i r_i^2}{r_o^2 - r_i^2}\left(1 - \frac{r_o^2}{r^2}\right) = \frac{p_i}{k^2 - 1}\left(1 - \frac{r_o^2}{r^2}\right) \tag{14.33}$$

$$\sigma_\theta = \frac{p_i r_i^2}{r_o^2 - r_i^2}\left(1 + \frac{r_o^2}{r^2}\right) = \frac{p_i}{k^2 - 1}\left(1 + \frac{r_o^2}{r^2}\right) \tag{14.34}$$

where $k = r_o/r_i$.

σ_r and σ_θ each have their maximum value at the inner surface, so that when $r = r_i$,

$$\sigma_r = -p_i \quad \text{(radial compressive stress)}$$

and

$$\sigma_\theta = \frac{r_0{}^2 + r_i{}^2}{r_0{}^2 - r_i{}^2} p_i \quad \text{(tensile hoop stress)}$$

$$= \frac{k^2 + 1}{k^2 - 1} p_i$$

At the outer surface, where $r = r_0$,

$$\sigma_r = 0 \quad \text{and} \quad \sigma_\theta = \frac{2p_i}{k^2 - 1}$$

It is important to note that stresses are dependent on the diameter or k ratio and not on the absolute dimensions.

Axial Stress and Strain

Now that expressions have been developed for the radial and circumferential stresses within the cylinder, the next step is to consider what conditions of stress and strain can exist axially along the cylinder. These will depend on the boundary conditions at the ends of the cylinder.

(a) *Cylinder built-in between rigid end supports—no change in length*

For this case $\varepsilon_z = 0$; in other words, plane strain exists.
   Now,

$$\varepsilon_z = \frac{\sigma_z}{E} - \frac{\nu}{E}(\sigma_r + \sigma_\theta)$$

Therefore

$$\frac{\sigma_z}{E} - \frac{\nu}{E}(\sigma_r + \sigma_\theta) = 0$$

$$\sigma_z = \nu(\sigma_r + \sigma_\theta) \tag{14.35}$$

Substituting for σ_r and σ_θ from eqns. (14.25) and (14.26),

$$\sigma_z = 2\nu A$$

$$\sigma_z = \frac{2\nu p_i r_i{}^2}{r_0{}^2 - r_i{}^2} = \frac{2\nu p_i}{k^2 - 1} \tag{14.36}$$

(b) *Cylinder with enclosed ends but free to change in length*

In this case there must be equilibrium between the force exerted on the end cover by the internal pressure and the force of the axial stress integrated across the wall of the vessel. Therefore

$$\sigma_z(\pi r_0{}^2 - \pi r_i{}^2) - p_i \pi r_i{}^2 = 0$$

so that

$$\sigma_z = \frac{p_i r_i^2}{r_0^2 - r_i^2} = \frac{p_i}{k^2 - 1} \qquad (14.37)$$

and

$$\varepsilon_z = \frac{\sigma_z}{E} - \frac{v}{E}(\sigma_r + \sigma_\theta) = \frac{(1 - 2v)p_i}{E(k^2 - 1)}$$

(c) *Pressure maintained by piston in each end of cylinder*

Since there is no connection between the piston and the cylinder, the axial force due to pressure is reacted entirely by the pistons, and therefore there can be no axial stress in the wall of the cylinder. Thus

$$\sigma_z = 0 \qquad (14.38)$$

and

$$\varepsilon_z = -\frac{v}{E}(\sigma_r + \sigma_\theta) = -\frac{2vp_i}{E(k^2 - 1)}$$

Example 14.1.   The cylinder of a hydraulic jack has a bore (internal diameter) of 150 mm and is required to operate up to 13·8 MN/m². Determine the required wall thickness for a limiting tensile stress in the material of 41·4 MN/m², and show by means of a diagram the distribution of radial and circumferential stress through the wall of the cylinder.

The given boundary conditions are that at $r = 75 \times 10^{-3}$, $\sigma_r = -13·8 \times 10^6$ and $\sigma_\theta = 41·4 \times 10^6$, since the maximum tensile hoop stress occurs at the inner surface. Therefore

$$-13·8 \times 10^6 = A - \frac{B}{5600 \times 10^{-6}}$$

and

$$41·4 \times 10^6 = A + \frac{B}{5600 \times 10^{-6}}$$

Adding the two equations,

$$2A = 27·6 \times 10^6$$

$$A = 13·8 \times 10^6 \, \text{N/m}^2$$

and

$$B = 154·5 \, \text{kN}$$

At the outside surface, $\sigma_r = 0$; therefore

$$0 = A - \frac{B}{r^2} = 13 \cdot 8 \times 10^6 - \frac{154 \cdot 5 \times 10^3}{r^2}$$

$$r^2 = 0 \cdot 0112 \, \text{m}^2$$

$$r = 0 \cdot 106 \, \text{m}$$

Therefore the required wall thickness is 31 m.

To plot the distribution of σ_r and σ_θ through the wall we have

$$\sigma_r = (13 \cdot 8 \times 10^6) - \frac{154 \cdot 5 \times 10^3}{r^2}$$

and

$$\sigma_\theta = (13 \cdot 8 \times 10^6) + \frac{154 \cdot 5 \times 10^3}{r^2}$$

and the distribution is shown in Fig. 14.3.

r, mm	75	80	85	90	95	100	106
σ_r, MN/m²	−13·8	−10·3	−7·7	−5·2	−3·25	−1·65	0
σ_θ, MN/m²	41·4	37·9	35·3	32·8	30·85	29·25	27·0

14.5. COMPOUND CYLINDERS

From Fig. 14.3 it will be observed that there is a great variation in the intensity of the stress in the wall of a thick cylinder subjected to internal fluid pressure, and thus the material is not used to the best advantage. In order to secure a more uniform stress distribution, one method is to build up the cylinder by shrinking one tube on the outside of another. The inner tube is brought to a state of compression by the contraction, on cooling, of the external tube; this tube will therefore be under a similar internal pressure, and hence be in a state of tension. When the compound tube is subjected to internal fluid pressure, the resultant stress is the algebraic sum of that due to the shrinking and that due to the internal pressure. The resultant tensile stress at the inner surface of the inner tube is not so large as if the cylinder were composed of one thick tube, and the final tensile stress at the inner surface of the outer tube is larger than if the cylinder consisted of one thick tube; thus a more even stress distribution is obtained.

In gun-making, it is not an easy matter to turn and bore long tubes to the degree of accuracy required for shrinking. It is usual in this case to wind around the outside of a tube a high tensile strength ribbon of a rectangular section with sufficient tension to bring the tube into a state of compression. Subsequent internal pressure then has to overcome the hoop compression before tensile stress can be set up in the tube.

A further method for creating hoop compression at the bore of a cylinder is known as *autofrettage*. This consists in applying internal pressure to a single cylinder until yielding and a prescribed amount of

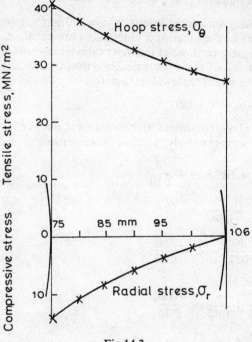

Fig 14.3

plastic deformation occur at the bore. Since the hoop stress falls in magnitude from the inner to the outer surface, a part of the wall from the bore will be in the plastic range of the metal while the remainder will still be elastic. On the release of pressure the elastic material cannot return to its original geometry owing to the permanent deformation at the inside of the vessel. Therefore the material that had been in the plastic range will be subjected to hoop compression, and the elastic outer material to hoop tension. This technique is discussed further in Chapter 18.

Example 14.2.   A vessel is to be used for internal pressures up to 207 MN/m². It consists of two hollow steel cylinders which are shrunk one on the other. The inner tube has an internal diameter of 200 mm and a nominal external diameter of 300 mm, while the outer tube is 300 mm nominal and 400 mm for the inner and outer diameters respectively. The interference at the mating surface of the two cylinders is 0·1 mm. Determine the radial and circumferential stresses at the bores and outside surfaces. The axial stress in the cylinders is to be neglected. $E = 207\,\text{GN/m}^2$.

The known conditions for the problem are as follows:

For the inner tube, at $r = 100$ mm, $\sigma_r = -207\,\text{MN/m}^2$
For the outer tube, at $r = 200$ mm, $\sigma_r = 0$

At the mating surface or interface, $r = 150$ mm nominally, and the radial stresses in the inner and outer tubes are equal, $\sigma_{ri} = \sigma_{ro}$. Also the radial displacement, u_i, of the inner cylinder inwards plus the radial displacement, u_0, of the outer cylinder outwards due to the shrink fit must equal the interference value; therefore

$$-u_i + u_0 = 0\cdot1\,\text{mm}$$

Using the above conditions and constants A, B and C, D for the inner and outer tubes respectively, we have four equations:

$$-207 \times 10^6 = A - \frac{B}{0\cdot01} \tag{14.39}$$

$$0 = C - \frac{D}{0\cdot04} \tag{14.40}$$

$$A - \frac{B}{0\cdot0225} = C - \frac{D}{0\cdot0225} \tag{14.41}$$

and

$$-\frac{u_i}{0\cdot15} + \frac{u_0}{0\cdot15} = \frac{0\cdot0001}{0\cdot15}$$

or

$$-\varepsilon_{\theta i} + \varepsilon_{\theta o} = \frac{0\cdot0001}{0\cdot15}$$

and substituting for the strains in terms of the stresses,

$$-\sigma_{\theta i} + \nu\sigma_{r_i} + \sigma_{\theta_o} - \nu\sigma_{r_o} = \frac{0\cdot0001}{0\cdot15} \times 207 \times 10^9$$

and since $\sigma_{r_i} = \sigma_{r_o}$,

$$\sigma_{\theta_o} - \sigma_{\theta_i} = 138 \times 10^6$$

Therefore

$$C + \frac{D}{0\cdot0225} - \left(A + \frac{B}{0\cdot0225}\right) = 138 \times 10^6 \tag{14.42}$$

The problem may be completely solved by determining the constants A, B, C and D from the above four equations. Substituting for C in eqns. (14.42) and (14.41),

$$\frac{D}{0 \cdot 04} + \frac{D}{0 \cdot 0225} - A - \frac{B}{0 \cdot 0225} = 138 \times 10^6$$

$$\frac{D}{0 \cdot 04} - \frac{D}{0 \cdot 0225} = A - \frac{B}{0 \cdot 0225}$$

Eliminating D between these two equations gives

$$-\frac{0 \cdot 0625}{0 \cdot 0175}\left(A - \frac{B}{0 \cdot 0225}\right) - A - \frac{B}{0 \cdot 0225} = 138 \times 10^6$$

or simplifying,

$$-4 \cdot 57A + 114B = 138 \times 10^6$$

Substituting for A from eqn. (14.39),

$$114B - 4 \cdot 57\left(-207 \times 10^6 + \frac{B}{0 \cdot 01}\right) = 138 \times 10^6$$

$$-343B = 138 \times 10^3 - 945\,000$$

from which

$$B = 2 \cdot 35\,\text{MN} \qquad \text{and} \qquad A = 28\,\text{MN/m}^2$$

Substituting for A and B,

$$-\frac{0 \cdot 0175D}{0 \cdot 04 \times 0 \cdot 0225} = (28 \times 10^6) - \frac{2 \cdot 35 \times 10^6}{0 \cdot 0225}$$

Therefore

$$D = 3 \cdot 93 \times 10^6$$

and

$$C = \frac{3 \cdot 93 \times 10^6}{0 \cdot 04} = 98 \cdot 2 \times 10^6$$

The required stresses are then as follows:

Inner Cylinder
Radial stress:

$$r = 100 \qquad \sigma_r = 28 \times 10^6 - \frac{2 \cdot 35 \times 10^6}{0 \cdot 01} = -207\,\text{MN/m}^2$$

$$r = 150 \qquad \sigma_r = 28 \times 10^6 - \frac{2 \cdot 35 \times 10^6}{0 \cdot 0225} = -76 \cdot 4\,\text{NM/m}^2$$

Circumferential stress:

$$r = 100 \qquad \sigma_\theta = 28 \times 10^6 + \frac{2\cdot35 \times 10^6}{0\cdot01} = +263\,\text{MN/m}^2$$

$$r = 150 \qquad \sigma_\theta = 28 \times 10^6 + \frac{2\cdot35 \times 10^6}{0\cdot0225} = +132\cdot4\,\text{MN/m}^2$$

Outer Cylinder
Radial stress:

$$r = 150 \qquad \sigma_r = 98\cdot2 \times 10^6 - \frac{3\cdot93 \times 10^6}{0\cdot0225} = -76\cdot3\,\text{MN/m}^2$$

$$r = 200 \qquad \sigma_r = 98\cdot2 \times 10^6 - \frac{3\cdot93 \times 10^6}{0\cdot04} = 0$$

Circumferential Stress

$$r = 150 \qquad \sigma_\theta = 98\cdot2 \times 10^6 + \frac{3\cdot93 \times 10^6}{0\cdot0225} = 272\cdot6\,\text{MN/m}^2$$

$$r = 200 \qquad \sigma_\theta = 98\cdot2 \times 10^6 + \frac{3\cdot93 \times 10^6}{0\cdot04} = 196\cdot4\,\text{MN/m}^2$$

If the cylinder had been made of one thick tube of the same overall dimensions as the compound vessel, then at the bore,

$$-207 \times 10^6 = E - \frac{F}{0\cdot01}$$

and at the outer surface,

$$0 = E - \frac{F}{0\cdot04}$$

Hence

$$E = 69 \times 10^6 \qquad \text{and} \qquad F = 2\cdot76 \times 10^6$$

from which the radial and circumferential stress distributions are as follows:

Radial Stress
$$r = 100 \qquad \sigma_r = -207\,\text{MN/m}^2$$
$$r = 150 \qquad \sigma_r = -53\cdot5\,\text{MN/m}^2$$
$$r = 200 \qquad \sigma_r = 0$$

Circumferential Stress
$$r = 100 \qquad \sigma_\theta = 345\,\text{MN/m}^2$$
$$r = 150 \qquad \sigma_\theta = 191\cdot5\,\text{MN/m}^2$$
$$r = 200 \qquad \sigma_\theta = 138\,\text{MN/m}^2$$

Fig. 14.4 shows diagrammatically the distribution of radial and hoop stresses through the wall of the compound and single cylinders, and illustrates the more efficient use of material in the former case.

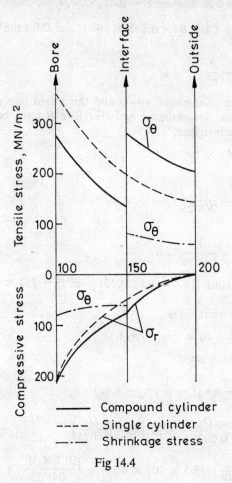

Fig 14.4

The stress distribution due to shrinkage only may be obtained directly as the difference between the curves for the single and compound cylinders. This approach would only apply if the compound tube was made of the one type of material throughout.

Shrinkage stresses alone could have been obtained by calculation from the four boundary conditions: $\sigma_r = 0$ at $r = 100$ and 200, and $\sigma_{r_i} = \sigma_{r_o}$ and $-u_i + u_o = 0 \cdot 1$ at the interface radius $r = 150$.

If the two tubes are of different materials then the appropriate elastic constants have to be used where they occur in the various equations.

Example 14.3. A bronze bush of 25 mm wall thickness is to be shrunk on to a steel shaft 100 mm in diameter. If an interface pressure of 69 MN/m² is required, determine the interference between bush and shaft. Steel: $E = 207 \, \text{GN/m}^2$; $\nu = 0.28$. Bronze: $E = 100 \, \text{GN/m}^2$, $\nu = 0.29$.

Using constants A and B for the shaft and C and D for the bush, then the radial stress is

$$\sigma_{r_s} = A - \frac{B}{r^2} \tag{14.43}$$

At the centre of the shaft $r = 0$ and this might imply that σ_{r_s} was infinite, but this cannot be so and therefore B must be zero; hence $\sigma_{rs} = A$. At the interface,

$$\sigma_{r_s} = -69 \times 10^6 = A$$

At $r = 50$,

$$\sigma_{r_b} = C - \frac{D}{0.0025} = -69 \times 10^6 \tag{14.44}$$

At $r = 75$,

$$\sigma_{r_b} = C - \frac{D}{0.0056} = 0 \tag{14.45}$$

Subtracting eqns. (14.44) and (14.45) gives $D = 312 \times 10^3$; therefore $C = 55.5 \times 10^6$.

Now, the interference is

$$\delta = -u_s + u_b = r(\varepsilon_{\theta_b} - \varepsilon_{\theta_s})$$

where $r = 50$; therefore

$$\frac{\delta}{50} = \frac{1}{E_b}(\sigma_{\theta_b} - \nu_b \sigma_{r_b}) - \frac{1}{E_s}(\sigma_{\theta_s} - \nu_s \sigma_{r_s})$$

$$= \frac{1}{E_b}\left[C + \frac{D}{0.0025} - 0.29\left(C - \frac{D}{0.0025}\right)\right] - \frac{1}{E_s}(A - 0.28A)$$

$$= \frac{1}{100 \times 10^9}\left[(55.5 \times 10^6 \times 0.71) + \left(\frac{312 \times 10^3}{0.0025} \times 1.29\right)\right]$$

$$\qquad + \frac{69 \times 10^6 \times 0.72}{207 \times 10^9}$$

$$= 0.002 + 0.000\,24$$

$$\delta = 0.112 \text{ mm}$$

which is the interference required at the nominal interface radius of 50 mm between the shaft and the bush.

14.6. THIN ROTATING DISC

A simplified example of a component such as a gas turbine rotor is a uniformly thin disc which, when rotating at a constant velocity, is subjected to stresses induced by centripetal acceleration. This is a problem which, like the thick cylinder, produces deformations symmetrical about the rotating axis. If the disc is thin in section then it is assumed that the radial and hoop stresses are constant through the thickness and that there is no stress in the z-direction.

The equation of equilibrium of an element, Fig. 14.5, is that derived

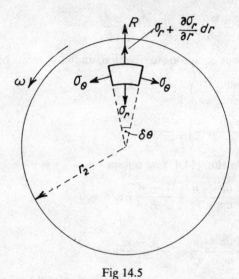

Fig 14.5

earlier for the axially-symmetrical stress system, but on this occasion a body force term must be included which is determined from the centripetal acceleration. Hence

$$\frac{d\sigma_r}{dr} + \frac{\sigma_r - \sigma_\theta}{r} + R = 0 \tag{14.46}$$

Now, R is the body force per unit volume; therefore

$$R r \delta\theta \, \delta r \, z = \rho \omega^2 r \, r \delta\theta \, \delta r \, z$$

which is the centrifugal force, where ρ is the density, ω is the steady rotational velocity in radians per second, and z is the thickness of the disc. Therefore

$$R = \rho \omega^2 r$$

and

$$\frac{d\sigma_r}{dr} + \frac{\sigma_r - \sigma_\theta}{r} + \rho\omega^2 r = 0 \qquad (14.47)$$

The strain–displacement equations are as previously for axial symmetry:

$$\varepsilon_r = \frac{du}{dr} \qquad \varepsilon_\theta = \frac{u}{r}$$

and so too are the stress–strain relationships:

$$\varepsilon_r = \frac{\sigma_r}{E} - \frac{\nu\sigma_\theta}{E} \qquad \varepsilon_\theta = \frac{\sigma_\theta}{E} - \frac{\nu\sigma_r}{E}$$

Using these four equations to obtain σ_r and σ_θ in terms of u,

$$\sigma_r = \left(\frac{du}{dr} + \frac{\nu u}{r}\right)\frac{E}{1 - \nu^2}$$

$$\sigma_\theta = \left(\frac{\nu \, du}{dr} + \frac{u}{r}\right)\frac{E}{1 - \nu^2}$$

Substituting in eqn. (14.47) we obtain

$$\frac{d^2u}{dr^2} + \frac{1}{r}\frac{du}{dr} - \frac{u}{r^2} + \frac{1 - \nu^2}{E}\rho\omega^2 r = 0$$

or

$$\frac{d^2u}{dr^2} + \frac{1}{r}\frac{du}{dr} - \frac{u}{r^2} = -\frac{1 - \nu^2}{E}\rho\omega^2 r \qquad (14.48)$$

This is a general linear differential equation of the second order. The general solution consists of the sum of two separate solutions known as the complementary function and the particular integral. The former is the solution of the left-hand side and the latter is obtained by considering the right-hand side of eqn. (14.48). Thus the complementary function is

$$u = Kr + \frac{K'}{r}$$

and the particular integral is

$$u = -\frac{1 - \nu^2}{E}\frac{\rho w^2 r^3}{8}$$

The complete solution is therefore

$$u = Kr + \frac{K'}{r} - \frac{1 - \nu^2}{E}\frac{\rho\omega^2 r^3}{8} \qquad (14.49)$$

in which K and K' are constants to be determined from the boundary conditions.

Substituting for u and du in the equations for σ_r and σ_θ, and simplifying the various constant terms by inserting new ones, A and B, we obtain

$$\sigma_r = A - \frac{B}{r^2} - \frac{3+v}{8}\,\rho\omega^2 r^2 \tag{14.50}$$

$$\sigma_\theta = A + \frac{B}{r^2} - \frac{1+3v}{8}\,\rho\omega^2 r^2 \tag{14.51}$$

The constants A and B are found from the appropriate boundary conditions of the problem.

(a) Solid Circular Disc of Uniform Thickness

If the disc is continuous from the centre to some outer radius $r = r_2$, then it is apparent that unless $B = 0$ when $r = 0$, the stresses would become infinite. To find A it is only necessary to use the condition that

$$\sigma_r = 0 \qquad \text{at} \qquad r = r_2$$

from which

$$A = \frac{3+v}{8}\,\rho\omega^2 r_2{}^2$$

and

$$\sigma_r = \frac{3+v}{8}\,\rho\omega^2(r_2{}^2 - r^2) \tag{14.52}$$

$$\sigma_\theta = \frac{3+v}{8}\,\rho\omega^2 r_2{}^2 - \frac{1+3v}{8}\frac{w}{g}\,\omega^2 r^2$$

$$= \frac{\rho\omega^2}{8}\,[(3+v)r_2{}^2 - (1+3v)r^2] \tag{14.53}$$

The maximum stress occurs at the centre, where $r = 0$, and then

$$\sigma_r = \sigma_\theta = \frac{3+v}{8}\,\rho\omega^2 r_2{}^2$$

(b) Circular Disc of Uniform Thickness with a Central Hole

In this case the boundary conditions are that the radial stress will be zero at the free boundaries, that is where $r = r_1$ and $r = r_2$, the radius of the hole and outside periphery of the disc respectively. Therefore

$$\sigma_r = A - \frac{B}{r_2{}^2} - \frac{3+v}{8}\,\rho\omega^2 r_2{}^2 = 0$$

and also

$$\sigma_r = A - \frac{B}{r_1^2} - \frac{3+\nu}{8} \rho \omega^2 r_1^2 = 0$$

Subtracting gives

$$B \left(\frac{1}{r_1^2} - \frac{1}{r_2^2} \right) = \frac{3+\nu}{8} \rho \omega^2 (r_2^2 - r_1^2)$$

Therefore

$$B = \frac{3+\nu}{8} \rho \omega^2 r_1^2 r_2^2$$

and substituting back for B in either of the above equations,

$$A = \frac{3+\nu}{8} \rho \omega^2 r_1^2 + \frac{3+\nu}{8} \rho \omega^2 r_2^2$$

$$= \frac{3+\nu}{8} \rho \omega^2 (r_1^2 + r_2^2)$$

Therefore

$$\sigma_r = \frac{3+\nu}{8} \rho \omega^2 \left[r_1^2 + r_2^2 - \frac{r_1^2 r_2^2}{r^2} - r^2 \right] \tag{14.54}$$

and

$$\sigma_\theta = \frac{3+\nu}{8} \rho \omega^2 \left[r_1^2 + r_2^2 + \frac{r_1^2 r_2^2}{r^2} - \frac{1+3\nu}{3+\nu} r^2 \right] \tag{14.55}$$

The maximum value of the hoop stress, σ_θ, is at $r = r_1$, and is given by

$$\sigma_{\theta\ max} = \frac{3+\nu}{4} \rho \omega^2 \left[r_2^2 + \frac{1-\nu}{3+\nu} r_1^2 \right] \tag{14.56}$$

σ_r is a maximum when $d\sigma_r/dr = 0$ or $r = \sqrt{(r_1 r_2)}$; therefore

$$\sigma_{r\ max} = \frac{3+\nu}{8} \rho \omega^2 (r_2 - r_1)^2 \tag{14.57}$$

When r_1 is so small that it is negligible compared to r_2, the maximum hoop stress becomes

$$\sigma_{\theta\ max} = \frac{3+\nu}{4} \rho \omega^2 r_2^2 \tag{14.58}$$

which is twice the value obtained for the maximum at the centre of a solid disc.

Example 14.4. A steel ring has been shrunk on to the outside of a solid steel disc. The interface radius is 250 mm and the outer radius of the assembly is 356 mm. If the pressure between the ring and the disc is not to fall below 34·5 MN/m², and the circumferential stress at the inside of the ring must not exceed 207 MN/m², determine the maximum speed at which the assembly can be rotated. What is then the stress at the centre of the disc? $\rho = 7\cdot75\text{Mg/m}^3$ $v = 0\cdot28$.

For the ring, when $r = 356$ mm, $\sigma_r = 0$, and when $r = 250$ mm, $\sigma_r = -34\cdot5 \times 10^6$; therefore

$$0 = A - \frac{B}{0\cdot126} - \left(\frac{3 + 0\cdot28}{8} \times 7\cdot75\omega^2 \times 0\cdot126 \times 10^3\right)$$

$$-(34\cdot5 \times 10^6)$$

$$= A - \frac{B}{0\cdot0625} - \left(\frac{3 + 0\cdot28}{8} \times 7\cdot75\omega^2 \times 0\cdot0625 \times 10^3\right)$$

from which

$$B = (4280 + 0\cdot025\omega^2)10^3 \qquad \text{and} \qquad A = (34\,000 + 0\cdot6\omega^2)10^3$$

Also when $r = 250$ mm, σ_θ must not exceed 207 MN/m²; therefore

$$207 \times 10^6 = A + \frac{B}{0\cdot0625} - \left[\frac{1 + (3 \times 0\cdot28)}{8} \times 7\cdot75\omega^2\right.$$

$$\left. \times 0\cdot0625 \times 10^3\right]$$

or

$$207 \times 10^3 = 34\,000 + 0\cdot6\omega^2 + 68\,500 + 0\cdot4\omega^2 - 0\cdot111\omega^2$$

After simplifying,

$$\omega^2 = 117\,500 \qquad \omega = 343 \text{ rad/s}$$

and

$$N = 343 \times \frac{60}{2\pi} = 3280 \text{ rev/min}$$

For the solid disc, using constants C and D, as shown previously, D must be zero; therefore

$$\sigma_r = C - \frac{3 + v}{8}\rho\,\omega^2 r^2$$

When $r = 250$, $\sigma_r = -34 \cdot 5 \times 10^6$; therefore

$$-34 \cdot 5 \times 10^6 = C$$

$$-\left(\frac{3 + 0 \cdot 28}{8} \times 7 \cdot 75 \times 10^3 \times 117\,500 \times 0 \cdot 0625\right)$$

and

$$C = (-34 \cdot 5 \times 10^6) + (23 \cdot 3 \times 10^6) = -11 \cdot 2 \times 10^6 \text{ N/m}^2$$

But at the centre of the disc, $r = 0$ and $\sigma_r = \sigma_\theta = C$; therefore

$$\sigma_r = \sigma_\theta = -11 \cdot 2 \text{ MN/m}^2$$

(c) Discs of Uniform Strength

It is possible to determine the required profile for a circular disc so that when it is rotating it is uniformly stressed throughout. Since both the radial and hoop stresses are to be equal at any point,* there is only one value of stress to be considered, and the whole problem can be uniquely determined by one equation, that of equilibrium.

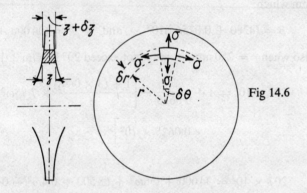

Fig 14.6

The element shown in Fig. 14.6 is subjected to radial and circumferential stresses as in Fig. 14.5, but here the thickness of the disc varies with radius. Considering the equilibrium of the element in the radial direction, then the net force outwards due to the radial stresses is given by

$$\sigma(z + \delta z)(r + \delta r)\delta\theta - \sigma z r \, \delta\theta$$

The radial force due to components of the circumferential stresses is

$$-2\sigma\left(z + \frac{\delta z}{2}\right)\delta r \sin\frac{\delta\theta}{2}$$

* The periphery shown is not a free boundary; thus $\sigma_r \neq 0$, since there would be a rim for blading or buckets which has not been shown. For a detailed discussion, see A. Stodola, *Steam and Gas Turbines*, Vol. 1 (McGraw-Hill, 1927).

and the force due to centripetal acceleration is

$$\rho \left(z + \frac{\delta z}{2} \right) \left(r + \frac{\delta r}{2} \right) \delta\theta \, \delta r \, \omega^2 r$$

Therefore, for equilibrium,

$$\sigma(z + \delta z)(r + \delta r)\delta\theta - \sigma z r \, \delta\theta - 2\sigma \left(z + \frac{\delta z}{2} \right) \delta r \sin \frac{\delta\theta}{2}$$

$$+ \rho \left(z + \frac{\delta z}{2} \right) \left(r + \frac{\delta r}{2} \right) \delta\theta \, \delta r \, \omega^2 r = 0 \qquad (14.59)$$

Simplifying, neglecting the higher orders of small quantities and letting $\sin \frac{1}{2}\delta\theta \to \frac{1}{2}\delta\theta$,

$$\sigma r \, \delta z \, \delta\theta + \sigma z \, \delta r \, \delta\theta - \sigma z \, \delta r \, \delta\theta + \rho z r \, \delta r \, \delta\theta \, \omega^2 r = 0$$

or, in the limit,

$$\frac{dz}{z} = -\rho \frac{\omega^2}{\sigma} r \, dr \qquad (14.60)$$

and integrating both sides,

$$\log_e z = -\rho \frac{\omega^2 r^2}{\sigma \, 2} + \text{constant}$$

Therefore

$$z = A e^{-\rho w^2 r^2 / 2\sigma} \qquad (14.61)$$

The constant A may be determined from a condition of geometry such as when $r = 0$, $z = z_0$; therefore $A = z_0$, and

$$z = z_0 e^{-\rho w^2 r^2 / 2\sigma} \qquad (14.62)$$

This expression gives the required thickness profile of the disc for limiting conditions of the uniform stress and speed of rotation.

14.7. THICK-WALLED SPHERICAL SHELL

This is a problem which has to be considered from first principles since the equilibrium equations so far derived are not applicable. It is convenient to have co-ordinates referred to the centre of the sphere since it is symmetrical about this point. Considering the applied forces in the form of uniformly distributed pressure, then the element within the wall of the vessel in Fig. 14.7 is subjected to the stress system shown. From symmetry there will be no shearing stress and the circumferential stress,

σ_θ, will be the same in all directions at any particular radius, r.* There will therefore be only one equation of equilibrium obtainable which will refer to the radial direction.

Along a radial line through the centre of the element, the equilibrium of forces will be as follows:

$$\left(\sigma_r + \frac{d\sigma_r}{dr}\,\delta r\right)(r + \delta r)\,\delta\theta\,(r + \delta r)\,\delta\theta - \sigma_r r\,\delta\theta\, r\,\delta\theta$$

$$-2\sigma_\theta\,\delta r\, r\,\delta\theta\sin\frac{\delta\theta}{2} - 2\sigma_\theta\,\delta r\, r\,\delta\theta\sin\frac{\delta\theta}{2} + Rr\,\delta\theta\, r\,\delta\theta\,\delta r = 0$$

Simplifying this equation and neglecting the higher-order terms gives

$$\frac{d\sigma_r}{dr} + \frac{2}{r}(\sigma_r - \sigma_\theta) + R = 0 \tag{14.63}$$

Partial derivatives are not required since σ_r is not dependent on θ, and the body force, R, can be ignored in this particular problem; therefore

$$\frac{d\sigma_r}{dr} + \frac{2}{r}(\sigma_r - \sigma_\theta) = 0 \tag{14.64}$$

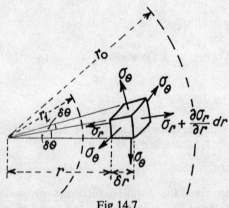

Fig 14.7

This equation may be compared with eqn. (14.12) for axial symmetry used for the thick-walled cylinder problem.

It will be readily observed that the strain–displacement relationships will be the same as previously derived for axial symmetry, and hence

$$\varepsilon_r = \frac{du}{dr} \quad \text{and} \quad \varepsilon_\theta = \frac{u}{r} \tag{14.65}$$

* In a more general problem which is not symmetrical, the spherical co-ordinates are r, θ and ϕ.

Since σ_θ is the same in all directions, the stress–strain relationships may be written in the following way:

$$\varepsilon_r = \frac{\sigma_r}{E} - \frac{2\nu\sigma_\theta}{E} \tag{14.66}$$

$$\varepsilon_\theta = \frac{\sigma_\theta}{E} - \frac{\nu}{E}(\sigma_r + \sigma_\theta) = -\frac{\nu\sigma_r}{E} + \frac{1-\nu}{E}\sigma_\theta \tag{14.67}$$

The first step of the solution is to obtain the stresses in terms of the displacement, u. Multiplying eqn. (14.66) by ν and adding to eqn. (14.67), we get

$$\nu\varepsilon_r + \varepsilon_\theta = \frac{1 - \nu - 2\nu^2}{E}\sigma_\theta$$

and substituting for σ_θ in eqn. (14.66),

$$\frac{\sigma_r}{E} = \varepsilon_r + \frac{2\nu}{1 - \nu - 2\nu^2}(\nu\varepsilon_r + \varepsilon_\theta)$$

Substituting for ε_r and ε_θ in terms of u,

$$\frac{\sigma_r}{E} = \frac{du}{dr} + \frac{2\nu}{1 - \nu - 2\nu^2}\left(\nu\frac{du}{dr} + \frac{u}{r}\right) = \frac{2\nu\dfrac{u}{r} + (1-\nu)\dfrac{du}{dr}}{1 - \nu - 2\nu^2} \tag{14.68}$$

$$\frac{\sigma_\theta}{E} = \frac{\nu\dfrac{du}{dr} + \dfrac{u}{r}}{1 - \nu - 2\nu^2} \tag{14.69}$$

Integrating σ_r with respect to r and inserting these values of $d\sigma_r/dr$, σ_r and σ_θ into the equilibrium equation (14.64) gives an expression in terms of u:

$$-\frac{2\nu}{1 - \nu - 2\nu^2}\frac{u}{r^2} + \frac{1}{r}\frac{2\nu}{1 - \nu - 2\nu^2}\frac{du}{dr} + \frac{1-\nu}{1 - \nu - 2\nu^2}\frac{d^2u}{dr^2}$$

$$+\frac{2}{r}\frac{du}{dr} - \frac{2\nu}{r(1 - \nu - 2\nu^2)}\frac{du}{dr} + \frac{4\nu^2}{r(1 - \nu - 2\nu^2)}\frac{du}{dr}$$

$$+\frac{2\nu - 1}{1 - \nu - 2\nu^2}\frac{2u}{r^2} = 0$$

On simplifying, this equation reduces to

$$\frac{d^2u}{dr^2} + \frac{2}{r}\frac{du}{dr} - \frac{2u}{r^2} = 0 \tag{14.70}$$

which may be compared in form to that for axial symmetry, eqn. (14.48). In the present problem, as there is no body force, the right-hand side of the equation is zero.

The general solution of this equation is

$$u = Kr + \frac{K'}{r^2}$$

which may be verified by direct substitution.

By substituting for u and du/dr in eqns. (14.68) and (14.69), the radial and hoop stresses can be expressed in terms of a single variable, r:

$$\sigma_r = \frac{E}{1 - v - 2v^2}\left[2v\left(K + \frac{K'}{r^3}\right) + (1 - v)\left(K - \frac{2K'}{r^3}\right)\right]$$

$$= \frac{E}{1 - v - 2v^2}\left[(1 + v)K - 2(1 - 2v)\frac{K'}{r^3}\right]$$

$$= \frac{E}{1 - 2v}K - \frac{2E}{1 + v}\frac{K'}{r^3} \tag{14.71}$$

and

$$\sigma_\theta = \frac{E}{1 - v - 2v^2}\left[v\left(K - \frac{2K'}{r^3}\right) + K + \frac{K'}{r^3}\right]$$

$$= \frac{E}{1 - v - 2v^2}\left[(1 + v)K + (1 - 2v)\frac{K'}{r^3}\right]$$

$$= \frac{E}{1 - 2v}K + \frac{E}{1 + v}\frac{K'}{r^3} \tag{14.72}$$

The above equations for σ_r and σ_θ can be rewritten, for simplicity, in the form

$$\sigma_r = A - 2\frac{B}{r^3} \tag{14.73}$$

$$\sigma_\theta = A + \frac{B}{r^3} \tag{14.74}$$

where A and B are constants determined from the boundary conditions of the problem. If these are internal pressure $- p_i$ at $r = r_i$ and external pressure $- p_o$ at $r = r_o$, then, from eqn. (14.73),

$$-p_i = A - \frac{2B}{r_i^3}$$

$$-p_o = A - \frac{2B}{r_o^3}$$

from which

$$B = \frac{r_i^3 r_o^3 (p_i - p_o)}{2(r_o^3 - r_i^3)}$$

and

$$A = \frac{p_i r_i^3 - p_o r_o^3}{r_o^3 - r_i^3}$$

Substituting for A and B in eqns. (14.73) and (14.74),

$$\sigma_r = \frac{p_i r_i^3 - p_o r_o^3}{r_o^3 - r_i^3} - \frac{r_i^3 r_o^3 (p_i - p_o)}{r^3 (r_o^3 - r_i^3)} \tag{14.75}$$

and

$$\sigma_\theta = \frac{p_i r_i^3 - p_o r_o^3}{r_o^3 - r_i^3} + \frac{r_i^3 r_o^3 (p_i - p_o)}{2r^3 (r_o^3 - r_i^3)} \tag{14.76}$$

For the case of internal pressure only, $p_o = 0$ and

$$\sigma_r = \frac{p_i r_i^3}{r_o^3 - r_i^3} - \frac{r_i^3 r_o^3 p_i}{r^3 (r_o^3 - r_i^3)}$$

$$= \frac{p_i r_i^3}{r_o^3 - r_i^3} \left(1 - \frac{r_o^3}{r^3}\right) \tag{14.77}$$

and

$$\sigma_\theta = \frac{p_i r_i^3}{r_o^3 - r_i^3} \left(1 + \frac{r_o^3}{2r^3}\right) \tag{14.78}$$

Example 14.5.   A spherical vessel of internal diameter 500 mm is required to withstand an internal pressure of 1035 bars. If the circumferential tensile stress at the inner surface is to be limited to 230 MN/m², determine the required wall thickness. Find also the change in internal diameter when under full pressure. $E = 207\,\text{GN/m}^2$, $v = 0.29$.

Since 1035 bars is 103·5 MN/m², then using eqns. (14.73) and (14.74) gives

$$\sigma_r = -103{\cdot}5 \times 10^6 = A - \frac{2B}{0{\cdot}25^3}$$

$$\sigma_\theta = 230 \times 10^6 = A + \frac{B}{0{\cdot}25^3}$$

from which

$$(230 \times 10^6) + (103{\cdot}5 \times 10^6) = \frac{3B}{0{\cdot}25^3}$$

or

$$B = 1.74 \times 10^6$$

and

$$A = (230 \times 10^6) - (111.2 \times 10^6) = 118.8 \times 10^6$$

At the outer surface, r_o, $\sigma_r = 0$; therefore

$$0 = (118.8 \times 10^6) - \frac{2 \times 1.74 \times 10^6}{r_o{}^3}$$

Therefore

$$r_o{}^3 = \frac{3.48}{118.8} = 0.0293 \quad \text{and} \quad r_o = 0.308 \, \text{m}$$

Hence the required wall thickness is 58 mm.

In order to find the change in internal diameter it is necessary to determine the displacement, u, at $r = 250$ mm. Now,

$$u = Kr + \frac{K'}{r^2}$$

and comparing eqns. (14.71) and (14.73) it is seen that the constants K and K' in terms of A and B are

$$K = \frac{1 - 2v}{E} A$$

and

$$K' = \frac{1 + v}{E} B$$

Thus

$$K = \frac{1 - (2 \times 0.29)}{207 \times 10^9} \times 118.8 \times 10^6 = 0.241 \times 10^{-3}$$

and

$$K' = \frac{1 + 0.29}{207 \times 10^9} \times 1.74 \times 10^6 = 10.85 \times 10^{-6}$$

At $r = 250$ mm,

$$u = (0.241 \times 10^{-3} \times 0.25) + \frac{10.85 \times 10^{-6}}{0.25^2}$$

$$= 0.234 \, \text{mm}$$

Hence the change in diameter is an increase of 0.468 mm.

Chapter 15
Theories of Yielding

15.1. All the theoretical analysis of the previous chapters has made use of a proportional stress–strain relationship. This is because Hooke's law established that metals have a linear elastic stress–strain range. However, if a metal specimen is subjected to simple axial loading, it is found that beyond a certain point, stress is no longer proportional to strain and the material is said to be *yielding*.* Knowing the stress at which this latter behaviour commenced, it would now be a simple matter to design a second specimen of the same material to withstand a particular load without any yielding occurring. This example is simple and there is only one principal stress to consider.

The problem of designing a pressure vessel, rotating disc, or some component containing a two or three principal stress system so that the material remains elastic, i.e. no yielding, when under full load is rather more complex. One could adopt a trial-and-error method of building a component and testing it to find when the deformations were no longer proportional to the applied load, but this would obviously be very un- economical. It is therefore essential to find some criterion based on stresses, or strains, or perhaps strain energy in the complex system which can be related to the simple axial conditions mentioned above. If a theoretical criterion can be established which agrees with complex material behaviour, it is then only necessary to establish experimentally the yield point in a simple tension or compression test.

A number of theoretical criteria for yielding have been proposed over the past century, and some of these will now be discussed in relation to ductile and brittle materials.

15.2. MAXIMUM PRINCIPAL STRESS THEORY

Theories of yielding are generally expressed in terms of principal stresses, since these completely determine a general state of stress. The element of material shown in Fig. 15.1(a) is subjected to three principal stresses, and the convention to be used is that $\sigma_1 > \sigma_2 > \sigma_3$.

The *maximum principal stress theory*, often attributed to Rankine, states that yielding will occur in a material under complex stress when σ_1 attains a value equal to the yield stress, σ_{Yt}, in a simple tension test on the same material. Yielding could also occur if the minimum stress, σ_3,

---

* See Chapter 19.

were compressive and reached the value of yield stress in a simple compression test. These statements may be written as

$$\sigma_1 = \sigma_{Yt}$$

or

$$\sigma_3 = \sigma_{Yc} \qquad (15.1)$$

for yielding to occur.

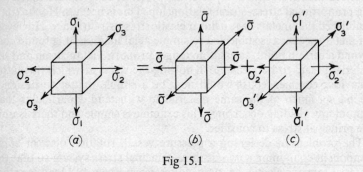

Fig 15.1

15.3. MAXIMUM PRINCIPAL STRAIN THEORY

St. Venant postulated that yielding commenced when the *maximum principal strain* (tensile), ε_1, was equivalent to the strain corresponding to the yield stress in simple tension. For yielding in compression the minimum principal strain, ε_3, would equal the yield strain in simple compression. If the strains are expressed in terms of stresses, then

$$\varepsilon_1 = \frac{\sigma_1}{E} - \frac{v}{E}(\sigma_2 + \sigma_3)$$

and yielding occurs when the expression equals σ_{Yt}/E:

$$\sigma_1 - v(\sigma_2 + \sigma_3) = \sigma_{Yt} \qquad (15.2)$$

or for compression,

$$\sigma_3 - v(\sigma_1 + \sigma_2) = \sigma_{Yc} \qquad (15.3)$$

15.4. MAXIMUM SHEAR STRESS THEORY

The assumption in this theory is that yielding is dependent on the maximum shear stress in the material reaching a critical value. This latter value is taken as the maximum shear stress in a simple tensile test, which is half the yield stress, or $\frac{1}{2}\sigma_{Yt}$. The maximum shear stress in the complex stress system will depend on the relative values and signs of the three principal stresses, always being half the difference between the maximum and minimum.

For a general three-dimensional stress system, or in the two-dimensional case with one of the stresses compressive and the other tensile, the maximum shear stress is

$$\hat{\tau} = \frac{\sigma_1 - \sigma_3}{2}$$

and for yielding

$$\frac{\sigma_1 - \sigma_3}{2} = \frac{\sigma_{Yt}}{2}$$

or

$$\sigma_1 - \sigma_3 = \sigma_{Yt} \qquad (15.4)$$

In a two-dimensional stress system when $\sigma_3 = 0$, i.e. for σ_1 and σ_2 tensile, the maximum difference between the principal stresses is

$$\hat{\tau} = \frac{\sigma_1 - 0}{2} = \frac{\sigma_1}{2}$$

and yielding occurs if

$$\frac{\sigma_1}{2} = \frac{\sigma_{Yt}}{2}$$

or

$$\sigma_1 = \sigma_{Yt} \qquad (15.5)$$

This theory is usually coupled with the names of Guest and Tresca.

15.5. TOTAL STRAIN ENERGY THEORY

The theories put forward so far have postulated a criterion for yielding in terms of a limiting value of stress or strain. The present theory, as proposed by Beltrami, and also attributed to Haigh, is based on a critical value of the *total strain energy* stored in the material, and this is a product of stress and strain.

It has been shown earlier that the work done in deformation or the stored elastic strain energy may be written as $\frac{1}{2}W\,\delta x$ or

$$\frac{\frac{1}{2}W\,\delta x}{Ax} = \frac{1}{2}\sigma\varepsilon \text{ per unit volume}$$

In a three-dimensional stress system, the total strain energy is

$$U_T = \tfrac{1}{2}\sigma_1\varepsilon_1 + \tfrac{1}{2}\sigma_2\varepsilon_2 + \tfrac{1}{2}\sigma_3\varepsilon_3$$

Now using the stress–strain relationships, the principal strains may be written as

$$\varepsilon_1 = \frac{\sigma_1}{E} - \frac{\nu}{E}(\sigma_2 + \sigma_3)$$

$$\varepsilon_2 = \frac{\sigma_2}{E} - \frac{\nu}{E}(\sigma_3 + \sigma_1)$$

$$\varepsilon_3 = \frac{\sigma_3}{E} - \frac{\nu}{E}(\sigma_1 + \sigma_2)$$

substituting for ε_1, ε_2 and ε_3 and rearranging,

$$U_T = \frac{1}{2E}(\sigma_1{}^2 + \sigma_2{}^2 + \sigma_3{}^2) - \frac{\nu}{2E}(2\sigma_1\sigma_2 + 2\sigma_2\sigma_3 + 2\sigma_3\sigma_1)$$

Yielding is said to occur when the above is equal to the total strain energy at yielding in simple tension, i.e. by putting $\sigma_2 = \sigma_3 = 0$ and $\sigma_1 = \sigma_{Yt}$,

$$U_T = \frac{\sigma_{Yt}{}^2}{2E}$$

Therefore

$$\sigma_1{}^2 + \sigma_2{}^2 + \sigma_3{}^2 - 2\nu(\sigma_1\sigma_2 + \sigma_2\sigma_3 + \sigma_3\sigma_1) = \sigma_{Yt}{}^2 \qquad (15.6)$$

In the two-dimensional system, $\sigma_3 = 0$ and

$$\sigma_1{}^2 + \sigma_2{}^2 - 2\nu\sigma_1\sigma_2 = \sigma_{Yt}{}^2 \qquad (15.7)$$

15.6. SHEAR OR DISTORTION STRAIN ENERGY THEORY

Huber, in 1904, proposed that the total strain energy of an element of material could be divided into two parts, that due to change in volume and that due to change in shape. These will be termed *volumetric strain energy*, U_v, and *distortion* or *shear strain energy*, U_s. It is rather more simple to determine the former quantity than the latter, and since the total strain energy has already been determined, the shear or distortion component can be determined as

$$U_s = U_T - U_v \qquad (15.8)$$

In order to show that the deformation of a material can be separated into change in volume and change in shape, consider the element in Fig. 15.1 subjected to the principal stresses σ_1, σ_2 and σ_3. These may be written in terms of the "average" stress in the element as follows:

$$\left.\begin{array}{l} \sigma_1 = \bar{\sigma} + \sigma_1' \\ \sigma_2 = \bar{\sigma} + \sigma_2' \\ \sigma_3 = \bar{\sigma} + \sigma_3' \end{array}\right\} \qquad (15.9)$$

where $\bar{\sigma}$ is the average or mean stress defined as

$$\bar{\sigma} = \frac{\sigma_1 + \sigma_2 + \sigma_3}{3} \qquad (15.10)$$

Now, when an element as in Fig. 15.1(b) is subjected to $\bar{\sigma}$ in all directions, this hydrostatic stress will produce a change in volume, but no distortion. Consider the effect of the σ' components of stress. Adding together eqns. (15.9) gives

$$\sigma_1 + \sigma_2 + \sigma_3 = 3\bar{\sigma} + \sigma_1' + \sigma_2' + \sigma_3'$$

but $\bar{\sigma} = \frac{1}{3}(\sigma_1 + \sigma_2 + \sigma_3)$; hence

$$\sigma_1' + \sigma_2' + \sigma_3' = 0 \qquad (15.11)$$

But from the stress–strain relationships,

$$\left. \begin{array}{l} \varepsilon_1' = \dfrac{\sigma_1'}{E} - \dfrac{v}{E}(\sigma_2' + \sigma_3') \\[2mm] \varepsilon_2' = \dfrac{\sigma_2'}{E} - \dfrac{v}{E}(\sigma_3' + \sigma_1') \\[2mm] \varepsilon_3' = \dfrac{\sigma_3'}{E} - \dfrac{v}{E}(\sigma_1' + \sigma_2') \end{array} \right\} \qquad (15.12)$$

Hence

$$\varepsilon_1' + \varepsilon_2' + \varepsilon_3' = \varepsilon_v' = \frac{(1 - 2v)}{E}(\sigma_1' + \sigma_2' + \sigma_3') \qquad (15.13)$$

and since the sum of the three stresses is zero, eqn. (15.11),

$$\varepsilon_1' + \varepsilon_2' + \varepsilon_3' = \varepsilon_v' = 0 \qquad (15.14)$$

Thus the stress components cause no change in volume but only change in shape.

The volumetric strain energy can now be determined from the hydrostatic component of stress, $\bar{\sigma}$.

$$\begin{aligned} U_v &= \tfrac{1}{2}\bar{\sigma}\bar{\varepsilon} \\[2mm] &= \tfrac{1}{2}\bar{\sigma}\frac{3\bar{\sigma}}{E}(1 - 2v) \\[2mm] &= \tfrac{1}{2}\left[\frac{\sigma_1 + \sigma_2 + \sigma_3}{3}\right]\left[\frac{3(1 - 2v)}{E}\right]\left[\frac{\sigma_1 + \sigma_2 + \sigma_3}{3}\right] \\[2mm] &= \frac{1 - 2v}{6E}(\sigma_1 + \sigma_2 + \sigma_3)^2 \qquad (15.15) \end{aligned}$$

But $U_s = U_T - U_v$; therefore

$$U_s = \frac{1}{2E}[\sigma_1{}^2 + \sigma_2{}^2 + \sigma_3{}^2 - 2\nu(\sigma_1\sigma_2 + \sigma_2\sigma_3 + \sigma_3\sigma_1)]$$
$$- \frac{1 - 2\nu}{6E}(\sigma_1 + \sigma_2 + \sigma_3)^2$$

which reduces to

$$U_s = \frac{1 + \nu}{6E}[(\sigma_1 - \sigma_2)^2 + (\sigma_2 - \sigma_3)^2 + (\sigma_3 - \sigma_1)^2]$$

per unit volume   (15.16)

or alternatively, using the relationship between E, G and ν,

$$U_s = \frac{1}{12G}[(\sigma_1 - \sigma_2)^2 + (\sigma_2 - \sigma_3)^2 + (\sigma_3 - \sigma_1)^2] \quad (15.17)$$

Now, the shear or distortion strain energy theory proposes that yielding commences when the quantity U_s reaches the equivalent value at yielding in simple tension. In the latter case σ_2 and $\sigma_3 = 0$ and $\sigma_1 = \sigma_{Yt}$; therefore

$$U_s = \frac{\sigma_{Yt}{}^2}{6G} \quad \text{per unit volume} \qquad (15.18)$$

and

$$\frac{1}{12G}[(\sigma_1 - \sigma_2)^2 + (\sigma_2 - \sigma_3)^2 + (\sigma_3 - \sigma_1)^2] = \frac{\sigma_{Yt}{}^2}{6G}$$

or

$$(\sigma_1 - \sigma_2)^2 + (\sigma_2 - \sigma_3)^2 + (\sigma_3 - \sigma_1)^2 = 2\sigma_{Yt}{}^2 \qquad (15.19)$$

In the two-dimensional system, $\sigma_3 = 0$ and

$$\sigma_1{}^2 + \sigma_2{}^2 - \sigma_1\sigma_2 = \sigma_{Yt}{}^2 \qquad (15.20)$$

for yielding to occur. This theory was also independently established by Maxwell, von Mises and Hencky, and is nowadays generally referred to as the *Mises criterion*.

The above analysis has been directly aimed at establishing a yield criterion on an energy basis. However, one might equally well propose that yielding occurs as a function of the differences between principal stresses. On this hypothesis it is evident that eqn. (15.19) is also obtained by considering the root mean square of the principal stress differences in the complex stress system in relation to simple tension. Thus

$$\sqrt{[\tfrac{1}{3}\{(\sigma_1 - \sigma_2)^2 + (\sigma_2 - \sigma_3)^2 + (\sigma_3 - \sigma_1)^2\}]} = \sqrt{[\tfrac{1}{3}(2\sigma_{Yt}{}^2)]}$$
$$(15.21)$$

The right-hand side of the equation is obtained for simple tension by putting $\sigma_1 = \sigma_{Yt}$ and $\sigma_2 = \sigma_3 = 0$. Squaring both sides of eqn. (15.21),

$$(\sigma_1 - \sigma_2)^2 + (\sigma_2 - \sigma_3)^2 + (\sigma_3 - \sigma_1)^2 = 2\sigma_{Yt}^2$$

which is the same as eqn. (15.19).

Summary

Many experiments have been conducted under complex stress conditions to study the behaviour of metals and test the validity of the foregoing theories. It has been shown that hydrostatic pressure, and by inference hydrostatic tension, does not cause yielding. Now, any complex stress system can be regarded as a combination of hydrostatic stress and a function of the difference of principal stresses, and therefore a yield criterion such as that of Tresca or von Mises which is based on principal stress difference would seem to be the most logical. It is now well established that for ductile metals, exhibiting yielding and subsequent plastic deformation, the shear strain energy theory correlates best with material behaviour.

The maximum shear stress theory, although not quite so consistent as the former, gives fairly reasonable predictions and is sometimes used in design by virtue of its simpler mathematical form. *The other theories are no longer used for ductile metals, some being positively unsafe.*

15.7. YIELD LOCI

The expressions for the five theories that have been discussed can be plotted graphically for the simplified conditions of $\sigma_3 = 0$ and equal yield stress in simple tension and compression. In Fig. 15.2 co-ordinate axes of the principal stresses σ_1 and σ_2 are drawn and the curves shown represent the five theories plotted on this basis. The maximum principal stress theory is represented by the square ABCD, and according to this theory any stress condition inside the boundaries of the square will be elastic, while outside the area, yielding will have occurred. The principal strain theory gives the rhombus EFGH, the co-ordinates of these points being

$$\left(\frac{\sigma_{Yt}}{1-\nu}, \frac{\sigma_{Yt}}{1-\nu}\right) \qquad \left(\frac{\sigma_{Yc}}{1+\nu}, \frac{\sigma_{Yt}}{1+\nu}\right)$$

$$\left(\frac{\sigma_{Yc}}{1-\nu}, \frac{\sigma_{Yc}}{1-\nu}\right) \qquad \left(\frac{\sigma_{Yt}}{1+\nu}, \frac{\sigma_{Yc}}{1+\nu}\right)$$

respectively.

For principal stress of opposite sign, the maximum shear stress theory has yield lines in the second and fourth quadrants denoted by RS and

QT. However, in the first and third quadrants, where σ_1 and σ_2 are of like sign, $\sigma_3 = 0$ has to be used to give the maximum shear stress, and the yield boundary coincides with the maximum principal stress theory at QAR and SCT. The two energy criteria plot as ellipses, and the ellipse due to the shear strain energy criterion circumscribes the maximum shear stress hexagon.

Each of these diagrams is termed a *yield locus*, and as stated previously, inside the locus elastic conditions prevail. The locus itself represents the onset of yielding, and outside the locus the material is in the plastic range. It is seen that, except for the small areas cut off at the corners A and C, the maximum shear stress criterion is the most conservative of the five loci.

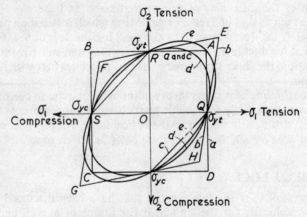

a = Maximum principal stress theory
b = Maximum principal strain theory
c = Maximum shear stress theory
d = Total strain energy theory
e = Shear strain energy theory

Fig 15.2

A diagram such as Fig. 15.2 serves several useful purposes. Experimental points can be plotted on it and compared with each of the theoretical curves as in Fig. 19.16. The latter can readily be assessed against each other, and ratios of σ_1 to σ_2 to cause yielding can be quickly determined. For example, in the case of equal biaxial tension (or compression), $\sigma_1/\sigma_2 = 1$, and proceeding along this line from the origin the loci are reached in the order of theories (*d*), (*a, c, e*) and (*b*), whereas for pure torsion, $\sigma_1/\sigma_2 = -1$, and the order of yield boundaries becomes (*c*), (*e*), (*d*), (*b*), (*a*).

For the case of three principal stresses, all non-zero, the yield locus becomes a yield envelope centred on co-ordinate axes σ_1, σ_2, σ_3. The maximum shear stress and shear strain energy theories are represented by hexagonal and circumscribing circular cylinders respectively as illustrated in Fig. 15.3. The axis of the cylinders is equally inclined to each of the three co-ordinate axes. Inside the cylinders an elastic condition exists, while outside the material is in the plastic range.

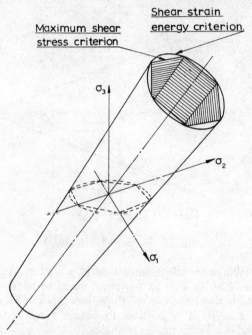

Fig 15.3

15.8. BRITTLE MATERIALS

Brittleness in a material may be defined as the absence of the ability to deform plastically. Materials such as flake cast-iron, concrete and ceramics when subjected to tensile stress will generally fracture at the elastic limit or only a very small strain beyond this point. This means that the term yield criterion used for ductile materials may often also imply a fracture criterion for a brittle material.

Experiment has shown that the maximum principal stress theory is the most satisfactory for predicting failure.

Some brittle materials are considerably stronger in compression than in tension, and Mohr put forward a construction, based on his stress circle, to allow for this in the application of the maximum shear stress theory. Mohr's stress circle for simple tension is represented in Fig. 15.4 by the circle on diameter OP, and for simple compression by the circle on OQ. If OP and OQ are made equal to the "yield" or fracture strengths, σ_{ut} and σ_{uc}, of a brittle material, then the tangents to these circles form

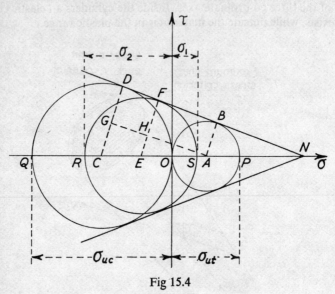

Fig 15.4

a fracture envelope for all combinations of σ_1 and σ_2. A general stress circle such as RSF in Fig. 15.4 is constructed to have a diameter of $(\sigma_1 + \sigma_2)$ and to touch the two envelope lines DN and MN. By a simple geometrical analysis it is possible to obtain a relationship between σ_1, σ_2, σ_{ut} and σ_{uc}. From similar triangles,

$$\frac{AE}{AC} = \frac{EH}{CG}$$

and expressing these lengths in terms of stresses in Fig. 15.4 we obtain

$$\frac{\sigma_{ut} - \sigma_1 + \sigma_2}{\sigma_{ut} + \sigma_{uc}} = \frac{\sigma_1 + \sigma_2 - \sigma_{ut}}{\sigma_{uc} - \sigma_{ut}}$$

and therefore

$$\frac{\sigma_1}{\sigma_{ut}} + \frac{\sigma_2}{\sigma_{uc}} = 1$$

From this equation, for a given ratio of the principal stresses, when one is tension and the other compression, σ_1 and σ_2 can be found.

As an example, for cast iron in simple shear, assuming a compressive strength four times the tensile strength, we have $\tau_u = \sigma_1 = -\sigma_2$ and $4\sigma_{ut} = -\sigma_{uc}$; therefore

$$\frac{\tau_u}{\sigma_{ut}} + \frac{\tau_u}{4\sigma_{ut}} = 1$$

or $\tau_u = 0 \cdot 8\sigma_{ut}$, which agrees well with experimental results. A "yield" or fracture locus similar to Fig. 15.2 can also be drawn for the Mohr theory as shown in Fig. 15.5.

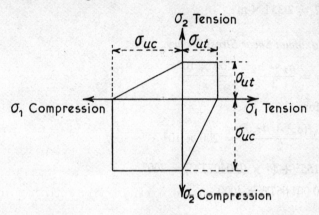

Fig 15.5

Example 15.1. A mild steel shaft of 50 mm diameter is subjected to a bending moment of $1 \cdot 9$ kN-m. If the yield point of the steel in simple tension if 200 MN/m² , find the maximum torque that can also be applied according to (*a*) the maximum principal stress, (*b*) the maximum shear stress, (*c*) the shear strain energy theories of yielding.

The maximum bending stress occurs at the surface of the shaft and is given by

$$\sigma_x = \frac{32M}{\pi d^3} = \frac{32}{\pi} \times \frac{1900}{125 \times 10^{-6}} = 155 \, \text{MN/m}^2$$

The maximum shear stress at the surface is

$$\tau_{xy} = \frac{16T}{\pi d^3} = \frac{16}{\pi} \times \frac{T}{125 \times 10^{-6}} = 40 \cdot 7 \times 10^3 T$$

(*a*) *Maximum Principal Stress Theory*

$$\sigma_1 = \frac{\sigma_x}{2} + \tfrac{1}{2} \sqrt{(\sigma_x{}^2 + 4\tau_{xy}{}^2)} = 200 \times 10^6$$

Therefore

$$\frac{155 \times 10^6}{2} + \tfrac{1}{2}\sqrt{[(155 \times 10^6)^2 + \{4 \times (40 \cdot 7 \times 10^3 T)^2\}]}$$

$$= 200 \times 10^6$$

$$\sqrt{[155^2 + \{4 \times (0 \cdot 0407 T)^2\}]} = 245$$

$$0 \cdot 001\,66 T^2 = 9000$$

and

$$T = 2 \cdot 33 \text{ kN-m}$$

(b) *Maximum Shear Stress Theory*

$$\tau = \frac{\sigma_1 - \sigma_2}{2} = \frac{200 \times 10^6}{2}$$

Therefore

$$\frac{\sqrt{(\sigma_x^2 + 4\tau_{xy}^2)}}{2} = 200 \times 10^6$$

$$155^2 + [4 \times (0 \cdot 0407 T)^2] = 200^2$$

$$0 \cdot 001\,66 T^2 = 4000$$

and

$$T = 1 \cdot 55 \text{ kN-m}$$

(c) *Shear Strain Energy Theory*

$$\sigma_1^2 + \sigma_2^2 - \sigma_1 \sigma_2 = (200 \times 10^6)^2$$

Putting $\sigma_x^2 + 4\tau_{xy}^2 = A$,

$$\tfrac{1}{4}(\sigma_x + \sqrt{A})^2 + \tfrac{1}{4}(\sigma_x - \sqrt{A})^2 - \tfrac{1}{4}(\sigma_x - \sqrt{A})(\sigma_x + \sqrt{A})$$

$$= (200 \times 10^6)^2$$

$$\tfrac{1}{4}[\sigma_x^2 + 2\sigma_x\sqrt{A} + A + \sigma_x^2 - 2\sigma_x\sqrt{A} + A - \sigma_x^2 + A]$$

$$= (200 \times 10^6)^2$$

Simplifying gives

$$\sigma_x^2 + 3\tau_{xy}^2 = (200 \times 10^6)^2$$

$$155^2 + [3 \times (0 \cdot 001\,66 T^2)] = 200^2$$

$$0 \cdot 001\,66 T^2 = 5330$$

Therefore

$$T = 1 \cdot 79 \text{ kN-m}$$

Example 15.2. In the compound pressure vessel example, page 339, determine where yielding would first occur according to the various theories already described.

An inspection of the values of circumferential and radial stresses shows that the maximum principal stress theory would predict yielding at the inside of the outer cylinder.

The maximum principal strain theory when $\sigma_3 = 0$ reduces to $(\sigma_1 - \nu\sigma_2)$. Assuming a value for ν of 0·3, then yielding is likely to occur first at the inside of either the inner or the outer cylinder. In the former case the equivalent stress is

$$263 - [0·3 \times (-207)] = 325 \, \text{MN/m}^2$$

and for the latter it is

$$272·6 - (0·3 \times 76·3) = 295·5 \, \text{MN/m}^2$$

Hence yielding will occur first at the bore of the inner cylinder.

Again by inspection it will be seen that the maximum shear stress anywhere in the vessel is at the inner surface of the inner cylinder, where

$$\frac{\sigma_1 - \sigma_2}{2} = \frac{263 - (-207)}{2} = 235 \, \text{MN/m}^2$$

The two energy theories are of very similar form and these also have their maximum value at the bore of the inner tube. The shear strain energy gives an equivalent stress $\bar{\sigma}$ as

$$\bar{\sigma} = \sqrt{[\tfrac{1}{2}\{263^2 + 207^2 - (263 \times -207)\}]} = 289 \, \text{MN/m}^2$$

and the total strain energy gives

$$\bar{\sigma} = \sqrt{[\tfrac{1}{2}\{263^2 + 207^2 - (2 \times 0·3 \times 263 \times -207)\}]}$$
$$= 269 \, \text{MN/m}^2$$

With the exception of the maximum principal stress theory, all the other theories agree on the position where yielding will first take place. If, however, a limiting tensile yield stress is given for the material, each theory gives a different permissible internal pressure (see Problem 15.1).

Example 15.3. In a cast-iron component the maximum principal stress is to be limited to one-third of the tensile strength. Determine the maximum value of the minimum principal stress using the Mohr theory. What would be the values of the principal stresses associated with a maximum shear stress of 390 MN/m²? The tensile and compressive strengths of the cast iron are 360 MN/m² and 1·41 GN/m² respectively.

Maximum principal stress $= 360/3 = 120\,\mathrm{MN/m^2}$ (tension)
According to Mohr's theory,

$$\frac{\sigma_1}{\sigma_{ut}} + \frac{\sigma_2}{\sigma_{uc}} = 1$$

Therefore

$$\frac{120}{360} + \frac{\sigma_2}{-1410} = 1 \quad \text{and} \quad \sigma_2 = -940\,\mathrm{MN/m^2}$$

The Mohr's stress circle construction for the second part of this problem is shown in Fig. 15.6. If the maximum shear stress is 390

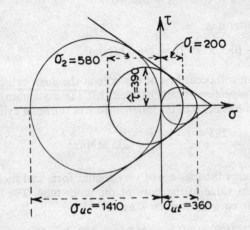

Fig 15.6

$\mathrm{MN/m^2}$, a circle is drawn of radius 390 units to touch the two envelope lines. The principal stresses can then be read off as $+200\,\mathrm{MN/m^2}$ and $-580\,\mathrm{MN/m^2}$.

Chapter 16
Thin Plates and Shells

16.1. The uses of plates and shells in engineering fall into two main categories of geometry, namely thin and thick. In the case of plates there are additionally two ranges of deformation, small and large, which influence the nature of the analysis. The solutions for thick plates and large deflections are beyond the scope of this book. "Thin" is a relative term which indicates that the thickness of material is small compared to the overall geometry. For example, a sphere of 15 m diameter and 150 mm wall thickness is of the same relative proportions as one of 1 m diameter and 10 mm thickness, and both are regarded as thin vessels for analysis. However, from a material and fabrication point of view there is no comparison between the problems of using 150 mm and 10 mm thicknesses.

Plates cover an area generally rectangular or circular and are used to support concentrated and distributed transverse loading. Shells generally contain a pressure due to a solid, liquid or gas. Some simple examples were introduced in Chapter 2 to illustrate statical determinacy.

16.2. ASSUMPTIONS FOR SMALL DEFLECTION OF THIN PLATES

(a) No deformation in the middle plane of the plate, i.e. a neutral surface.

(b) Points in the plate lying initially on a normal to the middle plane of the plate remain on the normal during bending.

(c) Normal stresses in the direction transverse to the plate can be disregarded.

Assumption (a) does not of course hold if there are external forces acting in the middle plane of the plate. Assumption (b) disregards the effect of shear force on deflection.

The deflection, w, is a function of the two co-ordinates in the plane of the plate, the elastic constants of the material, and the loading.

16.3. RELATIONSHIPS BETWEEN MOMENTS AND DEFLECTION

The first important step in the analysis of plates is similar to that for beams and is to relate the bending moments to curvature and hence deflection.

Consider an element of material as shown in Fig. 16.1 cut from a plate subjected to pure bending as in Fig. 16.2. The bending moments M_x and M_y *per unit length* are positive as drawn acting on the middle of the plate. This plane is undeformed and constitutes the *neutral surface*. The material above it is in a state of biaxial compression, and below it, in biaxial tension. The curvatures of the mid-plane in sections parallel to the xz

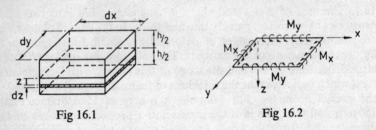

Fig 16.1   Fig 16.2

and yz planes are denoted by $1/r_x$ and $1/r_y$ respectively. At a depth z below the neutral surface the strains in the x- and y-directions of a lamina such as *abcd* are

$$\varepsilon_x = \frac{z}{r_x} \quad \text{and} \quad \varepsilon_y = \frac{z}{r_y} \tag{16.1}$$

using the same approach as for beams in Chapter 6.

The stress–strain relationships are

$$\varepsilon_x = \frac{\sigma_x}{E} - \frac{v\sigma_y}{E} \quad \text{and} \quad \varepsilon_y = \frac{\sigma_y}{E} - \frac{v\sigma_x}{E}$$

Combining with eqn. (16.1) and rearranging gives

$$\left. \begin{aligned} \sigma_x &= \frac{Ez}{1-v^2}\left(\frac{1}{r_x} + \frac{v}{r_y}\right) \\ \sigma_y &= \frac{Ez}{1-v^2}\left(\frac{1}{r_y} + \frac{v}{r_x}\right) \end{aligned} \right\} \tag{16.2}$$

The bending stresses are a function of plate curvatures and are proportional to distance from the neutral surface.

Next, we equate the required equilibrium between the internal moments, due to the bending stresses acting on the sides of the element, and the applied moments M_x and M_y.

$$\left. \begin{aligned} \int_{-h/2}^{h/2} \sigma_x z \, dy \, dz &= M_x \, dy \\ \int_{-h/2}^{h/2} \sigma_y z \, dx \, dz &= M_y \, dx \end{aligned} \right\} \tag{16.3}$$

Substituting from eqns. (16.2) for σ_z and σ_y in eqns. (16.3) and integrating gives

$$\left. \begin{array}{l} M_z = \dfrac{Eh^3}{12(1-v^2)}\left(\dfrac{1}{r_z}+\dfrac{v}{r_y}\right) = D\left(\dfrac{1}{r_z}+\dfrac{v}{r_y}\right) \\[3mm] M_y = \dfrac{Eh^3}{12(1-v^2)}\left(\dfrac{1}{r_y}+\dfrac{v}{r_z}\right) = D\left(\dfrac{1}{r_y}+\dfrac{v}{r_z}\right) \end{array} \right\} \tag{16.4}$$

where $D = \dfrac{Eh^3}{12(1-v^2)}$ is termed the *flexural rigidity*.

Since

$$\frac{1}{r_z} = -\frac{\partial^2 w}{\partial x^2} \quad \text{and} \quad \frac{1}{r_y} = -\frac{\partial^2 w}{\partial y^2}$$

where w is the deflection in the z direction, the relationships between applied moments and deflection are

$$\left. \begin{array}{l} M_z = -D\left(\dfrac{\partial^2 w}{\partial x^2}+v\dfrac{\partial^2 w}{\partial y^2}\right) \\[3mm] M_y = -D\left(\dfrac{\partial^2 w}{\partial y^2}+v\dfrac{\partial^2 w}{\partial x^2}\right) \end{array} \right\} \tag{16.5}$$

16.4. BENDING OF CIRCULAR PLATES FOR SYMMETRICAL LOADING

When the loading on the surface of a plate is symmetrical about a perpendicular central axis, the deflection surface is also symmetrical about that axis. Any diametral section may be used to indicate the deflection curve and the associated slope, ψ, and deflection, w, at any radius, r, as shown in Fig. 16.3.

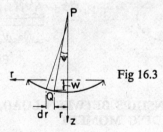

Fig 16.3

The curvature of the plate in the diametral section rz is

$$\frac{1}{R_r} = -\frac{d^2 w}{dr^2} \tag{16.6}$$

and for small values of w (positive downwards) the slope at any point is

$$\psi = -\frac{dw}{dr}$$

The second principal radius of curvature, R_θ, is in the plane perpendicular to rz and is represented by lines such as PQ which form a conical surface so that

$$\frac{1}{R_\theta} = \frac{\psi}{r} \tag{16.7}$$

Eqns. (16.4) relating bending moments and curvatures are now expressed for the circular plate as follows:

$$\left. \begin{aligned} M_r &= D\left(\frac{d\psi}{dr} + v\frac{\psi}{r}\right) \\ M_\theta &= D\left(\frac{\psi}{r} + v\frac{d\psi}{dr}\right) \end{aligned} \right\} \tag{16.8}$$

where M_r and M_θ are the bending moments per unit length acting as shown in Fig. 16.4 and in terms of deflections are expressed as

$$\left. \begin{aligned} M_r &= -D\left(\frac{d^2w}{dr^2} + \frac{v}{r}\frac{dw}{dr}\right) \\ M_\theta &= -D\left(\frac{1}{r}\frac{dw}{dr} + v\frac{d^2w}{dr^2}\right) \end{aligned} \right\} \tag{16.9}$$

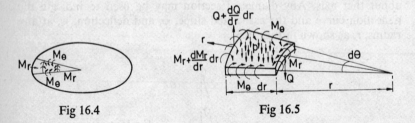

Fig 16.4   Fig 16.5

16.5. RELATIONSHIPS BETWEEN LOAD, SHEAR FORCE AND BENDING MOMENT

If an element, Fig. 16.5, is taken from the plate then it must be in equilibrium under the action of uniformly distributed loading, p per unit area and the resulting shear forces, Q per unit length, and bending moments, M per unit length. Owing to symmetry there are no shear forces on the radial sides of the element. For vertical equilibrium,

$$Qr\,d\theta + pr\,dr\,d\theta - \left(Q + \frac{dQ}{dr}\,dr\right)(r + dr)d\theta = 0$$

from which

$$\frac{dQ}{dr} + \frac{Q}{r} = p \tag{16.10}$$

For moment equilibrium,

$$\left(M_r + \frac{dM_r}{dr}\,dr\right)(r + dr)\,d\theta - M_r r\,d\theta - 2M_\theta\,dr\sin\frac{d\theta}{2}$$

$$+ Qr\,d\theta\,dr = 0$$

which may be simplified to

$$r\frac{dM_r}{dr} + M_r - M_\theta + Qr = 0 \tag{16.11}$$

16.6. RELATIONSHIPS BETWEEN DEFLECTION, SLOPE AND LOADING

Next we substitute eqns. (16.8) into eqn. (16.11) and after simplifying obtain

$$\frac{d^2\psi}{dr^2} + \frac{1}{r}\frac{d\psi}{dr} - \frac{\psi}{r^2} = -\frac{Q}{D} \tag{16.12}$$

which relates the slope at any radius to the shear force.

If eqns. (16.9) are substituted into eqn. (16.11) then

$$\frac{d^3w}{dr^3} + \frac{1}{r}\frac{d^2w}{dr^2} - \frac{1}{r^2}\frac{dw}{dr} = \frac{Q}{D} \tag{16.13}$$

expresses the variation of deflection with radius.

Eqns. (16.12) and (16.13) can be expressed in a form which makes a solution by integration rather more obvious, as follows:

$$\frac{d}{dr}\left[\frac{1}{r}\frac{d}{dr}(r\psi)\right] = -\frac{Q}{D} \tag{16.14}$$

and

$$\frac{d}{dr}\left[\frac{1}{r}\frac{d}{dr}\left(r\frac{dw}{dr}\right)\right] = \frac{Q}{D} \tag{16.15}$$

The shear force Q is a function of the applied loading p and may be related by simple statical equilibrium or by integrating eqn. (16.10) Multiplying through by $r\,dr$ gives

$$r\,dQ + Q\,dr = pr\,dr \qquad \text{or} \qquad d(Qr) = pr\,dr$$

$$Qr = \int_0^r pr\,dr$$

Substituting in eqn. (16.15),

$$r\frac{d}{dr}\left[\frac{1}{r}\frac{d}{dr}\left(r\frac{dw}{dr}\right)\right] = \frac{1}{D}\int_0^r pr\,dr \qquad (16.16)$$

If we know that p is equal to $f(r)$ or is constant then eqn. (16.16) can be integrated to find the deflection at any radius.

The constants of integration are evaluated from the boundary conditions for the particular problem being solved.

16.7. PLATE SUBJECTED TO UNIFORM LOADING, p PER UNIT AREA

In this problem the right-hand side of eqn. (16.16) reduces to $pr^2/2$, since p is constant; therefore

$$\frac{d}{dr}\left[\frac{1}{r}\frac{d}{dr}\left(r\frac{dw}{dr}\right)\right] = \frac{pr}{2D} \qquad (16.17)$$

Integrating,

$$\frac{1}{r}\frac{d}{dr}\left(r\frac{dw}{dr}\right) = \frac{pr^2}{4D} + C_1$$

Multiplying both sides by r and integrating again,

$$r\frac{dw}{dr} = \frac{pr^4}{16D} + \frac{C_1 r^2}{2} + C_2$$

$$\frac{dw}{dr} = \frac{pr^3}{16D} + \frac{C_1 r}{2} + \frac{C_2}{r} \qquad (16.18)$$

Finally

$$w = \frac{pr^4}{64D} + \frac{C_1 r^2}{4} + C_2 \log_e r + C_3 \qquad (16.19)$$

Case (a) Clamped Periphery

For a plate of radius a the boundary conditions are $dw/dr = 0$ at $r = 0$ and $r = a$, and $w = 0$ at $r = a$. Hence $C_2 = 0$ and

$$\frac{pa^3}{16D} + C_1\frac{a}{2} = 0 \qquad \text{so that} \qquad C_1 = -\frac{pa^2}{8D}$$

From eqn. (16.19),

$$0 = \frac{pa^4}{64D} - \frac{pa^4}{32D} + C_3 \quad \text{or} \quad C_3 = \frac{pa^4}{64D}$$

The deflection curve is thus

$$w = \frac{p}{64D}(a^2 - r^2)^2 \tag{16.20}$$

The maximum deflection occurs at the centre, so that

$$w_{max} = \frac{pa^4}{64D} \tag{16.21}$$

Bending stress may now be determined from the moment–slope relationships, eqns. (16.8) and (16.18):

$$M_r = \frac{p}{16}\left[a^2(1 + v) - r^2(3 + v)\right]$$

$$M_\theta = \frac{p}{16}\left[a^2(1 + v) - r^2(1 + 3v)\right]$$

At the periphery $r = a$ and

$$M_r = -\frac{pa^2}{8} \quad \text{and} \quad M_\theta = -v\frac{pa^2}{8} \tag{16.22}$$

At the centre $r = 0$ and

$$M_r = M_\theta = \frac{pa^2}{16}(1 + v) \tag{16.23}$$

From eqns. (16.2) and (16.4),

$$\sigma = \frac{12Mz}{h^3}$$

At the surface $z = h/2$ and

$$\sigma_{r\,max} = \pm\frac{6M_r}{h^2}_{max} = \pm\frac{3}{4}\frac{pa^2}{h^2} \tag{16.24}$$

Case (b) Simply Supported Periphery

For this case the boundary conditions are; $M_r = 0, r = a$; $dw/dr = 0$, $r = 0$; and $w = 0, r = a$ to determine the three constants of integration. The problem will instead be tackled by an alternative method using the principle of superposition. We can use the solution from the fixed edge case combined with that for a plate simply supported and subjected to edge moments equal but opposite in sense to the fixing moment, as illustrated in Fig. 16.6.

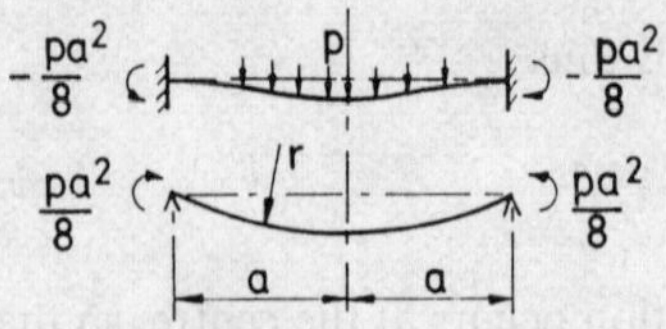

Fig 16.6

From eqn. (16.4) since, for $M_x = M_y$, $r_x = r_y$, then

$$\frac{1}{r} = \frac{M}{D(1 + \nu)}$$

The deflection at the centre of a spherical surface of radius a and curvature $1/r$ is

$$w = \frac{a^2}{2r}$$

Therefore

$$w = \frac{Ma^2}{2D(1 + \nu)} \tag{16.25}$$

But from case (a) the fixing moment is $M_r = -pa^2/8$; therefore

$$w = \frac{pa^4}{16D(1 + \nu)} \tag{16.26}$$

The resultant deflection at the centre for the simply supported plate is, by superposition of eqns. (16.21) and (16.26),

$$w = \frac{pa^4}{64D} + \frac{pa^4}{16D(1 + \nu)} = \frac{5 + \nu}{64(1 + \nu)D}pa^4 \tag{16.27}$$

The maximum bending moment occurs at the centre, and by superposition of eqn. (16.23) and $M = pa^2/8$ for the pure bending contribution,

$$M_r = M_\theta = \frac{pa^2}{16}(3 + \nu) \tag{16.28}$$

16.8. PLATE WITH CENTRAL CIRCULAR HOLE

Case (a) Edge Moments

A ring of moments is applied to the inner and outer boundaries as shown in Fig. 16.7. Since there is no shear force at the inner boundary, eqn.

(16.15) reduces to

$$\frac{d}{dr}\left[\frac{1}{r}\frac{d}{dr}\left(r\frac{dw}{dr}\right)\right] = 0 \tag{16.29}$$

Integrating twice and simplifying gives

$$\frac{dw}{dr} = \frac{C_1 r}{2} + \frac{C_2}{r} \tag{16.30}$$

and

$$w = \frac{C_1 r^2}{4} + C_2 \log_e r + C_3 \tag{16.31}$$

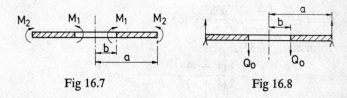

Fig 16.7   Fig 16.8

To find C_1 and C_2 we use the boundary conditions: at $r = b$, $M_r = M_1$; and at $r = a$, $M_r = M_2$; and by substituting eqn. (16.30) and its differential into eqn. (16.9) and solving, we obtain

$$C_1 = -\frac{2(a^2 M_2 - b^2 M_1)}{(1 + v)(a^2 - b^2)D}$$

$$C_2 = -\frac{a^2 b^2 (M_2 - M_1)}{(1 - v)(a^2 - b^2)D}$$

Since $w = 0$ at $r = a$ then, from eqn. (16.31),

$$C_3 = \frac{a^2(a^2 M_2 - b^2 M_1)}{2(1 + v)(a^2 - b^2)D} + \frac{a^2 b^2 (M_2 - M_1) \log_e a}{(1 - v)(a^2 - b^2)D}$$

The deflection at any radius may now be determined by substituting the values of C_1, C_2 and C_3 into eqn. (16.31).

Case (b) Edge Forces

In this case uniformly distributed transverse forces, Q_0, are applied at the inner and outer edges as shown in Fig. 16.8. The shear force per unit length at radius r is

$$Q = \frac{2\pi b Q_0}{2\pi r} = \frac{b Q_0}{r}$$

Substituting in eqn. (16.15) and integrating gives

$$\frac{dw}{dr} = \frac{bQ_0}{4D} r(2 \log_e r - 1) + \frac{C_1 r}{2} + \frac{C_2}{r} \tag{16.32}$$

$$w = \frac{bQ_0 r^2}{4D} (\log_e r - 1) + \frac{C_1 r^2}{4} + C_2 \log_e r + C_3 \tag{16.33}$$

For a simply supported periphery the boundary conditions are: at $r = a$, $w = 0$ and $M_r = 0$; at $r = b$, $M_r = 0$.

By substituting eqn. (16.32) and its differential into eqn. (16.9) and solving, we obtain

$$C_1 = -\frac{bQ_0}{2D} \left(\frac{1 - \nu}{1 + \nu} + \frac{2(a^2 \log_e a - b \log_e b)}{a^2 - b^2} \right)$$

$$C_2 = \frac{bQ_0}{2D} \frac{1 + \nu}{1 - \nu} \frac{a^2 b^2}{a^2 - b^2} \log_e \frac{b}{a}$$

Then, from eqn. (16.33),

$$C_3 = \frac{a^2 b Q_0}{4D} \left(1 + \frac{1}{2} \frac{(1 - \nu)}{(1 + \nu)} - \frac{b^2}{a^2 - b^2} \log_e \frac{b}{a} \right.$$
$$\left. - \frac{2(1 + \nu)}{(1 - \nu)} \frac{b^2}{a^2 - b^2} \log_e \frac{b}{a} \log_e a \right)$$

These constants may now be substituted into eqns. (16.32) and (16.33) to give the slope and deflection at any radius.

16.9. SOLID PLATE WITH CENTRAL CONCENTRATED LOAD

Considering the problem solved in the previous section and equating the total load around the inner periphery to the concentrated load, so that

$$2\pi b Q_0 = P$$

and taking the limiting case when b is infinitely small, $b^2 \log_e (b/a)$ tends to zero and the above constants of integration become

$$C_1 = -\frac{P}{4\pi D} \left\{ \frac{1 - \nu}{1 + \nu} + 2 \log_e a \right\} \qquad C_2 = 0$$

$$C_3 = \frac{Pa^2}{8\pi D} \left(1 + \frac{1}{2} \frac{1 - \nu}{1 + \nu} \right)$$

Substituting these values into eqn. (16.33),

$$w = \frac{P}{8\pi D} \left[\frac{3 + \nu}{2(1 + \nu)} (a^2 - r^2) + r^2 \log_e \frac{r}{a} \right] \tag{16.34}$$

which is the deflection at any radius for a solid plate simply supported and carrying a concentrated load at the centre.

For the fixed edge case we find the slope at the edge for the simply supported plate by differentiating eqn. (16.34); then this slope has to be made zero by the superposition of the appropriate ring of edge moments. The deflection due to these moments may then be superposed onto the deflection given by eqn. (16.34) to obtain a resultant.

Example 16.1. A cylinder head valve of diameter 38 mm is subjected to a gas pressure of 1·4 MN/m². It may be regarded as a uniform thin circular plate simply supported around the periphery by the seat, as shown in Fig. 16.9.

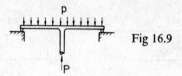

Fig 16.9

Assuming that the valve stem applies a concentrated force at the centre of the plate, calculate the movement of the stem necessary to lift the valve from its seat. The flexural rigidity of the valve is 260 N-m, and Poisson's ratio for the material is 0·3.

We have already derived solutions for a simply supported plate subjected to uniform loading, p, and a concentrated load, P, at the centre, hence the deflection at the centre is equal to the sum of the deflections due to the two separate load components. Therefore

$$w_{max} = \frac{Pa^2}{16\pi D}\frac{3+v}{1+v} + \frac{pa^4}{64D}\frac{5+v}{1+v}$$

but when the valve lifts from its seat, $P = -\pi a^2 p$; therefore

$$w_{max} = -\frac{7+3v}{1+v}\frac{pa^4}{64D}$$

$$= -\frac{7\cdot9 \times 1\cdot4 \times 10^6 \times 0\cdot19^4}{1\cdot3 \times 64 \times 260}$$

$$= -0\cdot0665\,\text{mm}$$

16.10. OTHER FORMS OF LOADING AND EDGE CONDITION

The solutions that have been obtained in Sections 16.7 and 16.8 can be used to advantage with the principle of superposition to analyse a number of other plate problems.

(a) Concentric Loading

A plate which is uniformly loaded transversely along a circle of radius b, as illustrated in Fig. 16.10 (a), can be split into the two components shown at (b), and the separate solutions which have been obtained previously can be superposed using the appropriate boundary conditions, namely that there must be continuity of slope at radius b.

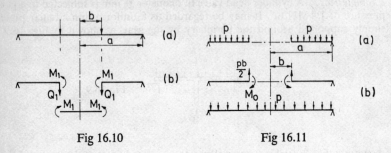

Fig 16.10   Fig 16.11

(b) Distributed Loading on Plate with Central Hole

The situation illustrated in Fig. 16.11(a) can be simulated by taking the deflection of the solid plate subjected to uniform loading and superposing that due to the appropriate moment and shear force at radius b, as in Fig. 16.11(b).

(c) Fixed Inner Boundary

This edge condition can be associated with several types of loading and merely entails applying the appropriate moment to give zero slope and shear force as a function of the applied loading.

A variety of configurations and loadings are illustrated in Fig. 16.12, all of which can be dealt with easily by superposition of the required components which give the appropriate boundary conditions. However, in all these cases the maximum deflection can be represented by the following relations

$$w_{max} = c' \frac{pa^4}{Eh^3} \quad \text{or} \quad w_{max} = c' \frac{Pa^2}{Eh^3}$$

where c' is a factor involving the ratio a/b and Poisson's ratio.

The maximum stresses can also be expressed by formulae as follows:

$$\sigma_{max} = c'' \frac{pa^2}{h^2} \quad \text{or} \quad \sigma_{max} = \frac{c''P}{h^2}$$

where c'' is also a factor as defined above. Values of c' and c'' for $v = 0.3$ and a/b in the range $1\frac{1}{4}$ to 5 for the cases in Fig. 16.12 are given in Table 16.1 (page 384).

16.11. BENDING OF RECTANGULAR PLATES

The theory of bending for rectangular plates is rather more involved than that for circular plates and is beyond the scope of this book. The reader who may require information on this topic is referred to one of the specialized texts on the theory of plates.

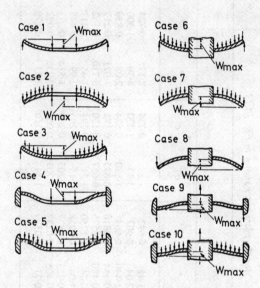

Fig 16.12

16.12. AXI-SYMMETRICAL THIN SHELLS

The analysis of thin shells of revolution subjected to uniform pressure was treated as a statically determinate problem in Section 2.9. The fundamental relationship between the principal membrane stresses in the wall and the principal curvatures of the shell to the applied pressure and wall thickness was shown to be

$$\frac{\sigma_1}{r_1} + \frac{\sigma_2}{r_2} = \frac{p}{t} \tag{16.35}$$

The simple applications of eqn. (16.35) to the cylinder and sphere under internal pressure were also dealt with in Chapter 2, together with an example on the self-weight of a concrete dome. The following example on liquid storage illustrates a further use of a thin shell.

Table 16.1

Coefficients c' and c'' for the plate cases shown in Fig. 16.12

Case	a/b = 1·25 c''	1·25 c'	1·5 c''	1·5 c'	2 c''	2 c'	3 c''	3 c'	4 c''	4 c'	5 c''	5 c'
1	1·10	0·341	1·26	0·519	1·48	0·672	1·88	0·734	2·17	0·724	2·34	0·704
2	0·66	0·202	1·19	0·491	2·04	0·902	3·34	1·220	4·30	1·300	5·10	1·310
3	0·592	0·184	0·976	0·414	1·440	0·664	1·880	0·824	2·08	0·830	2·19	0·813
4	0·194	0·00504	0·320	0·0242	0·454	0·0810	0·673	0·172	1·021	0·217	1·305	0·238
5	0·105	0·00199	0·259	0·0139	0·480	0·0575	0·657	0·130	0·710	0·162	0·730	0·175
6	0·122	0·00343	0·336	0·0313	0·74	0·1250	1·21	0·291	1·45	0·417	1·59	0·492
7	0·135	0·00231	0·410	0·0183	1·04	0·0938	2·15	0·293	2·99	0·448	3·69	0·564
8	0·227	0·00510	0·428	0·0249	0·753	0·0877	1·205	0·209	1·514	0·293	1·745	0·350
9	0·115	0·00129	0·220	0·0064	0·405	0·0237	0·703	0·062	0·933	0·092	1·13	0·114
10	0·090	0·00077	0·273	0·0062	0·71	0·0329	1·54	0·110	2·23	0·179	2·80	0·234

Example 16.2. The water storage tank illustrated in Fig. 16.13, of 20 mm uniform wall thickness, consists of a cylindrical section which is supported at the top edge and joined at the lower end to a spherical portion. An angle-section reinforcing ring of 5000 mm² cross-sectional area is welded into the lower joint as shown.

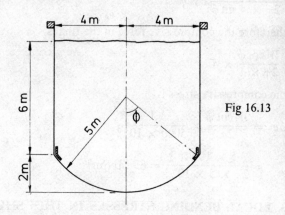

Fig 16.13

Calculate the maximum stresses in the cylindrical and spherical portions of the tank and the hoop stress in the reinforcing ring when the water is at the level shown. Density of water is 9·81 kN/m³.

The total force due to the weight of water in the tank is

$$W = 9{\cdot}81 \times 10^3 \left[(\pi \times 6 \times 4^2) + \left(\frac{2\pi}{3} \times 5^3 \right) + \left(\frac{\pi \times 3^3}{3} \right) \right.$$

$$\left. - (\pi \times 5^2 \times 3) \right]$$

$$= 3{\cdot}5 \, \text{MN}$$

The axial stress in the cylindrical part of the tank is

$$\sigma_a = \frac{3{\cdot}5}{2\pi \times 4 \times 0{\cdot}02} = 6{\cdot}95 \, \text{MN/m}^2$$

and the hoop stress is

$$\sigma_{h \, max} = \frac{6 \times 9{\cdot}81 \times 10^6 \times 4}{0{\cdot}02} = 11{\cdot}8 \, \text{MN/m}^2$$

The maximum stress in the spherical portion of the tank occurs at the bottom, where the pressure is $9{\cdot}81 \times 10^3 \times 8 \, \text{N/m}^2$; therefore

$$\sigma_{max} = \frac{9{\cdot}81 \times 10^3 \times 8 \times 5}{2 \times 0{\cdot}02} = 9{\cdot}81 \, \text{MN/m}^2$$

For the reinforcing ring, the tangential force per unit length at the edge of the spherical part is $W/(2\pi \times 4 \sin\phi)$. The inward radial component of that force is

$$\frac{W}{2\pi \times 4 \sin\phi} \cos\phi$$

and therefore the compressive force in the ring is

$$\frac{W \cot\phi}{2\pi \times 4} \times 4$$

and the compressive stress is

$$\sigma_c = \frac{W \cot\phi}{2\pi} \times \frac{1}{5000 \times 10^{-6}}$$

$$= \frac{3 \cdot 5 \times 10^6 \times \frac{3}{4}}{2\pi \times 5 \times 10^{-3}} = 83 \text{ MN/m}^2$$

16.13. LOCAL BENDING STRESSES IN THIN SHELLS

Whenever there is a change in geometry of the shell, particularly for discontinuities in the meridian such as in the above example, the membrane stresses cause displacements which give rise to local bending in the wall. The resulting bending stresses may be significant in comparison with the membrane stresses. This was the reason for introducing the reinforcing ring in the above problem.

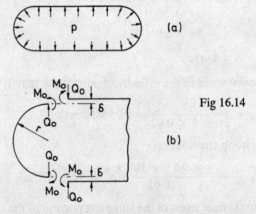

Fig 16.14

To illustrate the method of analysing local bending we will consider the elementary situation of a cylindrical vessel with hemispherical ends of the same thickness subjected to internal pressure as illustrated in Fig. 16.14.

The membrane stresses for the cylinder are

$$\sigma_1 = \frac{pr}{t} \quad \text{and} \quad \sigma_2 = \frac{pr}{2t}$$

and for the hemisphere,

$$\sigma = \frac{pr}{t}$$

as calculated in Chapter 2.

The corresponding radial displacements for the cylinder and hemisphere, respectively, are

$$u_c = \frac{r}{E}(\sigma_1 - \nu\sigma_2) = \frac{pr^2}{2tE}(2 - \nu)$$

$$u_h = \frac{r}{E}(\sigma_1 - \nu\sigma_2) = \frac{pr^2}{2tE}(1 - \nu)$$

The difference in deformation radially is

$$\delta = \frac{pr^2}{2tE}[(2 - \nu) - (1 - \nu)] = \frac{pr^2}{2tE} \tag{16.36}$$

In order to overcome this difference, shear and moment reactions are set up at the joint as shown in Fig. 16.14.

Since the cylindrical section is symmetrical with respect to its axis we may consider the reactions Q_0 and M_0 per unit length acting on a strip of unit width as illustrated in Fig. 16.15. The inward bending of the

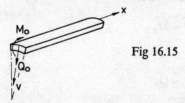

Fig 16.15

strip due to Q_0 sets up compressive circumferential strain. If the radial displacement is v then

$$\varepsilon_\theta = \frac{v}{r} \quad \text{and} \quad \sigma_\theta = \frac{Ev}{r}$$

The circumferential force due to this stress on the edge of the strip per unit length is Evt/r. The outward radial component of this force is

$$R = 2\frac{Evt}{r}\sin\frac{d\theta}{2} = \frac{Evt}{r}d\theta$$

$$= \frac{Evt}{r^2} \tag{16.37}$$

This force opposes the deflection of the strip and is distributed along the strip, being proportional to the deflection, v, at any point. This is a special case of the bending of a beam on an elastic foundation for which the deflection curve is

$$EI\frac{d^4v}{dx^4} = -w$$

where w is the distributed loading function. For the present problem EI is replaced by D, since the strip is restrained from distortion by adjacent material as for plates; therefore

$$D\frac{d^4v}{dx^4} = -\frac{Et}{r^2}v = -4D\beta^4v \qquad (16.38)$$

The solution of this equation is

$$v = e^{\beta x}(A\cos\beta x + B\sin\beta x) + e^{-\beta x}(C\cos\beta x + D'\sin\beta x) \qquad (16.39)$$

where A, B, C and D' are constants determined by the boundary conditions, and

$$\beta = \left[\frac{Et}{4Dr^2}\right]^{1/4} = \left[\frac{3(1-v^2)}{r^2h^2}\right]^{1/4}$$

As $x \to \infty$, $v \to 0$ and $M \to 0$, which gives $A = B = 0$. At $x = 0$ $M = M_0$ and $Q = Q_0$; therefore

$$D\frac{d^2v}{dx^2} = -M_0$$

and

$$D\frac{d^3v}{dx^3} = -Q_0$$

from which

$$C = \frac{1}{2\beta^3 D}(Q_0 - \beta M_0) \qquad \text{and} \qquad D' = \frac{M_0}{2\beta^2 D}$$

Substituting into eqn. (16.39), the deflection curve for the strip becomes

$$v = \frac{e^{-\beta x}}{2\beta^3 D}[Q_0\cos\beta x - \beta M_0(\cos\beta x - \sin\beta x)] \qquad (16.40)$$

This is a rapidly damped oscillatory curve, and thus bending of the cylinder and head is local to the joint.

When the cylinder and head are of the same material and wall thickness then the deflections and slopes at the joint produced by Q_0 are equal and

$M_0 = 0$. Therefore the boundary condition is at $x = 0$, $v = \delta/2$, and from eqn. (16.40),

$$\frac{\delta}{2} = \frac{Q_0}{2\beta^3 D}$$

or

$$Q_0 = \delta\beta^3 D = \frac{pr^2}{2tE}\frac{Et}{4\beta r^2} = \frac{p}{8\beta} \tag{16.41}$$

The deflection curve becomes

$$v = \frac{e^{-\beta x} p \cos \beta x}{16\beta^4 D} = \frac{pr^2}{4tE} e^{-\beta x} \cos \beta x \tag{16.42}$$

By differentiating this equation twice the bending moment and hence the bending stress can be calculated for any cross-section. These have to be added to the membrane stresses to get the resultant stress.

If the wall thickness of the head and cylinder are different then $M_0 \neq 0$ and the boundary conditions would be that (a) the sum of the edge deflections must be zero, and (b) the rotation of the edges must be the same.

The above solution is equally applicable for other shapes of head.

Example 16.3. Calculate the local bending stresses in the vessel shown in Fig. 16.14 if $p = 1$ MN/m^2, $r = 500$ mm, $t = 10$ mm, $v = 0.3$.

$$\beta = \left[\frac{3(1 - 0.3^2)}{0.5^2 \times 0.01^2}\right]^{1/4} = 0.182\,\text{m}^{-1}$$

$$Q_0 = \frac{p}{8\beta} = \frac{1 \times 10^6}{8 \times 0.182} = 6.87\,\text{kN/m}$$

Since the bending moment in the strip is $M = -D(d^2v/dx^2)$, and from eqn. (16.40) with $M_0 = 0$,

$$v = \frac{Q_0}{2\beta^3 D} e^{-\beta x} \cos \beta x$$

Therefore

$$M = -\frac{Q_0}{\beta} e^{-\beta x} \sin \beta x$$

This expression takes the largest value for $\beta x = \pi/4$, which gives

$$M_{max} = 0.121\,\text{kN-m}$$

This gives rise to a maximum bending stress of

$$\text{(axial)} \quad \sigma_b = \frac{6M_{max}}{t^2} = \frac{6 \times 0\cdot121 \times 10^3}{0\cdot01^2} = 7\cdot26 \, \text{MN/m}^2$$

The membrane stress is

$$\text{(axial)} \quad \sigma_m = \frac{pr}{2t} = \frac{1 \times 0\cdot5}{2 \times 0\cdot01} = 25 \, \text{MN/m}^2$$

The total axial stress is $\sigma_2 = 7\cdot26 + 25 = 32\cdot26 \, \text{MN/m}^2$.

The bending of the strip also produces circumferential stresses (a) due to prevention of strip from distorting, as in plates $= \pm 6\nu M/t^2$; (b) stresses due to shortening of the circumference $= -E\nu/r$.

Using the above values for ν and M and summing gives

$$\text{(hoop)} \quad \sigma_b = \frac{Q_0 e^{-\beta z}}{\beta t^2} \left(6\nu \sin \beta x - \frac{12(1 - \nu^2)}{2\beta^2 tr} \cos \beta x \right)$$

$$= 22\cdot6 e^{-\beta z}(0\cdot3 \sin \beta x - 0\cdot55 \cos \beta x) \, \text{MN/m}^2$$

The maximum value of this expression is 1.58 MN/m², which is small compared with the hoop membrane stress.

$$\text{(hoop)} \quad \sigma_m = \frac{pr}{t} = \frac{1 \times 0\cdot5}{0\cdot01} = 50 \, \text{MN/m}^2$$

Hence local bending does not have a serious influence in this particular case.

Example 16.4.   An upright cylindrical storage tank, of radius r, uniform wall thickness t, and height h, is filled to the top with liquid of density ρ. The base of the tank is built in to its foundation, Fig. 16.16, and since $t \ll h$ or r the shell may be regarded as infinitely long. Obtain an expression for the maximum bending moment due to discontinuity in the shell.

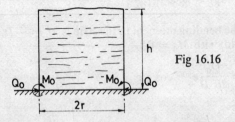

Fig 16.16

The governing equation is basically the same as eqn. (16.38) with an additional term which defines the variation of pressure loading due to the liquid:

$$D\frac{d^4v}{dx^4} + 4D\beta^4 v = -\rho(h - x) \tag{16.43}$$

The particular integral part of the solution is $-\rho(h-x)/4D\beta^4$ and the complete solution is

$$v = e^{\beta x}(C_1 \cos \beta x + C_2 \sin \beta x) + e^{-\beta x}(C_3 \cos \beta x + C_4 \sin \beta x)$$

$$-\frac{\rho(h-x)}{4D\beta^4} \tag{16.44}$$

The boundary conditions are as follows:

(i) The height of the cylinder can be regarded as "infinite", so that $M \to 0$ and $v \to 0$, giving $C_1 = C_2 = 0$.

(ii) At $x = 0$, $v = 0$ and $dv/dx = 0$; hence

$$C_3 = \frac{\rho h}{4D\beta^4} \quad \text{and} \quad C_4 = \frac{\rho}{4D\beta^4}\left(h - \frac{1}{\beta}\right)$$

Putting these values in eqn. (16.44) gives the deflection curve

$$v = -\frac{\rho h}{4D\beta^4}\left[1 - \frac{x}{h} - e^{-\beta x}\cos \beta x - e^{-\beta x}\left(1 - \frac{1}{\beta h}\right)\sin \beta x\right] \tag{16.45}$$

But $M = -D(d^2v/dx^2)$, so that differentiating eqn. (16.45) twice and substituting gives

$$M = \frac{\rho h}{2\beta^2}\left[-e^{-\beta x}\sin \beta x + \left(1 - \frac{1}{\beta h}\right)e^{-\beta x}\cos \beta x\right] \tag{16.46}$$

Now, the maximum value of M occurs at the discontinuity, where $x = 0$; therefore

$$M_{max} = \frac{\rho h}{2\beta^2}\left(1 - \frac{1}{\beta h}\right) = \frac{\rho}{2\beta^3}(\beta h - 1)$$

Chapter 17
Stress Concentration

17.1. In previous chapters the problems analysed have had stress distributions which were either uniform or varied smoothly and gradually over a significant area. However, in the vicinity of the point of application of a concentrated load there is a rapid variation in stress over a small area, in which the maximum value is considerably higher than the average stress in the full section of the material. This situation is known as a *stress concentration*.

Another way in which a stress concentration can be produced is at a geometrical discontinuity in a body, such as a hole, keyway, or other sharp change in sectional dimensions. If one imagines the stress in the components in Fig. 17.1 as represented by lines of flow, as in a fluid in a

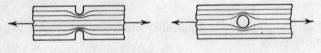

Fig 17.1

channel, then it is seen that the introduction of a notch or hole intercept and disturbs the flow locally.

Two questions come to mind when considering the above effect. If, as is the case, all points of support and load application disturb the uniformity of stress and cause stress concentration, it is surprising that one can obtain realistic results by, say, the simple bending and torsion theories considered earlier. This problem was studied theoretically by St. Venant, who stated the following principle. If the forces acting on a small area of a body are replaced by a statically equivalent system of forces acting on the same area, there will be considerable changes in the local stress distribution, but the effect on the stresses at distances large compared with the area on which the forces act will be negligible. For instance, in a bar gripped at each end and subjected to axial tension, the stress distribution at the ends will vary considerably according to whether gripping is by screw thread, button head, or wedge jaws. However, it has been shown that, at a distance of between one and two diameters from the ends, the stress distribution is quite uniform across the section. Similarly, it is immaterial how the couples are applied at the ends of a beam in pure bending. So long as the length is markedly greater than the cross-sectional dimensions, the assumptions and simple theory of bending will hold good at a distance of approximately one beam depth away from the concentrated force.

The second feature that is of interest is that, although a stress concentration is only effective locally (St. Venant), the peak stress at this point is sometimes far in excess of the average stress calculated in the body of the component. Why is it then normal practice under static loading to base design calculations on the main field of stress, and not on the maximum stress concentration value where a load is applied? Consider a point or line application of load, then theoretically the elastic stress would become infinite in the material under the load. This obviously cannot occur in practice since a ductile material will reach a yield point and plastic deformation will occur under the point of application of the load. The effect of the plastic flow is to cause a local redistribution of stress, which relieves the stress concentration slightly so that the peak value of stress does not continue to increase with increasing load at the same rate as in the elastic range. Eventually with still greater loading, general yielding in the body of the material will tend to catch up and encompass what was the stress concentration area.

The stress concentration set up by geometrical discontinuities, notches, holes, etc., is a function of the shape and dimensions of the discontinuity, and is expressed in terms of the *elastic stress concentration factor* denoted by K_t:

$$K_t = \frac{\text{Maximum boundary stress at the discontinuity}}{\text{Average stress at that cross-section of the body}}$$

This factor is obviously constant within the elastic range of the material. If loading is increased to the point where yielding occurs at the notch, there is a redistribution of stress similar to that at a concentrated load. The stress concentration factor is still defined in the same way as in the elastic range but is now denoted by K_p, the *plastic stress concentration factor*, and this is now also a function of the degree of plastic deformation that has occurred. It is found in practice that for ductile materials as $\sigma_{max} \to \sigma_{ult}$, $K_p \to 1$, i.e. the stress concentration does not reduce the static strength of the component. If anything the reverse is slightly the case owing to the triaxiality of stress and constraint of the notch. Brittle materials have little or no capacity for plastic deformation and therefore the stress concentration is maintained up to fracture. Whether or not there is an accompanying reduction in nominal strength depends largely on the structure of the material. Those such as glass and some cast-irons which have inherent internal flaws, which themselves set up stress concentration, show little reduction in strength over the unnotched condition. Others which have a homogeneous stress-free structure will show a considerable decrease in static strength for a severe notch.

From all of the foregoing it would appear that stress concentration does not present too serious a problem for components in service. However, there are two main aspects of material behaviour in which stress concentration plays the major part in causing failure. These are fatigue and brittle fracture (notch brittle reaction of a normally ductile metal), both of which topics are dealt with at length in later chapters.

The theoretical analysis of stress concentration is generally very complex, and sometimes the shape of a discontinuity is chosen for analysis not because it is in common usage, but rather that it is possible to solve the particular mathematical equations involved. Many classic theoretical solutions are due to Neuber, and a number of individual problems have been solved by other theoreticians and are available in published papers.

When theoretical solution is not possible or is very laborious, recourse is generally had to the technique of photoelasticity (see Chapter 26), which has provided many simple and accurate solutions to problems of stress concentration.

The majority of the theoretical solutions are beyond the scope of this book; therefore the discussion will be limited to the results of a few problems which are of practical interest in engineering.

17.2. CONCENTRATED LOAD ON THE EDGE OF AN INFINITELY LARGE PLATE

The local distribution of stress at the point of application of a concentrated load normal to the edge of an infinite plate was first studied in 1891 using photoelasticity. This led to the theoretical solutions a year later of Boussinesq and Flamant.

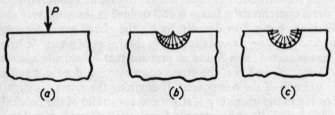

Fig 17.2

Consider the three systems of forces shown in Fig. 17.2 acting on the edge of an infinitely large plate of thickness b. The resultant force in each case is the same, and hence the systems are statically equivalent and therefore satisfy the principle of St. Venant. Now, case (a) is the one we wish to solve, but this will result in practice in a small volume of plastic flow as explained previously. To overcome this difficulty we replace the point load on the straight edge by a radial distribution of forces, as in (b) or (c), around a small semicircular groove. Experiment has shown that the forces in (c) give the better representation of the stress distribution due to a concentrated load on a straight edge. The solution by Flamant on this basis shows that the stress distribution is a simple radial one involving

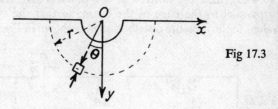

Fig 17.3

compression only. Using polar co-ordinates, and referring to Fig. 17.3, any element distant r from O at an angle θ to the normal to the edge of the plate is subjected to simple radial compression only of a magnitude

$$\sigma_r = -\frac{2P \cos \theta}{\pi b r}$$

An element of material at a depth y below the free edge of the plate and x from the point of loading is subjected to normal and shear stresses given in terms of the radial stress as

$$\sigma_x = \sigma_r \sin^2 \theta = -\frac{2P \sin^2 \theta \cos \theta}{\pi b r}$$

$$\sigma_y = \sigma_r \cos^2 \theta = -\frac{2P \cos^3 \theta}{\pi b r}$$

$$\tau_{xy} = \sigma_r \sin \theta \cos \theta = -\frac{2P \sin \theta \cos^2 \theta}{\pi b r}$$

But $\sin \theta = x/r$, $\cos \theta = y/r$ and $r = (x^2 + y^2)^{1/2}$. Therefore, in terms of rectangular co-ordinates,

$$\sigma_x = -\frac{2Px^2y}{\pi b(x^2 + y^2)^2}$$

$$\sigma_y = -\frac{2Py^3}{\pi b(x^2 + y^2)^2}$$

$$\tau_{xy} = -\frac{2Pxy^2}{\pi b(x^2 + y^2)^2}$$

17.3. CONCENTRATED LOAD BENDING A BEAM

The cross-section of the beam at which the load is acting is subjected to a complex stress condition composed of stress due to simple bending plus the stress due to the concentrated load itself.

Considering the radial pressure distribution on the small groove in Fig. 17.4 (a), then the horizontal components give rise to forces P/π acting parallel to the edge of the beam, so that the system of forces equivalent to the pressure distribution is as shown in Fig. 17.4 (b). In this

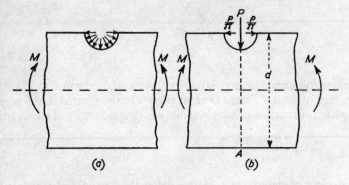

Fig 17.4

problem we are not considering an infinite plate, but a beam of finite depth, and consequently the horizontal forces, P/π, set up longitudinal tension and bending stresses. The former are given simply as load divided by area or $P/\pi \times (1/bd)$. The latter are determined by considering the bending moment about the axis of the beam given by $P/\pi \times \frac{1}{2}d$. The bending stresses are therefore

$$\sigma_x = \mp \frac{Pd}{2\pi} \frac{y}{I}$$

The total stress acting across the section OA of the beam is then obtained by the superposition of the various separate quantities:

$$\sigma_{x_{OA}} = \pm \frac{Pl}{4} \frac{y}{I} \mp \frac{Pd}{2\pi} \frac{y}{I} + \frac{P}{\pi bd}$$

$$= \pm \left(\frac{l}{4} - \frac{d}{2\pi} \right) \frac{12Py}{bd^3} + \frac{P}{\pi bd}$$

This expression is often referred to as the Wilson–Stokes solution.

The distribution of $\sigma_{x_{OA}}$ for long and short spans is compared with the simple bending distribution in Fig. 17.5, and it is seen that the more accurate solution gives rise to maximum longitudinal stresses which are *less* than those from simple bending theory.

Hence in this problem, although the stress concentration causes high normal compressive stresses, the tensile bending stress which would be expected to be the cause of failure is in fact reduced by the concentrated load.

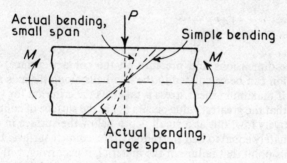

Fig 17.5

17.4. CONTACT STRESSES

Another important problem involving stress concentration is the condition of contact of two non-linear bodies under load. Typical examples may be found in the mating of gear teeth, a shaft in a bearing, and balls and rollers in bearings.

The mathematical solutions involve the principles of the theory of elasticity and are too complex and lengthy to be considered here, and discussion will be confined to the results of other investigators.

Two spherical bodies make contact at one point when unloaded. The application of load immediately creates a small surface of contact, owing to elastic deformation, which is circular in shape. The radius of this circular contact area is given by

$$r = 0\cdot88 \left[\frac{P}{2} \frac{(E_1 + E_2)D_1D_2}{E_1E_2(D_1 + D_2)} \right]^{1/3}$$

where P is the force between the balls of diameter D_1 and D_2 having elastic moduli E_1 and E_2. When the balls are of the same diameter and material

$$r = 0\cdot88 \left[\frac{PD}{2E} \right]^{1/3}$$

The distribution of pressure over the contact area is such that a maximum value occurs at the centre of the circle equal to

$$\sigma_{max} = 1 \cdot 5 \frac{P}{\pi r^2}$$

or, using the expression for r above,

$$\sigma_{max} = 0 \cdot 62 \left[\frac{4PE^2}{D^2} \right]^{1/3}$$

From the dimensions and pressure on the contact surface, the stress distribution can be calculated in the two balls along the axis normal to contact. If maximum shear stress is taken as the criterion for yielding, it is found that the greatest value occurs not at the surface of contact where it is relatively low, but at a small depth below the surface in each ball, approximately equal to half the radius of the contact surface. It is generally at this point that failure of the material, if occurring, will originate.

Maximum tensile stress occurs at the periphery of the contact area acting radially, and is given by

$$\sigma_r = \frac{1 - 2\nu}{2} \frac{P}{\pi r^2}$$

For a ball in a spherical seat of the same material, D_2 becomes negative and $E_1 = E_2 = E$ giving

$$r = 0 \cdot 88 \left[\frac{P}{E} \frac{D_1 D_2}{(D_2 - D_1)} \right]^{1/3}$$

$$\sigma_{max} = 0 \cdot 62 \left[\frac{PE^2 (D_2 - D_1)^2}{D_2{}^2 D_1{}^2} \right]^{1/3}$$

In the case of a ball on a plane surface, D_2 becomes infinite and

$$r = 0 \cdot 88 \left[\frac{PD}{E} \right]^{1/3}$$

$$\sigma_{max} = 0 \cdot 62 \left[\frac{PE^2}{D^2} \right]^{1/3}$$

Rollers in contact under load form a rectangular surface of deformation of a length equal to the length of the rollers and width given by

$$w = 2 \cdot 15 \left[\frac{PD_1 D_2}{El(D_1 + D_2)} \right]^{1/2}$$

for rollers of the same material, diameters D_1 and D_2 and length l. The maximum pressure occurs at the centre of the contact rectangle and is

$$\sigma_{max} = 0.59 \left[\frac{PE(D_1 + D_2)}{lD_1 D_2} \right]^{1/2}$$

For a roller on a plane surface, D_2 becomes infinite and

$$w = 2.15 \left[\frac{PD}{El} \right]^{1/2}$$

$$\sigma_{max} = 0.59 \left[\frac{PE}{lD} \right]^{1/2}$$

17.5. GEOMETRICAL DISCONTINUITIES

It was previously explained in the introduction that abrupt changes in geometry of a component give rise to stress concentration in a similar manner to those described in previous sections for loading. In most cases the failure of a component can be attributed to some form of geometrical stress raiser, either from bad design or misfortune.

Typical examples of stress raisers are oil holes, keyways and splines, threads, and fillets at changes of section. Generally one is concerned with the elastic stress concentration factor K_t (defined in the introduction). Since this is dependent on the geometrical proportions and shape of the notch and the type of stress system (tension, torsion, etc.), it is readily appreciated that to cover a wide range of parameters would require a great deal more space than part of one chapter. Furthermore, the limitations on theoretical solutions for stress concentration at notches have resulted in many analyses being obtained experimentally (generally photoelastically) and presented as charts of K_t against geometrical proportions. The subject has been treated very thoroughly by Neuber,[1] Peterson,[2] Frocht,[3] and others, and therefore attention will be confined here to a few special cases of interest.

(a) Circular Hole

The stress concentration caused by a small circular hole in a thin infinite plate subjected to tension was solved by Kirsch, who showed that the maximum stress occurred at the ends of the diameter of the hole transverse to the applied tension giving a value of $K_t = 3$.

1. Neuber, N., *Theory of Notch Stresses* (Michigan, Edwards, 1946).
2. Peterson, R. E., *Stress Concentration Design Factors* (New York, Wiley, 1953).
3. Frocht, M. M., *Photoelasticity*, Vols. I and II (New York, Wiley, 1941).

The solution for a hole in a strip of finite width under tension was obtained theoretically by Howland and confirmed photoelastically by many investigators.

The distribution of axial and transverse stress at the hole section is shown in Fig. 17.6, and the relationship between K_t and geometrical proportions is given in Fig. 17.7. The latter figure shows that the maximum value of K_t obtainable is 3·0, corresponding to the infinite plate, and the minimum value 2·0, when the strip is the same width as the hole.

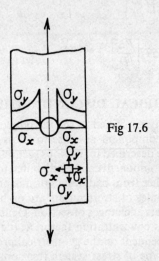

Fig 17.6

In pure bending of a finite width plate with a transverse hole, K_t is a function of plate thickness as well as of radius of hole r and width of plate w. For a very thick plate and a small hole, the K_t against r/w curve is identical with Fig. 17.7 for the tension strip. When the hole is large and the plate thin, K_t varies from 1·85 to 1·1 for decreasing width.

The case of a shaft with a transverse hole subjected to tension, bending or torsion is a common one. It introduces the feature of a three-dimensional stress system as against plane stress in the plate case. The stress concentration factor in a triaxial stress field, although defined in the same way as for a biaxial stress system, will be denoted by K_t'. Frocht has studied photoelastically a circular bar with a transverse hole under tension and has obtained a curve of K_t' against r/d as shown in Fig. 17.8. Also plotted here for comparison is Howland's curve for K_t for a plate under tension, and it is interesting to note that K_t' is noticeably higher.

A circular shaft with a transverse hole in pure bending has been examined by Frocht, and Peterson and Wahl. A curve of K_t' against r/d is given in Fig. 17.9. The interesting feature about this problem is that K_t' for bending can be related to K_t for a plate under tension by the hypothesis of laminar action. This means that the bent shaft can be

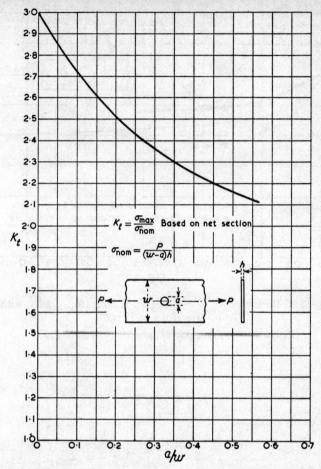

Fig 17.7   Stress Concentration Factor K_t for Axial Loading of a
Finite-width Plate with a Transverse Hole
(*Reference 2: by courtesy of John Wiley & Sons Ltd.*)

considered as consisting of a series of thin plates with a transverse hole
subjected to a tensile or compressive stress proportional to the distance
from the neutral axis; hence

$$K_t' = K_t \frac{y}{R}$$

where K_t is the value obtained from Fig. 17.7 at the a/w value corresponding to the laminar y from the neutral axis, and R is the radius of
the shaft.

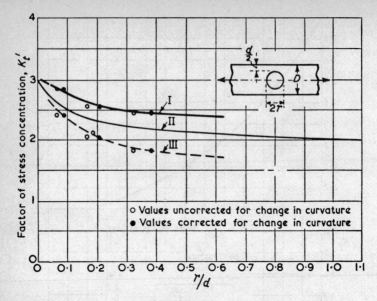

Fig 17.8 Summary of Experimental Results for Shafts with Transverse Holes in Tension

(*Reference 3: by courtesy of John Wiley & Sons Ltd.*)

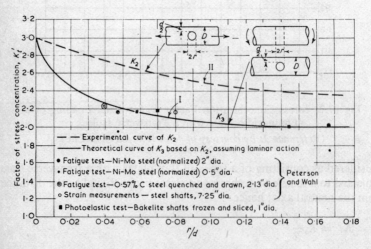

Fig 17.9 Comparison of Theoretical and Experimental Values of K_3 for Circular Shafts with Transverse Holes in Bending

(*Reference 3: by courtesy of John Wiley & Sons Ltd.*)

(b) Fillet Radius

Almost without exception cylindrical components do not have a uniform diameter from one end to the other. The journal bearings of a crankshaft have to mate into the web, and a motor shaft or railway axle requires shoulders to retain bearings or a wheel. At these changes of section a sharp corner would introduce an intolerable stress concentration which could lead to failure by fatigue. The problem is lessened by the introduction of a fillet radius, as in Fig. 17.10, to blend one section smoothly into

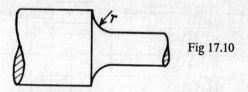

Fig 17.10

the next. Even a fillet radius will give rise to some stress concentration; however, this will be considerably less than with the sharp corner. The stress concentration at a fillet radius is not readily amenable to mathematical treatment and solutions have been obtained by photoelasticity and other experimental means.

Values of K_t for various geometrical proportions have been obtained for shafts in tension, bending and torsion. The latter two cases are probably the most common in practice and therefore only the charts for these, in Figs. 17.11 and 17.12, have been included, for reasons of space.

(c) Keyways and Splines in Torsion

Another design feature which requires careful attention is the keyway or spline in a shaft subjected to torsion. An approximation to the above is a shallow longitudinal groove with a semicircular root as in Fig. 17.13, and the mathematical solution by Neuber obtains the following expression for the shear stress concentration factor:

$$K_{ts} = \frac{\tau_{max}}{\tau_{nom}} = 1 + \sqrt{\frac{t}{\rho}}$$

where $\tau_{nom} = 16T/\pi d^3$, t = groove depth and ρ = radius of groove. In the special case of a semicircular groove where $t = \rho$, $K_{ts} = 2$.

The rectangular keyway of standard form having root fillets has been solved mathematically by Leven, and the curve for K_{ts} against the ratio of fillet radius to shaft diameter is shown in Fig. 17.14. The maximum shear stress occurs at a point 15° from the bottom of the keyway on the fillet radius.

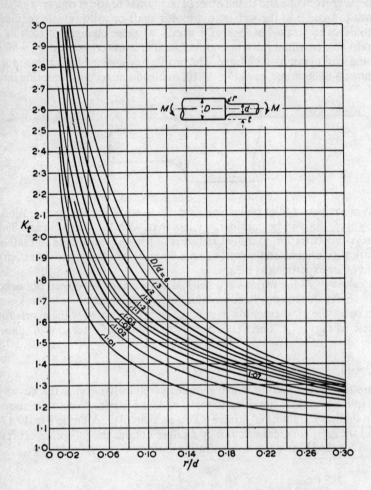

Fig 17.11   Stress Concentration Factor K_t for the Bending of a
Shaft with a Shoulder Fillet

(*Reference 2: by courtesy of John Wiley & Sons Ltd.*)

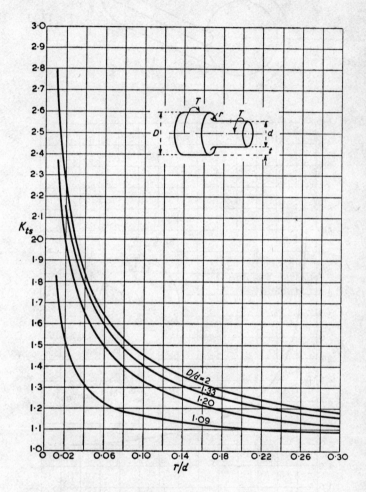

Fig 17.12   Stress Concentration Factor K_{ts} for the Torsion of a
Shaft with a Shoulder Fillet

(*Reference 2: by courtesy of John Wiley & Sons Ltd.*)

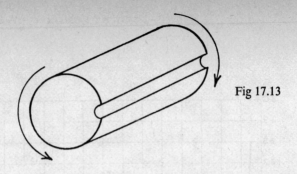

Fig 17.13

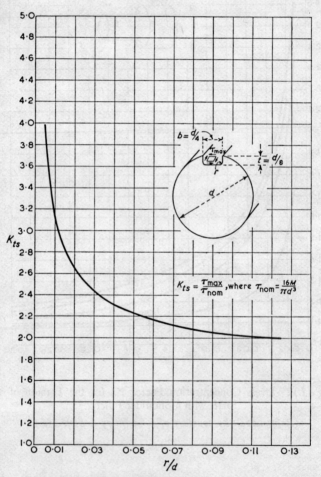

Fig 17.14 Stress Concentration Factor K_{ts} for the Straight
Portion of a Keyway in a Shaft in Torsion

(*Reference 2: by courtesy of John Wiley & Sons Ltd.*)

(d) Gear Teeth

The stress distribution in a loaded gear tooth is a complex problem. Stress concentration at the point of mating of two teeth due to contact load varies in position with rotation of the teeth, and further stress concentration occurs at the root fillets of a tooth, the latter being the more serious. The former can be analysed with the aid of the expressions given in the section under contact stresses. The fillet stress concentration

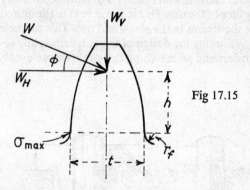

Fig 17.15

is a function of the load components, Fig. 17.15, causing bending and direct stress, and the gear tooth geometry. Empirical formulae in terms of the above have been derived from photoelastic experiments by Dolan and Broghammer as follows:

$$K_t = 0\cdot22 + \frac{1}{\left(\dfrac{r_f}{t}\right)^{0\cdot2}\left(\dfrac{h}{t}\right)^{0\cdot4}} \quad (14\tfrac{1}{2}° \text{ pressure angle})$$

$$K_t = 0\cdot18 + \frac{1}{\left(\dfrac{r_f}{t}\right)^{0\cdot15}\left(\dfrac{h}{t}\right)^{0\cdot45}} \quad (20° \text{ pressure angle})$$

A more recent investigation, very comprehensive in nature, was conducted by Jacobson,[4] who studied involute spur-gears (20° pressure angle) photoelastically and produced a series of charts of strength factor against the reciprocal of the number of teeth in the gear, which cover the whole possible range of spur-gear combinations.

4. Jacobson, M. A., "Bending stresses in spur gear teeth: proposed new design factors based on a photoelastic investigation," *Proc. Instn Mech. Engrs*, 1955.

(e) Screw Threads

One of the most common causes of machinery or plant having to be shut down is the fatigue of bolts or studs. This is principally due to the high stress concentration at the root of the thread. The problem has been tackled theoretically by Sopwith, and photoelastically by Hetenyi, and Brown and Hickson.[5] Each of the investigations confirmed that, for a bolt and nut of conventional design, the load distribution along the screw is far from uniform and reaches a maximum intensity at the plane of the bearing face of the nut. This is due in part to the unmatched and opposing signs of the strains in the screw and nut. This can be overcome to a large extent by altering the design of the nut, principally so that the nut thread is in tension and so matching the strains more evenly with the screw.

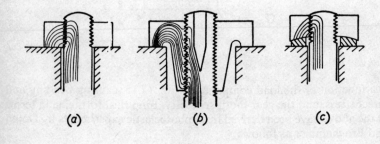

(a)   (b)   (c)

Fig 17.16

Various designs of special nut are shown in Fig. 17.16 and may be compared qualitatively with an ordinary nut in terms of the "flow lines" of stress and stress concentration.

Factors for the latter can reach very high values as is demonstrated in Fig. 17.17 (a) for the ordinary nut and screw. A maximum value of K_t of about 13 is obtained in the screw at $\frac{1}{3}$ of a pitch from the bearing face of the nut. On the other hand, a tension nut and hollowed-out screw, as in Fig. 17.16 (b), gives a much more uniform value of K_t over a greater length, as shown in Fig. 17.17 (b). Even in this case a peak value of stress concentration factor of about 9 is attained, which is quite considerable.

5. Brown, A. F. C. and Hickson, V. M., "A photoelastic study of stresses in screw threads," *Proc. Instn Mech. Engrs*, Vol. IB, 605, 1952–3.

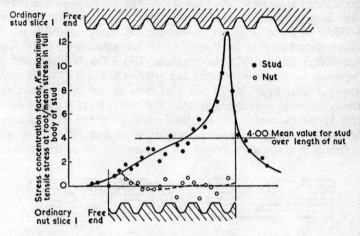

Fig 17.17(*a*) Stress Distribution in Ordinary Stud and Nut
(*Reference 5: by courtesy of the Institution of Mechanical Engineers*)

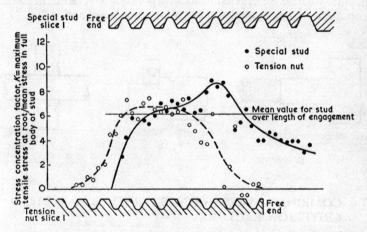

Fig 17.17(*b*) Stress Distribution in Special Stud with Tension Nut
(*Reference 5: by courtesy of the Institution of Mechanical Engineers*)

(f) Stress-relieving Holes and Grooves

A rather surprising feature about stress concentration is that a single discontinuity is more serious than closely-spaced multiple stress raisers of the same proportions. Neuber explains this phenomenon in terms of the "flow of force" through the member. Although multiple notches cause a disturbance in the "flow," there is a smoother transition and streamlining than with a single notch. This effect is illustrated diagrammatically for multiple holes and grooves in Fig. 17.18 and may be compared with Fig. 17.1. It has also been shown theoretically that, although there is an increase in stress at a notch, there is a reduction in stress at the extremities of the local area. Thus the introduction of additional notches, although causing their own stress concentration, has the effect of slightly relieving the adjacent stress concentration.

Fig 17.18

A stress-relieving groove is also useful when placed adjacent to a shoulder where it is not possible to incorporate a fillet radius of suitable proportions. Some examples of this application are shown in Fig. 17.19.

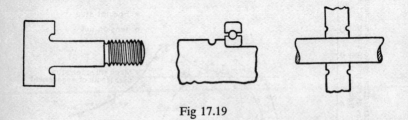

Fig 17.19

17.6. COMBINED STRESS CONCENTRATION AND YIELD CRITERION FACTOR

The usual definition of stress concentration factor, as given in the introduction, is in terms of the maximum stress at the discontinuity. There is of course a biaxial stress condition at the free surface of a notch, and therefore it might be more realistic to express stress concentration in terms of the maximum "equivalent stress," i.e. using one of the yield criteria discussed in Chapter 15. If the biaxial stresses can be determined theoretically at the free surface of a notch, then considering the shear strain energy criterion, which seems the most appropriate for both static and fatigue stress conditions in a ductile material, we have for $\sigma_1 > \sigma_2$ and $\sigma_3 = 0$

$$\sigma_{equiv} = \sqrt{(\sigma_1{}^2 - \sigma_1\sigma_2 + \sigma_2{}^2)}$$

Then

$$K_e = \frac{\sigma_{equiv}}{\sigma_{nom}}$$

In the elastic range there is a constant ratio between σ_1 and σ_2 for any particular point on the surface; therefore

$$\sigma_1 = k\sigma_2$$

and

$$K_e = \frac{\sigma_1\sqrt{[1 - (1/k) + (1/k^2)]}}{\sigma_{nom}}$$

It is seen from the above that K_e is always less than K_t, except when $k = 1$, and then $K_e = K_t$. This may partly explain why notched strength reduction in fatigue (Chapter 22) is nearly always less than would be expected from considerations of K_t.

Chapter 18
Elementary Plastic and Viscoelastic Analyses

18.1. PLASTICITY

The investigation into the behaviour of certain loaded items, in previous chapters, was based on the assumption that the material used obeyed Hooke's law. If, however, the material is stressed beyond the limit of proportionality a knowledge of the mechanical properties of the material, beyond this point, is required.

It is evident that the same principles must apply as for elastic deformation, namely equilibrium of forces, compatibility of deformation and the stress–strain law. In general, the problem is complicated since the latter is non-linear.

Materials such as mild steel possess clearly-defined upper and lower yield points in both tension and compression (see Chapter 19), but the problem is simplified if these are neglected and we assume that the material is elastic up to the value of yield stress σ_Y and that both this value of yield stress and Young's modulus are the same in tension and compression. A plastic range for the material then follows, during which yielding

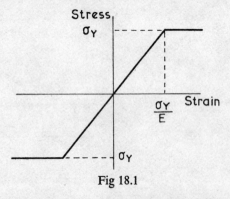

Fig 18.1

takes place at the constant stress σ_Y. This will result in an idealized diagram as shown in Fig. 18.1.

18.2. PLASTIC BENDING OF BEAMS: PLASTIC MOMENT

In considering the behaviour of beams subjected to pure bending which results in fibres being stressed beyond the limit of proportionality, the following assumptions will be made:

412

1. That the fibres are in a condition of simple tension or compression.
2. That any cross-section of the beam will remain a plane during bending as in elastic bending.

The axis, of cross-section, passing through the centroid remains a neutral axis during inelastic bending; thus a cross-section will bend about the neutral axis and the stresses will be greatest in the extreme fibres of the beam.

In elastic bending of a beam there is a linear stress distributed over a cross-section and the extreme fibres reach the yield stress, when the bending moment for this condition is given by

$$M_Y = \sigma_Y \frac{I}{y} \qquad (18.1)$$

where I is the second moment of area of the cross-section about the neutral axis, and y is the distance from the neutral axis to an extreme fibre. The value of the above bending moment will be found for beams of various cross-section.

(a) Rectangular Section
From eqn. (18.1),

$$M_Y = \sigma_Y \frac{bd^2}{6} \qquad (18.2)$$

and the stress distribution corresponding to this condition is shown in Fig. 18.2(a), all the fibres of the beam being in the elastic condition.

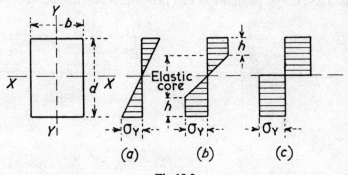

Fig 18.2

When the bending moment is increased above the value given in eqn. (18.2), some of the fibres near the top and bottom surfaces of the beam begin to yield and the appropriate stress diagram is given in Fig. 18.2(b). With further increase in bending moment, plastic deformation penetrates deeper into the beam. The total bending moment is obtained by consideration of both the plastic stress near the top and bottom of the beam and the elastic stress in the core of the beam. This moment is called the *elastic–plastic bending moment*. Therefore

$$M = \sigma_Y bh(d - h) + \sigma_Y \frac{b(d - 2h)^2}{6}$$

$$= \frac{\sigma_Y b}{6}[6h(d - h) + (d - 2h)^2]$$

$$= \frac{\sigma_Y bd^2}{6} \frac{6hd - 6h^2 + d^2 - 4dh + 4h^2}{d^2}$$

$$= \frac{\sigma_Y bd^2}{6}\left[1 + 2\frac{h}{d}\left(1 - \frac{h}{d}\right)\right] \tag{18.3}$$

At a distance $(\frac{1}{2}d - h)$ from the neutral axis, the stress in the fibres has just reached the value σ_Y; then, if R is the radius of curvature, we have

$$\sigma_Y = \frac{E(\frac{1}{2}d - h)}{R}$$

or

$$\frac{1}{R} = \frac{\sigma_Y}{E(\frac{1}{2}d - h)} \tag{18.4}$$

The values of M and $1/R$ calculated from eqns. (18.3) and (18.4) when plotted give the graph shown in Fig. 18.3. The connection between these

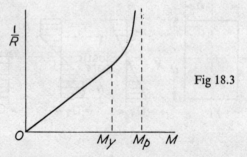

Fig 18.3

quantities is linear up to a value of $M = M_Y$. Beyond this point the relationship is non-linear and the curve becomes steeper with increase in depth, h, of the plastic state. When h becomes equal to $\frac{1}{2}d$, the stress distribution becomes that shown in Fig. 18.2 (c) and the highest value of bending moment is reached.

This *fully plastic moment* is given by eqn. (18.3), putting $h = \frac{1}{2}d$, as

$$M_p = \frac{3}{2}\sigma_Y \frac{bd^2}{6} = \sigma_Y \frac{bd^2}{4} \tag{18.5}$$

$$= \frac{3}{2}M_Y \tag{18.6}$$

The position of M_p is shown in Fig. 18.3 and is the vertical asymptote of the curve. Plastic collapse of the beam is shown, by eqn. (18.5), to occur at a moment $1\frac{1}{2}$ times that at initial yielding of the extreme fibres of the beam.

(b) I-section
When yielding is about to occur at the extreme fibres, the beam is still in the elastic condition and, in Fig. 18.4,

$$M_Y = \sigma_Y \left(\frac{bd^3}{12} - \frac{b_1 d_1^3}{12}\right)\frac{2}{d} \tag{18.7}$$

and the stress diagram is similar to that shown in Fig. 18.2(a).

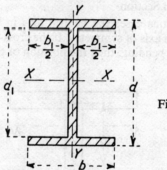

Fig 18.4

In the fully plastic condition, the stress diagram is similar to that in Fig. 18.2(c), and from eqn. (18.5) the fully plastic moment is given by

$$M_p = \sigma_Y \left(\frac{bd^2}{4} - \frac{b_1 d_1^2}{4}\right) \tag{18.8}$$

and the ratio M_p/M_Y is given by

$$\frac{M_p}{M_Y} = \frac{\dfrac{bd^2}{4} - \dfrac{b_1 d_1^2}{4}}{\dfrac{bd^3}{12} - \dfrac{b_1 d_1^3}{12}}\frac{d}{2}$$

or

$$\frac{M_p}{M_Y} = \frac{3}{2} \frac{1 - \dfrac{b_1 d_1{}^2}{bd^2}}{1 - \dfrac{b_1 d_1{}^3}{bd^3}} \qquad (18.9)$$

In an I-beam, 100 mm × 300 mm, with flanges and web 14 mm and 9 mm thick respectively,

$$\frac{M_p}{M_Y} = \frac{3}{2} \frac{1 - \dfrac{91 \times 272^2}{100 \times 300^2}}{1 - \dfrac{91 \times 272^3}{100 \times 300^3}}$$

$$= \frac{3 \times 0.255}{2 \times 0.33} = 1.16$$

Thus in this case, which is fairly representative of standard rolled I-section beams, the fully plastic moment is only 16% greater than that at which initial yielding occurs.

(c) Asymmetrical Section

In the two previous cases the neutral axis in bending of the section coincided with an axis of symmetry. If the cross-section is asymmetrical about the axis of bending, then the position of the neutral axis must be

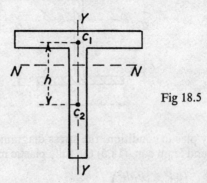

Fig 18.5

determined. Fig. 18.5 shows a T-bar section in which YY is the only axis of symmetry.

In the fully plastic condition the beam is bent about the neutral axis NN. If A_1 and A_2 are the areas of the cross-section above and below NN respectively, then since there can be no longitudinal resultant force in the beam during bending without end load,

$$A_1 \sigma_Y = A_2 \sigma_Y$$

or $A_1 = A_2 = \frac{1}{2}A$, where A is the total area of the cross-section. Thus the neutral axis divides the cross-section into two equal areas.

If C_1 is the centroid of the area A_1, C_2 the centroid of the area A_2, and h the distance between C_1 and C_2, then the fully plastic moment is given by

$$M_p = \tfrac{1}{2}A\sigma_Y h = \tfrac{1}{2}\sigma_Y Ah \tag{18.10}$$

Example 18.1. The flange and web of a T-bar section are each 12 mm thick, the flange width is 100 mm, and the overall depth of the section is 100 mm. The centroid of the section is at a distance of 70·6 mm from the bottom of the web, and the second moment of area, I_z, of the section about a line through the centroid and parallel to the flange is $22·4 \times 10^6$ mm^4. Determine the value of the ratio M_p/M_Y.

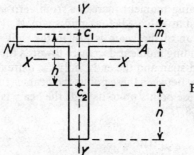

Fig 18.6

Let m be the distance of the neutral axis from the top of the flange (Fig. 18.6); then

$$100m = 100(12 - m) + (88 \times 12)$$

$$m = 11·3 \text{ mm}$$

If n is the distance of the centre of gravity of area A_2 from the bottom of the web, then

$$(100 \times 0·7)(88 + 0·35) + (12 \times 88 \times 44) =$$

$$[(100 \times 0·7) + (12 \times 88)]n$$

$$n = 46·8 \text{ mm}$$

Therefore h, the distance between C_1 (the centroid of A_1) and C_2 (the centroid of A_2), is

$$h = 88·7 - 46·8 + \frac{11·3}{2} = 47·55 \text{ mm}$$

so that

$$M_p = \frac{12 \times 188}{2} \sigma_Y \times 47 \cdot 55$$

$$M_Y = \frac{22 \cdot 4 \times 10^6}{70 \cdot 6} \sigma_Y$$

and

$$\frac{M_p}{M_Y} = 1 \cdot 7$$

18.3. PLASTIC HINGE IN BEAMS

Consider a beam simply supported at each end and carrying a central load W. The bending moment increases from zero at each end of the beam to a maximum value $\frac{1}{4}WL$ at mid-span. If W is such that the central cross-section of the beam is just in the fully plastic condition, then, since the bending moment decreases towards the ends, at a certain point between mid-span and the ends, only the extreme fibres will be in a plastic condition. Between these points the beam is in an elastic–plastic state, and from these points up to its ends the beam is in an elastic state as shown in Fig. 18.7.

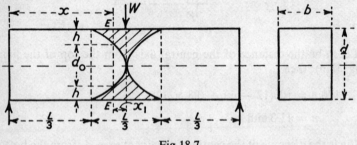

Fig 18.7

The value of W is found from $M_p = \frac{1}{4}WL$ or $W = 4M_p/L$, and from eqn. (18.6), $M_p = \frac{3}{2}M_Y$.

At a distance x from the free end of the beam, $M = \frac{1}{2}Wx$, and if x is such that $M = M_Y$, i.e. plasticity is just beginning, then

$$\frac{Wx}{2} = \frac{2}{3}M_p = \frac{2}{3}\frac{WL}{4}$$

or

$$x = \frac{L}{3} \tag{18.11}$$

The resisting moment at the section EE is given by eqn. (18.3) as

$$M = \sigma_Y \frac{bd^2}{6} \left[1 + \frac{2h}{d} \left(1 - \frac{h}{d} \right) \right]$$

Putting $h = \frac{1}{2}(d - d_0)$, then

$$M = \sigma_Y \frac{bd^2}{6} \left[1 - \frac{d_0{}^2}{3d^2} \right] \frac{3}{2}$$

$$= M_p \left[1 - \frac{d_0{}^2}{3d^2} \right] \tag{18.12}$$

The shape of the elastic–plastic boundary is readily found if M is calculated at a distance x_1 from mid-span; then

$$M = \frac{1}{2}W \left(\frac{L}{2} - x_1 \right) = \frac{1}{2} \frac{4M_p}{L} \left(\frac{L}{2} - x_1 \right)$$

$$= M_p \left(1 - \frac{2x_1}{L} \right) \tag{18.13}$$

Equating eqn. (18.12) to eqn. (18.13),

$$1 - \frac{d_0{}^2}{3d^2} = 1 - \frac{2x_1}{L}$$

or

$$d_0{}^2 = \frac{6d^2}{L} x_1 \tag{18.14}$$

Thus the depth of the elastic core varies parabolically with x_1 and has its maximum value when $d_0 = d$, i.e. at $x_1 = \frac{1}{6}L$.

When the load W is increased until the two regions of plasticity, at the top and bottom of the beam, meet, then the two halves of the beam will rotate with respect to each other, about the neutral axis of the cross-section at mid-span, as about a hinge. This is called the *plastic hinge*.

Example 18.2. A steel bar of square section 50 mm × 50 mm carries a uniformly distributed load w on a length of 2 m. If the bar is simply supported at each end, calculate the value of w when the central cross-section is in a fully plastic condition. Determine also the shape of the regions of plasticity and the length of that part of the beam which is wholly elastic. Assume a value for yield stress of 240 MN/m².

If the central cross-section is in the fully plastic state as in Fig. 18.8, then $M_{max} = \frac{1}{8}wL^2 = M_p$, from Section 1.17. Therefore

$$w = \frac{8M_p}{L^2}$$

$$= 8\sigma_Y \frac{bd^2}{4L^2}$$

$$= \frac{8 \times 240 \times 10^6 \times 0.05 \times 0.05^2}{4 \times 2^2}$$

$$= 15\,\text{kN/m}$$

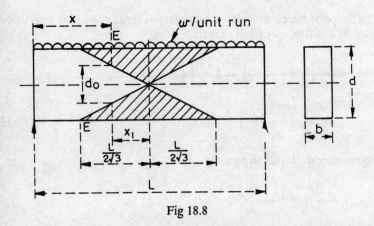

Fig 18.8

As in the previous case, the shape of the regions of plasticity is readily found by measuring x_1 from mid-span as in Fig. 18.8. The bending moment at a section EE is then given in terms of x_1 by

$$M = \frac{wL}{2}\left(\frac{L}{2} - x_1\right) - \frac{w}{2}\left(\frac{L}{2} - x_1\right)^2$$

$$= \frac{wL}{8L}(L^2 - 4x_1{}^2)$$

$$= \frac{M_p}{L^2}(L^2 - 4x_1{}^2)$$

$$= M_p\left[1 - 4\left(\frac{x_1}{L}\right)^2\right] \tag{18.15}$$

The resisting moment of the cross-section is given by eqn. (18.12); hence

$$M_p \left[1 - \frac{d_0^2}{3d^2} \right] = M_p \left[1 - 4 \left(\frac{x_1}{L} \right)^2 \right]$$

and

$$\frac{d_0^2}{3d^2} = 4 \left(\frac{x_1}{L} \right)^2$$

$$d_0^2 = 12d^2 \left(\frac{x_1}{L} \right)^2$$

Hence

$$d_0 = \frac{2\sqrt{3}\,d}{L}\,x_1 \qquad\qquad (18.16)$$

and in this particular problem,

$$d_0 = \frac{2\sqrt{3} \times 50}{2000}\,x_1 = 0 \cdot 0866 x_1$$

Thus when corresponding values of d_0 and x_1 are plotted, straight lines are obtained enclosing the regions of plasticity. x_1 has its maximum value when $d_0 = d$, and so from eqn. (18.16) $x_1 = L/2\sqrt{3} = 2000/2\sqrt{3} = 578$ mm. Hence the length of the beam which is wholly elastic is $2000 - (2 \times 578) = 844$ mm.

18.4. PLASTIC TORSION OF SHAFTS

In the following discussion it will be assumed that we have an ideal stress–strain relationship for the material as shown in Fig. 18.9, that a

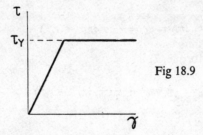

Fig 18.9

plane cross-section of the shaft remains a plane when in the plastic state, and that a radius remains straight. The shearing strain γ at a distance r from the axis of the shaft will still be given by $\gamma = r\theta/L$.

When the shaft has a twisting moment applied in the elastic range, the shear stress increases from zero at the shaft axis to a maximum value at the surface of the shaft, and

$$T = \frac{\tau\pi}{2} r^3$$

for a solid circular shaft. When the shear stress at the surface of the shaft has reached the value τ_Y the twisting moment required to give this stress is

$$T_Y = \frac{\tau_Y\pi}{2} r^3 \qquad (18.17)$$

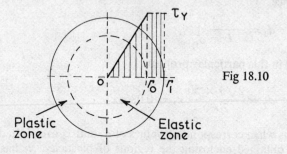

Fig 18.10

If the twisting moment is increased beyond this value, then plasticity occurs in fibres at the surface of the shaft and the stress diagram is as shown in Fig. 18.10. The torque carried by the elastic core is

$$T_1 = \frac{\tau_Y\pi}{2} r_0^3 \qquad (18.18)$$

where r_0 is the interface radius between elastic and plastic material; and that carried by the plastic zone is

$$T_2 = \int_{r_0}^{r_1} 2\pi r^2 \tau_Y \, dr = \frac{2\pi}{3} \tau_Y(r_1^3 - r_0^3) \qquad (18.19)$$

and the total torque, T, is

$$T_1 + T_2 = \tfrac{1}{2}\tau_Y\pi r_0^3 + \frac{2}{3}\pi\tau_Y(r_1^3 - r_0^3)$$

Therefore

$$T = \frac{2}{3} \pi r_1^3 \tau_Y \left[1 - \frac{r_0^3}{4r_1^3}\right] \qquad (18.20)$$

When the fully plastic condition is reached, the shear stress at all points in the cross-section is τ_Y, and it follows from eqn. (18.19) that the fully plastic torque is given by

$$T_p = \frac{2\pi}{3} \tau_Y r_1{}^3 \qquad (18.21)$$

and the ratio T_p/T_Y is

$$\frac{T_p}{T_Y} = \frac{4}{3}$$

When the fibres at the outer surface of the shaft are about to become plastic, the angle of twist is given by

$$\theta_Y = \frac{\tau_Y L}{Gr_1} \qquad (18.22)$$

and when the shaft is in the elastic–plastic condition, the angle of twist of the elastic core is given by

$$\theta = \frac{\tau_Y L}{Gr_0} \qquad (18.23)$$

Since we have assumed that radii remain straight, then the outer plastic region suffers the same angle of twist. From eqns. (18.22) and (18.23) it follows that

$$\frac{\theta_Y}{\theta} = \frac{r_0}{r_1} \qquad (18.24)$$

and eqn. (18.20) may be expressed in the form

$$T = \frac{2}{3} \pi r_1{}^3 \tau_Y \left[1 - \frac{1}{4} \left(\frac{\theta_Y}{\theta} \right)^3 \right] \qquad (18.25)$$

Example 18.3. A steel shaft, 100 mm in diameter and 3 m long, is in an elastic-plastic state under a twisting moment of 30 kN-m. Determine the diameter of the elastic core of the shaft and the angle of twist, given that $\tau_Y = 120 \text{ MN/m}^2$ and $G = 80 \text{ GN/m}^2$. What is the value of the fully plastic twisting moment for the shaft?

From eqn. (18.20) the elastic–plastic torque is

$$T = \frac{2}{3} \pi r_1{}^3 \tau_Y \left[1 - \frac{r_0{}^3}{4r_1{}^3} \right]$$

$$30\,000 = \tfrac{2}{3}\pi \times 0.05^3 \times 120 \times 10^6 \left[1 - \frac{r_0{}^3}{4 \times 0.05^3} \right]$$

from which

$$r_0{}^3 = 23 \times 10^{-6} \quad \text{and} \quad r_0 = 28.5 \text{ mm}$$

Hence the diameter of the elastic core is 57 mm.

The shear strain at r_0 is

$$\gamma = \frac{120 \times 10^6}{80 \times 10^9} = 0.0015$$

but γ is also $r\theta/L = 0.0285\theta/3$; therefore

$$\theta = \frac{0.0015 \times 3}{0.0285} = 0.158 \text{ rad}$$

From eqn. (18.21),

$$T_p = \frac{2\tau_Y\pi}{3} r_1^3 = \frac{2 \times 120 \times 10^6\pi}{3} \times 0.05^3 = 31.4 \text{ kN-m}$$

18.5. PLASTIC BENDING OR TORSION IN A STRAINHARDENING MATERIAL

Up to the present we have assumed that materials are perfectly plastic and have a stress–strain diagram as shown in Fig. 18.1. In the more general case, *strainhardening* occurs, increase in strain being accompanied by an increase in stress, and in the case of a beam a resultant stress–strain diagram, as shown in Fig. 18.11, is obtained. The material does not

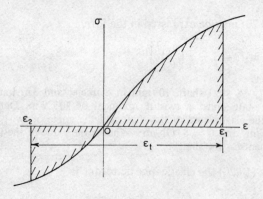

Fig 18.11

follow Hooke's law, but in order to investigate the pure bending of such beams, it will be assumed that cross-sections of the beam remain plane during bending, and thus extension and compression of the longitudinal fibres are proportional to their distance from the neutral axis. It will also be assumed that the same relationship exists between stress and strain during bending as in simple tension and compression.

(a) Beam of Rectangular Section Carrying a Bending Moment M

In Fig. 18.12, the strains ε_1 and ε_2 of the extreme fibres are given by y_1/R and $-y_2/R$ respectively. These quantities, however, can only be determined when the value of R and the position of the neutral axis are known, and these can be found from equilibrium conditions:

$$\int_A \sigma \, dA = b \int_{-y_2}^{y_1} \sigma \, dy = 0 \tag{18.26}$$

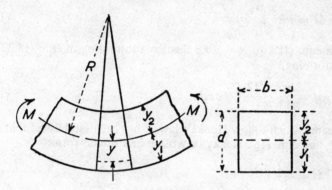

Fig 18.12

since the sum of the normal forces on a cross-section must be zero. Also

$$\int_A \sigma y \, dA = b \int_{-y_2}^{y_1} \sigma y \, dy = M \tag{18.27}$$

i.e. the sum of the moments of the above forces about the neutral axis is equal to the bending moment. But $y = R\varepsilon$, so that

$$dy = R \, d\varepsilon \tag{18.28}$$

and substituting in eqn. (18.26),

$$b \int_{-y_2}^{y_1} \sigma \, dy = bR \int_{\varepsilon_2}^{\varepsilon_1} \sigma \, d\varepsilon = 0 \tag{18.29}$$

Thus the position of the neutral axis must be such that $\int_{\varepsilon_2}^{\varepsilon_1} \sigma \, d\varepsilon$ is zero. In Fig. 18.11, let

$$\varepsilon_t = |\varepsilon_1 + \varepsilon_2| = \frac{y_1}{R} + \frac{y_2}{R} = \frac{y_1 + y_2}{R} = \frac{d}{R} \tag{18.30}$$

Then if ε_t is set out on the diagram in order to make the two shaded areas equal, ε_1 and ε_2 will be fixed since

$$\frac{y_1}{y_2} = \frac{\varepsilon_1}{\varepsilon_2} \tag{18.31}$$

Hence the position of the neutral axis is determined.

In order to find the magnitude of R, we have, by substituting the values of y and dy from eqn. (18.28) in eqn. (18.27),

$$M = bR^2 \int_{\varepsilon_2}^{\varepsilon_1} \sigma\varepsilon \, d\varepsilon \tag{18.32}$$

From eqn. (18.30), $R = d/\varepsilon_t$; then by substituting in eqn. (18.32) and transforming,

$$M = \frac{bd^3}{12} \frac{1}{R} \frac{12}{\varepsilon_t^3} \int_{\varepsilon_2}^{\varepsilon_1} \sigma\varepsilon \, d\varepsilon \tag{18.33}$$

In beams which follow Hooke's law, it has been established that $M = EI/R$, and eqn. (18.33) may be written in a similar form:

$$M = \frac{E_r I}{R} \tag{18.34}$$

where E_r is called the *reduced modulus* and is given by

$$E_r = \frac{12}{\varepsilon_t^3} \int_{\varepsilon_2}^{\varepsilon_1} \sigma\varepsilon \, d\varepsilon \tag{18.35}$$

Eqn. (18.34) can be used to calculate the bending moment for any given curvature of the beam when the prevailing conditions are those beyond the limit of proportionality of the material.

In order to find the value of bending moment corresponding to a particular curvature we require the appropriate value of E_r. This may be obtained from a curve of E_r against ε_t, which is constructed by assuming various values for ε_t, determining ε_1 and ε_2 from condition (18.29) and Fig. 18.11, and then evaluating the integral (18.35). It will be seen that the integral, and hence E_r, has units of force per unit area. The distribution of bending stress over the cross-section of the beam is the same as the stress–strain diagram in Fig. 18.11, where the depth of the beam corresponds to the strain axis.

(b) Solid Circular Shaft Subjected to Pure Torque T

Assuming as in Section 18.4 that, beyond the limit of proportionality, the cross-sections of the twisted shaft remain plane and the radii remain straight, then the shearing strain at a distance r from the shaft axis is given by $\gamma = r\theta/L$, where θ is the angle of twist.

The relationship between the shearing stress and the shearing strain for the material is given by Fig. 18.13. Owing to strain-hardening the stress increases with increase in strain. Then if the radius of the shaft is r_1, the maximum shear strain is $r_1\theta/L$ and the corresponding shear stress is the ordinate kl in the diagram.

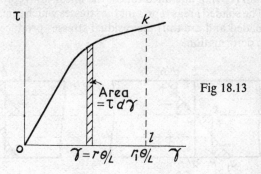

Fig 18.13

The torque, T, required to produce the angle of twist, θ, is given by the equilibrium relation

$$T = \int_0^{r_1} 2\pi r^2 \tau \, dr \tag{18.36}$$

and since $\gamma = r\theta/L$, then $d\gamma = \theta \, dr/L$, and substituting for r and dr in the above equation,

$$T = \frac{2\pi L^3}{\theta^3} \int_0^{r_1\theta/L} \tau\gamma^2 \, d\gamma \tag{18.37}$$

The integral is the second moment of the area $OklO$ about the vertical axis Oq. When this quantity has been calculated for a given value of $\gamma = r_1\theta/L$, the torque is found from eqn. (18.37). The distribution of shear stress across the shaft is of the same form as the shear stress–strain curve for the material in Fig. 18.13.

18.6. RESIDUAL STRESS DISTRIBUTION AFTER PLASTIC BENDING OR TORSION

(i) Elastic-plastic Material

Permanent set is produced when a beam is bent beyond the elastic limit and the deformation does not disappear when the load is removed. Those fibres which have suffered more permanent set prevent those which are less strained from recovering their initial length when the load is removed. Consequently some residual stresses are produced.

In a beam of rectangular cross-section, we will assume for simplicity that the beam has been bent to the fully plastic condition such that the stress distribution diagrams are the two rectangles *Ocdb* and *Oeka*, shown in Fig. 18.14(*a*). Assuming that when the material is stretched beyond the yield point and then unloaded it will follow Hooke's law during unloading, then the bending stresses which are superposed whilst the beam is being unloaded follow the linear law represented by the line a_1b_1. The shaded areas represent the stresses which remain after the beam is unloaded and are thus the residual stresses produced in the beam by plastic deformation.

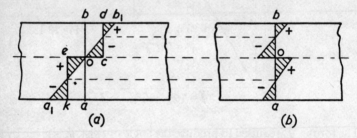

Fig 18.14

The rectangular and triangular stress distributions represent bending moments of the same magnitude; hence the moment of the rectangle *Ocdb* about the axis *eOc* is equal to the moment of the triangle *Obb₁* about the same axis. Since *bd* represents σ_Y it follows for equal moments about the axis *eOc* that the stress represented by bb_1 is equal to $1\frac{1}{2}\sigma_Y$, and thus the maximum residual tension and compression stresses after loading and unloading are db_1, and $a_1k = \frac{1}{2}\sigma_Y$. Near the neutral axis the residual stresses are equal to σ_Y.

In Fig. 18.14(*b*) the residual stresses are replotted on a conventional base of a cross-section, in order to show these residual tensile and compressive stresses more clearly. Diagrams showing the residual stresses in a partially yielded beam are shown in Fig. 18.15. In this case the

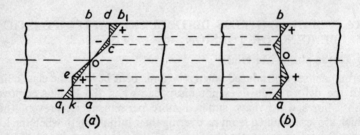

Fig 18.15

material has yielded to such an extent that the stress distribution diagrams are represented by the areas $Ocdb$ and $Oeka$ in Fig. 18.15(a). Again assuming the material to follow Hooke's law during unloading, the bending stress during this operation will follow the linear law represented by a_1b_1, such that the moments of $Ocdb$ and Obb_1 about the neutral axis are equal, and the shaded areas represent the residual stresses. These stresses are shown replotted on a conventional cross-section in Fig. 18.15(b).

The solution for residual stresses after plastic torsion follows the same arguments as above.

(ii) Strainhardening Material

If a beam has a strainhardening stress distribution given by the curve kOd, Fig. 18.16(a), then again assuming that the material follows Hooke's law during unloading, the residual stresses are distributed as shown by the shaded areas, so that the moments of Obd and Obc about the neutral axis are equal. Having obtained the curve kOd as in Section 18.5, the magnitude of the residual stress can be found for any given fibre. In

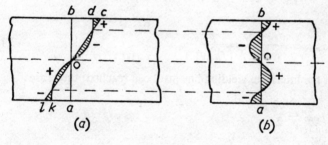

Fig 18.16

Fig. 18.16(b) the residual stresses are shown replotted on a conventional cross-section.

Similar reasoning may be applied when a shaft is subjected to torsion and yields a stress distribution curve similar to kOd.

18.7. AXIALLY SYMMETRICAL COMPONENTS

(i) Thick Cylinder under Internal Pressure

The problems analysed in the previous paragraphs only involved a uniaxial stress condition, and hence a simple tensile or shear yield stress was sufficient to define the onset of plastic deformation. In the axisymmetrical component, where there are at least radial and hoop stresses if not axial stress, a yield criterion of the type discussed in Chapter 15 has to be employed to determine the initiation of plastic flow.

The subsequent analysis will only consider an ideal elastic–plastic material, since the problem for a workhardening material is beyond the scope of this text. Furthermore, in order to simplify the mathematics the maximum shear stress theory of yielding will be adopted.

For a thick cylinder under internal pressure, the maximum shear stress occurs at the inner surface (Chapter 14), and therefore as the pressure is increased, plastic deformation will commence first at the bore and penetrate deeper and deeper into the wall, until the whole vessel reaches the yield condition.

At a stage when plasticity has penetrated partly through the wall, the vessel might be regarded as a compound cylinder with the inner tube plastic and the outer elastic. If the elastic–plastic interface is at a radius x and the radial pressure there is p_x, then from eqns. (14.33) and (14.34),

$$\sigma_r = \frac{p_x x^2}{r_0^2 - x^2}\left(1 - \frac{r_0^2}{x^2}\right) \tag{18.38}$$

$$\sigma_\theta = \frac{p_x x^2}{r_0^2 - x^2}\left(1 + \frac{r_0^2}{x^2}\right) \tag{18.39}$$

and

$$\tau_{max} = \frac{\sigma_\theta - \sigma_r}{2} = \frac{p_x r_0^2}{r_0^2 - x^2} \tag{18.40}$$

But at the interface, yielding has just been reached; therefore

$$\tau_{max} = \frac{\sigma_Y}{2}$$

and

$$\frac{p_x r_0^2}{r_0^2 - x^2} = \frac{\sigma_Y}{2}$$

Therefore

$$p_x = \frac{\sigma_Y}{2r_0^2}(r_0^2 - x^2) \tag{18.41}$$

From this value of p_x the stress conditions in the elastic zone can be determined using eqns. (14.33) and (14.34). It is now necessary to consider the equilibrium of the plastic zone in order to find the internal pressure required to cause plastic deformation to a depth of $r = x$. The equilibrium equation is

$$r\frac{d\sigma_r}{dr} + \sigma_r - \sigma_\theta = 0 \quad \text{(eqn. (14.12))} \tag{18.42}$$

Hence

$$r \frac{d\sigma_r}{dr} - 2\tau_{max} = 0$$

or

$$\frac{d\sigma_r}{dr} = \frac{\sigma_Y}{r} \qquad (18.43)$$

Integrating this equation gives

$$\sigma_r = \sigma_Y \log_e r + K \qquad (18.44)$$

Now, when $r = x$, $\sigma_r = -p_x$; therefore

$$-p_x = \sigma_Y \log_e x + K$$

or

$$K = -p_x - \sigma_Y \log_e x$$

Substituting for K in eqn. (18.44),

$$\sigma_r = \sigma_Y \log_e r - p_x - \sigma_Y \log_e x$$

$$= \sigma_Y \log_e \frac{r}{x} - p_x \qquad (18.45)$$

Therefore, using eqn. (18.41),

$$\sigma_r = \sigma_Y \log_e \frac{r}{x} - \frac{\sigma_Y}{2r_0^2}(r_0^2 - x^2) \qquad (18.46)$$

which gives the distribution of radial stress in the plastic zone; and when $r = r_i$, $\sigma_r = -p_i$; therefore

$$p_i = -\sigma_Y \log_e \frac{r_i}{x} + \frac{\sigma_Y}{2r_0^2}(r_0^2 - x^2) \qquad (18.47)$$

where p_i is the internal pressure to cause yielding to a depth of $r = x$.
    The hoop stress is

$$\sigma_\theta = \sigma_Y + \sigma_r$$

Therefore

$$\sigma_\theta = \sigma_Y \left(1 + \log_e \frac{r}{x}\right) - \frac{\sigma_Y}{2r_0^2}(r_0^2 - x^2) \qquad (18.48)$$

    The internal pressure, p_{max}, required to cause yielding right through the wall is found by putting $x = r_0$ in eqn. (18.47):

$$p_{max} = -\sigma_Y \log_e \frac{r_i}{r_0} \qquad (18.49)$$

and

$$\sigma_r = \sigma_Y \log_e \frac{r}{r_0} \tag{18.50}$$

$$\sigma_\theta = \sigma_Y \left(\log_e \frac{r}{r_0} + 1\right) \tag{18.51}$$

The stress distribution of σ_r and σ_θ for the cases considered above are illustrated in Fig. 18.17 in terms of σ_Y for a vessel where $r_0 = 2r_i$.

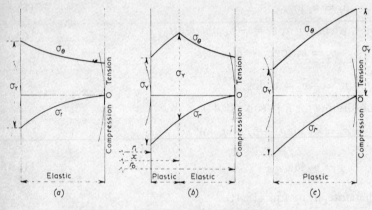

Fig 18.17

For a more detailed analysis of thick-walled cylinders, the reader is referred to *High-pressure Engineering*, by W. R. D. Manning, Bulleid Memorial Lectures 1963, University of Nottingham.

(ii) Thin Rotating Disc with a Central Hole

From Section 14.6 (*b*), it will be seen that the maximum hoop stress occurs on the surface of the hole, and both hoop and radial stresses are positive throughout the disc, the former being the greater of the two. Therefore maximum shear stress is given by

$$\tau_{max} = \frac{\sigma_\theta - 0}{2} \tag{18.52}$$

since $\sigma_z = 0$.

Yielding occurs first at the hole and gradually spreads outwards with increasing rotational speed. Using the maximum shear stress criterion, the speed ω_Y at which yielding first commences is given by

$$\tau_{max\ (r=r_1)} = \frac{\sigma_\theta}{2} = \frac{\sigma_Y}{2} \tag{18.53}$$

From eqn. (14.55), putting $r = r_1$,

$$\sigma_\theta = \frac{\rho \omega_Y{}^2}{4} [r_2{}^2(3 + v) + r_1{}^2(1 - v)] = \sigma_Y \qquad (18.54)$$

Therefore

$$\omega_Y{}^2 = \frac{4\sigma_Y}{\rho [r_2{}^2(3 + v) + r_1{}^2(1 - v)]}$$

or

$$\omega_Y = \sqrt{\frac{4\sigma_Y}{\rho \{r_2{}^2(3 + v) + r_1{}^2(1 - v)\}}} \qquad (18.55)$$

To determine the rotational speed, ω, at which there is a plastic zone for $r_1 < r < c$, consider the equilibrium of an element. The equilibrium equation is

$$r\frac{d\sigma_r}{dr} + \sigma_r - \sigma_\theta + \rho \omega^2 r^2 = 0 \quad \text{(from Section 14.6)}$$

and, since $\sigma_\theta = \sigma_Y$ in the plastic zone,

$$r\frac{d\sigma_r}{dr} + \sigma_r = \sigma_Y - \rho \omega^2 r^2 \qquad (18.56)$$

Integrating both sides,

$$r\sigma_r = r\sigma_Y - \frac{\rho}{3} \omega^2 r^3 + K$$

Therefore

$$\sigma_r = \sigma_Y - \frac{\rho}{3} \omega^2 r^2 + \frac{K}{r} \qquad (18.57)$$

When $r = r_1$, $\sigma_r = 0$; therefore

$$0 = \sigma_Y - \frac{\rho}{3} \omega^2 r_1{}^2 + \frac{K}{r_1}$$

so that

$$K = -r_1\sigma_Y + \frac{\rho}{3} \omega^2 r_1{}^3$$

Substituting for K in eqn. (18.57),

$$\sigma_r = \sigma_Y - \frac{\rho}{3} \omega^2 r^2 - \frac{r_1}{r} \sigma_Y + \frac{\rho}{3} \omega^2 \frac{r_1{}^3}{r} \qquad (18.58)$$

At $r = c$, eqn. (18.58) becomes

$$\sigma_r = \frac{\sigma_Y}{c}(c - r_1) - \frac{\rho\,\omega^2}{3\,c}(c^3 - r_1^{\,3}) \tag{18.59}$$

But this value of σ_r must be the same as σ_r at $r = c$ for the elastic zone. It is firstly necessary to determine the constants A and B in eqns. (14.50) and (14.51). The boundary conditions are: at $r = r_2$, $\sigma_r = 0$; and at $r = c$, $\sigma_\theta = \sigma_Y$. Therefore

$$0 = A - \frac{B}{r_2^{\,2}} - \frac{3 + v}{8}\rho\omega^2 r_2^{\,2}$$

and

$$\sigma_Y = A + \frac{B}{c^2} - \frac{1 + 3v}{8}\rho\omega^2 c^2$$

from which

$$A = \frac{c^2\sigma_Y + \dfrac{\rho}{8}\omega^2\{(1 + 3v)c^4 + (3 + v)r_2^{\,4}\}}{c^2 + r_2^{\,2}}$$

$$B = \frac{c^2 r_2^{\,2}\sigma_Y + \dfrac{\rho}{8}\omega^2 c^2 r_2^{\,2}\{(1 + 3v)c^2 - (3 + v)r_2^{\,2}\}}{c^2 + r_2^{\,2}}$$

Therefore for the elastic zone at $r = c$,

$$\sigma_r = \frac{c^2\sigma_Y + \dfrac{\rho}{8}\omega^2\{(1 + 3v)c^4 + (3 + v)r_2^{\,4}\}}{c^2 + r_2^{\,2}}$$

$$- \frac{r_2^{\,2}\sigma_Y + \dfrac{\rho}{8}\omega^2 r_2^{\,2}\{(1 + 3v)c^2 - (3 + v)r_2^{\,2}\}}{c^2 + r_2^{\,2}}$$

$$- \frac{3 + v}{8}\rho\omega^2 c^2 \tag{18.60}$$

Equating eqns. (18.59) and (18.60) and simplifying,

$$\omega^2 = \frac{12\sigma_Y\{2cr_2^{\,2} - r_1(c^2 + r_2^{\,2})\}}{\rho\{3(3 + v)cr_2^{\,4} - (1 + 3v)(2r_2^{\,2} - c^2)c^3 - 4r_1^{\,3}(r_2^{\,2} + c^2)\}} \tag{18.61}$$

where ω is the angular speed to cause plasticity to a radial depth of $r = c$.

If ω_p is the speed at which the disc becomes fully plastic, then substituting $c = r_2$ in eqn. (18.61) and simplifying,

$$\omega_p{}^2 = \frac{\sigma_Y(r_2 - r_1)}{\dfrac{\rho}{3}(r_2{}^3 - r_1{}^3)} = \frac{3\sigma_Y}{\rho(r_2{}^2 + r_1 r_2 + r_1{}^2)}$$

and

$$\omega_p = \sqrt{\frac{3\sigma_Y}{\rho(r_2{}^2 + r_1 r_2 + r_1{}^2)}}$$

The stress distributions of σ_r and σ_θ for various degrees of plastic deformation are shown in Fig. 18.18 for $r_2 = 10r_1$. A more complete analysis of this problem has been made by Heyman, *Proc. Instn. Mech. Engrs*, **172**, 1958.

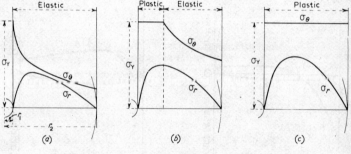

Fig 18.18

Example 18.4. A metal disc of uniform thickness is 300 mm in diameter and has a central hole 50 mm in diameter. Determine the increase in rotational speed, over that for initial yielding at the hole, necessary to cause plastic deformation throughout the disc. Poisson's ratio = 0·3.

$$\frac{\omega_p}{\omega_Y} = \sqrt{\frac{3\{r_2{}^2(3 + \nu) + r_1{}^2(1 - \nu)\}}{4(r_2{}^2 + r_1 r_2 + r_1{}^2)}}$$

$$= \sqrt{\frac{3\{(150^2 \times 3{\cdot}3) + (25^2 \times 0{\cdot}7)\}}{4(150^2 + 3750 + 25^2)}} = \sqrt{2{\cdot}08}$$

Therefore

$$\omega_p = 1{\cdot}445\omega_Y$$

An increase of speed of 44·5% is required for full plasticity.

18.8. RESIDUAL STRESSES IN AXIALLY SYMMETRICAL BODIES

If plastic deformation is caused in one part of a body and not in the remainder, then, on removal of external load, there still exists a stress system in the body, owing to the strain gradient and hence interaction between parts of the body not being able to return to the unstrained state. Residual stresses have important implications in engineering practice.

In the thick cylinder under internal pressure in Section 18.7(i) it is assumed that plastic deformation has partly penetrated the wall. On release of pressure the elastic outer zone tries to return to its original dimensions, but is partly prevented by the permanent deformation of the inner plastic material. Hence the latter is put into hoop compression and the former is in hoop tension, such that equilibrium exists. The residual stress distributions are then as shown diagrammatically in Fig. 18.19. They may be calculated by using eqns. (14.33) and (14.34) to

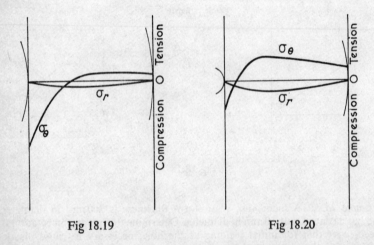

Fig 18.19   Fig 18.20

determine the elastic unloading stresses from the internal pressure p_i (eqn. (18.47)) and superposing these on to the loaded stress distribution. The residual compressive hoop stress at the bore has to be nullified first on repressurizing, thus allowing a greater elastic range of hoop stress and therefore a greater internal pressure. This process of obtaining favourable residual stresses in thick cylinders is known as *autofrettage* (see also Section 14.5).

When a rotating disc is stopped after partial plasticity has occurred, a similar condition of residual stresses is obtained as for the thick cylinder above. Compressive hoop stress is obtained at the central hole, which may be calculated using the elastic stress equations and the appropriate rotational speed, followed by superposition on to the loaded stress pattern. The above action, known as *overspeeding*, helps to increase the elastic stress range, and hence the speed limit available under working conditions. The residual stress distributions for this case are shown in Fig. 18.20 for the same conditions as in Fig. 18.18(*b*).

18.9. VISCOELASTICITY

Viscoelasticity is the term used to describe the behaviour of a material which combines the characteristics of a viscous liquid and an elastic solid. The model which may represent a Hookean solid is a spring for which stress is linearly related to strain. However, in the case of a Newtonian liquid stress is proportional to *strain rate* denoted by $\dot{\varepsilon}$ and the constant of proportionality is η, the coefficient of viscosity of the liquid.

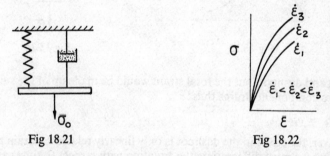

Fig 18.21   Fig 18.22

A suitable model representation is the dashpot, in which a piston is moved through the Newtonian fluid. A simple combination in parallel of the above two models is shown in Fig. 18.21, which is known as a *Voigt–Kelvin solid*. Its stress–strain–time response is represented by

$$\sigma = E\varepsilon + \eta\dot{\varepsilon} \tag{18.62}$$

The essential difference between this type of solid and a purely elastic solid is the introduction of time dependence. With three interrelated functions, stress, strain and time, there is no longer one particular stress–strain relationship as with a Hookean solid. If one conducted a simple tension test on a viscoelastic material, the resulting stress–strain curve would vary in shape and position depending on the *rate* of straining as shown in Fig. 18.22. This diagram also shows that stress is *not* proportional to strain even in the "elastic"* range.

Returning to the Voigt–Kelvin model and applying a step input of stress, σ_0, held constant for time t, then the resulting strain response is

---

* The material may recover on unloading to the origin but not necessarily along the same path as when loading.

illustrated in Fig. 18.23. The deformation that occurs with time at constant stress is termed *creep* and is discussed in detail in Chapter 23. On removal of the stress there is a strain–time response termed *recovery*.

Another model which involves viscoelastic behaviour was proposed by Maxwell and is simply a spring in series with a dashpot as shown in Fig. 18.24. A stress applied to this model would have the same value in both

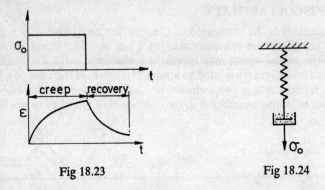

Fig 18.23   Fig 18.24

spring and dashpot, but the total strain would be made up of the sum of the two component strains thus:

$$\varepsilon = \varepsilon_s + \varepsilon_d$$

However, the stress in the dashpot is only linearly related to strain rate; therefore we must differentiate the equation with respect to time to give

$$\dot{\varepsilon} = \dot{\varepsilon}_s + \dot{\varepsilon}_d$$

For the spring

$$\dot{\varepsilon}_s = \frac{\dot{\sigma}}{E}$$

and for the dashpot

$$\dot{\varepsilon}_d = \frac{\sigma}{\eta}$$

Thus

$$\dot{\varepsilon} = \frac{\dot{\sigma}}{E} + \frac{\sigma}{\eta} \tag{18.63}$$

The strain response to a step input of stress, σ_0, is shown in Fig. 18.25. The above models are representative of *linear* viscoelasticity. *This does not mean that stress is proportional to strain.* If, in a series of creep experiments at different constant stress levels, the ratios of strain to stress (creep compliance) plotted against time form one curve, then linear viscoelastic behaviour has been exhibited for which the constitutive law is

$$\varepsilon = \sigma f(t) \tag{18.64}$$

which occasionally may be expressed as

$$\varepsilon = g(\sigma) h(t) \tag{18.65}$$

Non-linear viscoelastic behaviour is represented by the constitutive equation

$$\varepsilon = j(\sigma, t) \tag{18.66}$$

Perhaps the most interesting groups of materials to exhibit viscoelastic behaviour are rubbers and plastics, generally termed high polymers (see Chapter 24).

The simple Maxwell and Voigt–Kelvin models cannot in themselves represent the stress–strain–time response of polymers. However, combinations of them, as for example in Fig. 18.26, do give a fair representation

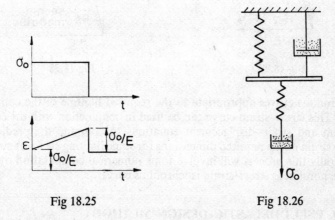

Fig 18.25   Fig 18.26

of the small range of linear-viscoelastic behaviour exhibited by polymers. Non-linear viscoelasticity can only be modelled by including non-linear spring and dashpots.

Stress and strain analysis for viscoelastic materials follows exactly the same steps as for elastic or plastic problems in respect of equilibrium of forces and compatibility of deformations. The essential difference arises in the third step, that of a constitutive or stress–strain law. In particular, the inclusion of time is different and complicating, and unfortunately the majority of plastics and rubbers are only linearly viscoelastic up to relatively insignificant strain values. Thus, although eqns. (18.64) and (18.65) are not too difficult to handle analytically, they are not too relevant to engineering applications. However, the application of the non-linear viscoelastic relation, eqn. (18.66), results in extremely complex analysis even for the more simple engineering components. To overcome these difficulties the following approach is now being widely adopted in the design of polymer load-bearing components.*

Firstly a family of strain–time creep curves for various stresses is required in tension and/or compression (generally available from the plastics manufacturer). A number of constant time sections taken through the family of curves enables *isochronous* stress–strain curves to be plotted as shown in Fig. 18.27. It is now possible to select one of these

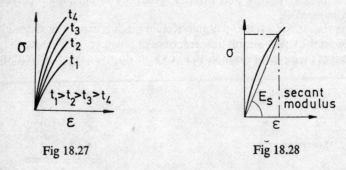

Fig 18.27   Fig 18.28

isochronous curves appropriate to the required lifetime of the component. This stress–strain curve can be used in conjunction with the equilibrium and strain–displacement equations to compute the predicted creep strain or the required dimensions for some limiting strain or stress. Naturally this process will involve some numerical computation owing to the non-linear stress–strain isochronous curve.

18.10. PSEUDOELASTIC DESIGN METHOD

In order to avoid the computation described above it has been shown that a satisfactory approximation to the isochronous curve can be made by the use of a *secant modulus*, $E_s(t)$. This is the slope of a line drawn from the origin to some specific point on the isochronous curve, Fig. 18.28. This point can be defined either by (*a*) the expected maximum stress or (*b*) a limiting value of strain, which is generally taken as 1%. Having determined for the appropriate time the secant tension, compression or shear modulus as required, this value is then inserted in the established

---

* P. P. Benham and D. McCammond, "Approximate creep analysis for thermoplastic beams and struts", *J. Strain Analysis*, **6**, No. 1 (1971).

linear elastic formulae for stress or deflection for the particular component being designed. For some problems, e.g. a plate or rotating disc*
where in elasticity Poisson's ratio, v, occurs we must use for viscoelasticity the time-dependent equivalent known as the *creep contraction* or
lateral strain ratio, $v(t)$. Although this method is approximate it does
provide in a simple manner predictions which are quite reasonable and
generally on the safe side. Further discussion on the use of creep data for
polymers will be found in Chapter 24.

Example 18.5. A circular diaphragm 200 mm in diameter is simply supported
around the edge. It is to be made of polypropylene homopolymer and subjected to a uniform pressure of 10 N/m^2. Determine a suitable thickness so
that a stress of 4 kN/m^2 is not exceeded. What will be the central deflection
after 1000 h at 20°C? Creep contraction ratio is 0·45 and the following values
have been taken from creep curves at 1000 h:

σ (kN/m²)	1·38	2·76	4·14	5·52	6·9	8·28
ε (%)	0·32	0·63	0·95	1·35	1·81	2·33

The maximum tensile stress at the centre of the diaphragm is obtained
from eqn. (16.28):

$$\sigma_r = \sigma_\theta = \frac{3[3 + v(t)]pr^2}{8h^2}$$

Hence

$$h^2 = \frac{3(3 + 0·45) \times 10 \times 0·1^2}{8 \times 4000}$$

and $h = 5·7 \text{ mm}$.

By plotting the values of stress and strain given, the isochronous
curve is obtained, from which, at 4 kN/m^2 and 0·92 % strain, the secant
modulus is found to be 435 kN/m^2 at 1000 h.

The deflection at the centre of the plate is obtained from eqn. (16.27):

$$w = \frac{3[1 - v(t)][5 + v(t)]pr^4}{16E_s(t)h^3}$$

$$= \frac{3(1 - 0·45)(5 + 0·45) \times 10 \times 0·1^4}{16 \times 435 \times 10^3 \times 0·57^3 \times 10^{-6}}$$

$$= 7 \text{ mm}$$

* D. McCammond and P. P. Benham, "A study of design stress analysis procedures for thermoplastic products using time dependent data", *Plastics and Polymers*, October 1969.

Part 2

Mechanical Properties and Testing of Materials

Chapter 19

Tension, Compression and Torsion

19.1. Before a designer can complete the theoretical stress analysis of a projected component he requires some information about the properties of strength and ductility of the proposed material from which the component will be made. A manufacturer engaged in producing a metal from the raw material into finished bars and sections has to know whether the mechanical properties are up to specification. Investigation of the failure of a component necessitates a study of the condition of the material to determine possible causes of failure. Thus a knowledge of mechanical tests and properties of materials is essential to many engineers.

19.2. TENSION

(i) Elastic Range of a Metal

The most widely used method of obtaining the information required above is by means of a tensile test and the tensile stress–strain relationship

It was demonstrated by Hooke in 1678 that, up to a certain limit, the extension of a piece of metal is proportional to the load producing that extension. This behaviour, known as *Hooke's law*, plays an important part in the mechanics of solids.

Straining a metal bar in tension and measuring the resulting force is termed a *tension test*. The extension of a bar a few inches long in the elastic range amounts to only a few thousandths of an inch, and delicate instruments, known as *extensometers*, have to be employed to measure the extension. Plotting a graph of load as ordinate and extension as abscissa results in a straight line up to a certain point as in Fig. 19.1 (*a*). If the load axis is divided by the original cross-sectional area of the bar, stress is obtained, and when the extension values are divided by the original gauge length from which the extensions were measured, then strain is obtained and a stress–strain graph results, Fig. 19.1 (*b*), which is also linear. Fig. 19.1 (*c*) shows the typical shape of a flat or round bar specimen for tension testing. The enlarged ends are for gripping in the jaws of the testing machine and the reduced parallel portion contains the *gauge length*.

Returning to Figs. 19.1 (*a*) and (*b*), the points marked *P* and *E* are termed the *limit of proportionality* and the *elastic limit* respectively. The former is the last point of proportionality between load and extension, or stress and strain. The elastic limit, often indistinguishable from the proportionality limit, is the highest point from which, after complete removal of load from the test piece, there is no permanent strain remaining. The term *elasticity* relating to material behaviour is defined as a coincident value of strain for a load during loading and the same load during unloading. It should be noted that this does not imply proportionality and therefore a material can behave elastically without having a linear stress–strain relationship, although there are very few materials that exhibit non-linear elasticity.

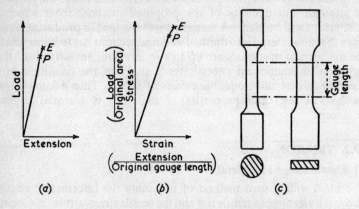

Fig 19.1 (*a*) Elastic Load-Extension Curve. (*b*) Elastic Stress–Strain Curve. (*c*) Tension Test Pieces

One of the most important mechanical constants for a linear elastic material is that relating elastic stress and strain or the slope of the stress–strain diagram. The constant is known as *Young's modulus of elasticity*, *E*, and some examples of typical values for various metals are given in Table 19.1. Also given in the table are values of *Poisson's ratio*, another of the elastic constants, which can also be determined in a tensile test by measuring lateral, as well as longitudinal, strain.

(ii) Plastic Range of a Metal

When straining of a metal is continued beyond the elastic limit, yielding commences and the material is in the plastic range. There are two characteristically different types of transition from elastic to plastic behaviour. These are illustrated in Fig. 19.2 (*a*) and (*b*): the first is principally found in low- and medium-carbon steels and the second in alloy steels and non-ferrous metals. In the former, the point U, at which there is a sudden drop in load with further strain, is termed the *upper yield point*. This is followed by L, the *lower yield point*, from which there is a marked extension at almost constant load.

Table 19.1

Material	Modulus of elasticity, E, GN/m²	Poisson's ratio, v
SAE 1045 steel, annealed	200	0·287
SAE 1045 steel, hot-rolled	197	0·292
SAE 2330 steel, as rolled	197	0·291
SAE 6150 steel, as rolled	202	0·289
Cold-rolled shafting	204	0·287
Cutlery stainless steel	210	0·278
18–8 Stainless steel	190	0·305
36% Ni–8% Cr steel	182	0·333
Cast-iron (no alloying elements)	100	0·211
Cast-iron (15% Ni–2·05% Cr)	93	0·299
Malleable iron	163	0·265
Copper (99·9%)	108	0·355
Brass (70–30)	110	0·331
Phosphor bronze (Cu 90%, Zn 6%, Sn 4%)	111	0·349
17ST Duralumin (Cu 3·85%, Mn 0·57%, Mg 0·05%, Al 95·08%)	70	0·330
Aluminium alloy (Si 1·07%, Mg 0·56%, Al 98·37%)	68	0·334
Monel metal	172	0·315

(Adapted from M.S. Thesis by R. W. Vose, M.I.T., 1936)

The upper yield point of low- and medium-carbon steels is a complex phenomenon, still not fully understood, which is a function of strain rate, temperature, type of testing machine and geometry of the specimen. In some cases the upper yield point does not appear and the curve departs from the elastic range immediately into the horizontal lower yield range.

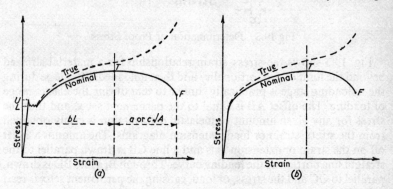

Fig 19.2 Nominal and True Tensile Stress–Strain Curves

The latter, however, may be regarded as a material property and is used as the yield point stress for design purposes. It is this rather unique shape of stress–strain curve to which one approximates when using an ideal elastic–plastic relationship in plastic bending or torsion theory as in Chapter 18, since mild steel is widely used for structural work.

The mechanism of deformation during yielding is such that the inter-atomic forces maintained during elasticity are exceeded, allowing sliding of rows of atoms over one another. The behaviour is analogous to a pack of cards being subjected to shear on the outer surfaces. The deformation in the crystal structure is termed *slip* and occurs on planes of maximum shearing stress. In mild steel specimens, in the initial yield region any surface scale flakes off along lines between 45° and 50° to the axis of the bar and specimens with a polished surface show shadowy bands at approximately the same angles as above. These surface markings are known as *Lüders lines*, and are useful as a qualitative indication of initial yielding.

Many materials, particularly the light alloys, do not exhibit a clearly-defined elastic limit, limit of proportionality or yield point, and several methods of indicating these stresses are in use. The most widely used is that involving a measure of permanent set and is called the *offset method*.

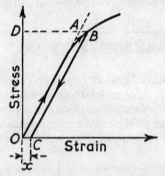

Fig 19.3   Determination of Proof Stress

Fig. 19.3 shows the stress–strain relationship for a material stressed beyond the limit of proportionality and then unloaded. The slope during the unloading stage is practically similar to that during the elastic range of loading. The offset AB is equal to the permanent set x, and thus the stress for any given amount of inelastic deformation is easily obtained from the stress–strain or load–extension diagrams. The amount x is set off on the strain or extension axis and a line CB is drawn parallel to the straight line portion of the loading curve. Through B, a line BD is drawn, parallel to OC and the stress, or load, causing the permanent set x is read off at D.

In defining a particular stress by this method, the magnitude of x is chosen from experience, and is that value which it is considered will give a stress of practical value to the designer. The values of various offsets and the corresponding stresses, applicable to different materials, are referred to in various British Standards, and these should be consulted for fuller details.

Part of the stress–strain curve for an aluminium alloy is shown in Fig. 19.4, and typical values of proof stress and strain offset are indicated on the diagram. The elastic limit is taken to be that stress at which there

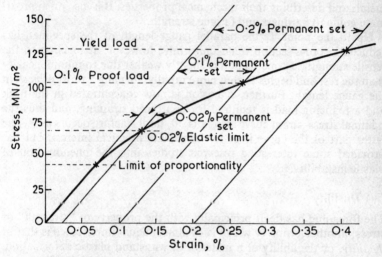

Fig 19.4   Proof Stress and Permanent Set Values during Continuous Yielding
(From Lessels, *Strength and Resistance of Metals:*
by courtesy of John Wiley & Sons Ltd.)

is a permanent set of 0·02%; this is generally called the 0·02% elastic limit. The proof stress is that stress at which there is a permanent set of 0·1% and is called the 0·1% proof stress. The yield stress is that stress at which there is a permanent set of 0·2% of the gauge length.

On a 50 mm gauge length, the extensions corresponding to the above stresses are 0·01, 0·05 and 0·1 mm respectively, and thus extensometer measurements are required for the accurate production of the load–extension diagram.

Once the initial yield region has been passed in Figs. 19.2 (*a*) and (*b*), the stress–strain curves have a common form. An increasing stress is required to cause continued straining, and this behaviour is known as *workhardening* or *strainhardening* and the metal does in fact become harder.

The property of workhardening is quite important and will be discussed further, later on. The specimen has a limit to which it can be workhardened uniformly in a simple tension test, and this is reached when the slope of the nominal stress–strain curve becomes zero as at T, in Figs. 19.2 (*a*) and (*b*). This point is known as the *tensile strength* (in the past, the ultimate tensile strength, U.T.S.) and is given by the maximum load carried by the specimen divided by the original cross-sectional area. This is a very important quantity as it is sometimes used in design in conjunction with a safety factor, and is always quoted when comparing metals and describing their mechanical properties. It is also an approximate guide to hardness and fatigue strength.

Up to the point T, the parallel gauge length of the specimen has reduced in cross-section quite uniformly; however, at or close to the tensile strength, T, one section is slightly weaker due to inhomogeneity than the rest and begins to thin more rapidly, forming a waist or *neck* in the gauge length. Further, extension is now concentrated in the neck and a reducing load is required for continued straining, and thus the nominal stress–strain curve also falls off, until fracture occurs at F. The latter part of the curve is of no immediate practical interest, but has provided some interesting research studies into the phenomenon of plastic instability.

(iii) Ductility

The foregoing has dealt principally with the property of "strength" or stress; another property which is of almost equal importance is that of *ductility*, or the ability of a material to withstand plastic deformation. In a tensile test this is expressed in two ways, by the percentage elongation of the gauge length after fracture and by the percentage reduction in cross-sectional area referred to the neck or minimum section at fracture.

The second quantity is expressed algebraically as

$$\text{Reduction in area} = \frac{A_0 - A_F}{A_0} \times 100\%$$

where A_0 and A_F are the original and fracture areas respectively.

Elongation is the increase in the gauge length divided by the original gauge length, or algebraically

$$\text{Elongation} = \frac{L_F - L_0}{L_0} \times 100\%$$

where L_0 and L_F are the original and final gauge lengths respectively. Elongation is only partly a material property since it is also dependent on the geometrical form of the test piece. It was pointed out by Barba that comparison of elongation could only be made for geometrically similar test bars. Therefore for cylindrical bars the ratio (length/diameter) must be constant. The values adopted by various countries are shown in Table 19.2.

Table 19.2

	L/D	L
Great Britain	3·54 ($L = 50$ mm $D = 14\cdot3$ mm)	$4\sqrt{\text{Area}}$
Europe	10	$11\cdot3\sqrt{\text{Area}}$
	5	$5\cdot65\sqrt{\text{Area}}$
U.S.A.	4 ($L = 50$ mm $D = 12\cdot7$ mm)	$4\cdot51\sqrt{\text{Area}}$

The elongation at fracture includes the necked region, and this is relatively independent of original gauge length and is more a function of the material and shape of cross-section.

The elongation of the gauge length, apart from the neck, is proportional to the original length, and therefore elongation for geometrically similar specimens of a material can be written as $e = a + bL$ as proposed by Barba and illustrated in Fig. 19.2 (a). Unwin suggested that the constant term, a, relating to the neck elongation was a function of the cross-section and could be written as $c\sqrt(A)$ to relate specimens of different shape, where A is the original area. Hence, elongation $\varepsilon = c\sqrt(A) + bL$, and c and b have become known as *Unwin's constants*. Typical values of these constants for various metals are given in Table 19.3.

Table 19.3

Material	Unwin's constants	
	b	c
Brass	0·486	0·091
Low-carbon steel	0·279	0·73
Medium-carbon steel	0·16	0·454
Nickel steel	0·107	0·57
Nickel-chrome steel	0·109	0·722

(*From Lessels*, Strength and Resistance of Metals:
by courtesy of John Wiley & Sons Ltd.)

The distribution of strain over the gauge length of a specimen can vary considerably from one metal to another, as illustrated in Fig. 19.5, depending on the grain size and microstructure.

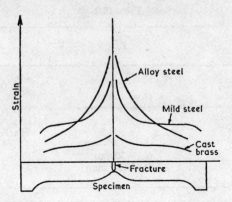

Fig 19.5   Strain Distribution along a Tensile Test Piece

(iv) True Stress in the Tensile Test

The lower stress–strain curves in Fig. 19.2 were based on nominal values of stress, that is the load at all points was divided by the original cross-sectional area. This is a convenient arrangement for most practical purposes and is the generally adopted procedure. However, as the specimen is strained, so the cross-sectional area reduces, and hence the true stress is higher than the nominal stress. The true stress is obtained by dividing the load by the current area of the specimen corresponding to that load. The current area of the bar may be obtained by direct measurement at various stages during the test or by calculation as follows. It is generally true to assume that the volume of the gauge length remains constant in the plastic range; therefore

$$A_0 L_0 = A_1 L_1$$

where subscripts refer to original and current conditions. Thus

$$A_1 = \frac{A_0 L_0}{L_1} = \frac{A_0 L_0}{L_0 + dL_0}$$

$$= \frac{A_0}{1 + (dL_0/L_0)}$$

$$= \frac{A_0}{1 + \varepsilon}$$

and

$$\text{True stress} = \frac{W_1}{A_1} = \frac{W_1}{A_0}(1 + \varepsilon)$$

or the nominal stress × (1 + current strain).

The shape of the true stress curve is shown in Fig. 19.2. It is only strictly valid up to the point where necking commences, since the change in geometry at the neck sets up a complex stress system which cannot be determined simply from the load divided by the area of the neck.

(v) Logarithmic or Natural Strain

It was suggested by Ludwik that strain defined as dL_0/L_0 was only really applicable to small values and that for larger strains it was more logical to use a definition of an increase in strain as being the ratio of an increment of length over the current gauge length, i.e. $d\bar{\varepsilon} = dL/L$. Therefore the total strain is given by

$$\int_{L_0}^{L} d\bar{\varepsilon} = \int_{L_0}^{L} \frac{dL}{L}$$

or

$$\bar{\varepsilon} = [\log_e L]_{L_0}^{L} = \log_e \left(\frac{L}{L_0}\right) = \log_e \left(\frac{L_0 + dL}{L_0}\right)$$

Therefore

$$\bar{\varepsilon} = \log_e(1 + \varepsilon)$$

Hence *natural or logarithmic strain* is the Napierian logarithm of unity plus the ordinary strain.

From considerations of constant volume,

$$\frac{L}{L_0} = \frac{A_0}{A}$$

Therefore natural strain can also be written as

$$\bar{\varepsilon} = \log_e \left(\frac{A_0}{A}\right)$$

or

$$\bar{\varepsilon} = 2 \log_e \left(\frac{d_0}{d}\right) \quad \text{(for a cylindrical bar)}$$

Expressing strain in this way enables the tensile stress–strain curve to be continued beyond the point where necking begins, since measurements of A can be taken in the neck to determine the equivalent natural strain.

The form of the true-stress/log-strain curve is shown in Fig. 19.6 and it is seen that from the point N where necking begins to F at fracture there is a linear relationship. This is a measure of strainhardening, and the slope in this region gives what is termed the *strainhardening coefficient*. This is mostly of interest when it is necessary to compare the condition of two different metals.

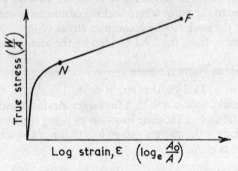

Fig 19.6   A True-stress/Log-strain Curve

(vi) Fracture in Tension

Two basic types of fracture can be obtained in tension, depending on the material, temperature, strain rate, etc., and these are termed *brittle* and *ductile*. The main features of the former are that there is little or no plastic deformation, the plane of fracture is normal to the tensile stress and separation of the crystal structure occurs. In the second type, ductile failure is preceded by a considerable amount of plastic deformation and fracture is by shear or sliding of the crystal structure on microscopic or general planes at about 45° to the tensile stress. Probably the most well-known type of failure is called the *cup and cone* in the cylindrical bar. An example of this along with other typical failures is shown in Plate I. The flattish central area is where failure is initiated due to the triaxial stress system set up in the neck of the specimen. However, a microscopic examination of the central region shows that it is not brittle separation as it at first might appear, but a mass of microscopic shear planes at 45° to the axis of the bar. The lip around the edge of the specimen is obviously shear, and it varies in size depending on the metal and its condition, and is not always formed continuously around the periphery, parts being torn out and left on the "cone" end.

A cylindrical bar even though relatively ductile may not produce a sufficient neck to set up marked triaxiality of stress, and failure is found to occur on a single shear plane right across the specimen. When the two parts of a fractured flat bar are placed together it is seen that there is a gap in the middle region, showing the initiation of fracture there, while the outer regions continued to extend.

(vii) Overstrain, Hysteresis and Bauschinger Effects

If a metal is taken into its plastic range and at some point the load is removed, then the unloading line, although having slight curvature, approximates to the slope of the original elastic range. On reapplication of load the line diverges only slightly from that for unloading, and yielding occurs only when the load has reached the previous point of commencement of unloading. This effect is shown diagrammatically in Fig. 19.7 and is known as *overstraining*. It should be emphasized that this word does not mean damage to the static properties of the metal although it may convey that impression.

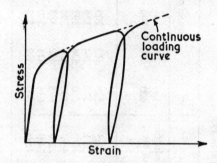

Fig. 19.7   Hysteresis Loops during Overstrain

Overstrain is in fact a very useful property in enabling a higher yield range to be obtained, although at the expense of elasticity and proportionality. The latter can be regained by a rest period or mild heat-treatment (boiling water). Instances of the usefulness of overstraining on components are illustrated in Sections 18.7 and 18.8.

In Fig. 19.7 the "loops" formed by the unloading and reloading lines during overstrain are caused by mechanical hysteresis, that is, there is a lag between stress relaxation and strain recovery, and similarly on reloading. In unidirectional loading (tension only) hysteresis loops are generally quite narrow and in some metals, virtually non-existent. Hysteresis is also the term used to describe the loop obtained when reversed loading is conducted on a metal, i.e. through yield in tension followed by compression or vice versa. Fig. 19.8 shows the form of a tension–compression loop. A most interesting feature about this form of deformation is what is known as the *Bauschinger effect*, in which yielding in compression occurs at a slightly lower value than previous yielding in tension, and vice versa if compression were applied first. This effect is

Table 19.4

Steel	Tensile tests, cold				Tensile strength, hot (°C), MN/m²							
	Elastic limit, MN/m²	Tensile strength, MN/m²	Elongation on 50 mm, %	Reduction of area, %	600	650	700	750	800	850	900	950
3% Nickel-chrome	745	874	23·0	62·5	525	365	209	173	142	107	70	—
Stainless steel	726	746	24·0	58·1	374	264	158	278	94	87	121	—
Silicon-chrome	790	975	21·0	40·0	650	534	392	275	165	111	60	46
Chrome steel	650	835	24·5	55·0	560	464	294	201	108	111	113	80
Cobalt-chrome	650	896	13·0	22·0	695	472	383	210	155	90	125	93
High-speed steel	711	943	15·0	24·0	634	411	330	260	151	119	130	87
High-nickel chrome	650	1050	27·0	45·0	660	592	523	442	371	300	232	193

due to the residual stress after the first yield adding to the reversed applied stress, and thus lowering the required value of the latter for the second yield.

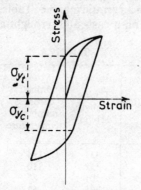

Fig 19.8   Hysteresis Loop for Reversed Loading

(viii) Effect of Temperature

At low temperatures, the strength of a mild steel is increased; the material, however, may become very brittle under certain circumstances, its ductility being almost negligible (see Chapter 21).

With increase in temperature, the elastic limit decreases, and is accompanied by a corresponding decrease in the modulus of elasticity. The tensile strength of the material decreases until a temperature of approximately 150°C is reached. From this temperature the strength increases, reaching a maximum value in the neighbourhood of 300°C, after which it decreases as the temperature is further increased. Table 19.4 shows the effect of high temperature on the tensile strengths of various steels.

Table 19.5

Temperature (°C)	Tensile strength, MN/m²		Brinell hardness number at temperature		Brinell hardness number after cooling	
	Sand-cast	Die-cast	Sand-cast	Die-cast	Sand-cast	Die-cast
20	286	371	129	138	—	—
200	248	340	101	115	129	138
250	216	302	78	80	121	129
300	201	240	50	55	101	105
350	136	139	27	30	70	85

The ductility decreases with increase in temperature until a temperature of about 150°C is reached. After this temperature the ductility increases with further increase of temperature. The increase is not regular, however, but takes place in an erratic manner. Tables 19.5 and 19.6 show the effect of temperature on a cast- and wrought-aluminium alloy respectively.

Table 19.6

Temperature (°C)	Tensile strength, MN/m²	Brinell hardness number at temperature	Brinell hardness number after cooling
20	433	134	—
200	332	110	134
250	301	87	125
300	201	52	90
350	124	27	75
400	—	12	70

(ix) Influence of Composition and Heat-treatment

The chemical composition of a material and the heat-treatment to which it has been subjected have a great effect on the strength and ductility of the material.

The mechanical properties of steel are very largely influenced by the amount of carbon in the steel, and Table 19.7, consisting of tests on a complete family of steels, shows the variations very clearly.

Table 19.7
Mechanical Properties of Steels

Analysis							
Carbon, %	0·130	0·190	0·290	0·380	0·450	0·570	0·630
Silicon, %	0·028	0·037	0·037	0·066	0·167	0·121	0·111
Sulphur, %	0·048	0·034	0·046	0·030	0·040	0·034	0·038
Phosphorus, %	0·056	0·058	0·057	0·025	0·058	0·058	0·055
Manganese, %	0·400	0·620	0·750	0·690	0·610	0·600	0·560
Tensile test							
Elastic limit, MN/m²	195	203	247	213	270	316	348
Yield stress, MN/m²	266	266	315	337	371	375	405
Tensile strength, MN/m²	382	448	538	609	660	721	795
Elongation, %	30·00	29·90	25·00	22·50	18·80	15·20	12·25
Contraction, %	69·00	60·80	57·50	54·40	36·00	30·60	14·62
Hardness Number	116	127	157	171	191	229	233

(From J. H. Smith, "Some experiments on fatigue of metals," *J. Iron St. Inst.*, No. II, 1910)

It will be noted that the strength of the steel increases with increase in the carbon content, but the ductility decreases as the carbon content is increased. In Fig. 19.9 stress is shown plotted against percentage elongation, and in Fig. 19.10 all the results in the table are shown plotted to a base of carbon content. It is worth noting that there are apparently simple linear relationships between a number of the quantities.

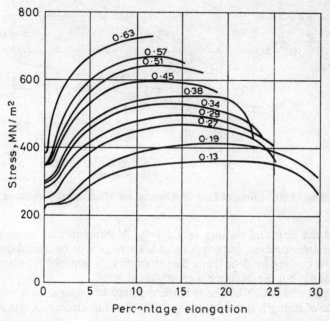

Fig 19.9   Effect of Carbon Content on the Tensile Stress–Strain Curve

A steel may be hardened by heating it to a high temperature, and then rapidly cooling it in a cold liquid. The strength of the steel is greatly increased by this process, but its ductility is greatly reduced, and it becomes very brittle. Usually further heat-treatment is required before the steel is of commercial use. The amount of hardness is influenced by the rate at which quenching takes place, and also by the carbon content. Table 19.8 contains a list of alloy steels, and shows the variations of strength and ductility with hardening. Each steel was tested before hardening, that is in the rolled condition. It was then hardened by heating to about 800°C, soaked for three-quarters of an hour and quenched.

A steel which has been hardened, due to rapid cooling or to repeated straining processes, may be brought to its original state by heating to a

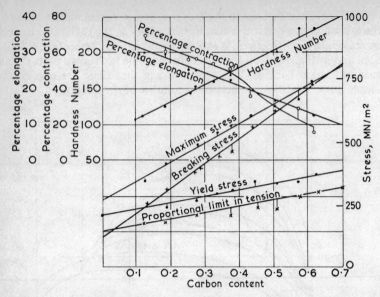

Fig 19.10   Effect of Carbon Content on Mechanical Properties

blood-red heat and cooling very slowly. Aluminium may be annealed by the same process, but copper and brass require to be cooled quickly, as slow cooling hardens them. The strength of a material is reduced by annealing, but the ductility is greatly increased.

A material which has been hardened may be brought to a required degree of strength and ductility by further heat-treatment, or tempering.

Table 19.8

	Yield point, MN/m²		Tensile strength, MN/m²		Elongation on 50 mm gauge length, %		Reduction in area, %	
	Before	After	Before	After	Before	After	Before	After
Mild steel	330	360	451	569	31·5	31·5	55·3	62·6
3% Ni-Cr steel	399	650	560	833	27·5	22·5	52·2	53·4
5% Ni-Cr steel	416	843	607	1020	28·5	17·0	61·2	51·0
3% Ni-Cr steel	556	1050	742	1205	23·0	14·0	50·0	46·5
Special Ni-Cr	486	1050	682	1200	26·0	14·0	52·0	40·0
Ni-Cr-V steel	595	1190	850	1380	14·0	14·5	31·1	52·0

Table 19.9 shows the effect of tempering on a nickel-chrome steel—a steel which is of great use in automobile and aircraft work, and large connecting rods, etc. The steel was hardened by heating to 820°C, soaked for three-quarters of an hour and quenched in oil. It was then tempered for one hour at the temperatures stated, and cooled in oil.

Table 19.9

Tempering heat °C	Yield point, MN/m²	Tensile strength, MN/m²	Elongation on 50 mm gauge length, %	Reduction in area, %	Brinell hardness number	Izod impact, J
Hardened only	1700	1930	10·0	30·0	532	21·7
150	1700	1917	11·0	36·0	532	25·5
200	1647	1840	11·0	39·0	512	29·9
250	1575	1730	11·0	42·0	477	25·5
300	1475	1608	11·5	45·0	460–444	19·0
350	1390	1500	12·0	47·5	430	17·6
400	1260	1375	12·5	49·5	402	20·3
450	1160	1269	13·5	51·0	375	28·5
500	1082	1160	15·5	52·5	351	40·7
550	1005	1075	18·0	54·0	321	54·2
600	920	990	20·5	56·5	293	71·9
650	804	911	23·0	61·5	269	101·0
700	741	1080	18·5	30·0	302	49·5

The results are shown plotted to a base of tempering temperatures in Fig. 19.11, and it will be easily seen that increase in tempering temperature reduces the strength of the steel, but increases its ductility. It should also be noted that the strength of this steel is much above that of a mild steel. The addition of nickel, chromium, vanadium, and manganese adds greatly to the strength of a steel, but decreases ductility.

19.3. COMPRESSION

The mechanical properties of a ductile metal are generally obtained from a tension test. However, compression behaviour is of interest in the metal-forming industry, since most processes, rolling, forging, etc., involve compressive deformations of the metal, and also often of the forming equipment (rolling mill).

In compression an elastic range is exhibited as in tension and the

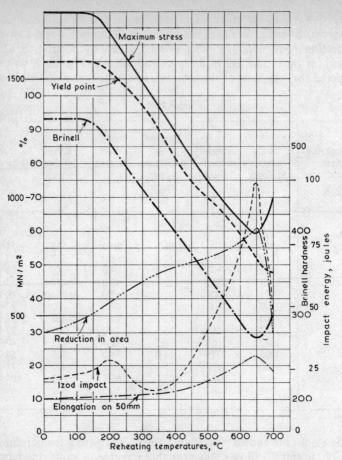

Fig 19.11   Effect of Tempering Temperature on Mechanical
Properties

elastic modulus, proportional limit and yield point or proof stresses have closely corresponding values for the two types of deformation. The real problem arises in a compression test when the metal enters the plastic range. The test piece has to be relatively short ($L/D \not> 4$) to avoid the possibility of instability and buckling. The axial compression is accompanied by lateral expansion, but this is restrained at the ends of the specimen owing to the friction between the machine platens and the end faces, and consequently on a shortish specimen marked barrelling occurs as in Fig. 19.12. This causes a non-uniformity of stress distribution, and conical sections of material at each end are strained and hardened to a lesser degree than the central region. The effect on the load–compression curve, after the smaller values of plastic strain have been achieved, is a fairly rapid rise in the load required to overcome friction and cause further compression.

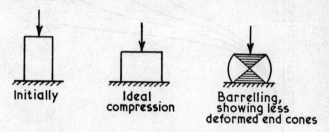

Initially   Ideal compression   Barrelling, showing less deformed end cones

Fig 19.12   Deformation during a Compression Test

Owing to the barrelling effect, only an average stress can be computed from the load–compression curve, based on an average area determined from considerations of constant volume. Hence, as was derived earlier for tension,

$$A = \frac{A_0}{(1 + \varepsilon)}$$

remembering that compressive strains are negative, thus giving A larger than A_0. Hence the average stress is

$$\sigma = \frac{W}{A_0}(1 + \varepsilon)$$

Various methods have been attempted to overcome the effects of barrelling, none of which is completely successful. The most satisfactory appears to be the technique of using several cylinders of the same metal having different diameter-to-length ratios. Incremental compression tests are conducted on the set of cylinders at a series of loads of increasing magnitude, measuring the strain for each of the cylinders at each load. Extrapolation of curves of D/L against strain with load as parameter to a value of $D/L = 0$, representing an infinitely long specimen where barrelling would be negligible, enables the true compressive stress–strain curve to be determined, Fig. 19.13. Failure of a ductile metal in compression only occurs owing to excessive barrelling causing axial splitting around the periphery.

For brittle materials, such as flake cast-iron, concrete, etc., which would not normally be used in tension, the compression test is used to give quantitative mechanical properties. Although end friction still occurs, which affects the stress values somewhat, owing to the absence

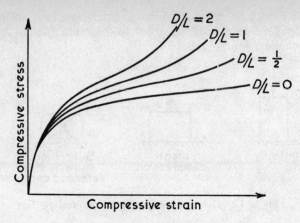

Fig 19.13   Compressive Stress–Strain Curves for Various Diameter/
Length Ratios

of ductility in these materials the barrelling condition is barely achieved. Fracture takes place on planes of maximum shear stress as illustrated in Fig. 19.14. Further details of compression of concrete, granite, wood, etc., are given in Chapter 24.

Fig 19.14   Modes of Compression Failure in Various Materials

19.4. TORSION

The usual method of obtaining a relationship between shear stress and shear strain for a material is by means of a torsion test. This may be conducted on a circular-section solid or tubular bar. By applying a torque to each end of the test piece by a testing machine and measuring the angular twist over a specified gauge length, a torque–twist diagram can be plotted. This is the equivalent in torsion to the load–extension diagram in tension. It was shown theoretically in Chapter 9 that, under elastic conditions in torsion, the applied torque is proportional to the angle of twist on the assumption that shear stress is proportional to shear strain. This is found to be true experimentally, and the linear torque–

twist relationship obtained enables the shear or rigidity modulus to be determined, since

$$G = \frac{T}{\theta}\frac{L}{J}$$

where the symbols are as specified in Chapter 9.

The torsion test is not like the tension or compression tests in which the stress is uniform across the section of the specimen. In torsion there is a stress gradient across the cross-section, and hence at the end of the elastic range yielding commences in the outer fibres first while the core is still elastic, whereas in the direct stress tests yielding occurs relatively evenly throughout the material. With continued twisting into the plastic

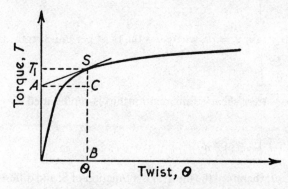

Fig 19.15   Torque–Twist Curve and Nadai Construction

range, more and more of the cross-section yields until there is penetration to the axis of the bar. The torque–twist diagram, Fig. 19.15, appears of much the same form as a load–extension diagram, and workhardening will occur at a gradually decreasing rate as straining proceeds, but of course there is no fall off in the curve, as in tension, since necking cannot take place. In fact, ductile metals can absorb extremely high values of shear strain (200%) before failure occurs.

Although the shear-stress/shear-strain relationship can be determined easily in the elastic range, difficulties are introduced for a solid bar in the plastic range due to the stress variation mentioned above. One solution to this problem is to conduct the torsion test on a thin-walled tubular specimen in which the shear stress in the plastic range may be assumed to be constant through the wall thickness, and is given by

$$\tau = \frac{T}{2\pi r^2 t}$$

where t is the wall thickness and r is the mean radius.

Shear strain is obtained in the plastic range from the same assumptions that apply in the elastic range; hence

$$\gamma = \frac{r\theta}{L}$$

(The tangent of the angle γ must be used for large strains.)

The solid bar in torsion has been considered by Nadai,* so that the shear-stress/shear-strain curve can be derived from the torque–twist curve as follows.

Consider a solid circular bar of radius r_0 subjected to a torque T; then for equilibrium

$$T = 2\pi \int_0^{r_0} \tau r^2 \, dr$$

Now, shear strain $\gamma = r\theta$, where θ is the twist per unit length; therefore

$$T = \frac{2\pi}{\theta^3} \int_0^{\gamma_0} \tau \gamma^2 \, d\gamma$$

If the shear-stress/shear-strain relationship is represented as $\tau = f(\gamma)$, then

$$T = \frac{2\pi}{\theta^3} \int_0^{\gamma_0} f(\gamma)\gamma^2 \, d\gamma$$

But $\gamma_0 = r_0\theta$; therefore the integral is a function of θ, and differentiating both sides above with respect to θ,

$$\frac{d(T\theta^3)}{d\theta} = 2\pi f(r_0\theta)r_0^3\theta^2 = 2\pi r_0^3\theta^2\tau_0$$

where $\tau_0 = f(\gamma_0) = f(r_0\theta)$; hence

$$\tau_0 = \frac{1}{2\pi r_0^3\theta^2} \frac{d}{d\theta}(T\theta^3)$$

Now, the term on the right-hand side of the above equation, $(1/\theta^2)(d/d\theta)(T\theta^3)$, may be written as

$$\theta \frac{dT}{d\theta} + 3T$$

Therefore

$$2\pi r_0^3\tau_0 = \theta \frac{dT}{d\theta} + 3T$$

* *Flow and Fracture of Solids* (Wiley).

Referring to the torque–twist diagram in Fig. 19.15, at the point S the torque is T_1 and the angle of twist per unit length is θ_1. Drawing a tangent to the curve at S, the intercept on the ordinate is A, and from A drawing a horizontal line to cut SB at C, then

$$\theta\frac{dT}{d\theta} + 3T = CS + 3BS$$

Hence the shear stress τ_0 can be computed from the T–θ curve for various values of θ or γ_0/r_0.

Fracture in torsion for ductile metals generally occurs in the plane of maximum shear stress perpendicular to the axis of the bar, whereas for brittle materials failure occurs along a 45° helix to the axis of the bar due to tensile stress across that plane.

19.5. COMPLEX STRESS TESTS

The various theoretical criteria of yielding, discussed in Chapter 15, are based on assumed critical values of stress, strain or strain energy at which points yielding will commence. In order to substantiate one or more of those theories it is necessary to study material behaviour at yielding under complex stress conditions. There have been many experimental investigations of this nature to establish the "yield locus" for various metals and, as was mentioned earlier, the shear strain energy or maximum shear stress theories seem to give the best correlation with experiment for ductile metals. Experiments on brittle materials agree more with the maximum principal stress or Mohr theories. There are several forms of loading suitable for producing complex stress in a specimen; a tube subjected to internal pressure, combined with axial tension or compression, and a tube in torsion, combined with axial loading or bending, are the principal methods employed. It is evident that a conventional testing machine is unsuitable for any of the above without modification and additional loading facilities. It is therefore usual to build a special testing rig for these types of experiment.

The object of the test is to determine a series of values of principal stresses σ_1 and σ_2 at which yielding occurs, so that a yield locus can be plotted, Fig. 19.16. Of the various forms of combined loading mentioned above, internal pressure and axial loading give principal stresses directly, while in the case of torsion combined with axial loading, or bending, the principal stresses have to be calculated from the values of shear and normal stresses at yield. For sensitive and accurate results it is essential to use a thin-walled rather than a solid specimen owing to the stress gradient over the cross-section of the latter affecting the determination of initial yielding.

The test is carried out preferably by keeping a fixed ratio between the principal stresses or between the shear and direct stresses as loading is increased until extensometers indicate yielding is commencing. However, that arrangement may be difficult or impossible with some apparatus and the alternative is to put on a steady value for one stress system and gradually increase the other until yield is reached.

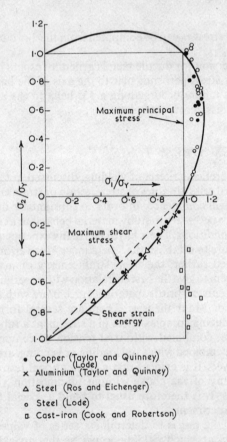

● Copper (Taylor and Quinney) (Lode)
× Aluminium (Taylor and Quinney)
△ Steel (Ros and Eichenger)
○ Steel (Lode)
□ Cast-iron (Cook and Robertson)

Fig 19.16   Experimental Results for Complex Yielding

Some experimental results plotted on a yield locus are shown in Fig. 19.16. The majority of experiments have been conducted on nominally isotropic material; however, there is often a considerable degree of anisotropy in metals used in service and therefore more information on yield criteria under conditions of anisotropy would be very valuable.

19.6. TESTING MACHINES

A testing machine for applying tension, compression or bending to a component or specimen consists of two units. One is a straining mechanism and the other a load weighing system.

For tension the test piece is gripped at each end in jaws. These are in the form of a pair of wedges, which are serrated on the faces in contact with the specimen, as shown in Plates II and III, and are housed in a wedge box. The smooth ends of the specimen, either a round or flat bar, are gripped by the serrations as the strain is applied, pulling the wedges firmly together. An alternative gripping arrangement for very hard or brittle materials or shorter specimens is to have threaded collars screwed on to each end of the test bar which are held in self-centring chucks with spherical seatings to avoid applying bending to the specimen. One end of the latter is attached either to the weighing mechanism or a fixed cross-head of the frame of the machine. The other end is attached to the straining unit, which may also incorporate a weighing system.

Straining is effected either by a manually- or power-operated screw or pair of screws or by a hydraulic ram attached to a moving crosshead holding one of the wedge boxes. A gear box with the mechanical system, or variable output pump with the hydraulic arrangement, allows for a wide range of straining speeds.

The original load measuring or applying system used in testing machines was a single lever (steelyard) balanced on knife-edges and connected through further knife-edges to another crosshead holding the specimen. The load was measured by moving a jockey weight along the lever until, at balance, the latter "floated" between two stops, and the value of load was indicated from a scale mounted along the lever. This type of machine although very accurate has the disadvantages that for large capacities, say 500 kN, it becomes very cumbersome and, requiring such a long steelyard, takes up much space. The next development was to employ a compound lever system, so that the magnitude of the force transmitted from the specimen was reduced to a value which could be balanced by a small steelyard and jockey weight.

With increasing routine industrial testing of materials the above types of machine are somewhat inconvenient, since for a continuous test to fracture the jockey weight has to be moved smoothly along the steelyard to keep it balanced, and rapid reading of the load is difficult. The next improvement to be introduced was the self-indicating mechanism as used on most modern testing machines. This system makes use of a multiple lever system, but instead of the steelyard and jockey weight, there is a pendulum or spring, to react the force on the specimen, which in turn is linked to a pointer. This moves around a calibrated dial, giving readings of load on the specimen directly in newtons.

The modern method in load measuring systems is that of the load cell employing electrical resistance strain gauges (see Chapter 25). The cell is mounted in series with the upper wedge box in the stationary crosshead. It is advantageous in being compact and its electrical principle enables a wide range of control to be employed by automatic and permanent recording of load variation on a pen recorder, X–Y plotter, or oscilloscope.

With one or two exceptions, all makes of testing machine will only strain in one direction, and therefore tension and compression tests have to be conducted in different "compartments" of the straining frame. Usually tension is applied on one side of the moving crosshead and compression on the other. Compression tests do not require especially

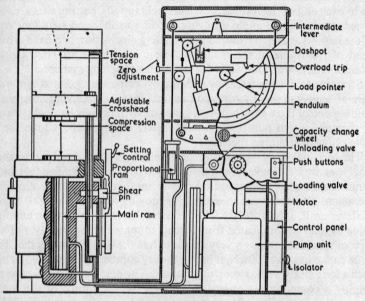

Fig 19.17 Mechanism of Avery 250 kN Universal Hydraulic
Testing Machine

shaped jaws, the specimen being loaded through flat ends via hardened platens and a spherical seating. Other attachments are generally available so that the straining and loading mechanisms can be employed on a variety of other types of test, including bending of a beam and hardness indentation. Two of the types of testing machine described above are shown in Plate IV and Figs. 19.17 and 19.18.

It is often very useful to have a continuous record during a test of load plotted against deformation, and this is termed an *autographic record*. The principles of autographic recording are the same on the various types of testing machine although the actual technique differs widely. In one system, an extensometer attached to the specimen rotates a drum either by mechanical or electrical means, in proportion to the deformation of the specimen. A pen resting on the drum is linked to the load measuring system and is moved along a generator of the drum in proportion to the load on the test piece. Consequently the combined movements of drum and pen result in a tracing of the load–deformation diagram as illustrated in Plate V.

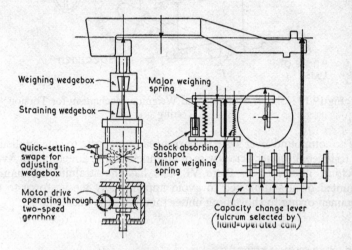

Fig 19.18   Mechanism of Denison 67 kN Universal Mechanical
Testing Machine

Torsion testing machines are much less frequently required than tension/compression machines and there are few manufacturers of the former type. Torsion of a test piece is most conveniently effected by a mechanical system. The specimen is usually a cylindrical solid or hollow bar and is gripped at each end by jaws supported in bearings in the frame of the machine. The input twist at one end of the test piece can be achieved by either a manually-operated worm and gear wheel of a suitable ratio, or an electric motor drive through a gear box. At the other end of the test piece the machine has to react and measure the torque. In principle this may be done by the simple lever and steelyard system shown in Fig. 19.19. A modern torsion machine employs a self-indicating torque-measuring mechanism, and one arrangement is shown in Fig. 19.20 in which a compound lever system reduces the magnitude of the torque and changes it to a direct force at point A in the dial mechanism. The small counter-balance weights react the force through the flexible tapes, T, which in turn causes the rotation of the indicating pointer via the rack and pinion.

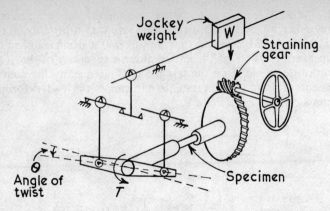

Fig 19.19   Simple Straining and Weighing Mechanism for Torsion
Testing

The compound lever system is arranged to measure torque applied to
the test piece in either a clockwise or an anticlockwise direction. An Avery
machine, illustrated in Plate VI, also has the straining mechanism
mounted on a sliding bed to avoid applying any tensile force to the
specimen owing to shortening under torsion in the plastic range.

19.7. EXTENSOMETERS

These are devices for measuring the deformations accompanying direct
stress, tension or compression. Since they are principally required for
measurement in the elastic and initial yield ranges, when determining the
elastic modulus or proof stresses, the strains are very small ($<0.2\%$
mild steel, $<0.6\%$ aluminium and steel alloys). Consequently, extensometers are very delicate and sensitive instruments. There are three
principles which can be employed, mechanical, optical or electrical.
Some examples are described briefly below.

The principal element of the Johansson extensometer, having a range
of gauge lengths from 50 mm to 3 mm, is a twisted thin metal strip.
Application or release of tension in the strip, due to movement of the
gauge points, causes untwisting or further twisting of the strip. A very
light pointer attached to the centre of the strip and perpendicular to its
axis moves across a calibrated scale under the action of the twisting
motion. On the most sensitive gauge, deformations of 0.000 25 mm over a
range of 0.0125 mm can be measured.

The Huggenberger extensometer relies simply on compound lever
magnification of specimen deformation. On gauge lengths of 12.5 or
25 mm, magnifications of 300 to 2000 can be obtained on the various

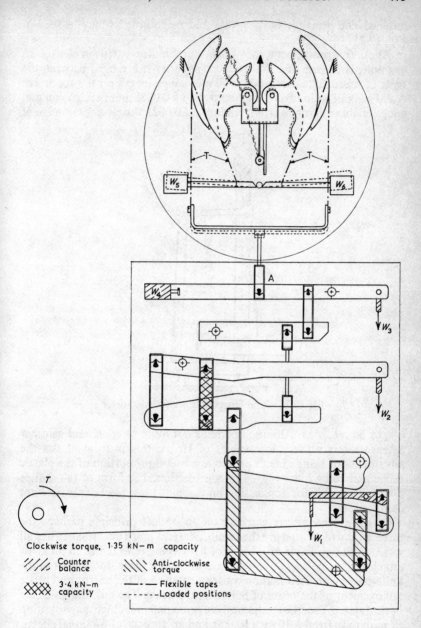

Clockwise torque, 1·35 kN–m capacity

///// Counter   \\\\\ Anti-clockwise
      balance   torque

XXXX 3·4 kN–m   —·— Flexible tapes
      capacity   ------- Loaded positions

Fig 19.20 W. & T. Avery Self-indicating Torque-measuring
Mechanism

models. The principle of operation and magnification may be seen in Fig. 19.21.

Another arrangement, of which the Lindley extensometer is an example, is the use of a dial gauge. The latter is arranged between the ends of scissor-like clamps on a 50 mm gauge length on the specimen. A 2:1 magnification from the clamps and a 0·0025 mm per division dial gauge enable minimum strain readings of 0·0025 % over a total range of

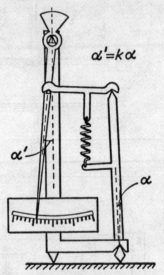

Fig 19.21   The Huggenberger Extensometer

6 % to be recorded. Although perhaps not quite so consistent as other extensometers at very low values of strain, this instrument has the advantages of being able to penetrate a useful proportion of the plastic range and being quite rugged in construction. The form of the clamps makes it suitable for application on machined test pieces only, and not components.

Of the extensometers working on an optical principle perhaps the three most widely used are the Lamb, Martens and Tuckerman. They all work on the principle of a beam of light being reflected from a mirror attached in some way to knife-edges on the test piece. Movement of the knife-edges with deformation causes a rotation of the mirror and hence a movement of the beam of light across a scale. The advantages of the optical system over the mechanical are principally very high magnification (if required), from × 10 to × 10 000, and an absence of frictional effects.

The Lamb extensometer can be arranged for tension or compression tests, and the principle is also applied for the measurement of lateral expansion and contraction. The instrument is shown in one form in

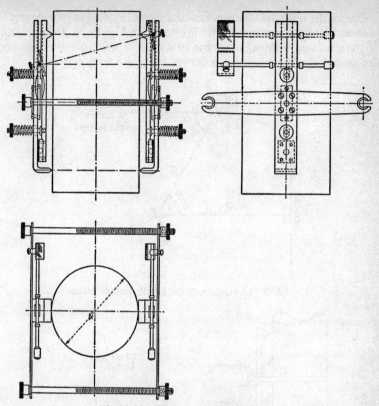

Fig 19.22   The Lamb Extensometer

Fig. 19.22, in which it is arranged for a compression test on a cement or concrete cylinder. It consists of two elements, which are clamped on opposite sides of the specimen. Each element consists of two hardened and ground steel plates having a knife-edge on each; between these plates a hardened and lapped roller is inserted, and as the specimen alters in length the rollers move in opposite directions. Each roller carries a mirror, the angular movement of which is measured by a telescope and scale in the usual way. The specimen does not require to be marked off, as the knife-edges are "gauge" when the ends of the plates are flush—a gauge is supplied to check this. When using two mirrors, the scale reading indicates the mean elongation of both sides of the specimen, which automatically corrects any errors due to non-axial loading. It can be shown that the deformation of the specimen is given by

$$\delta l = \frac{d}{4L + 2a} \delta x$$

where δx is the scale reading and d is the mean roller diameter, Fig. 19.23.

One of the best known examples of a single rotating mirror system is that of the Martens extensometer illustrated in Fig. 19.24. Deformation of the specimen rotates the rhomb to which the mirror is attached and the change in scale reading is observed through a telescope.

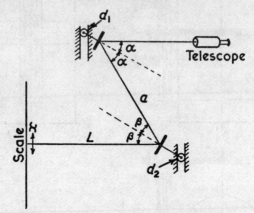

Fig 19.23   Optical Path for Lamb Extensometer

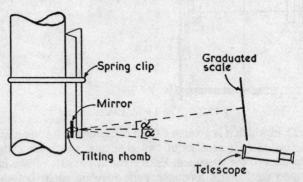

Fig 19.24   Principle of the Martens Extensometer

The Tuckerman gauge employs a collimated beam of light, which passes through a fixed roof-edge prism before being reflected from a lozenge and mirror as shown, Fig. 19.25. The great advantage of this arrangement is that it eliminates the possible errors due to tilt and linear displacement of a mirror relative to the telescope in the single mirror system such as the Martens.

Extensometers employing electrical devices to monitor displacement have become popular in the last decade. The sensing element is usually a linear variable differential transformer (LVDT) or one involving variation of inductance or capacitance proportional to displacement. These devices can be made quite small and light and can cover a wide range of

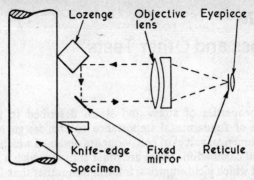

Fig 19.25(a) Simplified Optical Diagram of the Tuckerman
Extensometer

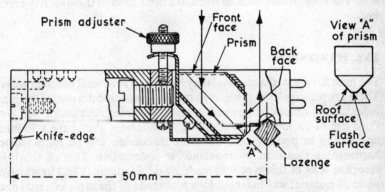

Fig 19.25(b) Sectional View of Standard Tuckerman
Extensometer

strain. They have a particular advantage over mechanical and optical
extensometers when it is required to provide a feedback signal in a servo-
controlled testing facility.

Further discussion of strain measurement will be found in Chapter 27.

BIBLIOGRAPHY

Beaumont, R. A., *Mechanical Testing of Metallic Materials* (Pitman, 1954).
British Standard No. 18: 1962.
Hetenyi, M., *Handbook of Experimental Stress Analysis*, Chapters 1–4
(Wiley, 1950).
Judge, A. W., *Engineering Materials;* Vol. III: *The Testing of Materials*
(Pitman, 1947).
Lessels, J. M., *Strength and Resistance of Metals* (Wiley, 1954).
Marin, J., *Engineering Materials: their Mechanical Properties and Applica-
tions* (Prentice-Hall, 1952).

Chapter 20

Hardness and Other Tests

20.1. The properties of stress and strain described in the previous chapter are of fundamental importance to the design and research engineer. Numerical values of behaviour are obtained which are required in theoretical calculation. There are other tests to which a material can be subjected which yield empirical information rather than basic properties. The results cannot be employed quantitatively for design purposes, but principally serve as a check that a material is in some desired condition. The present chapter is devoted to a discussion of three such types of test.

20.2. HARDNESS

The term *hardness* of a material may be defined in several different ways. These are principally in relation to the resistance to deformation such as indentation, abrasion, scratching and machining. Although the last three are of importance in certain circumstances, they have a limited application in practice, and therefore discussion will be restricted to hardness as measured by resistance to indentation. One of the first recorded tests of this type was made by de Réaumur (1722) in which a piece of material was indented by a tool made of the same material and the volume of the resulting indentation measured. Since then there have been several variations of the principle of this type of test used and widely adopted in engineering practice. The most popular are the Brinell, Vickers and Rockwell methods and these are discussed in the following sections.

(i) Brinell Test

Brinell published the details of the indentation hardness test he devised in 1901. The principle involved is that a hardened steel ball is pressed under a specified load into the surface of the metal being tested. The hardness, which is quoted as a number, is then defined as

$$\text{Hardness number} = \frac{\text{Load applied to indenter in kg}}{\text{Contact area of indentation in mm}^2}$$

or Brinell Hardness Number (B.H.N.) $= P/A$. It is noted that the number has in fact units of pressure. The contact area A is given by

$$A = \pi D h$$

$$= \frac{\pi D}{2}[D - \sqrt{(D^2 - d^2)}]$$

where h = depth of impression in mm, D = ball diameter in mm d = surface projected diameter of impression in mm.

If the hardness number of a metal is determined at several different values of load it is found that the number is not constant. The results give all or part of a curve of the form shown in Fig. 20.1 depending on

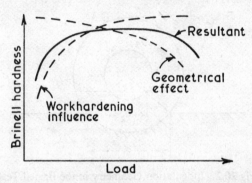

Fig 20.1   Variation of Hardness with Load in the Brinell Test

the condition of the material. The curve is attributed to two separate effects. The rising portion with increase in load is caused by the non-proportional effect of workhardening on the size of the impression. Thus a soft metal will show a marked apparent rise in hardness while a heavily cold-worked material will show none.

The falling part of the curve in Fig. 20.1 with increasing indenter load is caused by ungeometrical similarity between the spherical areas of successive impressions. This feature is very important and is worthy of further analysis. Consideration of Fig. 20.2 shows that similarity can only be obtained for different loads if different ball sizes are used, since the total angle subtended by the centre of a ball and the indentation must be equal in each case. Hence the condition for similarity is that

$$\frac{d_1}{D_1} = \frac{d_2}{D_2} = \text{constant}$$

For a given angle of indentation the mean pressure is $P/\frac{1}{4}\pi d^2$, but since $d = \text{constant} \times D$ it follows that for similarity $P/D^2 = \text{constant}$.

The highest value on the curve of Fig. 20.1 is known as the optimum hardness number and is the figure quoted for a material when hardness is required. Brinell obtained the optimum hardness for steels using a 10 mm diameter ball at a load of 3000 kg, and these conditions have become a British Standard for hardness testing. Using the above values it is seen that $P/D^2 = 30$, and it is therefore possible to obtain comparable hardness numbers on a metal for different sizes of ball if the load is chosen to satisfy the above relationship. For softer metals and thin sheet it is found that other values are required for the constant equal to P/D^2 in

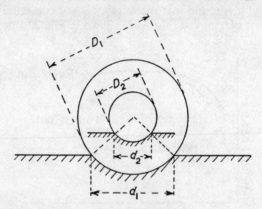

Fig 20.2 Indentation Geometry in the Brinell Test

order to give optimum hardness. Values of $P/D^2 = 10$ (non-ferrous alloys), 5 (copper, aluminium) and 1 (lead, tin) have been adopted as standard. The peak of the hardness curve generally occurs for a value of d/D between 0·25 and 0·5 and this fact can be used to assist in choosing the correct value of P/D^2. It is important when quoting a hardness number to state the conditions used, ball size, load, etc. It is also essential to space successive impressions adequately and keep them clear of the edge of the material, owing to the plastic deformation caused in the area around the indentation ("ridging" for hard metals, "sinking" for soft metals).

Another very useful feature of the Brinell method is that an empirical relationship has been found to exist between hardness number and tensile strength for steels. Thus, $K \times$ B.H.N. (kg/mm²) = tensile strength (MN/m²), where K lies between 3·4 and 3·9 for the majority of steels. Hence, for checking of correct heat-treatment, or a fractured component, it is only necessary to carry out a Brinell test to obtain an approximate value for the tensile strength.

The size of the Brinell indentation is such that the test is generally employed for checking raw stock or unmachined components rather than finished products. Typical hardness values for various materials are given in Table 20.1.

Table 20.1

Material	Condition	Hardness		Tensile strength, MN/m^2
		Brinell	Vickers	
Pure aluminium	Annealed	25	(25)	54
	Cold rolled (hard)	55	(55)	185
Duralumin alloy (3·5–4·5 Cu, 0·4–0·8 Mg, 0·4–0·7 Mn, 0·7 Si)	Solution- and pre-cipitation-treated	120	(120)	433
6% Al-Zn, 4% Mg alloy	Solution- and pre-cipitation-treated	180	(181)	587
Pure copper	Annealed	42	(42)	221
	Cold rolled (hard)	119	(119)	371
Brass (60–40)	Cold drawn	178	(179)	659
Mild steel (0·19 C)	Annealed	127	(127)	451
	Cold rolled	(192)	192	595
Si-Mn spring steel	Quenched and tempered	(415)	435	1180
4% Ni, 1·5% Cr, 0·3% C steel	Quenched and tempered	(434)	460	1640
Ball bearing steel	—	—	700	—
Tungsten carbide	Sintered	—	1200	—

( ) equivalent values

(ii) Meyer Analysis

Apart from the considerations of geometrical similarity in the Brinell tests discussed earlier, it was also shown by Meyer that a relationship between P and d exists as follows:

$$P = ad^n$$

where a and n are material constants. The above is an exponential function, and taking the logarithms of both sides, we have

$$\log P = \log a + n \log d$$

Plotting $\log P$ against $\log d$ gives a linear relation from which the slope determines n. This constant is a function of the ability of the metal to workharden and ranges from 2·0 for a metal which has no capacity for workhardening to 2·5 for a material which is able to workharden easily.

The constant a represents the resistance to first penetration and is a function of the ball size employed, which n is not.

Meyer proposed that the hardness number in the Brinell-type test should be defined in terms of the projected area of indentation rather than the spherical area so that

$$\text{Meyer hardness number} = \frac{P}{\frac{1}{4}\pi d^2} = \frac{4P}{\pi d^2}$$

Results in Fig. 20.3 show that, although the curves are still dependent on the workhardening condition of the material, dissimilarity of impression appears to be absent as indicated by the lack of fall-off in the curves as compared with the Brinell results for the same metals.

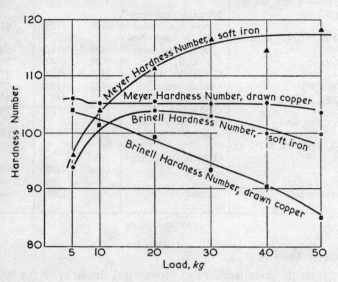

Fig 20.3   Brinell and Meyer Hardness Numbers for Drawn Copper and Soft Iron (1 mm Diameter Ball)

(*After H. O'Neill and Lessels: by courtesy of John Wiley & Sons Ltd.*)

The Brinell test can be conducted on a conventional tension/compression testing machine using a suitable holder for the ball indenter. However, for simplicity and speed of operation the majority of commercial tests are done using a Brinell hardness tester. In principle this consists of a frame supporting a pivoted lever. Weights are hung on one end of the lever on which, close to the pivot, is attached the ball indenter. A table under the latter supports the specimen, and a simple hydraulic system is used to lower and raise the indenter and lever on and off the test piece. The diameter of the impression is measured with a separate microscope.

(iii) Vickers Test

This test was devised about 1920 and employs a square-based diamond pyramid as the indenting tool. The angle between opposite faces of the pyramid is 136° and this was chosen so that close correlation can be obtained between Vickers and optimum Brinell Hardness Numbers. The angle of 136° corresponds to the geometry of an impression given by a d/D ratio of 0·375.

In the Vickers test, hardness number is defined in the same way as for Brinell, that is indenter load, kg, divided by the contact area of the impression, mm². If l is the average length of the diagonal of the impression, the contact area is given by

$$\frac{l^2}{2 \sin \frac{1}{2}(136)} = \frac{l^2}{1·854}$$

Therefore

$$\text{V.P.N.} = 1·854 \frac{P}{l^2}$$

There are two features of this test which are essentially different and advantageous over the Brinell method. Firstly, there is geometrical similarity between impressions under different indenter loads, and hardness number is virtually independent of load as shown in Fig. 20.4, except

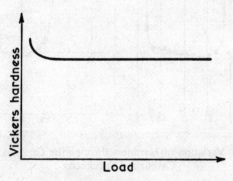

Fig 20.4   Variation of Hardness with Load in the Vickers Test

at very low loads where there is often a higher hardness owing to a "skin" effect on the test piece. The standard loads recommended in B.S. 427 are 1, 2·5, 5, 10, 20, 30, 50 and 100 kg.

The second advantage of the Vickers test is that the upper limit of hardness number is controlled by the diamond, therefore allowing values up to 1500 to be determined which is far in excess of that possible with the steel ball in the Brinell test.

The extremely small size of the impression necessitates a very good surface finish on the test sample, but means that it is advantageous in checking the hardness of finished components without leaving a severe mark.

The Vickers test is also very useful for surveying a specimen which has a variation in hardness through the cross-section. The most common examples of this are the surface hardening treatments such as nitriding, case hardening, etc., in which the surface layer of the metal, to a depth of, say, 1 mm, is much harder than the core. The small size of indentation of

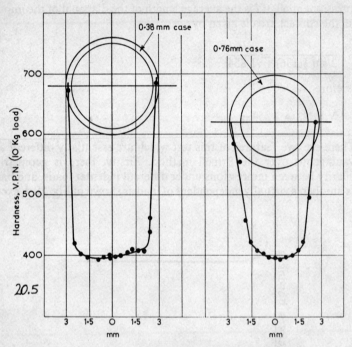

Fig 20.5 Variation of Hardness through the Cross-section of Carburized Specimens

the diamond pyramid enables the hardness to be determined at a series of points over a cross-section, from which the variation can be plotted as in Fig. 20.5, which shows the situation in a heat-treated and case-hardened (carburized) steel bar. The Vickers hardness for various materials is given in Table 20.1.

A special machine is used for Vickers testing which has the same principle as described previously for the Brinell machine, except that a mechanically-driven cam controlled by an oil dashpot is used to lower and raise the weighted lever, the load being automatically applied for

15 seconds. There is also a built-in microscope and shutter device for measuring the diagonals of the impression to an accuracy of 0·001 mm. A Brinell-type test can also be conducted using a different holder with either 1 mm or 2 mm diameter balls.

(iv) Rockwell Test

This test was introduced in the U.S.A. at about the same time as the Vickers test in England. It is quite popular as it has a wide range of versatility, is rapid and useful for finished parts.

Two types of penetrator are employed for different purposes, a diamond cone with rounded point for hard metals and a $\frac{1}{16}$ in (1·6 mm) diameter hardened steel ball for metals of medium and softer hardness values.

In this test hardness is defined in terms of the depth of the impression rather than the area, and the hardness number is read directly from an indicator on the machine having three scales, A, B and C.

The procedure for applying load to the specimen is rather different from the Vickers and Brinell and is illustrated in Fig. 20.6. Initially a

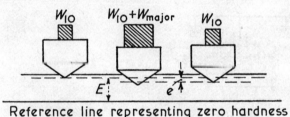

Fig 20.6   Load Application in the Rockwell Test

minor load of 10 kg is applied, which is followed by a major load, being an additional 50, 90 or 140 kg depending on the indenter and type of metal. The major load is now removed, but the minor load is retained while the hardness number is read, where

$$H_R = E - e$$

and H_R = Rockwell Number; e = depth of penetration due to the major load only but while the minor load is operating; E = arbitrary constant which is dependent on the type of penetrator. Table 20.2 gives the various relevant details of the test (B.S. 891).

(v) Hardness Comparator

For a machine part which is too large to be placed in any of the machines just described, an instrument called a hardness comparator, Fig. 20.7, may be used. This instrument is of a size which slips into a pocket and is

Table 20.2

Indicator scale	Penetrator	Minor load, kg	Major load, kg	Total load, kg	E	Materials
A	Diamond cone	10	50	60	100	Thin, hardened, steel strip, other extremely hard materials
B	$\frac{1}{16}$ in (1·6 mm) diameter steel ball	10	90	100	130	All mild and medium-carbon steels, hard non-ferrous alloys
C	Diamond cone	10	140	150	100	Hardened steels, alloy steels, materials harder than B100

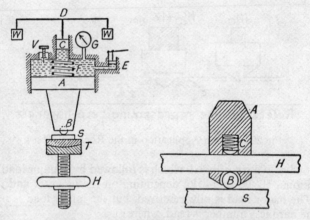

Fig 20.7   Brinell Hardness Comparator

thus useful for testing the hardness of an object which it is not desirable to move. It consists of a steel cylinder A, carrying a ball B resting on the specimen and also touching a bar H of standard Brinell hardness, which can be inserted in A. A spring C holds H in position. The top of the cylinder is struck with a hammer and an indentation is made both in the bar and in the specimen. Let

P = force of the blow
A_1 = area of indentation in standard bar
A_2 = area of indentation in specimen

B_1 = Brinell hardness number for bar
B_2 = Brinell hardness number for material
$P = k$ (B.H.N. × area of indentation) $= kA_1B_1 = kA_2B_2$

Therefore $\qquad B_2 = \dfrac{A_1B_1}{A_2}$

(vi) Comparison of Hardness Values

Owing to the wide use of the Brinell, Vickers and Rockwell methods and the varying preferences for any one of these tests, there are occasions when the same material or component is hardness tested by different methods in different laboratories. This has led to a demand for some correlation between hardness values determined by the three tests. It has been shown that there is no general relationship between the hardness scales, and empirical formulae only hold good for materials of closely similar composition and condition.

However, based on experimental results, the British Standards Institution has issued a table (B.S. 860:1939) of *approximately* comparative values for the three tests, but it is emphasized that it is not intended that the table shall be used as a conversion system for standard values from one hardness scale to another.

20.3. DUCTILITY TESTS

There is a wide range of tests included in several British Standards, for example B.S. 1639:1964 dealing with bend tests for bar and sheet material, which are concerned with the ductility of bar material, wire, rivets, tubes, chain links, etc. These tests are only qualitative in nature and generally involve some specified form of extreme plastic deformation after which the specimen must not show any visible sign of cracking. The tests provide crude information supplementary to the ductility values found from a normal tensile test, but related more specifically to the geometry and possible forming deformation of an article, e.g. rivets or tubes.

It is not appropriate here to describe all of these specialized tests; however, one example which is applied more generally to *samples of material*, known as the bend test, is perhaps worthy of note.

Bend Test

In this type of test an initially straight bar of round or rectangular section is plastically deformed by bending through some large angle in one direction only.

The angle, usually 120° to 180°, through which the test piece may be bent without visible cracking or fracturing occurring, gives a measure of the severity of the test and quality of the material. It has been found that this type of test is more severe on a rectangular than on a circular section bar having a diameter equal to the thickness of the rectangular section.

When results are being quoted, the final internal minimum radius of curvature is given as well as the final angle of bend achieved. When tests are made on a range of thicknesses, the final minimum radius of curvature is expressed as a function of the initial thickness or diameter of cross-section.

One common arrangement for a bend test is where the test piece is simply supported at each end and is subjected to a point load (initially) at the centre. The first stage is that of elastic curvature, after which occurs plastic curvature and hence permanent deformation (see Chapter 18). The test piece bends either uniformly into an arc of a circle, or else more acutely at the central region, allowing the ends to remain sensibly straight. This latter condition is termed *peaking*. The application of axial tension to the ends of the test piece reduces the curvature at the centre and hence tends to reduce any trends towards peaking.

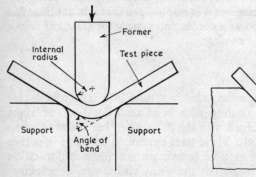

Fig 20.8   Bend Test

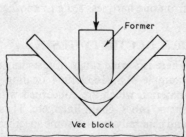

Fig 20.9   Bending Test Piece into a Vee Block

The most simple arrangement for a bend test is shown in Fig. 20.8 in which a radiused former forces the specimen to bend between two supports. It is important to radius the corners of the supports to obviate excessive damage to the test piece. On the other hand, the minimum radius compatible with little damage is desirable, so that the frictional force along the test piece, which helps to lessen peaking, is not reduced too much.

It is sometimes desirable to put a definite limit on the angle of bend attained, and this may be done conveniently by using vee-blocks, as illustrated in Fig. 20.9. Another method of reducing peaking during a test is to press the specimen into a block of lead or other suitably soft material. The latter then acts as a plastic foundation, or distributed support, assisting the former to transmit its shape to the test piece, as in Fig. 20.10.

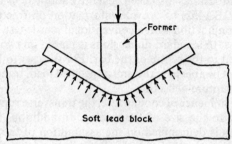

Fig 20.10   Bending Test Piece into a Soft Lead Block

In all the above tests the former is the moving member against which the test piece reacts. In another arrangement of the bend test the former is fixed and the specimen is wrapped around it by a separate tool. As seen in Fig. 20.11, the test piece is initially clamped at one end like a

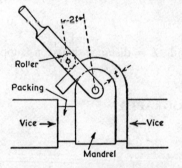

Fig 20.11   Wrapping Test Piece
round Mandrel

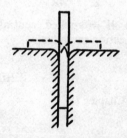

Fig 20.12   Reversed Bend
Test of Strip

cantilever between a mandrel (former) and vice, and rotation of the wrapping tool about its bearing bends the test piece around the curved end of the mandrel.

A variation of the above principle is found in the reverse bend test in which part of a strip is clamped in a vice with suitably rounded shoulders, and the protruding end of the test piece is alternately bent through a right angle to each side of the axis as in Fig. 20.12. The criterion of acceptability takes the form of a specified number of bend reversals without visible cracking or fracture.

20.4. TRANSVERSE TEST FOR CAST IRON

In contrast to the bend tests described in the previous section for checking ductility, a bend test has also been generally adopted as a standard test for cast iron (B.S.1452) to provide information on fracture stress and deflection, in conjunction with a conventional tensile test.

A test bar cast to standard dimensions is placed on two supports and a load is applied to the centre of the bar. For the cast to be acceptable, the test bar must be able to withstand a specified load, transverse rupture stress and pre-rupture deflection.

Owing to the influence of cooling rate, the transverse rupture stress will vary according to the size of the casting. In addition, the maximum transverse stress is determined on the assumption of the simple elastic theory of bending, which of course will not hold since the test piece will exceed the elastic limit. From the above it is apparent that the transverse rupture stress is not wholly independent of size and shape of test bar; however, the elastic calculation of stress has been standardized as a convenient way of expressing results without having to give full details of bar dimensions.

The transverse rupture stress, or modulus of rupture, the alternative term, is given by

$$\sigma = \frac{M}{Z} = \frac{WL}{4Z}$$

where W = applied central load, L = distance between supports, Z = section modulus.

BIBLIOGRAPHY

See Chapter 19.

Chapter 21
Toughness and Unstable Fracture

21.1. INTRODUCTION

There are a number of causes for material or structural failure other than simply exceeding the tensile (or compressive or shear) yield or maximum strength as determined by the tests described in Chapter 19. If a material has a low resistance to crack propagation, catastrophically rapid failure can occur initiating from metallurgical or manufacturing flaws in the material. This type of unstable fracture results from low energy absorption characteristics of a material and is often described as *brittle fracture*. The capacity of material to absorb energy is described as *toughness*.

Brittle fracture of steel was reported as long ago as 1886; however, the importance of this type of failure was not appreciated then and fractures were not always reported. Structures were almost entirely of mild steel of riveted construction, and failures occurred principally in storage tanks at ambient or lower temperatures.

Welded structures developed from 1912, but the first reported failure appears to be a light railway and roadbridge at Hasselt, Belgium, in 1938. Two more bridges collapsed in the next two years. From 1940 onwards, welding was used widely in shipping to speed up production and there was a corresponding sharp rise in the number of brittle fractures, Liberty ships and tankers being the main source. In recent times the *World Concord* oil tanker, and an oil storage tank at Fawley, Southampton, have provided spectacular examples of brittle fracture. Three contrasting service failures are illustrated in Plates VII (top and bottom) and VIII (top), the former being that of a chain link, for which periodic annealing is required to minimize the possibility of fracture. The oil tanker, *World Concord*, Plate VII (bottom), broke in two in cold severe weather in the Irish Channel. Thirdly, an oil storage tank split and collapsed while undergoing a test filling with a water temperature of 4°C. Some of the fractured plate showing both ductile and brittle fractures is illustrated in Plate VIII (top).

The principal features of these failures are as follows: poor design, faulty material, or a bad repair job acting as a notch, coupled with a sudden drop in temperature causes initiation of a crack. A large structure has a lot of strain energy available to assist in the propagation of a crack, and in this respect a welded structure is more "continuous" than a riveted structure and is therefore slightly worse from the point of view of crack arrest conditions. Fractures occur catastrophically at very low nominal stress values (generally below the yield stress of the steel), and there is a complete absence of macroscopic plastic deformation. Distinct chevron patterns appear on the fracture surface as illustrated in Plate VIII (centre). The above low energy or brittle fractures were in low-carbon steels which in general are regarded as ductile.

In recent times, progress in the aerospace industry has necessitated the development of ultra-high-strength alloys for rocket motors and space vehicles. Some of these low-ductility materials were found to be susceptible to unstable fracture from small flaws owing to low toughness. This resulted in the development by Irwin of the theory of linear elastic fracture mechanics which was accompanied by the establishment of tests to measure the fracture toughness of materials.

This chapter will briefly introduce the essential features of unstable fracture and toughness measurement. For convenience the first part of the chapter will deal with traditional brittle fracture of tough ductile steel and the second part will introduce the modern concepts of fracture mechanics.

21.2. FRACTURE TERMINOLOGY

There are four aspects to consider, (a) the behaviour prior to fracture, (b) the fracture mechanism, (c) the appearance of the fracture, and (d) fracture resistance. Firstly, consider the stress–strain curves in Fig. 21.1: in curve (1), fracture occurs immediately after the elastic range

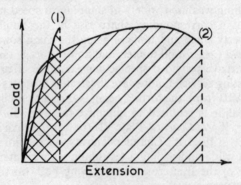

Fig 21.1   Load–Extension Curves for a Brittle and a Ductile Material

and is said to be brittle; curve (2), which is typical for most metals, exhibits a plastic range, which may vary from 10 to 50% strain before fracture. The material in this case is termed ductile. Hence a brittle material does not exhibit much plastic deformation before fracture while a ductile material does.

The second aspect is the fracture mechanism. There are only two paths for a crack passing through a metal, either transcrystalline or intercrystalline. The latter only occurs in a few particular circumstances (creep, stress corrosion, etc.). The former is the more general mechanism, of which there are two types related to the crystallographic planes known as *shear* and *cleavage*. Shear is the result of certain crystal planes sliding over one another, and is associated with a great deal of local plastic deformation. Cleavage occurs on different crystallographic planes caused by a normal (tensile) stress and involves very little local plastic deformation.

The third aspect of fracture, that of appearance, is a useful way of describing fracture in a qualitative way, but is not precise as to mechanism. Shear has a dull grey appearance and is often described as fibrous or silky. Cleavage, on the other hand, reveals smooth reflecting planes and is described as bright, and crystalline or granular.

Finally, fracture resistance falls into two main categories, (a) above-yield fracture stress with high energy absorption which is classed as tough, and (b) below-yield and low-energy fracture described as frangible.

Summary of Terms

Prior deformation	Ductile	Brittle
Crystallographic mode	Shear	Cleavage
Appearance	Fibrous	Crystalline
Resistance	Tough	Frangible

21.3. BRITTLE FAILURE OF A DUCTILE METAL

Under certain conditions a ductile metal which normally fractures by shear will fail either partly or totally by cleavage. This phenomenon is now commonly termed brittle fracture; however, the use of the word "brittle" in this sense does not imply complete absence of plastic deformation. In the same way, the term "brittle" used to describe a fatigue fracture does not imply a cleavage mechanism.

21.4. TOUGHNESS

In Chapter 8 the concepts of elastic strain energy were discussed and problems were solved in which, during elastic deformations both of static and dynamic form, the amount of stored strain energy was determined. When a material is deformed beyond the proportional limit, accurate theoretical predictions of energy absorbed would be much more difficult to obtain owing to the non-linearity between stress and strain. However, a knowledge of the capacity of a material to absorb energy before fracture can be determined experimentally; and this quantity is termed the *toughness* of the material. Toughness is a property which can be used quantitatively in design specification, and has a very important place in mechanical testing and properties. This is because it gives a useful indication of the condition of the material, correct heat-treatment, metallurgical defects, etc., and also because there are many components and structures which in service are required to absorb *elastic* strain energy and, in the event of accidental over-strain, should be able to absorb some *plastic* strain energy without fracture.

In a normal tensile test the energy to fracture is given by the area under the stress–strain curve. Illustrated in Fig. 21.1 are two curves,

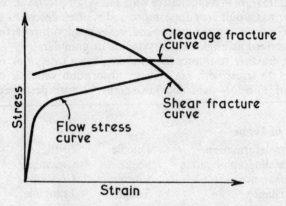

Fig 21.2   Diagrammatic Representation of Flow and Fracture Stress Curves

(1) for a brittle and (2) for a ductile material having approximately the same tensile strength, showing the difference in energy to fracture. However, the assessment of toughness is not simply determined from a tensile stress–strain curve since fracture energy is considerably influenced by factors such as stress conditions and geometry of material, velocity of loading, and temperature of test.

21.5. FRACTURE STRESS CURVES

To explain certain features of the phenomenon of fracture, Davidenkov and others have proposed the hypothesis of relationships between stress and strain which uniquely determine the fracture point on the true stress–strain curve for a material. The relationships are termed *fracture stress curves*, and it is suggested that there is one applicable for the cleavage mechanism and another for shear, as shown in Fig. 21.2. The type of fracture then depends on the particular fracture curve with which the flow stress curve first intersects. Different test conditions will cause various relative positions of the curves and hence one or other mode of fracture with various amounts of prior plastic flow.

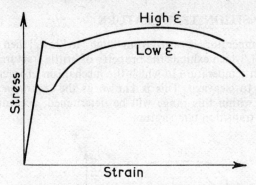

Fig 21.3   Effect of Strain Rate on the Tensile-stress/Strain Curve for Mild Steel

21.6. EFFECT OF TEMPERATURE, STRESS CONDITION AND STRAIN RATE

If a normally ductile metal failing by shear is to be made to fracture in a brittle manner by cleavage, then the flow stress curve must be raised to intersect first with the cleavage fracture stress curve.

Lowering the test temperature has the effect of raising the yield point, tensile strength and fracture stress. However, it also raises the shear and cleavage fracture curves, though not by as much as the flow stress curve. Consequently, ductility may or may not be affected, and fracture may be changed from shear to cleavage in some metals, or remain in shear at very low temperatures in other metals.

So far, the discussion has been concerned with simple uniaxial stress, but in a complex stress system the situation can be changed in the following way. In a triaxial stress state where $\sigma_1 > \sigma_2 > \sigma_3$, the maximum shear stress is $\frac{1}{2}(\sigma_1 - \sigma_3)$, but as $\sigma_3 \to \sigma_1$, $\hat{\tau} \to 0$. In the extreme case of hydrostatic tension and compression, $\sigma_1 = \sigma_2 = \sigma_3$ and $\hat{\tau} = 0$. It is evident that shear cannot occur and hence cleavage fracture will result. The introduction of a discontinuity or notch into a piece of material causes a stress concentration and triaxiality of stress to a degree which depends on notch geometry and loading condition. There is therefore the tendency to raise the flow stress curve and induce cleavage fracture.

Most non-ferrous metals and alloy steels are not particularly affected by strain rate. On the other hand, mild steel is highly strain rate dependent, and the yield point can be considerably raised, far more in proportion than the tensile strength, as is illustrated by Fig. 21.3. This is therefore another factor which can contribute towards a change from shear to cleavage fracture.

21.7. TRANSITION TEMPERATURE

If the test temperature is varied from "high" to "low," then a metal such as mild steel, which exhibits the property of brittle fracture, will have a range of test temperature in which the mechanism of fracture changes from shear to cleavage. This is known as the *transition temperature* range, and within this range will be determined, according to some criterion, a transition temperature.

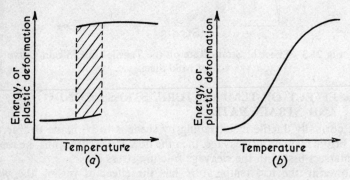

Fig 21.4   Brittle-Ductile Energy vs. Temperature Transition Curves

Typical transition curves are shown in Fig. 21.4 (*a*) and (*b*) in which plastic deformation at, or energy to, fracture is plotted against test temperature. The first shows what is known as bimodal behaviour in which there are two distinct branches, one at high energy and temperature and the other at low energy and temperature joined by a narrow region (10 to 20°C) of scattered points at upper and lower energy values. The second diagram shows continuous behaviour where there is a gradual fall from high to low energy with decrease in temperature. The transition range in this case can be as much as 100°C from complete shear to entire cleavage.

21.8. BRITTLE FRACTURE LABORATORY TESTING

Laboratory tests may be divided into two categories, firstly standard tests for quality control of material, and secondly research tests which aim at reproducing the features of service failure and studying crack initiation and propagation phenomena.

Between 1900 and 1908 several investigators arrived at the conclusion that an impact bend type of test gave the most effective and convenient measure of the toughness of a material for quality control. Although details differ from one investigator to another, the principle of the test is much the same. A small bar of material having a notch cut into some part to initiate fracture is struck by a heavy pendulum having a known initial potential energy. The bar is bent to fracture, and the resistance offered to the pendulum is a measure of the energy absorbed by the test piece. This type of test is simple, rapid and appears to indicate lack of toughness in a material under certain conditions when a conventional tensile test would not do so. The notched impact bend test has now been universally adopted for checking toughness of a material.

The two forms of the test most widely used are the Charpy and Izod. The principle of the former is shown in Fig. 21.5. The test piece is a square bar of material, 10 mm × 10 mm × 55 mm, containing a notch cut in the middle of one face. In this country the notch is a 45° vee, 2 mm deep, with a root radius of 0·25 mm. The original notch and that still used frequently in Europe is of keyhole form. The test piece is simply supported at each end on anvils 40 mm apart. A heavy pendulum is supported at one end in a bearing on the frame of the machine, and a striker is situated at the other end. The pendulum in its initially raised position has an available energy of 300 J and on release swings down to strike the specimen immediately behind the notch, bending and fracturing it between the supports. A scale and pointer indicate the energy absorbed during fracture.

In the Izod test, Fig. 21.6, the specimen is of circular section, 11·43 mm in diameter and 71 mm long, or square section, 10 mm × 10 mm × 75 mm, and the Izod notch is a vee as described above for the Charpy test, 3·33 mm and 2 mm deep for the round and square specimens respectively. The specimen is supported as a vertical cantilever "built-in" to jaws up to the notched cross-section. A pendulum and striker having an initial energy of 166 J is arranged to swing and strike the free end of the test piece with a velocity of 3–4 m/s on the same side as the notch and 22 mm above it. The energy absorbed by the test piece is again recorded on a scale.

It is seen that in both tests the notch is on the tension side of bending, thus initiating fracture. The Charpy and Izod tests have been adopted as British Standards and details of alternative types of specimen and other conditions are given in B.S. 131:1961. There is no direct correlation between energy values given in each of these tests; however, experimental results on a wide range of steels show that there is a linear relationship over the range from 20 to 95 J.

A typical machine on which both Izod and Charpy tests can be conducted is illustrated in Plate IX.

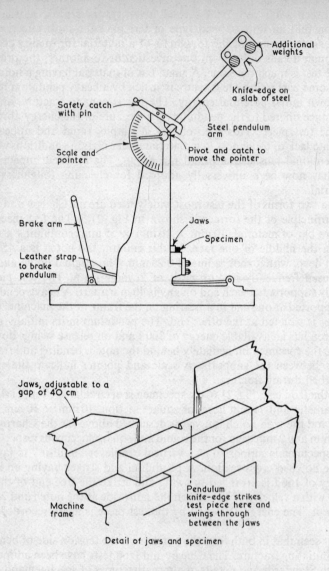

Additional weights

Knife-edge on a slab of steel

Safety catch with pin

Steel pendulum arm

Scale and pointer

Pivot and catch to move the pointer

Brake arm

Jaws

Specimen

Leather strap to brake pendulum

Jaws, adjustable to a gap of 40 cm

Pendulum knife-edge strikes test piece here and swings through between the jaws

Machine frame

Detail of jaws and specimen

Fig 21.5   Diagram of Charpy-Type Impact Testing Machine
(*Losenhausen*)

Some typical values of strength, ductility and toughness are given in Table 21.1. The most notable feature is the effect of the type of heat-treatment given, whether normalizing, quenching or quenching and tempering. The reason for this is the difference in metallurgical structure in each case influencing the ductility of the material, and the ease

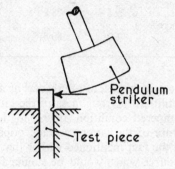

Fig 21.6 Arrangement of Test Piece in the Izod Test

with which a crack can propagate through the notched bar. The quenched structure is hard and brittle and can absorb little energy. In the quenched and tempered structure, a higher tempering temperature gives greater ductility and therefore higher impact energy. If the tempering temperature is arranged to give the same tensile strength as a

Table 21.1

Material	Condition	Tensile strength, MN/m²	Reduction in area, %	Izod impact energy at room temperature, J
0·1 C, 0·3 Mn steel	Annealed	377	65	54·2
0·21 C, 0·82 Mn steel	Annealed	505	58	40·6
0·5 C steel	Normalized	787	63	29·8
Ni-Cr-Mo steel	Quenched 840°C, tempered 650°C	895	64	114·0
	Quenched 840°C, tempered 500°C	1472	47	29·8
3·0 Ni, 1·0 Cr steel	Quenched 840°C	1668	35	19·0
	Quenched and tempered 550°C	865	60	93·5
Stainless steel (18–8)	Cold rolled	987	—	46·1

Table 21.2

0·5 C steel	Izod, J	Tensile strength, MN/m²	Elonga-tion, %	Hardness, V.P.N.
Normalized	29·8	78·8	20	230
Quenched 810°C, tempered 680°C	96·2	80·8	23	247

normalized structure as in the case of a 0·5% carbon steel in Table 21.2, in spite of similar ductilities, the toughness of the latter is not nearly so great as for the tempered condition, owing to the resistance of the refined grain structure of the latter to crack popagation. The above case also illustrates the fact mentioned earlier that the area under the tensile-stress/strain curve which would be similar for these two conditions is not necessarily the best guide to toughness, as the impact values show.

Another case in which satisfactory tensile strength and ductility are obtained, but in which very low impact toughness occurs, is in the behaviour known as *temper brittleness*. This phenomenon arises in certain alloy steels if slow cooling is allowed after tempering. If the cooling is rapid, i.e. quenching, then expected energy values are obtained.

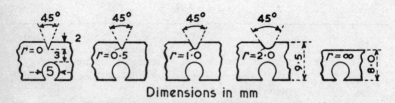

Dimensions in mm

Fig 21.7   Notch Geometry for the Schnadt Impact Test

An impact bend test which has aroused much interest and controversy in recent years is that due to Henri Schnadt. In principle it is similar to the Charpy test except that five different forms of notch are employed as shown in Fig. 21.7, ranging from a sharp-rooted vee through increasing root radius to simply the straight edge of the specimen, thus giving a complete range of stress triaxiality at the notch root. A further and most important feature is that the compression material behind the notch is drilled out and a hardened-steel pin inserted. This has the effect of eliminating the tension side bending stress gradient, since when the specimen is struck it fractures by hinging about the pin. The results of the test have to be interpreted in the light of a theory proposed by Schnadt in which triaxiality of stress at the notch is related to the maximum principal stress in the material.

An impact test can also be conducted in tension, but it is not as simple or rapid as the bend test, and is therefore generally confined to research studies. With careful instrumentation it can then be employed for examining such effects as strain rate on the tensile-stress/strain curve, and this information can make an important contribution to the general field of fracture mechanics.

There are far too many specialized research tests for them all to be described in this chapter. The four main classifications are notch-tensile, notched slow and impact bend, combined bending and tension of a notched test piece, and explosion tests.

Tension tests are conducted on either a cylindrical bar containing a circumferential vee-notch, or a plate specimen which at one section has a vee-notch cut in each edge. Energy to fracture cannot be simply determined in this test; however, the nominal maximum stress can be obtained, and in the case of brittle fracture this is also the fracture stress. It is found that with decreasing test temperature the tensile strength rises uniformly, there being no discontinuity at the transition temperature. Also, whereas service brittle fractures generally occur at stress levels below virgin yield, this effect can only be achieved in the notch-tensile test under very special circumstances.

The slow bend test, of which examples are the Van der Veen and Lehigh tests, consists in bending a piece of plate, simply supported at each end, by a central gradually applied load which is on the opposite side of the plate to a small vee-notch. Load–deflection curves are plotted from which the energy absorbed is determined. Plastic deformation close to the notch, deflection at maximum load and fracture appearance are used as criteria for the determination of brittle fracture. A set of load–deflection curves at different temperatures is shown in Fig. 21.8 and it is interesting to note the sharp drop in load at the onset of brittle fracture, even though there is some prior plastic deformation as in the curve for +12°C.

One of the most interesting research tests of recent years has been developed by T. S. Robertson. The principle is illustrated in Fig. 21.9. The test plate, which is of service thickness, is welded along its edges to extension pieces which are allowed to yield under load to avoid bending of the test plate and through which a transverse stress is applied to the test piece. A hole is cut in the end of the plate and a small sawcut is made in the edge of the hole along the axis of the plate. The hole is filled with liquid nitrogen and the opposite end of the plate is heated, thus providing a temperature gradient along the

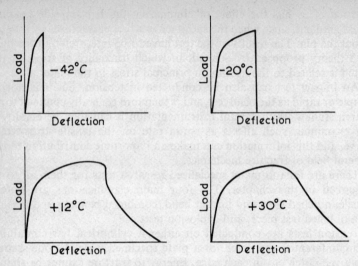

Fig 21.8   Load-Deflection Curves at Different Temperatures in a
Slow Bend Test

axis. An impact force on the outer edge of the hole serves to initiate a
crack from the sawcut, which will then propagate along the axis of the
plate until it is arrested when reaching material at a higher temperature.
The results are plotted in the form of transverse stress against crack

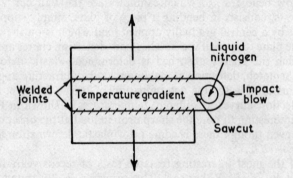

Fig. 21.9   Test Piece Arrangement in the Robertson Test

arrest temperature. The curves in Fig. 21.10 show that at a particular
arrest temperature there is a sharp rise in stress. This value is generally
lower than the transition temperature determined in other types of test.
It is also important to note that the "knee" in the curve occurs at
transverse stress values well below the yield stress of the material.

Although this is rather a complicated and expensive test it is perhaps the nearest approach so far to reproducing service-type brittle fractures in the laboratory. In addition, fundamental information can be obtained on crack initiation and propagation.

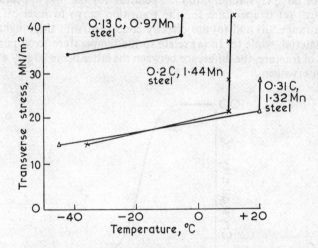

Fig 21.10   Transverse Stress vs. Crack Arrest Temperatures
for Several Steels in the Robertson Test

(*From T. S. Robertson, "Brittle fracture of mild steel," Engineering,* **172**, *pp. 445–8, 1951*)

21.9. FRACTURE APPEARANCE

It has already been mentioned that visual observation of the appearance of fractured surfaces can help in assessing the fracture mode. An examination of specimens associated with various points along a transition curve shows a marked change in appearance from one end of the curve to the other. At high temperature and energy associated with shear, the specimen is distorted owing to plastic deformation, and the structure is dull and fibrous. At the other end of the scale, there is negligible plastic deformation and the fracture surface is bright and crystalline. In between these two extremes there is a mixture of the conditions described above. A typical set of Charpy specimen fractures is shown in Plate VIII (bottom). Another method of determining transition temperature is by an estimation of the area of the fracture surface which is crystalline in nature, and plotting this against test temperature. This yields a curve of the form shown in Fig. 21.11, where 100% crystallinity corresponds to low temperature and complete cleavage, and 0% crystallinity to a fully ductile shear fracture.

A number of criteria have been adopted by different countries and investigators for defining transition temperature or acceptance levels of energy, and these may be listed as follows: (*a*) energy at a selected temperature, (*b*) temperature for a specified energy value, (*c*) temperature for 50% crystallinity, (*d*) temperature for the first appearance of cleavage, (*e*) temperature for the transition curve to level off in the brittle range. (*a*) and (*b*) are directly concerned with the toughness of the material, while (*c*) to (*e*) relate to the temperature for a particular mode of fracture, the difference between the latter being largely a matter of conservatism.

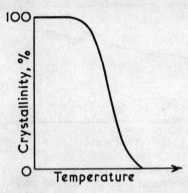

Fig 21.11   Relationship between Crystallinity and Fracture
Temperature

21.10. METALLURGICAL FACTORS

Probably the two most important factors necessary to avoid brittle fracture are a careful consideration of the design of a structure and improved properties of a metal to resist the onset of cleavage. The latter aspect is bound up with metallurgical considerations.

Although transition behaviour cannot be eliminated in carbon steels, the actual transition temperature can be lowered by various means. The effect of carbon content is such that decreasing amounts of carbon result in lowering the transition temperature. Manganese is a very important element in improving transition properties, and a Mn/C ratio of 3 to 1 is considered to be very desirable. Phosphorus, nitrogen and hydrogen are all detrimental and increase the tendency towards brittle failure, while nickel improves toughness, and aluminium plays an important part during steel making by removing oxygen and nitrogen.

Grain size has a marked effect on notch toughness: the coarser the grain size the higher will be the transition temperature. It is therefore desirable to have rapid cooling, when normalizing say, for thick plates to promote a refined grain structure and keep transition temperature down.

21.11. FRACTURE MECHANICS

In 1920, A. A. Griffith considered the growth of unstable cracks in glass, being an entirely brittle elastic solid, and proposed that a crack would propagate if the incremental release of stored elastic strain energy in the material exceeded the increase in surface energy due to the creation of a new crack surface. This may be expressed as

$$\frac{\partial(U_E)}{\partial c} \geqslant \frac{\partial(U_S)}{\partial c} \tag{21.1}$$

for the extension of a crack of length $2c$.

For a plate of unit thickness subjected to plane stress, σ, normal to the crack, the stored elastic energy, $U_E = \pi\sigma^2 c^2/E$, and the surface energy, $U_S = 4cS$, where S is the surface tension; thus eqn. (21.1) becomes

$$\frac{\partial}{\partial c} \frac{\pi\sigma^2 c^2}{E} \geqslant \frac{\partial}{\partial c} 4cS$$

and if $\sigma = \sigma_f$ at fracture,

$$\sigma_f = \sqrt{\frac{2ES}{\pi c}} \tag{21.2}$$

For the case of plane strain where thickness is much greater than crack length,

$$\sigma_f = \sqrt{\frac{2ES}{(1 - v^2)\pi c}} \tag{21.3}$$

Eqns. (21.2) and (21.3) relate the flaw size and stress for unstable fracture in a completely brittle elastic solid.

Since no metal is ideally brittle and there is always a small zone of plastic yielding at the tip of a flaw or crack, Orowan suggested that eqn. (21.2) should be modified to include a plastic work term, p:

$$\sigma_f = \sqrt{\frac{2E(S + p)}{\pi c}} \tag{21.4}$$

but we now find that S is smaller than p by a factor of 10^{-2} to 10^{-3}; thus S can be neglected so that

$$\sigma_f = \sqrt{\frac{2Ep}{\pi c}} \tag{21.5}$$

Thus a *small* plastic zone and its associated plastic work can promote unstable fracture. If there is more extensive plasticity a stable situation exists between the increasing stress required for propagation accompanied by further yielding, and so on.

21.12. CRACK TIP STRESS DISTRIBUTION

The elastic stress distribution in the region near the crack tip is given by

$$\left. \begin{aligned} \sigma_x &= \frac{K}{\sqrt{(2\pi r)}} \cos\frac{\theta}{2} \left(1 - \sin\frac{\theta}{2}\sin\frac{3\theta}{2}\right) \\ \sigma_y &= \frac{K}{\sqrt{(2\pi r)}} \cos\frac{\theta}{2} \left(1 + \sin\frac{\theta}{2}\sin\frac{3\theta}{2}\right) \\ \tau_{xy} &= \frac{K}{\sqrt{(2\pi r)}} \cos\frac{\theta}{2} \sin\frac{\theta}{2}\cos\frac{3\theta}{2} \end{aligned} \right\} \qquad (21.6)$$

where the geometry at the crack tip is defined as in Fig. 21.12. The factor K was termed by Irwin the *stress intensity factor*. For an infinitely sharp crack in an infinitely wide plate,

$$K = \sigma\sqrt{(\pi c)} \qquad (21.7)$$

where σ is the general plate stress.

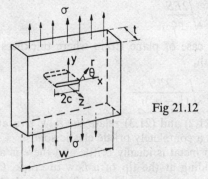

Fig 21.12

It has been shown that for various types of flaw in finite structures the stress intensity factor has the form $K = \sigma\sqrt{(\alpha\pi c)}$, where α is a parameter dependent on geometry. For example, a finite plate subjected to tension containing a central crack of length $2c$ has

$$K = \sigma\sqrt{(\pi c)} \left[\frac{W}{\pi c} \tan\frac{\pi c}{W}\right]^{1/2} \qquad (21.8)$$

where σ is the applied tensile stress.

The inclusion of the effect of a plastic zone at the crack tip can only be treated analytically so far for the case of plane stress in a non-hardening material.

The extent of the plastic zone beyond the crack tip is

$$R \approx 2r_y = \frac{K^2}{\pi\sigma_y^2}$$

as shown in Fig. 21.13.

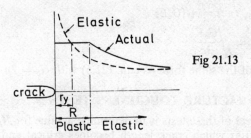

Fig 21.13

Irwin has proposed that stress relaxation in the plastic zone is equivalent, as far as K is concerned, to an additional length of "elastic" crack $r_y = K^2/2\pi\sigma_y^2$.

Thus using $2(c + r_y)$ instead of $2c$ in eqn. (21.8) we obtain

$$K = \sigma\sqrt{W}\left[\tan\frac{\pi}{W}\left(c + \frac{K^2}{2\pi\sigma_y^2}\right)\right]^{1/2} \qquad (21.9)$$

The plastic zone size for fully plane strain conditions is considered to be about one-quarter to one-third of that for plane stress owing to the restraint of triaxiality.

21.13. CRACK EXTENSION FORCE

It is considered that crack growth is controlled by a *crack extension force* denoted by G (units of length-force/unit area) which increases with stable crack growth until the transition occurs from slow to rapid growth. At this point G has reached the value of the critical crack extension force G_c or critical strain energy release rate.

Now, by definition,

$$G = \frac{K^2}{E} \quad \text{(plane stress)} \qquad (21.10)$$

where K is the stress intensity factor and E is the elastic modulus.

When $G = G_c$, $K = K_c$ (units of force/length$^{3/2}$), a critical value of the stress intensity factor which is termed the *fracture toughness*, and for plane stress

$$K_c = \sqrt{(G_c E)} \tag{21.11}$$

and for plane strain

$$K_{Ic} = \sqrt{\left(G_{Ic}\frac{E}{1 - v^2}\right)} \tag{21.12}$$

From eqns. (21.7) and (21.11),

$$\sigma_f\sqrt{(\pi c)} = \sqrt{(G_c E)}$$

$$\sigma_f = \sqrt{\frac{G_c E}{\pi c}} \tag{21.13}$$

This is of the same form as eqn. (21.5) with $G_c = 2p$.

21.14. FRACTURE TOUGHNESS TESTS

The object of these tests is to determine a value for K_{Ic} (which is lower than K_c) at which crack length becomes critical and propagates fast. To evaluate K_{Ic} it is necessary to measure load and crack length at the onset of instability. The load is determined from the testing machine

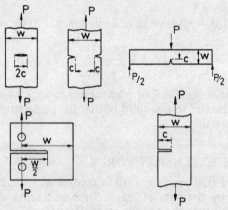

Fig 21.14

autographic recorder or load cell, while the crack length at instability may be determined by cinematography, acoustically or by displacement or continuity gauges. Some examples of commonly used specimens are illustrated in Fig. 21.14. In order that the initial crack shall have a natural sharpness it is general practice to grow a short fatigue crack from the end of a manufactured notch or slit. An example of fracture toughness data for a high-strength steel is illustrated in Fig. 21.15.

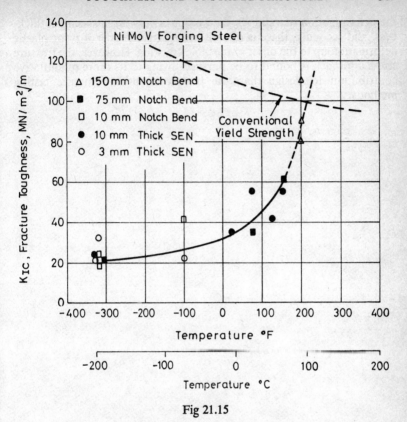

Fig 21.15

21.15. FRACTURE STRESS ANALYSIS

The procedure for designing against unstable fracture is firstly to establish the K relationship for the likely defect geometry in the component or structure. Next, having selected a material, a suitable K_{Ic} test is conducted (if its value is unknown) to simulate the likely fracture state (temperature, strain rate, size, etc.) in the structure. It is also necessary to estimate or measure on a prototype the largest likely flaw size. Finally, with the foregoing information and the relationships derived in earlier sections, the allowable applied load (stress) on the component or structure can be determined. Conversely, to withstand a particular load a material must be chosen having adequate fracture toughness for the largest expected size of defect.

The above method has been very successfully applied to high-strength alloys of medium and low toughness, since largely elastic conditions prevail with a small plastic zone size. There are problems in using the method for high-toughness/low-strength materials. Firstly, the determination of valid K_{Ic} data requires very large specimen thickness, and secondly, there is often a much larger degree of prior plastic deformation up to the onset of unstable cracking. However, the fracture stress approach is bound to be pursued owing to its more quantitative contribution to design than the traditional brittle fracture testing methods.

Chapter 22
Fatigue

22.1. In the early part of the 19th century the failure of some mechanical components subjected to nominal stresses well below the tensile strength of the material aroused some interest among a few engineers of that time. There was no obvious defect in workmanship or material, and the only feature common to these failures was the fact that the stresses imposed were not steady in magnitude, but varied in a cyclical manner. This phenomenon of failure of a material when subjected to a number of varying stress cycles became known as *fatigue*, since it was thought that fracture occurred due to the metal weakening or becoming "tired".

The first real attack on this problem was made by the German engineer Wöhler in 1858. Since then a great deal of research has been conducted on fatigue of metals, and in more recent times other materials also, and although this work has resulted in an ever increasing understanding of the problem there is as yet no complete solution. It has been estimated from time to time that at least 75% of all machine and structural failures have been caused by some form of fatigue. It is therefore evident that every engineer should be aware of this phenomenon, and have some idea of its mechanics and what can be done to minimize or avoid the risk of this type of failure.

22.2. FORMS OF STRESS CYCLE

Throughout the working life of a component subjected to cyclical stress the magnitude of the upper and lower limits of cycles may vary considerably. Similarly the shape of the stress–time cycle may differ in various circumstances. However, it is general practice when considering fatigue behaviour to assume or employ a sinusoidal cycle having constant upper and lower stress limits throughout the life. Fig. 22.1 (*a*) shows a general type of stress cycle, which is termed fluctuating, in which an alternating stress is imposed on a mean stress. This cycle can consist of any combination of upper and lower limits, within the static strength, which are both positive, or both negative. When one stress limit in a cycle is positive and the other negative as shown in Fig. 22.1(*b*) it is known as a reversed cycle. There are two particular cases of the reverse and fluctuating cycles which arise frequently in engineering practice. The first is a symmetrical reversed cycle in which the mean stress is zero and the upper and lower limits are equal positive and

511

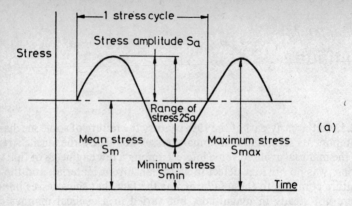

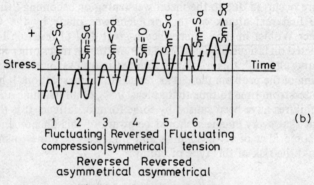

Fig 22.1 Some Stress Cycles

negative. The second is a fluctuating cycle in which the mean stress is half the maximum, the minimum being zero.

The relationship between the various stress values is of some importance. The mean stress, S_m,* is half the algebraic sum of the maximum stress, S_{max}, and the minimum stress, S_{min}. The range of stress, $2S_a$, is the algebraic difference between S_{max} and S_{min}. The ratio of the minimum to the maximum stress is termed the *stress ratio*, R. Hence

$$S_m = \frac{S_{max} + S_{min}}{2}$$

$$2S_a = S_{max} - S_{min}$$

$$R = \frac{S_{min}}{S_{max}}$$

* The symbol, S, is being used for stress to conform with B.S. 3518: Part I: 1962; however, σ is also a recommended symbol (I.S.O.).

The foregoing has only considered stresses as being positive or negative in a general sense. In practice, fatigue can be generated in direct stress due to axial loading or bending, or shear stress due to cyclic torsion, or any combinations of these.

22.3. S–N CURVE

The most readily obtainable information on fatigue behaviour is the relationship between the applied cyclic stress, S, and the number of cycles to failure, N. When plotted in graphical form the result is known as an S–N curve. Fig. 22.2 shows the three ways of plotting the

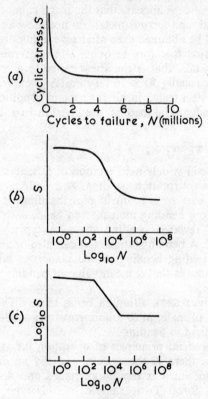

Fig 22.2 Typical Methods for Presenting Fatigue Curves

variables: (a) S vs. N, (b) S vs. $\log_{10} N$, (c) $\log_{10} S$ vs. $\log_{10} N$. The number of cycles to failure at any stress level is termed the *endurance*, and this may vary between a few thousand cycles at high stress and as much as 100 million cycles at low cyclic stresses, for a complete S–N curve. It is immediately apparent that a linear scale for endurance, N, over the whole range is impractical. It is for this reason that the semi-logarithmic or double logarithmic plots (b) and (c) are employed. The former, (b), is the most widely used method of presentation.

The S–N curve as plotted in Figs. 22.2 (b) and (c) reveals three distinctly different regions. There is the shallow sloping portion at very low endurances (discussed in a later section), followed by a more rapid decrease in stress with cycles. At the end of this second region there is a "knee" after which, depending on the material, the curve either becomes parallel to the N-axis or continues with a steadily decreasing slope. Most steels and ferrous alloys exhibit the former types of curve, and the stress range at which the curve becomes horizontal is termed the *fatigue limit*. Below this value it appears that the metal cannot be fractured by fatigue. In general, non-ferrous metals do not show a fatigue limit and fractures can still be obtained even after several hundred million cycles of stress. It is usual therefore to quote what is termed an *endurance limit* for these metals, that is the stress range to give a specific large number of cycles, usually 50×10^6. Typical S–N curves for an aluminium alloy and a steel tested in air for a plain condition, i.e. no stress concentration, under axial loading, are given in Figs. 22.3 (a) and (b).

22.4. FATIGUE TESTING

The earliest and still widely used method of fatigue testing of laboratory specimens is by rotation bending. A cylindrical bar is arranged either as a cantilever or a beam in pure bending; it is then rotated while subjected to a bending moment and hence each fibre of the bar suffers cycles of reversed bending stress. A typical arrangement is shown in Plate X. A bending fatigue test can also be arranged, without rotation, by alternating bending in one plane. An advantage of this test over the former is that a mean value of bending moment can be introduced.

Axial load fatigue tests, although being more difficult to perform, subject a volume of material to a uniform stress condition as opposed to the stress gradient in bending.

There are three main principles of operation for axial load fatigue machines. Firstly, there is the system whereby an electromagnet in series with the specimen is rapidly energized and de-energized, thus applying cyclical force. Two examples of this type of machine are the Amsler "Vibrophore" and the Haigh. In both of these machines, to reduce the amount of power input to the electromagnet to achieve the required load amplitude, a resonant vibration system is employed using springs.

Another method of achieving a resonant condition of springs in series with a specimen is by mechanical means. A small out-of-balance rotating mass is usually employed as in Schenck-type machines.

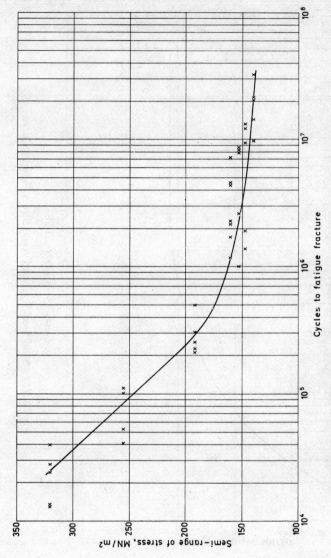

Fig 22.3(a) Aluminium Alloy 24S-T3 Reversed Axial Stress Fatigue Curve

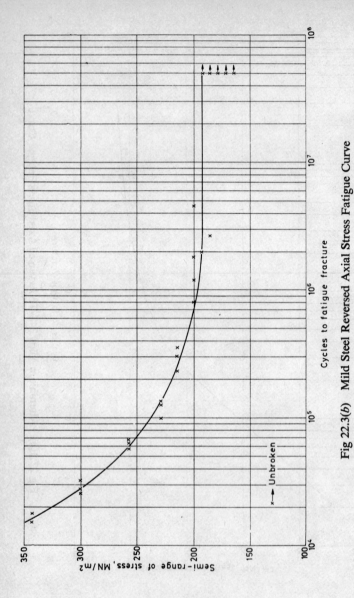

Fig 22.3(b) Mild Steel Reversed Axial Stress Fatigue Curve

The third arrangement for producing fluctuating forces on a specimen is by means of hydraulic pulsation. One, or two, counter-acting pistons in cylinders in series with the test piece are supplied with oil from a pump via a pulsating unit. This consists of a motor-driven piston of variable stroke, which pulsates the oil passing to the main rams, inducing fluctuating pressure and hence force on the specimen. The Losenhausen fatigue machines employ the foregoing principle.

Figs. 22.4 to 22.6 show diagrammatically the arrangement of the three types of fatigue machine described above.

Frequency is an important factor in fatigue testing owing to the large number of cycles it is necessary to achieve at lower stress ranges and the consequent time factor in obtaining an S–N curve. Bending fatigue machines generally run in the range from 30 to 80 Hz. The electrical excitation push–pull machines operate between 50 and 300 Hz, while the mechanical excitation is usually restricted to about 50 Hz. The hydraulic machines are inevitably somewhat slower due to the limitations on pulsating the oil to the required pressures, and this type has a frequency, often variable, in the range from about 3 to 16 Hz. The testing capacity of axial load fatigue machines at present commercially available varies from 20 to 2000 kN.

Cyclic torsion or combined bending and torsion fatigue machines generally operate on the principle of direct mechanical displacement of the specimen by a variable eccentric, crank and connecting rod system. Depending on the capacity of the machine, frequencies vary from 16 to 50 Hz.

22.5. FATIGUE MECHANISM

A great deal of research has been devoted to a study of the mechanism of fatigue, and yet there is still not a complete understanding of the phenomenon. It is not an easy problem to handle theoretically or experimentally, since the process commences within the atomic structure of the metal crystals and develops from the first few cycles of stress, extending over thousands or millions of subsequent cycles to eventual failure.

The fatigue mechanism has two distinct phases, initiation of a crack and the propagation of this crack to final rupture of the material. One of the earliest and classic studies of initiation was made by Ewing and Humfrey, who metallurgically examined the polished surface of a rotating bending specimen at intervals during its fatigue life. They observed that above a certain value of cyclic stress (the fatigue limit) some crystals on the surface of the specimen developed bands during cycling. These bands are the result of sliding or shearing of atomic planes within the crystal and are termed *slip bands*. With continued cyclic action these slip bands broaden and intensify to the point where separation

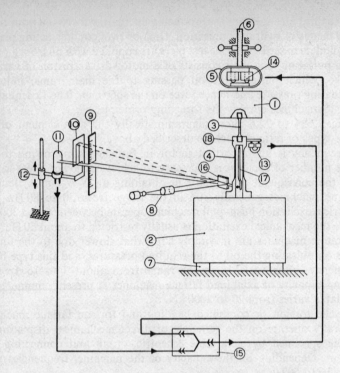

Fig 22.4   Working Principle of the High-frequency Amsler
Vibrophore Fatigue Machine

1. Main moving mass
2. Opposing mass
3. Specimen
4. Dynamometer
5. Pre-load spring
6. Adjusting spindle
7. Vibration insulators
8. Optical projector
9. Dynamometer scale
10. Diaphragm
11. Photoelectric cell
12. Slides of photoelectric cell
13. Impulse generator
14. Driving magnet
15. Amplifier
16. Oscillating mirror
17. Comparison strip
18. Specimen holder

occurs within one of the bands and a crack is formed. The appearance
of fatigue slip bands is illustrated in Plate XI. By a more careful
study using an X-ray technique, Gough showed that the mechanism
of deformation or slip under cyclic stress was the same as for static
stress, and yet a crack could develop at low stresses in fatigue slip which
would not develop under the equivalent static condition. Orowan
suggested that the repeated plastic deformation within slip bands
caused a strainhardening process which, on reaching a limiting value
for the material, had raised the stress in the slip planes to a point where
a crack would form. However, more recently it has been demonstrated

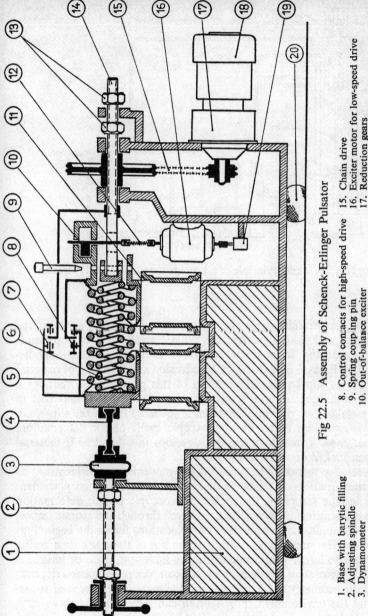

Fig 22.5   Assembly of Schenck-Erlinger Pulsator

1. Base with barytic filling
2. Adjusting spindle
3. Dynamometer
4. Test piece
5. Oscillation spring (outside)
6. Pre-load spring (inside)
7. Control contacts for low-speed drive
8. Control contacts for high-speed drive
9. Spring coup ing pin
10. Out-of-balance exciter
11. Cross spring joints
12. Flexible drive shaft
13. Lock-nuts, lecked for high-speed drive
14. Spindle for low-speed drive
15. Chain drive
16. Exciter motor for low-speed drive
17. Reduction gears
18. Electric motor for low-speed drive
19. Transmitter to cycle counter for high-speed drive
20. Rubber buffers

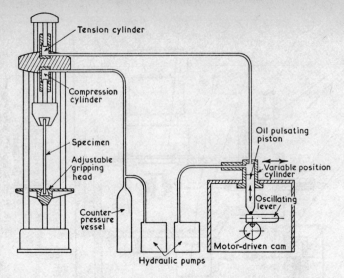

Fig 22.6 Working Principle of a Losenhausen Universal Hydraulic
Fatigue Machine

by Wood that strainhardening occupies only a short initial period of the
endurance, and that thereafter the fatigue mechanism in slip bands
is non-hardening in character. Further interesting behaviour discovered by Forsyth is that known as extrusion and intrusion of material
from fatigue slip bands. Observation of this phenomenon requires a
delicate technique using an electron microscope. It is found that in
some metals a fine ribbon of material is squeezed out of the slip band,
which is extrusion under the microscopic cyclic plastic deformation.
In other cases a trough (intrusion) develops in a band as if material
has been sucked down from the surface.

Present-day theories are based on the movement of dislocations as
being the initiation of fatigue. A dislocation is a fault or misplacement
in the atomic lattice of the metal. Microscopic plastic deformation
allows dislocations or vacancies to "move" through the atomic lattice,
and it is thought that coagulation of dislocations forms the beginnings
of a crack. The stress concentration at the tips of the crack then assists
its growth to larger proportions, and the initiation phase of fatigue is
over. Propagation of a fatigue crack is a complex phenomenon depending on the geometry of the component, the material, the type of stress-
ing, and the environment. Propagation can occupy as much as 90%
of the total endurance, hence movement is relatively slow. Experimental
evidence reveals a discontinuous progress of the crack; it is moving for
some cycles and stationary for some. This effect is also borne out by the

appearance of the fractured specimen, which shows graded markings over the fatigue area as in Plate XII. Other features common to fatigue fractures are the absence of macroscopic plastic deformation in the area of the crack and the relatively smooth surface of the cross-section leading into the deformed region of the final rupture.

22.6. MEAN STRESS

It is quite common to think of fatigue in terms of the range of cyclic stress; however, the mean stress in the cycle has an important influence on fatigue behaviour. There are two obvious limiting conditions for the mean and range of stress. One is for the mean stress equal to the static strength in tension or compression, whence the range of stress must be zero. The other condition is for zero mean stress and a stress range equal to twice the static strength. Between these boundaries there is an infinite number of combinations of mean and range of stress. It is obviously

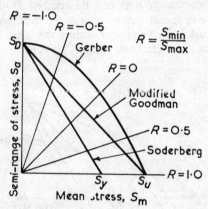

Fig 22.7   Diagrams for Mean Stress vs. Range of Stress

impossible to study the problem experimentally completely; consequently, empirical laws have been developed to represent the variation of mean and range of stress, in terms of static strength values and the fatigue curve for reversed stress (zero mean). This latter condition is the most widely used for obtaining experimental data, principally because of the simplicity of the rotating beam test.

If a diagram is plotted of the semi-range of stress as ordinate and the mean stress as abscissa, known as an S_a–S_m diagram, as in Fig. 22.7, then the limiting conditions are $S_m = 0$ and $S_a = S_D$, the fatigue limit for reversed stress, and $S_a = 0$ and $S_m = S_u$. A straight line joining these pairs of co-ordinates represents one empirical law known as the modified Goodman relationship. Algebraically this is given by

$$S_a = S_D \left(1 - \frac{S_m}{S_u}\right)$$

Another relationship is obtained by joining the limiting coordinates by a parabola known as the Gerber parabola. This is expressed algebraically in the above symbols as

$$S_a = S_D \left[1 - \left(\frac{S_m}{S_u}\right)^2\right]$$

A more conservative line for design purposes was proposed by Soderberg using a yield stress instead of the tensile strength in the Goodman relationship:

$$S_a = S_D \left(1 - \frac{S_m}{S_y}\right)$$

The simplest conclusion that can be reached from the foregoing is that an increase in tensile mean stress in the cycle reduces the allowable range of stress for a particular endurance. This applies similarly for direct stress or shear stress (torsional) fatigue.

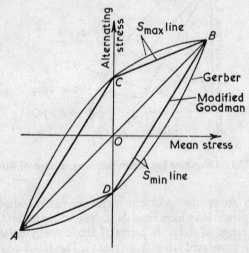

Fig 22.8   Alternative Presentation of S_m–S_a Diagrams

Compressive mean stresses appear to cause little or no reduction in stress range, some materials even show an increase; consequently the S_a–S_m diagram is not symmetrical about the zero mean stress axis.

Another way of plotting a mean-stress/alternating-stress relationship, favoured by the German Standards, is as shown in Fig. 22.8.

The ordinates represent the alternating S_{max} and S_{min} values and the abscissa the mean stress condition. The 45° line AOB joins the tensile and compressive strength passing through the origin. C and D are the S_{max} and S_{min} values for reversed stress at the fatigue limit. The locus is then completed either by parabolas or straight lines. Any stress falling inside the locus will be safe, while outside it will be liable to fatigue failure, when the diagram relates to the fatigue or endurance limit region.

Further S_a–S_m diagrams could be drawn having different boundary values relating to lower endurances, the limiting case being when the ordinate CD equals the ordinate distance between A and B for one cycle endurance. (See B.S. 3518: Part I for details of the Haigh, Smith and Ros diagrams.)

22.7. STRESS CONCENTRATION

Probably the most serious effect in fatigue is that of stress concentration. It is virtually impossible to design any component in a machine without some discontinuity such as a hole, keyway, or change of section. These features are known as *stress raisers* or sources of stress concentration. This concept was introduced in Chapter 17, and it was explained that under static loading in the elastic range the local peak stress at a notch or discontinuity is raised in magnitude above the nominal stress on a cross-section away from the notch. The theoretical elastic stress concentration factor, K_t, for a notch is defined as the maximum stress at the notch divided by the average stress on the minimum area of cross-section at the notch.

For ductile metals, static stress concentration does not reduce the strength owing to redistribution of stress when the material at the notch enters the plastic range. However, under fatigue loading the position is very different, since a fatigue limit tends to correspond with the static elastic limit of the material. Consequently, the fatigue limit for the material having the peak stress at the discontinuity would correspond to the static elastic limit, and hence the fatigue limit based on the nominal stress would be reduced by a factor dependent on the elastic concentration factor. In practice this is generally not so and the notched fatigue strength is rather better than the "plain" fatigue strength divided by K_t. This has led to what is termed the *fatigue strength reduction factor*, which is defined as

$$K_f = \frac{\text{Plain fatigue strength at } N \text{ cycles or fatigue limit}}{\text{Notched fatigue strength at } N \text{ cycles or fatigue limit}}$$

and generally $K_t > K_f > 1$. Fig. 22.9 illustrates the variation in K_f

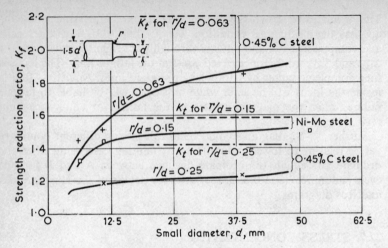

Fig 22.9 Relation between K_t and K_f for Fillets, Specimens
Geometrically Similar
(*Adapted from Lessels: by courtesy of John Wiley & Sons Ltd.*)

with size of component and radius of fillet in relation to K_t. It is seen
that K_f approaches K_t as the size of the bar is increased.

Another way of representing material reaction to stress concentration
in fatigue is by means of a *notch sensitivity factor q*, which is defined in
terms of the theoretical stress concentration factor and the fatigue
strength reduction factor. Thus

$$q = \frac{K_f - 1}{K_t - 1}$$

and for a component or specimen which yields K_f in fatigue equal to
K_t, the factor $q = 1$ shows maximum sensitivity. Where there is no
strength reduction, $K_f = 1$ and q becomes zero showing no sensitivity.
Peterson has produced curves of sensitivity factors for various notch
geometries and radii in different conditions of steel, and these are
reproduced in Fig. 22.10.

22.8. TEMPERATURE

Many components are subjected to fatigue conditions while working
at temperatures other than ambient. Components in aeroplanes at
high altitude may experience many degrees of frost, while a steam or
gas turbine will be running at several hundred degrees Celsius.

Tests on a number of aluminium and steel alloys ranging down to
$-50°C$, and lower in some cases, have shown that the fatigue strength
is as good as and often a little better than at normal temperature $+20°C$.

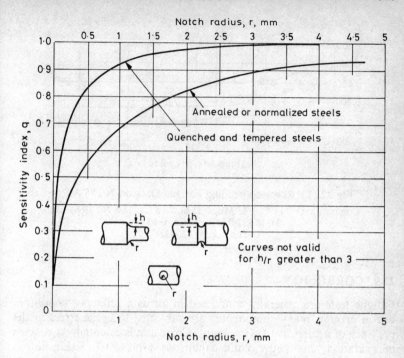

Fig 22.10   Sensitivity Indices for Various Structural Conditions
of Steel
(*After R. E. Peterson*)

On the other hand, fatigue tests at higher temperatures show little
or no effect up to about 300°C, after which for steels to about 400°C
there is an increase in fatigue strength to a maximum value, followed by
a rapid fall to values well below that at +20°C. A further interesting
feature is that a material having a fatigue limit characteristic at ambient
temperature will lose this at high temperatures, and the *S–N* curve will
continue to fall slightly even at high endurances, and an endurance
limit has to be quoted. *S–N* curves for a nickel-chrome alloy at various
temperatures are shown in Fig. 22.11.

Above a certain temperature there is interaction between fatigue
and creep effects (see Chapter 23), and it is found that up to, say,
700°C for heat-resistant alloy steels, fatigue is the criterion of fracture,
whereas at higher temperatures, creep becomes the cause of failure.
This has led to the use of combined fatigue–creep diagrams, where the
cyclic stress is plotted against the steady or creep stress for various
temperatures and endurances (hours).

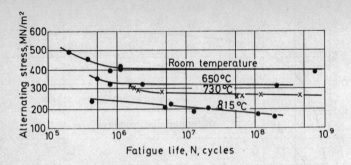

Fig 22.11   Reversed-bending Fatigue Data on N-155 Alloy
Composition: 0·14 C, 1·43 Mn, 0·35 Si, 20·8 Cr, 18·8 Ni, 19·65 Co,
3·0 Mo, 2·0 W, 0·99 Cb, 0·13 Sm,
Test speed: 120 Hz

22.9. CORROSION

Fatigue tests are generally conducted in air as a reference condition, but in practice many components are subjected to cyclic stress in the presence of a corrosive environment. Corrosion is essentially a process of oxidation, and under static conditions a protective oxide film is formed which tends to retard further corrosion attack. In the presence of cyclic stress the situation is very different, since the partly protective oxide film is ruptured in every cycle allowing further attack. A rather simplified explanation of the corrosion fatigue mechanism is that the microstructure at the surface of the metal is attacked by the corrosive, causing an easier and more rapid initiation of cracks. The stress concentration at the tips of fissures breaks the oxide film and the corrosive in the crack acts as a form of electrolyte with the tip of the crack becoming an anode from which material is removed, thus assisting the propagation under fatigue action. It has been shown that the separate effects of corrosion and fatigue when added do not cause as serious a reduction in strength as the two conditions acting simultaneously.

One of the important aspects of corrosion fatigue is that a metal having a fatigue limit in air no longer possesses one in the corrosive environment, and fractures can be obtained at very low stress after hundreds of millions of cycles.

Gough and Sopwith and also McAdam conducted the pioneer work in corrosion fatigue, and some results of the former are given in Table 22.1, which illustrates the very considerable lowering of fatigue strength in the presence of a salt spray. The increased resistance of the two bronzes is probably due to the cooling of the spray dissipating hysteresis effects, since these metals are inherently very corrosion resistant.

Table 22.1

Corrosion Fatigue of Metals

Material	Tensile strength, MN/m²	Fatigue limit, MN/m² (air)	Corrosion fatigue at 50×10^6 cycles, MN/m² (salt spray)
Mild steel	(470)	262	172
0·5% C steel	980	386	604
15% Cr steel	669	380	142
17% Cr, 1% Ni steel	841	507	142
18% Cr, 8% Ni steel	1020	369	245
Beryllium bronze	649	252	268
Phosphor bronze	428	152	180
Aluminium bronze	552	221	152
Duralumin	435	141	52
Manganese alloy	255	105	negligible

Corrosion fatigue is time as well as cycle dependent and therefore frequency of testing is an important factor, since there is less time involved at a high frequency to reach a particular endurance, thus giving a greater fatigue strength.

A considerable study has been made of the value of protective surface coatings in increasing corrosion fatigue strength. The two categories of coating are metallic and non-metallic. For many years the former has proved the more successful, but in recent times the development of plastics and synthetic rubbers has provided new possibilities for protection. In a recent programme one of the authors has found that nylon, phenolic-epoxy-silicone resin and other polymers achieved complete protection up to 50×10^6 cycles, at a reversed torsional stress just below the air fatigue limit, for a spring steel in a brine atmosphere. Some of Gough's results for a variety of coatings are reproduced in Table 22.2.

22.10. CUMULATIVE DAMAGE

Although most fatigue tests and some components are subjected to a constant amplitude of cyclic stress during the life to fracture, there are many instances of machine parts and structures which receive a load spectrum, that is, the load and cyclic stress vary in some way under working conditions. To establish any difference between fatigue under varying and constant amplitude conditions, tests are conducted in

Table 22.2
Corrosion Fatigue on Coated 0·5% Carbon Steel

Coating	Fatigue strength at 10^7 cycles, MN/m²	
	Air	Salt spray
None	253	62
Enamel	265	172
Galvanized	228	255
Sherardized	228	234
Zinc-plated	248	228
Cadmium-plated	234	212
Cadmium and enamel	244	207
Cadmium and oil	244	207
Phosphate-treated	273	200

which a certain number of cycles is done at one stress level followed by a number at a higher or lower stress, and this sequence is repeated till failure occurs. It is suggested that damage by fatigue action accumulates and that a certain total damage line is represented by the *S–N*

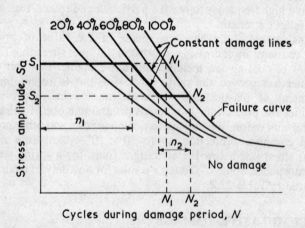

Fig 22.12   Diagrammatic Representation of Cumulative Damage

curve. One way of representing this algebraically was proposed by Miner. If n_1 cycles are conducted at a stress level S_1, at which the fracture endurance would be N_1, and if this is followed by n_2 cycles out of N_2 at a second stress level S_2, as in Fig. 22.12, and so on, then

$$\frac{n_1}{N_1} + \frac{n_2}{N_2} + \frac{n_3}{N_3} + \ldots = 1$$

or

$$\Sigma \frac{n}{N} = 1$$

Test results often show the sum of the cycle ratios (n/N) differing widely from the value of unity, generally covering a range from about 0·6 to 2·0 with, in a few cases, extreme values well outside this range. However, the average of series of tests is often quite close to 1·0, and with the present state of knowledge, more complex laws have yet to be developed and satisfactorily proved. The problem is complicated by the apparent dependence of damage on stress history, mean stress, stress concentration, etc. For example, it is often found that for a two-stress level test, in which one stress is applied for a number of cycles and then run to failure at a second stress, if $S_1 < S_2$, then $\Sigma(n/N) > 1$, and for $S_1 > S_2$, $\Sigma(n/N) < 1$. In addition, the variation from unity is greater for larger differences between S_1 and S_2. In spite of these difficulties, Miner's cumulative damage hypothesis still provides a useful starting point for the analysis of fatigue under variable loading spectra.

22.11. FREQUENCY

It has been shown that for cycle frequencies between about 3 and 100 Hz there is no variation in fatigue strength. At higher frequencies ranging up to 10 kHz there appears to be some increase in fatigue strength of the order of 10 to 20%. There is evidence that at very low frequencies fatigue strength in the finite range falls about 10%.

22.12. LOW ENDURANCE FATIGUE

The discussion of fatigue so far has been concerned with endurances in the range of 10^5 to 10^8 cycles. However, the advances in science and engineering during the last two decades have required the use of metals under very much more severe working conditions of stress, temperature, etc. Owing to high stress and strain conditions, designers now have to consider the possibilities of fatigue occurring in the range from 100 to 10 000 cycles. These low numbers of cycles may not imply a short life in time—a pressure vessel may work for 25 years before achieving 1000 cycles of cleaning and inspection.

In low endurance fatigue there is marked plastic deformation (hysteresis) in every cycle, and due to strainhardening (or softening) effects it becomes of some significance whether the cycle is of constant load (stress) amplitude or constant strain amplitude, a problem which does not arise at long endurance and hence low stress. A good example of strain cycling, which has come into prominence with the nuclear power and rocket era, is the effect of repeated thermal changes in a component.

At low endurances the $S-N$ and $\varepsilon-N$ curves take the form shown in Fig. 22.13. It is usual to consider the $S-N$ curve as starting from a point at a $\frac{1}{4}$ cycle representing a tension test or single pull to fracture. Likewise the $\varepsilon-N$ curve is started at a $\frac{1}{4}$ cycle using the true fracture ductility in simple tension as the ε value. Much recent work has been

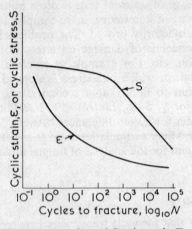

Fig 22.13   Typical Stress-cycle and Strain-cycle Fatigue Curves at Low Endurance

concentrated on strain cycle testing, and it has been found that, if the plastic strain range, ε_p, or width of the hysteresis loop, is determined during the life of the specimen, then a relationship of the form, $\varepsilon_p N^\alpha =$ constant, holds for most metals up to 10^5 cycles. It is seen that on a log ε_p–log N graph a straight line results. It has further been demonstrated that α is between 0·5 and 0·6 for many metals at room temperature, and the constant is broadly related to the ductility of the material, i.e. the greater the latter, the larger the constant term. This is illustrated in log ε_p–log N curves in Fig. 22.14 for a variety of metals.

Thermal strain cycling, that is, cyclic strains induced by changes in temperature, appears to yield the same results as if the specimen or component were kept at a constant temperature of a value equal to the upper limit of the previous temperature cycle, and the cyclic strains induced mechanically.

Discontinuities in form, notches, holes, etc., have only a little strength reduction effect at low endurances for stress cycling conditions, far less than at long endurances.

Under strain cycle testing the position is very different; strength reduction is at least maintained, or becomes slightly worse at short endurances. The different behaviour in these two types of test is probably attributable to stress redistribution during plastic deformation in stress cycling reducing the stress concentration effect. In strain cycling, there appears to be no relief with plastic deformation, the strain concentration being maintained.

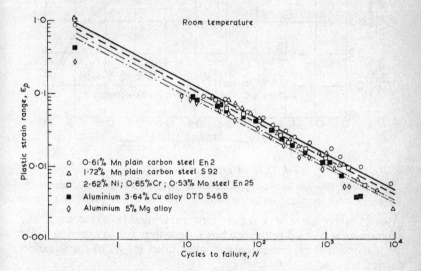

Fig 22.14   Plastic Strain Range vs. Cycles to Failure for Various Materials

(A. C. Low)

22.13. THEORIES OF FATIGUE UNDER COMBINED STRESS

In discussion of fatigue so far, only uniaxial stresses have been considered, but in practice biaxial or triaxial cyclic stresses may often be encountered. Most of the experimental evidence to date from combined stress fatigue tests, generally in bending and torsion, shows that for ductile metals the shear strain energy is the criterion of failure. Some results for two steels under combined cyclic bending and torsion are given in Fig. 22.15.

The equation (15.20) for shear strain energy developed in Chapter 15 may be used for fatigue, if the fatigue limit in reversed bending is substituted for the yield stress in simple tension, so that

$$\sigma_1{}^2 + \sigma_2{}^2 - \sigma_1\sigma_2 = S_D{}^2$$

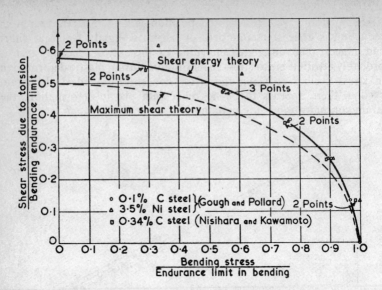

Fig 22.15 Combined Stress Tests

(From R. E. Peterson: by courtesy of the Society for Experimental Stress Analysis)

In cases where there is a mean stress present, then either an S_a–S_m diagram must be used or the bending fatigue limit found experimentally for that mean stress condition and substituted for S_D in the equation above.

BIBLIOGRAPHY

Forrest, P. G., *Fatigue of Metals* (Pergamon, 1962).
Heywood, R. B., *Designing against Fatigue* (Chapman & Hall, 1962).
Lessels, J. M., *Strength and Resistance of Metals* (Wiley, 1954).
Pope, J., *Metal Fatigue* (Chapman & Hall, 1959).

Chapter 23
Creep

23.1. Consider the room temperature tensile-stress/strain curve shown in Fig. 23.1 in which, at some point in the plastic range of the material, loading is held constant at A. It is observed that the metal continues to stretch slightly at this constant load until after a few seconds or minutes equilibrium is reached at B. This time-dependent increase in strain AB at constant load is known as *creep*.

If, on the other hand, instead of load, the strain is held constant at A as in Fig. 23.1, it is found that there is a drop in load with time to equilibrium at point C. This behaviour is termed *creep-stress relaxation*.

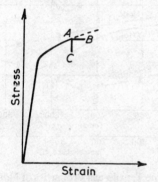

Fig 23.1   Room-temperature Creep and Relaxation Effects

Both of the above phenomena are very much more pronounced at high temperature and can operate continuously over very long periods (years). In creep, fracture may occur eventually, and although stress relaxation cannot result directly in fracture it can cause serious side effects which may be detrimental to the running of plant.

The increasing number and range of thermal problems in fields such as steam plant, gas turbine, nuclear and chemical engineering and kinetic heating in aeronautics necessitate a considerable knowledge of the creep properties of materials.

23.2. INFLUENCE OF HIGH TEMPERATURE ON THE TENSILE TEST

Experiments have shown that, if a continuous tensile test at normal straining rates is conducted at temperatures above ambient, there is a considerable change in strength and elasticity properties. Some metals, notably carbon steels, show an initial increase in tensile strength up to 300°C, after which there is a rapid fall to about 50% of the ambient strength at 500 to 600°C. In general for most metals, including carbon steels, yield point, proportional limit, and modulus of elasticity decrease immediately with rise in temperature while ductility may increase somewhat. Fig. 23.2 shows the relationships between the above properties and temperature for a medium-carbon steel.

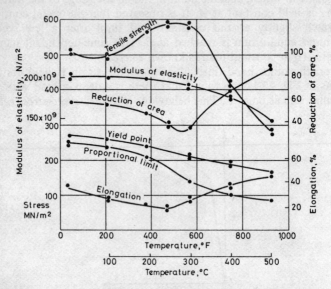

Fig 23.2   Short-time Tensile-test Properties of Normalized Medium-carbon (0·37 per cent Carbon) Steel at Elevated Temperatures
(After P. G. McVetty)

(From Lessels: by courtesy of John Wiley & Sons Ltd.)

If tests are conducted at one temperature, but the straining rate though continuous is decreased in successive tests, it is again found that strength is reduced much as with increasing temperature in the previous case. Table 23.1 illustrates this effect for various metals tested by Tapsell.

From the foregoing it is seen that the combined effects of high temperature and time of load application are very detrimental to the tensile properties of metals. Since metals in service are subjected to stress at high temperature for periods of up to 100 000 hours, long-term creep tests are necessary to assess stress and temperature limits for a given allowable creep strain.

Table 23.1

Material	Temperature, °C	Tensile strength, MN/m²		
			Length of test	
		2–10 min	5 days	25 days
0·3% C (cast)	555	262	—	60
0·5% C	400	548	362	338
0·5% C	500	376	155	124
3·0% Ni, 0·25% Cr	400	593	370	324
3·0% Ni, 0·25% Cr	500	424	138	108
3·0% Ni, 0·25% Cr	600	255	46	25
Ni-Cr-Si	600	510	—	138
Ni-Cr-Si	700	441	—	69
Ni-Cr-Si	800	225	—	26
Duralumin	150	335	262	224
Duralumin	250	162	77	57
Duralumin	350	51	22	19
Brass (60–40)	150	359	324	293
Brass	250	276	108	62
Lead	20	26 (1¼ mm)	8	—

23.3. LONG-TERM CREEP TESTS

As long ago as 1910, Andrade conducted creep experiments on lead at room temperature which produced the same effects as high temperature on other metals. The procedure in this type of test is to apply a constant tensile load to a cylindrical specimen, which is surrounded by a furnace to maintain accurately a constant temperature. Measurements are taken of extension at frequent intervals of time until the specimen fractures, or the test is stopped after a sufficiently lengthy period. Typical curves of creep strain against time are plotted in Fig. 23.3 for various nominal stress levels at constant temperature. After the initial strain due to the application of load, there is a gradually diminishing creep rate, which in the case of higher stresses is followed by a rapid increase in creep rate to fracture. At lower stresses the final stage is preceded by a region of constant creep rate, the length of which depends on the stress value. It appears that there might be a limiting creep stress, below which $d\varepsilon/dt = 0$, having the same sort of significance as the fatigue limit under cyclic stress. It has been shown that there is no reliable criterion of this form for creep, and a "limiting creep stress" is nowadays based on a permissible creep strain after a given time.

There are four principal aspects of each of the curves shown in Fig. 23.3: (a) initial strain, elastic and sometimes partly plastic, due to the first application of load; (b) primary stage, a period of decreasing creep rate; (c) secondary stage, where the creep rate is sensibly constant; (d) tertiary stage, an increasing rate of strain leading to fracture.

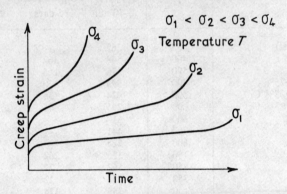

Fig 23.3 Typical Creep Curves for Various Stresses at Constant Temperature

A physical interpretation of the above phases may be summarized as follows:

(b) Primary stage: strainhardening is occurring more rapidly than softening by the high temperature, resulting in a decreasing rate of creep. Slip bands can be seen in the crystal structure, taking a form similar to that in room temperature plastic flow.

(c) Secondary stage: the constant rate of creep appears to take place through equilibrium being obtained between strainhardening and thermal softening. Structural deformation is quite different from that of the primary stage, severe distortion occurring in the intercrystalline material.

(d) Tertiary stage: by this time, the stress level has increased slightly owing to reduction in area, and stress concentrations at cracks opening up in the grain boundaries contribute to an accelerating creep condition resulting in fracture.

One important feature to note at this stage is that the elongation to fracture in creep, even for a ductile metal, is only a fraction of that obtained for continuous loading to fracture at high temperature.

Andrade was the first to propose a relationship between creep strain and time to describe primary and secondary creep, and this took the form

$$\varepsilon = \alpha t^{1/3} + \beta t \qquad (23.1)$$

where t is time and α and β are material constants which are a function of stress and temperature. Graham and Walles extended the analysis to cover the tertiary stage by proposing a relationship of the form

$$\varepsilon = \alpha t^{1/3} + \beta t + \gamma t^3 \qquad (23.2)$$

where γ is a further material constant dependent on stress and temperature.

The family of curves shown in Fig. 23.3 was based on stress as the parameter and temperature constant; however, a similar family of curves would be obtained if a constant stress was considered and temperature became the parameter. Thus a complete picture of the creep behaviour of a metal would necessitate the construction of several families of creep curves for various stresses and temperatures.

23.4. INTERPRETATION OF CREEP DATA

Typical of the standards of creep strength required for metals are the stresses to give minimum creep rates of 1 % strain in 10 000 hours, or 1 % strain in 100 000 hours.

A 10 000-hour test occupies approximately 1 year, and it is therefore evident that, although a few creep tests may be conducted over periods of this length or longer, it is a very slow and inconvenient process for obtaining a range of data at different stresses and temperatures. As a result, methods have been sought whereby long-life data can be extrapolated from short-term tests. Great care has to be exercised in this approach, since very small variations in temperature or metallurgical condition can cause completely erroneous extrapolated values.

The four basic quantities which must be analysed against each other are creep strain, ε, time, t, stress, σ, and temperature, T. Interest centres principally on the secondary stage of creep, where the rate is constant for very long times, producing the major contribution to the total creep strain to fracture. Consequently in eqn. (23.2) the tertiary term is neglected and the primary is replaced by constant ε_0, being the intercept of the extrapolated secondary stage on to the strain axis, Fig. 23.4. Thus

$$\varepsilon_t = \varepsilon_0 + \dot{\varepsilon} t \qquad (23.3)$$

where $\dot{\varepsilon} = d\varepsilon/dt$, the minimum creep rate.

It has been found experimentally that the minimum creep rate and stress are related as

$$\dot{\varepsilon} = K\sigma^n \qquad (23.4)$$

Hence a log-log plot of $\dot{\varepsilon}$ against σ yields a straight line. Initial estimates

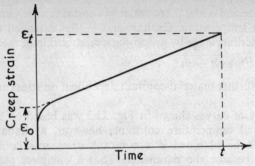

Fig 23.4 Simplified Creep Curve

can be made of the stress to give a particular creep rate by extrapolation. Alternatively, the constants K and n can be determined from the straight line of $\log \dot{\varepsilon}$ against $\log \sigma$, and the time to reach a specified value of total creep strain in the secondary stage is given by

$$t = \frac{\varepsilon_t - \varepsilon_0}{K\sigma^n} \tag{23.5}$$

using eqns. (23.3) and (23.4). The difficulty of this approach is that tertiary creep may commence in practice before the predicted value of secondary creep can be achieved.

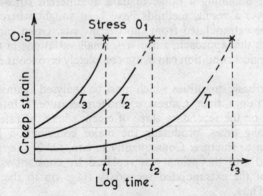

Fig 23.5 Creep Strain vs. Log Time Curves, Extrapolated to a Specified Strain

Another approach to short-time testing is to conduct creep tests at various stresses and temperatures such that a part of the secondary stage is obtained. Graphs of creep strain against log time are then extrapolated to the limiting value of creep strain required, say 0·5%, as in Fig. 23.5, for constant temperature T. Similar graphs obtained at other temperatures then enable cross-plotting of temperature against log time with stress as parameter, all related to the required creep strain of 0·5% as illustrated in Fig. 23.6. These curves can then be extrapolated to a required life of, say, 100 000 hours, and finally a relationship can be plotted, using Fig. 23.6, for stress against temperature which gives 0·5% creep strain in 100 000 hours. Although several extrapolations are involved in this method they tend to be rather more reliable than the simple extrapolation of a creep–time curve.

As there is as yet no real substitute for a few long-term tests to ensure reliable creep knowledge, it is useful to have a quick sorting test to enable the best material from a group to be selected for long-term tests.

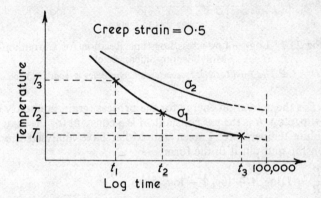

Fig 23.6 Temperature vs. Log Time Curves, Extrapolated to a Specified Time

The *stress-rupture test* is widely used for the above purpose and also as a guide to the rupture strength at very long endurances. The principle is to apply various values of stress, in successive tests at constant temperature, of a magnitude sufficient to cause rupture in times from a few minutes to several hundred hours. Plotting log stress against log time as in Fig. 23.7 yields a family of straight lines with temperature as parameter. These appear to extrapolate back to the respective hot tensile strengths. Extrapolation forward to longer times is possible, but care has to be exercised in that oxidation owing to the high temperature can cause a marked increase in the slope, and hence reduction in stress for a required life.

One of the physical approaches to the problem of creep suggests that viscous flow in fluids is analogous to secondary creep in metals and hence a rate-process theory is applicable, relating creep rate and temperature, of the form

$$\dot{\varepsilon} = \frac{d\varepsilon}{dt} = Ae^{-E/RT} \qquad (23.6)$$

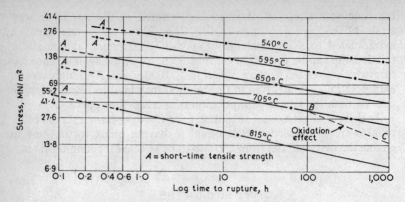

Fig 23.7   Log-rupture-stress/Log-time Relation for Chromium-Molybdenum-Silicon Steel

(Adapted from Lessels: by courtesy of John Wiley & Sons Ltd.)

where E is the activation energy for the process (grain boundary movement, slip, etc.), R is the gas constant, A is a constant for the material, T is absolute temperature. Larsen and Miller have analysed the above relationship and put it in the form

$$\frac{E}{R} = T(\log_e A + \log_e t - \log_e \varepsilon) \qquad (23.7)$$

or

$$\left(\frac{E}{R}\right)_\sigma = T(B + \log_e t) \qquad (23.8)$$

for a given value of strain. B is a constant for a given strain and $(E/R)_\sigma$ is a function of the stress level, σ. The right-hand side of eqn. (23.8) is known as the *Larsen–Miller parameter*, $T(B + \log_e t)$, and plotting $\log_e \sigma$ against the parameter often yields a family of straight lines for different creep strains, known as *master creep curves*, which correlate well over a wide range of times, temperatures and different metals. From the above curves a general relationship may be written in the form

$$\log_e \sigma = C_1 + C_2 T(B + \log_e t) \qquad (23.9)$$

where C_1 and C_2 are constants. Combining the strain-rate/stress equation

$$\dot{\varepsilon} = K\sigma^n$$

with that above for stress, temperature and time gives

$$\dot{\varepsilon} = C\sigma^{n'} e^{-\alpha/T} \qquad (23.10)$$

where C, α and n' are material creep constants.

23.5. CREEP TESTING EQUIPMENT

The most common form of creep test is conducted in simple tension, and since constant loading on the specimen is required over very long periods a dead weight loading system is usually employed. A typical arrangement is shown in Fig. 23.8, where the lower end of the test piece is gripped in screwed or colleted chucks attached to a positioning screw. The upper end of the test piece is hung from a beam pivoted on the frame of the machine. The load is suspended from the free end of the beam, and below the weight pan is located a microswitch so that, when

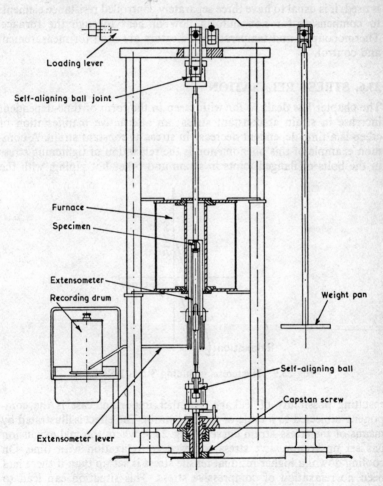

Fig 23.8   Arrangement of a Tension Creep Machine

(Electronic and Mechanical Engineering Co. Ltd.)

the specimen is fractured or has suffered a specified extension, the magnified movement of the load operates the microswitch and cuts out the furnace. This item together with the control circuit is perhaps the most important part of the apparatus. In previous sections it has been stated that slight temperature variation has a great influence on the minimum creep rate; it is therefore essential to have uniform and constant temperature along the length of the specimen. The relevant British Standard specifies a maximum variation along the gauge length of 2°C and a variation of mean temperature of not more than ±1°C up to 600°C and ±2°C from 600°C to 1000°C. If a resistance furnace is used, it is usual to have three separately controlled resistance elements to compensate for non-uniform flow of heat through the furnace. Thermocouples and temperature indicators are used for measurement and control.

23.6. STRESS RELAXATION

The chapter has dealt so far with creep in the form of time-dependent increase in strain at constant stress; an alternative manifestation of creep is a time-dependent decrease in stress at constant strain. A common example of this phenomenon is the relaxation of tightening stress in the bolts of flanged joints in steam and other hot piping, with the

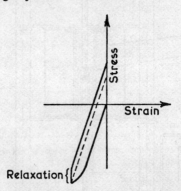

Fig 23.9   Hysteresis including Stress Relaxation

resulting possibility of leakage. Another important case is the component subjected to a cycle of thermal strain. The effect is illustrated by means of the stress–strain curve in Fig. 23.9, where thermal expansion has set up compressive stress followed by relaxation with time. On cooling down, a higher residual tensile stress is set up than if there had been no relaxation of compressive stress. This situation can lead to failure eventually in a metal such as a flake cast-iron which is weak in tension. Thermal strain concentration around nozzle openings in pressure vessels is another example where reversal of stress after relaxation could in time lead to a thermal fatigue failure.

A stress-relaxation/time curve is similar to a mirror image of a creep-strain/time curve as shown in Fig. 23.10. Just as much care has to be taken with relaxation testing as with conventional creep testing since results are very sensitive to temperature variation. The procedure usually adopted is to load the specimen to an initial stress which will give a specified strain of, say, 0·15%. The stress is then adjusted with time so that the specified strain is maintained.

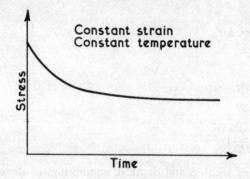

Fig 23.10   Relaxation of Stress with Time, at Constant Strain and Temperature

In the Barr and Bardgett types of test the decrease in stress is noted after 48 hours for various initial stresses at constant temperature. If the decrease in stress is plotted against initial stress, the intercept of the curve on the latter axis gives the initial stress required for "zero" decrease in stress, i.e. no relaxation. However, this is a very short-term test and can only safely be used for sorting materials.

23.7. ELEMENTARY DESIGN PROBLEMS INVOLVING CREEP

Consideration will now be given to the application of the creep equation (23.4) in simple design problems for high-temperature service.

(i) Creep during Pure Bending of a Rectangular Bar

The assumptions that will be made in this problem are as follows: (a) plane sections remain plane under creep deformation; (b) longitudinal fibres experience only simple axial stress; (c) creep behaviour is the same in tension as in compression.

If the radius of curvature of the neutral axis is R, then the strain at a distance y from the axis is

$$\varepsilon = \frac{y}{R} \tag{23.11}$$

and for creep in the secondary stage,

$$\frac{d\varepsilon}{dt} = K\sigma^n$$

Therefore

$$\frac{d(y/R)}{dt} = K\sigma^n$$

or

$$\frac{y}{R} = K\sigma^n t$$

and

$$\sigma = \left(\frac{y}{RKt}\right)^{1/n} \tag{23.12}$$

If the bar is of width b and depth d, equilibrium of the external and internal moments is given by

$$M = 2\int_0^{d/2} \sigma b y \, dy \tag{23.13}$$

Substituting for σ, we have

$$M = 2\int_0^{d/2} \left(\frac{y}{RKt}\right)^{1/n} b y \, dy = \frac{2b}{(RKt)^{1/n}} \int_0^{d/2} y^{1+(1/n)} \, dy$$

Integration gives

$$M = \frac{2b}{(RKt)^{1/n}} \frac{n}{2n+1} \left(\frac{d}{2}\right)^{2+(1/n)} \tag{23.14}$$

Substituting for $(RKt)^{1/n}$ from eqn. (23.12) gives

$$M = \frac{2b\sigma}{y^{1/n}} \frac{n}{2n+1} \left(\frac{d}{2}\right)^{2+(1/n)} \tag{23.15}$$

Inserting the second moment of area $I = \frac{1}{12}bd^3$ and rearranging, we have the bending stress at distance y from the neutral axis as

$$\sigma = \frac{My}{I} \frac{2n+1}{3n} \left(\frac{2y}{d}\right)^{(1/n)-1} \tag{23.16}$$

It has been put in this form to show the difference from the simple linear elastic distribution of stress, My/I.

The stress distribution across the section is shown in Fig. 23.11 for $n = 1$, which corresponds to no creep and simple elastic conditions, and $n = 10$, where secondary-stage creep is occurring. It is seen that the effect of creep is to relax the outer fibres' stress, but to increase the core bending stress. This is because minimum creep rate must be proportional to the distance from the neutral axis in order to satisfy eqn. (23.12), and this automatically adjusts the stresses to the distribution shown.

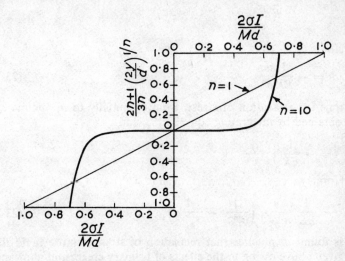

Fig 23.11   Distribution of Bending Stress for the Cases of Zero Creep and Secondary-stage Creep

(ii) Creep Relaxation in a Bolt

Bolted flanged joints at high temperature represent an example of the problem of stress relaxation. A bolt is initially tightened to a stress σ_0, producing an elastic strain ε_0, and to simplify the problem it is assumed that the flange is not deformed by the bolt stress. After a period of time, t, the effect of creep is to induce plastic deformation or creep strain, ε, which allows a relaxation of stress, σ, and elastic strain. Now, the total strain must remain the same if the flange is rigid; therefore

$$\varepsilon_0 = \varepsilon + \frac{\sigma}{E} \qquad (23.17)$$

or differentiating with respect to time,

$$0 = \frac{d\varepsilon}{dt} + \frac{1}{E}\frac{d\sigma}{dt}$$

or

$$\dot{\varepsilon} = -\frac{1}{E}\frac{d\sigma}{dt} \tag{23.18}$$

Substituting for $\dot{\varepsilon}$ in terms of stress, we have

$$K\sigma^n = -\frac{1}{E}\frac{d\sigma}{dt}$$

Therefore

$$dt = -\frac{1}{EK}\frac{d\sigma}{\sigma^n} \tag{23.19}$$

The time for relaxation of stress, from σ_0 initially to σ_t at time t, is then obtained by integrating eqn. (23.19), and

$$t = -\frac{1}{EK}\int_{\sigma_0}^{\sigma_t}\frac{d\sigma}{\sigma^n}$$

Therefore

$$t = \frac{1}{EK}\ \frac{1}{n-1}\left(\frac{1}{\sigma_t^{n-1}} - \frac{1}{\sigma_0^{n-1}}\right) \tag{23.20}$$

It is found in practice that relaxation of stress is more rapid than that given above owing to the effects of primary creep, and allowances for this can only be made by using a more complex creep–rate–stress–time function.

(iii) Creep in a Thin Tube under Internal Pressure

The case of creep in a biaxial stress system is an important one and there has been only a very limited number of investigations in this field. The present example will serve to illustrate the theoretical assumptions and approach that have been made so far.

In order to relate uniaxial creep data to a biaxial problem, some of the laws of plasticity are invoked, namely that (a) the principal strains and stresses are coincident in direction, (b) plastic deformation occurs at constant volume, (c) the maximum shear stresses and shear strains are proportional.

For constant volume the sum of the three principal strains is zero; therefore

$$\varepsilon_1 + \varepsilon_2 + \varepsilon_3 = 0 \tag{23.21}$$

Expressing maximum shear stress and strain in terms of the difference in principal stress and strains,

$$\frac{\varepsilon_1 - \varepsilon_2}{\sigma_1 - \sigma_2} = \frac{\varepsilon_2 - \varepsilon_3}{\sigma_2 - \sigma_3} = \frac{\varepsilon_3 - \varepsilon_1}{\sigma_3 - \sigma_1} = \beta \qquad (23.22)$$

Rearranging the above equations to give the individual principal strains in terms of the principal stresses,

$$\left.\begin{array}{l} \varepsilon_1 = \dfrac{2\beta}{3}\left[\sigma_1 - \tfrac{1}{2}(\sigma_2 + \sigma_3)\right] \\[2mm] \varepsilon_2 = \dfrac{2\beta}{3}\left[\sigma_2 - \tfrac{1}{2}(\sigma_3 + \sigma_1)\right] \\[2mm] \varepsilon_3 = \dfrac{2\beta}{3}\left[\sigma_3 - \tfrac{1}{2}(\sigma_1 + \sigma_2)\right] \end{array}\right\} \qquad (23.23)$$

These may be expressed as a constant creep rate by writing

$$\left.\begin{array}{l} \dot{\varepsilon}_1 = \alpha[\sigma_1 - \tfrac{1}{2}(\sigma_2 + \sigma_3)] \\ \dot{\varepsilon}_2 = \alpha[\sigma_2 - \tfrac{1}{2}(\sigma_3 + \sigma_1)] \\ \dot{\varepsilon}_3 = \alpha[\sigma_3 - \tfrac{1}{2}(\sigma_1 + \sigma_2)] \end{array}\right\} \qquad (23.24)$$

where α is a function relating the three principal stresses to the simple uniaxial stress creep condition.

Using the von Mises yield criterion to obtain the equivalent uniaxial stress, σ_e, gives

$$\sigma_e = \frac{1}{\sqrt{2}}\sqrt{[(\sigma_1 - \sigma_2)^2 + (\sigma_2 - \sigma_3)^2 + (\sigma_3 - \sigma_1)^2]}$$

From the simple secondary-stage creep law,

$$\dot{\varepsilon} = K\sigma_e{}^n$$

and for simple tension, σ_2 and $\sigma_3 = 0$ and $\sigma_e = \sigma_1$; therefore

$$\dot{\varepsilon} = \alpha\sigma_e$$

and hence

$$\alpha = K\sigma_e{}^{n-1} \qquad (23.25)$$

Therefore the three principal creep rates may be written as

$$\left.\begin{array}{l} \dot{\varepsilon}_1 = K\sigma_e{}^{n-1}[\sigma_1 - \tfrac{1}{2}(\sigma_2 + \sigma_3)] \\ \dot{\varepsilon}_2 = K\sigma_e{}^{n-1}[\sigma_2 - \tfrac{1}{2}(\sigma_3 + \sigma_1)] \\ \dot{\varepsilon}_3 = K\sigma_e{}^{n-1}[\sigma_3 - \tfrac{1}{2}(\sigma_1 + \sigma_2)] \end{array}\right\} \qquad (23.26)$$

In the thin tube under internal pressure where σ_1 is the hoop stress and σ_2 the axial stress, $\sigma_1 = 2\sigma_2$ and $\sigma_3 = 0$. Hence

and

$$
\left.
\begin{aligned}
\sigma_e &= \frac{\sqrt{3}}{2}\,\sigma_1 \\[1em]
\dot{\varepsilon}_1 &= \left(\frac{\sqrt{3}}{2}\right)^{n+1} K\sigma_1{}^n \\[1em]
\dot{\varepsilon}_2 &= 0 \\[1em]
\dot{\varepsilon}_3 &= -\left(\frac{\sqrt{3}}{2}\right)^{n+1} K\sigma_1{}^n
\end{aligned}
\right\}
\tag{23.27}
$$

where $\sigma_1 = pr/t$ (eqn. (2.15)).

It is interesting to observe that there is no creep in the axial direction, which has also been verified experimentally. It is evident that to extend the solution to internal pressure combined with axial tension or torsion only requires the insertion of a different ratio of σ_1 to σ_2.

23.8. CREEP UNDER CONDITIONS OF CYCLIC STRESS AND TEMPERATURE

It is traditional to think of creep in terms of constant stress and temperature over long periods of time, and there are many examples in practice where this is true or very nearly so.

In the last two decades certain engineering developments such as gas and steam turbines, rockets and supersonic aircraft have involved the use of metals not only at very high temperatures but also with dynamic fluctuating stresses. In short, the problem is one in which the mean or steady component of stress can induce creep, and the alternating component of stress may lead to fatigue failure. The earliest investigations into this problem were made between 1936 and 1940 by various German investigators, and since then there have been many interesting studies both in this country and the U.S.A.

The problem of fatigue at high temperature was discussed in Chapter 22, and this phenomenon can be unaccompanied by creep for reversed or zero mean stress. Therefore, when considering a material for high-temperature service it is usual to think of the behaviour in terms of a diagram such as Fig. 23.12, in which fatigue failure is the criterion within certain stress and temperature limits, and beyond these creep is the predominant factor. If it is a question not of one or other phenomenon acting on its own, but of both influences operating simultaneously, then the solution becomes rather more involved. Fatigue is essentially a cycle-dependent mechanism whereas creep is time-dependent. It is therefore both desirable and convenient to express fatigue behaviour at high temperature also in terms of time to rupture as suggested by Tapsell. One of the reasons for this is because of the greater dependence of fatigue on cyclic frequency at high temperature.

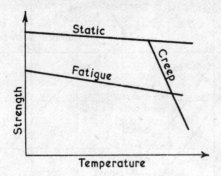

Fig 23.12   Strength Limitations with Increasing Temperature

The most useful way of presenting data for combined creep and fatigue conditions is in the form of a diagram of alternating stress against steady or mean stress, which is similar in most respects to the S_a–S_m diagram in normal fatigue, Fig. 22.7. Test results are plotted as the combination of alternating and mean stress to produce either rupture or a specified creep strain after a particular number of hours at constant temperature. Points along the abscissa axis represent creep conditions only and points along the ordinate axis are for fatigue only. Some results obtained by Tapsell on 0·26% carbon steel are given in Fig. 23.13 for various amounts of total creep strain occurring in 100 hours at 400°C under different combinations of cyclic and steady stress. Theoretically predicted curves are also shown for creep strains of 0·002 and 0·005. Lazan has derived a theoretical expression, neglecting any effects of stress history, relating the alternating and steady stress components for secondary-stage creep in a given time. Using the relationship (23.4) for creep rate and stress,

$$\dot{\varepsilon} = K\sigma^n$$

and an expression for the instantaneous stress σ at any time during a cycle,

$$\sigma = \sigma_m + \sigma_a \sin \omega t \tag{23.28}$$

where σ_m = cyclic mean stress, σ_a = cyclic amplitude of stress, ω = cyclic frequency, t = time, then combining the above,

$$\dot{\varepsilon} = K\sigma_m^n \left(1 + \frac{\sigma_a}{\sigma_m} \sin \omega t\right)^n \tag{23.29}$$

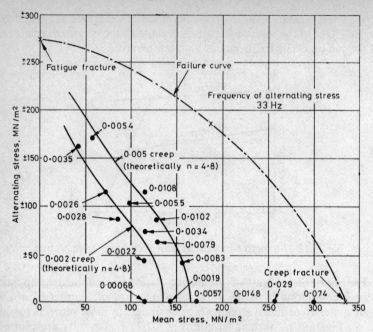

Fig 23.13 0·26% Carbon Steel: Total Creep occurring in 100 h
at 400°C

During a time interval t' the creep strain will be

$$\int_0^{t'} \dot{\varepsilon}\, dt = \int_0^{t'} K\sigma_m{}^n \left(1 + \frac{\sigma_a}{\sigma_m}\sin \omega t\right)^n dt \qquad (23.30)$$

If σ_e is the equivalent static stress to cause the same creep strain in time t', then

$$\int_0^{t'} \dot{\varepsilon}\, dt = \int_0^{t'} K\sigma_e{}^n\, dt \qquad (23.31)$$

Therefore

$$\int_0^{t'} K\sigma_e{}^n\, dt = \int_0^{t'} K\sigma_m{}^n \left(1 + \frac{\sigma_a}{\sigma_m}\sin \omega t\right)^n dt \qquad (23.32)$$

or

$$\frac{\sigma_m}{\sigma_e} = \left[\frac{\displaystyle\int_0^{t'} dt}{\displaystyle\int_0^{t'} \left(1 + \frac{\sigma_a}{\sigma_m}\sin \omega t\right)^n dt}\right]^{1/n} \qquad (23.33)$$

The above expression is plotted in Fig. 23.14 and yields curves for
$n = 7, 9$ and 11, which show the combinations of mean and alternating
stress, that give the same amount of secondary creep per unit time as
an equivalent static stress.

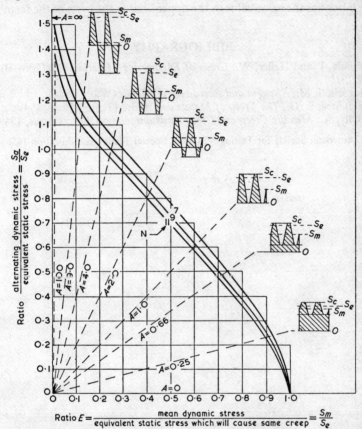

Fig 23.14   Theoretical Curves showing the Effect of Alternating
Stress on Creep, assuming Negligible Stress History Effect
(*From Lazan: by courtesy of the American Society for Testing Materials*)

The influence of alternating stress on the minimum creep rate and
time to rupture varies considerably with temperature, material, and
length of time. At higher temperatures or long life, the alternating
stress appears to have little effect on creep rate; in fact, there are
cases where creep strengthening has resulted. On the other hand, at
lower temperatures or shorter rupture times, fatigue appears to play
a more detrimental part, giving a higher creep rate. The rupture strain
is also somewhat reduced by the presence of cyclic stress.

Creep behaviour under constant load, but with a cyclical variation of temperature, has received less attention than cyclic loading. However, some information* indicates that, at the same stress, time to rupture is increased for cyclic temperature as against steady temperature, where the latter has the same value as the upper limit in the cycle of the former.

BIBLIOGRAPHY

Finnie, I. and Heller, W., *Creep of Engineering Materials* (McGraw-Hill, 1959).
Lessels, J. M., *Strength and Resistance of Metals* (Wiley, 1954).
Stanford, E. G., *The Creep of Metals and Alloys* (Temple Press, 1949).
Sully, A., *Metallic Creep and Creep Resistant Alloys* (Butterworths, 1949).

* American Society for Testing Materials, Special Technical Publication 165.

Chapter 24
Non-metallic Materials

24.1. The theoretical analyses in the first part of this book have been almost exclusively related to isotropic linear-elastic materials. These are generally assumed to be metals, which have the most widely used role as engineering materials. However, in the field of civil engineering and in particular building construction, concrete is just as important even though it is frequently combined with a metal framework or reinforcement. It may be regarded as isotropic and linear elastic for design purposes, but has vastly different strengths in tension and compression.

Plastics and rubber have an increasing use in engineering not only because of their special properties, but in some circumstances as effective replacements for metals. They are the class of materials which exhibit viscoelastic properties (see Chapter 18) and may be used in an unreinforced or fibre-reinforced state.

Finally, this chapter describes some of the properties of timber, which again is an important structural material, but is vastly different from all those above owing to its distinctive grain structure. This makes its properties markedly anisotropic, which must be allowed for in design. In addition, there is probably a larger proportion of significant defects, e.g. knots, in timber than in other engineering materials.

Although in one chapter only brief justice can be done to the properties of a few of the most important non-metallic materials, it may perhaps encourage the student to read further and broaden his outlook on engineering materials.

24.2. CONCRETE

Concrete consists of a mixture of cement, water and inert materials such as sand, gravel and crushed stone. The cement and water form a paste which on hardening holds the mass firmly together. The mixture of lime, sand and water has been used through the ages as an air-drying lime mortar to retain building blocks. However, apart from pozzuolana volcanic ash which was imported from Italy, it was not until 1824 that a cement was manufactured which would harden under water. This of course is essential for harbours, dams, bridges, etc. Aspdin, a Leeds builder, fired, at about 1400°C, a mixture of clay and chalk and after grinding to a powder obtained a cement which he called Portland cement. There are two alternative processes used today to make Portland cement, known as "wet" and "dry". In the former, lime and clay are excavated and mixed in ball mills, which crush them to a slurry with water. After final blending the mixture is fired in a kiln, the clinker is cooled and coarsely crushed. Finally, the granules are ground to a very fine powder in a ball mill and gypsum is added to regulate setting time. The "dry" process, as the name suggests, excludes water at the mixing stage, but is otherwise the same. It is slightly less popular, it being more difficult to maintain close control of the mix. Portland cement is seldom used alone for structural purposes since it would be too costly and would crack on hardening owing to excessive shrinkage. There are a number of standard tests required to check the chemical composition, fineness, strength and setting time, details will be found in B.S. 12:1958.

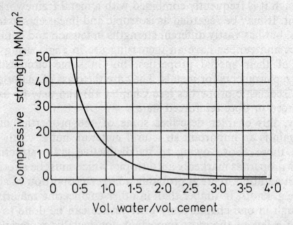

Fig 24.1   Relation between Compressive Strength and Water/
Cement Ratio

We shall now return to the discussion of concrete, having introduced its most important constituent. When concrete is used without reinforcement, or reinforced only for shrinkage and temperature changes, it is known as *plain concrete*. This is used for massive work sustaining only compression such as walls, dams, piers, etc. If tensile stresses are likely to arise in concrete structural members, resistance is provided by steel bars embedded in the concrete. This is then known as *reinforced concrete*. If the reinforcement is placed under an initial tensile stress prior to placement of the concrete so as to introduce precompression in the concrete when set, then this form of construction is termed *prestressed concrete*. It has the advantage of economy in the use of concrete and steel by reducing stresses and deflections.

The properties which are generally most important in concrete are durability, watertightness, strength and wear resistance. These properties are very dependent on the "design" and "cure" of a concrete mix. The

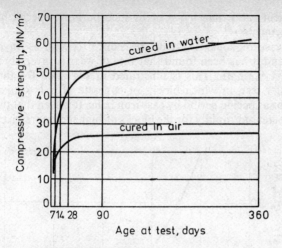

Fig 24.2 Effect of Ageing and Curing on Compressive Strength

influence of water/cement ratio and curing time on compressive strength are illustrated in Figs. 24.1 and 24.2. The standard test is conducted on a cube of 150 mm side, and the conditions of preparation and loading rate are given in B.S. 1881. Strength is also measured in tension and bending tests, and typical values are given in Table 24.1. It is sometimes desirable to produce concrete having a comparatively high strength within a few days after placing, in order to get early usage. High early strength is achieved either by a normal Portland cement and a lower

Table 24.1

*Comparison of Compressive, Bending and
Tensile Strength of Plain Concrete*

Strength of plain concrete, MN/m^2			Ratio, per cent		
Compressive	Bending modulus of rupture	Tensile	Modulus of rupture to compressive strength	Tensile strength to compressive strength	Tensile strength to modulus of rupture
6·9	1·59	0·76	23·0	11·0	48
20·7	3·35	1·90	16·2	9·2	57
34·5	4·66	2·76	13·5	8·0	59
48·3	5·90	3·59	12·2	7·4	61
62·1	6·97	4·35	11·2	7·0	63

water-cement ratio than usual or by incorporating special high-early-strength cements.

In the case of concrete roads, resistance to wear is an important property and it has been found that rate of wear is inversely related to compressive strength. This is illustrated in Fig. 24.3 for the Talbot–Jones rattler test, in which blocks of concrete of a 1:4 mixture were abraded for a specific period by cast-iron balls. It is seen that the amount of wear decreased rapidly for the blocks of higher compressive strength.

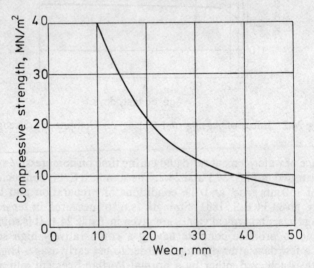

Fig 24.3   Relation between Compressive Strength and Wear of Concrete

The shear strength of rich concrete is found to average about half the compressive strength, while the fatigue limit under fluctuating (zero to maximum) cycles of stress is also about 50% of the maximum compressive strength. Since the stress–strain curves for a concrete may not be truly linear elastic, the modulus of elasticity can be defined either by an initial tangent modulus (at the origin) or a secant modulus at one-third of the maximum strength. Typical values for the initial tangent modulus range from $16\,GN/m^2$ at three days to $39\,GN/m^2$ after five years.

24.3. PLASTICS

As engineering materials, plastics are now third in importance after metals and concrete, and their usage and applications are expanding rapidly year by year. As with concrete, plastics are used both in the unreinforced and reinforced states. The latter is effected by glass or carbon fibres.

The term *plastics* describes a large and varied group of materials which consist of, or contain as an essential ingredient, an organic substance. These materials, although solid in the finished state at ordinary temperatures, at some stage in their manufacture have been or can be formed into various shapes by flow, usually by the application, singly or together, of heat and pressure.

Plastics may be divided into two general classes:

(a) Those which harden on cooling from the basic polymerization process and which may be resoftened and rehardened by successive heating and cooling. These are called *thermoplastics*.

(b) Those which are hardened by chemical changes. They remain hardened without cooling and do not soften when reheated. These are known as *thermosetting plastics*.

Thermoplastics have generally been adopted in the unreinforced state for batch runs of semi-automatically moulded components. Thermosetting plastics have been widely used with glass-fibre reinforcement for one or a few off structural elements. The essential differences between the unreinforced and reinforced states are strength and economy of production.

Unreinforced Plastics

Plastics are low-density high-molecular-weight materials consisting of long entangled molecular chains which generally have a hydrocarbon structure. Deformational behaviour is viscoelastic (see Chapter 18) in nature, that is to say combining the properties of an elastic solid with a viscous fluid. This means that stress is not simply related to strain, as say in a Hookean material, but is time and temperature dependent and has "memory" of stress and strain history. Apart from the dependence of properties on small variations in temperature there are also certain temperature regimes for polymers where there is a radical change in structure. These are termed *main* and *secondary transition temperatures*, of which the most important is the *glass-rubber transition*. Above this temperature the polymer has a low stiffness and is soft and flexible, and below it the material is relatively rigid or glass-like. Normal room temperature may be above or below the transition for various polymers. For example, polymethylmethacrylate, PMMA (Perspex), is well below its transition and is relatively brittle, whereas polythene is above its transition and so is soft and has a high elongation to break.

All the mechanical properties and tests described in the previous five chapters have some relevance to the use of plastics, but before discussing any particular property in more detail it is worth summarizing the main feature of plastics as a class of materials:

(a) General. Low density, in the range 0·83 to 2·5 g/cm³, low surface hardness (at best comparable to that of aluminium).

(b) Thermal. High coefficient of expansion (many times those for metals); low conductivity.

(c) Electrical. Good electrical insulation.

(d) Chemical. Resistant to acids, alkalis and water, but susceptible to some organic liquids; resistance to sunlight and weather varies from good to poor depending on the particular plastic.

(e) Optical. Transparent, translucent, or opaque; some are birefringent (see Chapter 26).

(f) Mechanical. High strength-to-weight ratio and low stiffness compared with metals; toughness varies from very low (brittle) in some, to high in other plastics; time, temperature and history have previously been mentioned.

The mechanical properties of plastics fall into two principal categories: deformation, i.e. creep, relaxation and recovery; and strength, i.e. tensile, impact, fatigue.

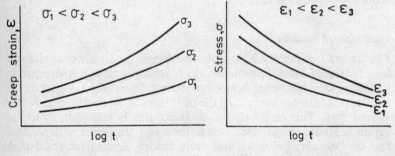

Fig 24.4   Creep Curves   Fig 24.5   Isometric Stress–Time Curves

Creep tests are conducted generally under simple tensile constant load and temperature, and strain is measured with time, resulting in *creep curves* typically as shown in Fig. 24.4. From these basic curves can be plotted *isometric stress–time curves*, Fig. 24.5, which is a useful alternative presentation. The creep rupture curve is also included for comparison. *Creep modulus* is defined as the ratio of stress to strain at a particular time during a creep test, and the variation of creep modulus with time and stress is illustrated in Fig. 24.6. The most useful way of characterizing stress versus strain (which is *always* time dependent) is by means of *isochronous curves*, Fig. 24.7, which derive from constant time sections through creep curves.

In applications where strain is controlled rather than stress then data are required from stress relaxation tests in which the decay of stress with time at constant strain is determined. *Relaxation modulus* is defined as the ratio of stress to strain at a particular time in a relaxation test.

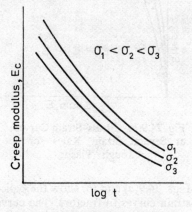

Fig 24.6   Creep Modulus Curves

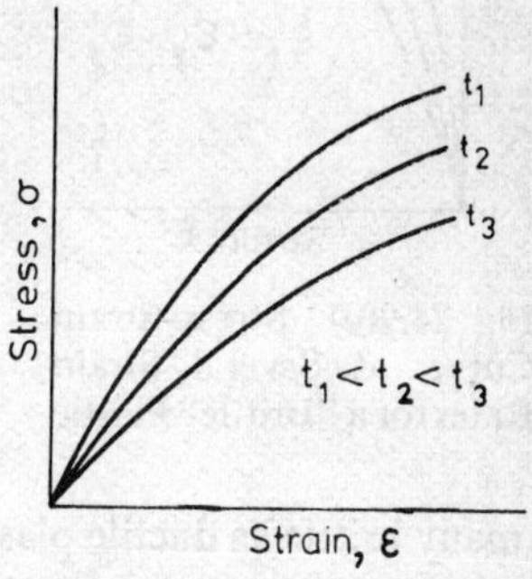

Fig 24.7   Isochronous Stress–
Strain Curves

The effect of temperature is much the same relatively as for metals, that is as temperature is increased so does the rate of creep. In other words, stiffness or creep modulus decreases with increasing temperature as shown in Fig. 24.8.

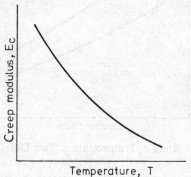

Fig 24.8   Creep Modulus v. Temperature

The first aspect of strength which we will consider is the conventional tensile test to fracture. The principal difference from metals is the marked dependence on strain rate and temperature and the existence of

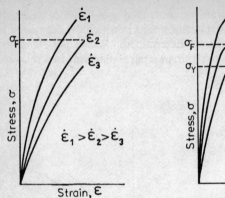

Fig 24.9(a) Stress–Strain Curves at Several Strain Rates for a "Brittle" Plastic

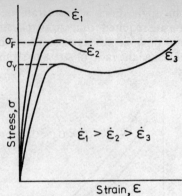

Fig 24.9(b) Stress–Strain Curves at Several Strain Rates for a "Tough" Plastic

as many brittle as ductile plastics. Figs. 24.9(a) and (b) show the typical form of continuous tensile stress–strain curves to fracture. The curves in (a) are for a brittle plastic, and although non-linear do not exhibit a yield point before fracture at σ_F. On the other hand, the ductile or

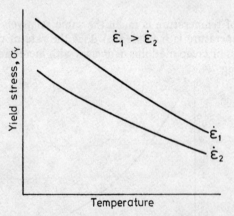

Fig 24.10   Yield Stress v. Temperature at Two Different Strain Rates

tough plastic in (b) exhibits yielding, which is the onset of large semi-permanent deformation, at σ_Y, and after continued straining (in some plastics to over 100%) fracture occurs at σ_F, which may be above or below the value at σ_Y. The variation of yield stress with temperature is illustrated in Fig. 24.10, which brings out the reduction in strength as temperature increases.

Before proceeding to the next aspect of strength it is essential to appreciate that, whereas in the case of metals one carefully conducted tensile test can provide a repeatable meaningful value, plastics on the other hand exhibit considerable scatter in test results. It is therefore always necessary to conduct a number of tests under nominally identical conditions and to take the average of the results.

Impact strength is studied using a Charpy principle test, but with reduced striking energy and velocity and a notched specimen of about 6 mm × 3 mm cross-section. Results of tests are plotted as impact strength (energy per unit area) against temperature for standard notch

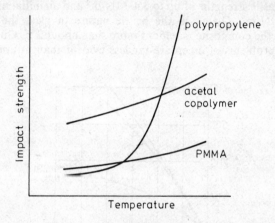

Fig 24.11   Impact-strength/Temperature Curves for Several Thermoplastics (Notch tip radius, 0·2 mm)

geometries, and some typical results are shown in Fig. 24.11. The curve for polypropylene shows a sharp change in impact strength due to the rubber–glass transition, whereas the data for PMMA has not reached its transition.

There is relatively little fundamental data on fatigue of unreinforced thermoplastics to date. Stress-cycles-to-failure plots exhibit the same general shape as for metals; i.e. the lower the cyclic stress the longer is the life. At the present time there are perhaps not very many engineering applications of plastics which involve large numbers of stress cycles. However, this situation will change with the increasing use of plastics. The most important difference compared with metals is the frequency dependence of plastics on fatigue. Owing to high hysteresis and poor conductivity, at other than fairly low frequencies (< 5 Hz), failure is due to thermal softening rather than the slow propagation of a fatigue crack.

Reinforced Plastics

There are three principal ways in which glass fibre is used to reinforce a resin matrix (polyester, epoxy or phenolic thermosets). Firstly, short chopped strands which are randomly mixed into the resin give a composite which is fairly isotropic. Secondly, the reinforcement can take the form of a series of woven mats sandwiched by resin, which results in uniform properties parallel to the 0° and 90° weave directions. Finally, if it is required to have superior strength in a certain direction to carry specific loads, batches of single filaments can be specifically oriented. Since glass has a high strength and modulus, the resulting composite achieves tensile strengths of up to 350 MN/m² and an initial modulus of 20 to 25 GN/m². Owing to the brittle nature of glass the strain to fracture of the composite is seldom more than about 2%. On the other hand, the problems of creep are far less evident than in unreinforced plastics.

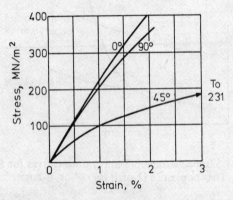

Fig 24.12   Tensile Stress–Strain Curves for Glass Cloth Reinforced Epoxy Resin

Some examples of tensile stress–strain curves are given in Fig. 24.12. Fracture of composites is a complex process of debonding at the resin–fibre interface followed by resin and fibre cracking. However, on a macroscopic scale, cracks can be initiated and propagated under cyclic stress resulting in conventional fatigue failure. More fatigue data is available for reinforced plastics, and typical fatigue curves are shown in Fig. 24.13.

24.4. RUBBER

This material possesses four salient characteristics which give it a wide field of usefulness in engineering practice. These are (a) its value as an electrical insulator, (b) its impermeability to water, (c) its ability to withstand great deformation without damage, (*d*) its ability to absorb shock by mechanical hysteresis.

As an electrical insulator it is in wide demand for covering wires, bushings, etc., and its waterproof qualities make it suitable for hoses conveying water. The ability to withstand great deformation makes it especially useful for pneumatic tyres for vehicles, and its ability to absorb energy gives it a wide use in the manufacture of members which take up shock in machines and structures.

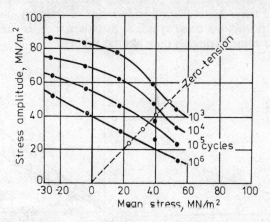

Fig 24.13 Fatigue Data for Chopped Strandmat Polyester Resin
in Axial Loading

(*Owen and Smith*)

One constituent of rubber is sulphur, and the percentage of this substance in the material determines whether the resultant product is hard or soft. A high percentage will result in a hard brittle material, whilst about 7·5% sulphur will give a good grade of soft rubber which will stretch many times its initial length without fracture.

The stress–strain diagrams for both a hard and a soft rubber, in tension, are shown in Fig. 24.14. The shapes of the curves show that the modulus of elasticity in tension is not constant; also since in the case of soft rubber the curvature is reversed at the higher stress range, when the material is very greatly extended it will exert an increasing resistance to further strain.

Rubber is usually said to be highly elastic but it does not fulfil the definition of elasticity already adopted. The stress–strain diagram for a cycle of stress is shown in Fig. 24.15. The shaded area shows the amount of energy lost in mechanical hysteresis. This loss is considerable, even if the loading and unloading are carried out at a slow rate, and increases considerably when the applied cycles are carried out rapidly. The energy lost in this manner in pneumatic tyres running at high speed tends to produce heat and has a weakening effect on the rubber.

The high mechanical hysteresis of soft rubber makes it an outstanding material for damping vibration and absorbing the energy of rapidly applied shock loads. For compression members, such as anti-vibration mountings, volume for volume, rubber will absorb about four times as much energy as spring steel at the same stress.

Owing to the shortage of natural rubber during the last war a number of synthetic rubbers, such as Neoprene, were provided by the petro-chemical industry. Apart from a wide range of strength and distensi-bility being possible in these synthetic materials, they also have a better resistance to some chemicals than natural rubber.

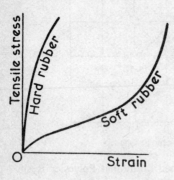

Fig 24.14   Tensile Stress–Strain Curves for Hard and Soft Rubber

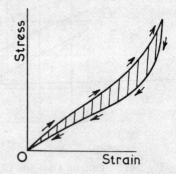

Fig 24.15   Mechanical Hysteresis in Rubber

The types of universal testing machine described in Chapter 19 for tensile, compressive and bending tests were generally designed to test metals. Since these materials are not particularly sensitive to straining or loading rate and overall extensions to fracture are not very large, the testing machines are limited in these respects. A more sophisticated range of testing machines has been developed over the past ten years specifically catering for the needs of testing rubbers and plastics. Generally, these are closed-loop servo-controlled machines of either mechanical or hydraulic drive with lower load ranges than are required for metals. An example, illustrated in Plate XIII, is the Instron 50 kN capacity testing machine, which provides cross-head speeds ranging from 0·5 to 500 mm/min, loading or straining rate held constant at any setting of full range taking 0·2 to 500 min, automatic cycling, accurate load or strain holding for long periods and various other control systems. There are, of course, a number of other manufacturers who provide similar testing machine facilities as described above.

24.5. TIMBER

Timber has been used as a structural material since prehistoric times. One of its advantages today over other structural materials such as metal, concrete, brick, etc., is its lower cost. Wood has a relatively low density and is easily shaped by sawing. It is subject to attack by insects and to decay which is analogous to the problem of corrosion of metals. Although wood is easily inflammable, substantial load-bearing members will retain a considerable proportion of their strength even though badly charred on the surface.

There are two general classes of timber known as *softwood* and *hardwood*. The former relates to all cone-bearing trees such as cedar, fir, pine and spruce, while the latter term covers all broad-leaf trees such as oak, elm, ash, beech, teak, walnut, mahogany and poplar. These classes relate to hardness and strength in general, but there are particular exceptions where a "softwood" is harder and stronger than a "hardwood". For general structural purposes the softwoods are more generally used, while hardwoods are reserved for interior finishing and furniture.

Green timber has a moisture content of about 32%, and this is reduced to about 12% by exposure to the atmosphere, which is known as *seasoning* and takes several weeks to months depending on the type of wood. When it is required to reduce the moisture content still further, timber is dried in a kiln at a temperature between 70° and 75°C.

The softwoods are made up for the most part of an aggregation of elongated tubular cells which extend in a direction parallel to the axis of the tree trunk. These cells are closed at the ends and absorb water through their porous walls. At right angles to these tubular cells are numerous groups of cells extending in a radial direction and called *rays*. Through these rays, plant food is distributed across the tree section. In some softwoods there are longitudinal tubes called *resin ducts* in which resin is formed. The hardwoods have a much more complex structure than the softwoods; for example, in hardwoods the rays are much larger and more numerous. The elongated tubular cells are of minor importance, and the principal structural elements of the hardwoods are longitudinal wood fibres made up of elongated sharp-pointed cells with very thick walls. The general longitudinal direction of the tubular cells which make up the greater part of wood gives a distinctive "grain" to its structure. There are three types known as straight-grained, cross-grained and twisted, one of which develops in any tree as it grows.

The principal defects found in timber which reduces its strength are knots and shakes. *Knots* are caused by the trunk end of a branch being enclosed by successive annual growth rings of the main trunk. *Shakes* is a term used for cracks which may occur radially or circumferentially.

In the case of wooden structural members such as beams or columns, the weakening effects of shakes and knots depends on their location in respect of the stress system set up and also on their size.

Owing to the grain structure of wood its stress–strain properties are anisotropic, which means that it has different strength and elasticity in various directions. For example, wood is much weaker in tension and compression across the grain than along the grain; in shear the reverse obtains. For this reason in a wooden beam the horizontal shear stress is a much more important factor than in a steel beam. Similarly it is

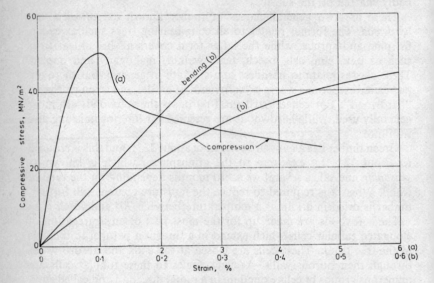

Fig 24.16   Pure Bending and Compression of Silver Spruce along the Grain

easier to transmit compressive rather than tensile forces in axially loaded timber members. As a consequence the standard tests (B.S. 373) that are used to assess the elastic and strength properties of wood are by three- or four-point bending and by compression parallel and perpendicular to the grain. Wood does not exhibit a truly linear elastic range, but up to a strain of about 0·05 to 0·1 % there is sufficient linearity to establish a value for a modulus of elasticity which is in the range 8 to 12 GN/m² for most timber. An increasing degree of non-linearity occurs beyond the elastic limit, which may be regarded as yielding. There is no sharp yield point and a steady breakdown of the structure occurs up to the "ultimate" load at which major fracturing occurs. Examples of stress–strain curves in pure bending and compression for silver spruce are shown in Fig. 24.16. For this particular spruce the moisture content was 14·8%, the specific gravity was 0·48 and the density was 485 kg/m³.

Owing to considerable scatter in values obtained from tests on wood it is essential to conduct a number of tests and assess the data statistically. It is perhaps worth mentioning one term which is particular to the bend test, namely *modulus of rupture*. This is defined as the maximum bending stress determined from the linear elastic formula using the bending moment at fracture of the wooden beam. This ignores the non-linearity of the load-deflection curve.

Wood has a high resilience or capacity to absorb energy owing to relatively low modulus values combined with proof stresses in the range 20 to 50 MN/m²; thus it compares favourably with some metals in resisting shock or impact loading.

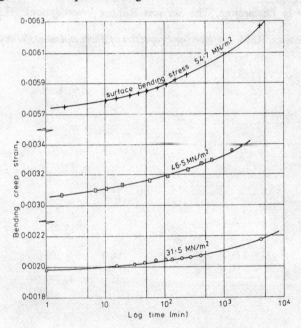

Fig 24.17   Bending Creep Curves for Silver Spruce

As is the case with plastics, timber also exhibits time dependence of stress and strain or creep at ambient temperature. Typical creep curves in pure bending for the spruce mentioned above at 20°C are shown in Fig. 24.17. This is one of the reasons why working stresses in timber structural members have to be maintained at 50% or less than the maximum strength in a continuous short-term test in order to ensure structural stability.

Cyclical stresses affect timber by a process of fatigue and fracture, as is the case with all materials, although naturally the microscopic and macroscopic mechanisms are particular to the structure of timber.

BIBLIOGRAPHY

Blanks, R. F. and Kennedy, H. L., *The Technology of Cement and Concrete* (Wiley, 1955).

Jastrzebski, Z. D., *Nature and Properties of Engineering Materials* (Wiley, 1959).

Kinney, S. F., *Engineering Properties and Applications of Plastics* (Wiley, 1957).

Meredith, R., *Mechanical Properties of Wood and Paper* (Interscience Publishers, 1953).

Rubber in Engineering, The Services Rubber Investigations (H.M.S.O., 1946).

Ogorkiewicz, R. M. *Engineering Properties of Thermoplastics* (Wiley, 1970).

Part 3

Experimental Analysis of Stress and Strain

Chapter 25
Resistance Strain Gauges

25.1. In 1856 it was noted by Lord Kelvin that, if a metal wire was strained, its electrical resistance changed. However, it was not until 1938 that this principle was employed in the U.S.A. as a means of measuring the surface strain on a component. Since then the wire resistance strain gauge has become the most widely used technique in the world of those available for experimental stress analysis.

Although it is quite suitable for measurement on small components, one of its great advantages of being electrical in principle enables very large numbers of gauges to be used on a structure such as an aeroplane with the readings being recorded automatically, permanently and quickly.

25.2. PRINCIPLE AND CONSTRUCTION OF A WIRE RESISTANCE STRAIN GAUGE

The majority of strain gauges consist of a fine metal wire, approximately 0·025 mm in diameter, formed into a grid and bonded to a thin paper backing as shown in Fig. 25.1. This complete unit is then cemented

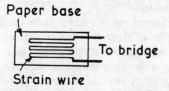

Fig 25.1   Wire Strain Gauge

on to the surface of the component or structure in which the strain is to be measured. A later development of the strain gauge is illustrated in Fig. 25.3B and is termed a metal foil gauge in which the grid of wire filaments is obtained by "printing" on to a metal foil.

As was stated in the introduction, a strained wire has its electrical resistance altered. A simple method of measuring change in resistance is by means of a Wheatstone bridge, Fig. 25.2, in which the four "arms" of the bridge contain resistances, one of which is the "active" strain gauge. A battery supplies the electromotive force and a galvanometer detects the potential difference across two apex points.

Strain in the component causes the gauge wire cemented to it to be strained and the consequent resistance change is detected and measured on the Wheatstone bridge. A proportional relationship between strain and change in resistance enables the strain in the component to be determined.

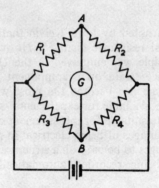

Fig 25.2   Strain Gauge Bridge Circuit

25.3. SIZE AND FORM OF GAUGES

There are nowadays a great many manufacturers of and applications for strain gauges, and it would be quite impossible in one chapter to give full details; however, it is of interest to appreciate the range of gauges that is available.

A number of different metals have been used for the gauge wire, the two most popular of these being a copper-nickel alloy of approximately 55–45% composition respectively and an alloy of nickel, chromium and iron with other minor elements.

The range of gauge resistance available is between 60 and 10 000 ohms, although the majority are produced with a resistance of a few hundred ohms. The safe current that can be carried by the gauge for long periods is around 25 mA, although an overload to twice this figure is generally acceptable for short periods.

Gauge size and shape are important factors. On areas of a component where it is expected that the strain pattern will vary only slightly, a long narrow gauge, 10 to 25 mm, can be employed. In regions of likely stress concentration and steep strain gradients a much shorter gauge length must be used, and for this application gauges as small as 2 mm effective length are produced; the width now has to be greater than the length in order to achieve a suitable resistance value.

A gauge will only react to strain parallel to the direction of the wire. Consequently, unless directions of strain in a component are known and gauges cemented-on to correspond with these directions, they will only record the component of strain parallel to their length. Hence, for completely unknown strain directions at a point on a surface, it becomes necessary to have three gauges cemented at the point, or in a small area, and set at known relative angles in order to obtain three strain components from which to determine a complete solution. A set of gauges as described above is termed a rosette, two arrangements, for ease of analysis, being either 60° or 45° between each of the three gauges. Although it is often possible to cement three individual gauges on a small surface area to form a rosette, manufacturers can supply a made-up rosette gauge. Some examples of gauge shape and form are given in Figs. 25.3A and 25.3B.

25.4. AXIAL SENSITIVITY OF A STRAIN GAUGE

The governing equation for the operating principle of a strain gauge is

$$R = \rho \frac{L}{A}$$

where R is the resistance of a conductor, whose resistivity is ρ, and L and A are the length and cross-sectional area of the conductor respectively. The above expression indicates that, if the conductor is changed in length by strain, which also implies a change in area, the resistance will change. However, experiment has shown that change in resistance is not simply a function of the change in geometry of the conductor. It appears that resistivity itself varies in proportion to strain.

Treating all the terms in the equation above as variables, and differentiating accordingly gives

$$dR = \frac{\rho A \, dL + LA \, d\rho - L\rho \, dA}{A^2} \tag{25.1}$$

$$\frac{dR}{R} = \frac{\rho A \, dL + LA \, d\rho - L\rho \, dA}{A\rho L}$$

$$= \frac{dL}{L} + \frac{d\rho}{\rho} - \frac{dA}{A} \tag{25.2}$$

Dividing by dL/L,

$$\frac{dR/R}{dL/L} = 1 + \frac{d\rho/\rho}{dL/L} - \frac{L \, dA}{A \, dL} \tag{25.3}$$

Now, the volume of the conductor is $V = AL$; therefore

$$dV = A \, dL + L \, dA \tag{25.4}$$

But also it can be shown that

$$\frac{dV}{V} = \varepsilon(1 - 2v) \tag{25.5}$$

Therefore

$$\frac{dL}{L}(1 - 2v)AL = A\,dL + L\,dA \tag{25.6}$$

so that

$$\frac{L\,dA}{A\,dL} = -2v \tag{25.7}$$

and

$$\frac{dR/R}{dL/L} = 1 + \frac{d\rho/\rho}{dL/L} + 2v \tag{25.8}$$

$(dR/R) \div (dL/L)$ is termed the *axial sensitivity* or *gauge factor* and is denoted by K_a, i.e.

$$K_a = 1 + \frac{d\rho/\rho}{dL/L} + 2v$$

Suppliers of strain gauges quote a gauge factor determined for each batch, and the most commonly used gauges have a sensitivity ranging from 2 to about 3·5. The gauge factor is the constant directly relating unit change in resistance to strain; hence

$$\varepsilon = \frac{1}{K_a}\frac{dR}{R} \tag{25.9}$$

25.5. CROSS-SENSITIVITY

The diagrams of various strain gauge grids, Figs. 25.3A and B, show that a certain amount of the grid wire is unavoidably placed transverse to the gauge axis. This wire will react to strains perpendicular to the gauge direction and gives rise to the term *cross-sensitivity*, K_c. The ratio K_c/K_a is generally about 2%.

The manufacturer's calibration is carried out on a uniaxial strain field, so that if a gauge is used in service in the same manner, cross-sensitivity can be ignored. If a gauge is situated in a two-dimensional strain field, then readings will be affected by the transverse strain, and a correction should be applied to obtain the true strain along the gauge axis.

Thus if $K_c/K_a = n$, then it can be shown that the true strains in two

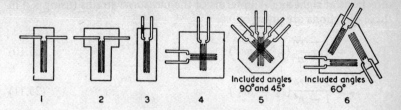

Fig 25.3A Single and Rosette Wire Strain Gauges

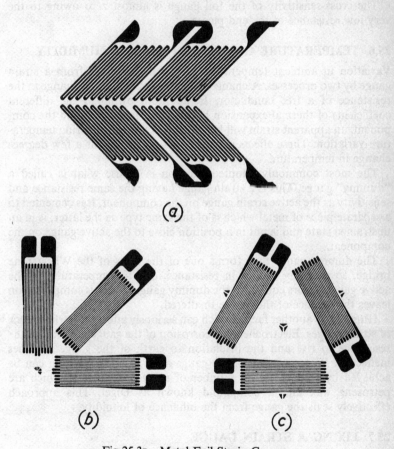

Fig 25.3B Metal Foil Strain Gauges
(a) Torque gauge. (b) 45°–90° rosette gauge. (c) Delta rosette gauge.
(By courtesy of Westland Aircraft Ltd., Saunders-Roe Division)

directions at right angles in terms of the measured strains (using K_a) in these directions are given by

$$\varepsilon_1 = \frac{1 - vn}{1 - n^2}(\varepsilon_1' - n\varepsilon_2') \tag{25.10}$$

$$\varepsilon_2 = \frac{1 - vn}{1 - n^2}(\varepsilon_2' - n\varepsilon_1') \tag{25.11}$$

where ε_1 and ε_2 are the true axial strains and ε_1' and ε_2' are the apparent strains from the change in resistance.

The cross-sensitivity of the foil gauge is almost zero owing to the very low resistance of the end pieces.

25.6. TEMPERATURE COMPENSATION AND HUMIDITY

Variation in ambient temperature affects the readings from a strain gauge by two processes. A change in temperature causes a change in the resistance of a free conductor. In addition, owing to the different coefficients of thermal expansion between the gauge wire and the component, an apparent strain will be induced in the gauge during temperature variation. These effects cannot be ignored even for a few degrees change in temperature.

The most commonly adopted solution is to use what is called a "dummy" gauge. This is a strain gauge having the same resistance and sensitivity as the active strain gauge on the component. It is cemented to a separate piece of metal which is of the same type as the latter, is in an unstrained state and is put in a position close to the active gauge on the component.

The dummy gauge then forms one of the arms of the Wheatstone bridge, so that any change in resistance, due to temperature, of the active gauge occurs equally in the dummy gauge, and this compensation leaves the balance of the bridge unaltered.

Humidity is another factor which can seriously affect the performance of strain gauges. Electrochemical corrosion of the gauge wire causes the resistance to rise and the insulation to earth of the gauge becomes ineffective. Waterproofing of the gauge after fixing in position can be achieved by coating with a number of substances, two of which are petrosene wax and a compound known as Digel. This approach effectively seals the gauge from the influence of humidity.

25.7. FIXING A STRAIN GAUGE

This operation is quite straightforward but requires care and patience if the gauge is to give reliable results. The most important feature is to ensure that the gauge is stuck evenly and securely all over to the metal surface. The first essential is that the latter should be clean of rust, scale and grease but left slightly roughened to secure keying of the cement. After lightly cleaning the back of the gauge, and applying a thin coating of the cement to the metal surface, the gauge is placed in position and pressed evenly and firmly to remove excess cement and air bubbles. The nitrocellulose type of cement will dry at room temperature in one to two days, the epoxy-resin type can be treated also in this way, or the curing can be speeded by a soaking period at temperatures between 55° and 71°C. Before moisture proofing, final drying-out for about 2 hours under radiant heat is advisable, followed by a check on the insulation resistance to earth. The gauge is now ready for wiring into the bridge circuit.

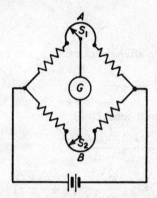

Fig 25.4   Null Method Bridge Circuit

25.8. BRIDGE CIRCUITS FOR STATIC STRAIN

The basic Wheatstone bridge circuit was illustrated in Fig. 25.2. Consideration of Kirchhoff's law, for the flow of current in a circuit, and Ohm's law shows that for zero galvanometer deflection, i.e. zero potential difference across AB,

$$\frac{R_1}{R_3} = \frac{R_2}{R_4} \tag{25.12}$$

If we now replace R_1 by the dummy gauge resistance R_d and R_2 by the active gauge resistance R_a, it is apparent that a change in R_a due to strain requires an alteration in the resistance value R_4 to balance the bridge. The necessary change in R_4 is a measure of the strain in the gauge. This usage of the Wheatstone bridge is known as the *null method* of determining strain.

In practice, a change in R_4 is not achieved by using a variable resistor owing to possible variations in contact resistance being of the same order as the change in R_a. Fig. 25.4 shows a typical bridge circuit for measuring static strains. Slide wires S_1 and S_2 are connected in at the apex points A,B. S_1 serves to obtain an initial balance of the bridge before the component is strained. Slide wire S_2 is calibrated and, instead of changing R_4, allows the bridge to be balanced without variation in contact resistance. In commercial equipment S_2 is related to a scale from which percentage change in resistance can be read directly.

An analysis of the resistance conditions in the bridge is as follows. After initially balancing the bridge by the slide wire S_1, and including the portions of each slide wire resistance in the arms of the bridge, then R_d becomes R_d', R_a becomes R_a', etc.:

$$\frac{R_d'}{R_3'} = \frac{R_a'}{R_4'} \tag{25.13}$$

Rebalancing the bridge after strain using S_2, we have

$$\frac{R_d'}{R_3' - \Delta S_2} = \frac{R_a' + \Delta R_a'}{R_4' + \Delta S_2} \tag{25.14}$$

Substituting for R_d' from eqn. (25.13),

$$\frac{R_a' R_3'}{R_4'(R_3' - \Delta S_2)} = \frac{R_a' + \Delta R_a'}{R_4' + \Delta S_2} \tag{25.15}$$

$$= \frac{R_a'}{R_4' + \Delta S_2}\left(1 + \frac{\Delta R_a'}{R_a'}\right)$$

Therefore

$$\frac{R_3'(R_4' + \Delta S_2)}{R_4'(R_3' - \Delta S_2)} = 1 + \frac{\Delta R_a'}{R_a'} \tag{25.16}$$

Simplification gives

$$\frac{\Delta R_a'}{R_a'} = \frac{\Delta S_2(R_3' + R_4')}{R_4'(R_3' - \Delta S_2)} \tag{25.17}$$

where the strain is given by

$$\frac{1}{K_a}\frac{\Delta R_a'}{R_a'} \quad \text{or} \quad \frac{1}{K_a}\frac{\Delta S_2(R_3' + R_4')}{R_4'(R_3' - \Delta S_2)}$$

25.9. ARRANGEMENT OF GAUGES

(i) Direct Strain

A useful way of measuring tensile or compressive strain in a member and eliminating from the readings any bending effect that may be occurring is illustrated in Fig. 25.5. Two active gauges placed on opposite faces of the member are wired in series to form one arm of the bridge, and two dummy gauges on another piece of metal are wired in the same way for a second arm. It is apparent that the tensile and compressive components of bending strain will cause the respective resistance changes to cancel out, leaving the resistance change due to direct strain only.

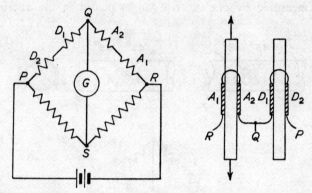

Fig 25.5   Gauge Arrangement and Bridge Circuit for Direct Strain

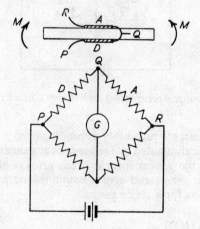

Fig 25.6   Gauge Arrangement and Bridge Circuit for Bending Strain

(ii) Bending

The presence of direct strain during the measurement of bending strain can be inconvenient, and this can be overcome by having two gauges cemented on opposite surfaces of the member as in Fig. 25.6. Each gauge forms one arm of the bridge and acts as a dummy for the other. Hence both arms are also active and this doubles the sensitivity of the bridge to bending, while eliminating direct strain.

(iii) Torsion

A cylindrical bar subjected to torsion has directions of principal strain at 45° to the longitudinal axis of the bar. Although torque can be measured by one or two gauges placed in the appropriate

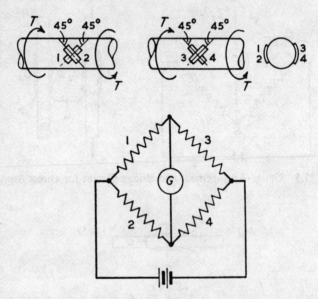

Fig 25.7   Gauge Arrangement and Bridge Circuit for Torsion

directions on the bar, the possibility of bending or axial strains affecting the readings can be eliminated by the use of four gauges as shown in Fig. 25.7. In this case the gauges form the four arms of the bridge, act as dummies for each other, and give a combined output which is four times as sensitive as for a single gauge.

25.10. CALIBRATION

A manufacturer's gauge factor is quoted as being accurate to 1%. However, it is generally desirable for an investigator to make quite sure of the sensitivity of the gauges being used by conducting a calibration on a known strain field. Direct tension or compression can be used if the elastic modulus of the material is known accurately. Alternatively, gauges cemented to a bar subjected to pure bending entail only a measurement of the curvature of the bar (from the deflection) to obtain the surface bending strain. This method of calibration is often employed since it only requires a simple rig and no knowledge of the elastic constants.

25.11. MEASUREMENT OF DYNAMIC STRAIN

Many machine components are subjected to dynamic strain; it is therefore desirable to be able to use resistance strain gauges under these circumstances. Although the principles already described for measuring static strain also apply for dynamic strain, the galvanometer is no longer a suitable instrument for detecting the fluctuating potential across the Wheatstone bridge. The galvanometer has to be replaced by an electrically-operated pen recorder or a cathode-ray oscilloscope, which has the necessary rate of response to accommodate fluctuations of strain of up to several thousand hertz.

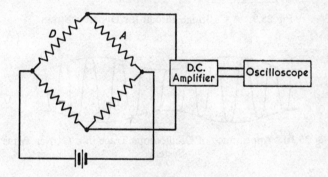

Fig 25.8   D.C. Bridge Circuit for Dynamic Strain

Unfortunately the sensitivity of strain measurement usually required is such that the potential difference across the bridge is too low to operate an oscilloscope or pen recorder, and one or more stages of amplification have to be introduced into the circuit.

It would be inappropriate in a single chapter to enter upon a discussion of the various electronic circuits which can be used in conjunction with the basic Wheatstone bridge. The reader who wishes or has to study the matter further is referred to the more detailed texts on the subject. However, it is worth pointing out here that there are two basic systems employing either direct current or alternating current. The former system, Fig. 25.8, employs a d.c. amplifier, and in its simplest form is only suitable for measuring dynamic strains of low frequency. For high-frequency measurement it is more usual to employ an a.c. circuit, Fig. 25.9. The battery of the d.c. circuit is now replaced by an oscillator, to energize the bridge, having a fixed frequency and a steady output. It is now possible to use an a.c. amplifier, which has certain advantages over the d.c. equivalent, and the image on the cathode-ray tube appears as a waveform of frequency determined by the

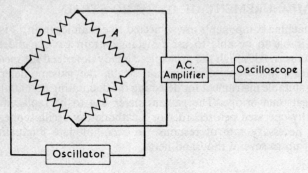

Fig 25.9   A.C. Bridge Circuit for Dynamic Strain

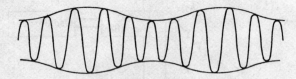

Fig 25.10   Appearance of Oscilloscope Trace in a Carrier Wave System

oscillator, and enveloped or modulated by the signal wave from the strain gauge as illustrated in Fig. 25.10. This is termed a carrier wave system and can be used equally well for static or high-frequency strain measurement up to about one-fifth of the carrier wave frequency.

25.12. SLIP RINGS

It is sometimes required to make strain measurements on reciprocating or rotating components and it then becomes necessary to include some form of sliding contact in the electrical circuit. An arrangement of brushes in contact with a metal disc or circumferential ring on the moving component is termed a slip ring. The chief problem in the use of slip rings is that the change in contact resistance at the brushes during rotation is of the same order as the signal from the strain gauge. The choice of contact materials is important, and silver wire or silver graphite brushes are commonly used. The effect of change in contact resistance can be reduced to negligible proportions by having the slip ring connections in the oscilloscope or battery parts of the circuit. If this cannot be arranged a variable resistor in parallel with one of the fixed resistance arms of the bridge also reduces the influence of contact resistance.

25.13. LOAD CELLS AND TORQUE BARS

In recent years the electrical resistance strain gauge has been used in a capacity other than for strain analysis. In some testing machines and structures, the forces transmitted are determined by an elastically strained block of metal, to which strain gauges are attached, which is in series with the specimen or structural member. The strain gauge block of metal is termed a *load cell*. There are, of course, other arrangements of the same principle such as an elastically deformed ring on which strain gauges are cemented. This is known as a *proving ring* or *loop dynamometer*. Strain gauge techniques are now so well developed that manufacturers can guarantee a very high accuracy and long-term stability for load cells and dynamometers.

A simple variation of the principle of the load cell for measuring torque is an elastically twisted bar, to which strain gauges are fixed, arranged in series with the member or specimen being subjected to torsion.

The circuits previously described for direct strain, bending and torsion measurement are applicable in the case of load cells and torque bars.

25.14. ROSETTE STRAIN COMPUTATION

In Section 25.3 it was explained that for the complete determination of strain at a point on the surface of a component, it was necessary to measure the strain in three directions at the point using a rosette gauge arrangement.

Let the three measured strains be ε_l, ε_m and ε_n and the angle between the l and m, and m and n, directions be $45°$ in each case. Then this arrangement is known as a $45°$ rosette as shown in Fig. 25.11. If the angle between ε_l and the principal strain ε_1 is θ, then from eqn. (13.32), Chapter 13, the principal strains are related to the measured strains as follows:

$$\left.\begin{aligned}
\varepsilon_l &= \varepsilon_1 \cos^2 \theta + \varepsilon_2 \sin^2 \theta \\
\varepsilon_m &= \varepsilon_1 \cos^2 (\theta + 45°) + \varepsilon_2 \sin^2 (\theta + 45°) \\
\varepsilon_n &= \varepsilon_1 \cos^2 (\theta + 90°) + \varepsilon_2 \sin^2 (\theta + 90°)
\end{aligned}\right\} \quad (25.18)$$

These equations may be rewritten as

$$\left.\begin{aligned}
\varepsilon_l &= \tfrac{1}{2}(\varepsilon_1 + \varepsilon_2) + \tfrac{1}{2}(\varepsilon_1 - \varepsilon_2)\cos 2\theta \\
\varepsilon_m &= \tfrac{1}{2}(\varepsilon_1 + \varepsilon_2) - \tfrac{1}{2}(\varepsilon_1 - \varepsilon_2)\sin 2\theta \\
\varepsilon_n &= \tfrac{1}{2}(\varepsilon_1 + \varepsilon_2) - \tfrac{1}{2}(\varepsilon_1 - \varepsilon_2)\cos 2\theta
\end{aligned}\right\} \quad (25.19)$$

Solving the above equations simultaneously gives

$$\left.\begin{array}{l} \varepsilon_1 = \tfrac{1}{2}(\varepsilon_l + \varepsilon_n) + \dfrac{\sqrt{2}}{2}\sqrt{[(\varepsilon_l - \varepsilon_m)^2 + (\varepsilon_m - \varepsilon_n)^2]} \\[2mm] \varepsilon_2 = \tfrac{1}{2}(\varepsilon_l + \varepsilon_n) - \dfrac{\sqrt{2}}{2}\sqrt{[(\varepsilon_l - \varepsilon_m)^2 + (\varepsilon_m - \varepsilon_n)^2]} \end{array}\right\} \tag{25.20}$$

$$\tan 2\theta = \frac{2\varepsilon_m - \varepsilon_l - \varepsilon_n}{\varepsilon_l - \varepsilon_n} \tag{25.21}$$

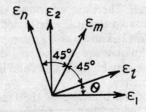

Fig 25.11   Rectangular or 45° Rosette

To obtain principal stresses from principal strains we use the stress–strain relationships, equations (13.57):

$$\left.\begin{array}{l} \varepsilon_1 = \dfrac{\sigma_1}{E} - \dfrac{\nu\sigma_2}{E} \\[2mm] \varepsilon_2 = \dfrac{\sigma_2}{E} - \dfrac{\nu\sigma_1}{E} \end{array}\right\} \tag{25.22}$$

from which

$$\left.\begin{array}{l} \sigma_1 = \dfrac{E}{1 - \nu^2}(\varepsilon_1 + \nu\varepsilon_2) \\[2mm] \sigma_2 = \dfrac{E}{1 - \nu^2}(\varepsilon_2 + \nu\varepsilon_1) \end{array}\right\} \tag{25.23}$$

25.15. CIRCLE CONSTRUCTION FOR STRAIN ROSETTE READINGS

Consider the general case of three strain gauges a, b and c, having arbitrary orientation as shown in Fig. 25.12 and strain readings ε_a, ε_b and ε_c in the given directions. As an example to illustrate the method assume $\varepsilon_a > \varepsilon_c > \varepsilon_b$. The procedure is as follows:

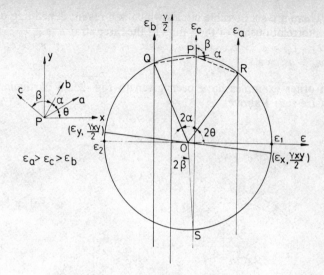

Fig 25.12

1. Set up a vertical axis to represent $\varepsilon = 0$ (which will subsequently be the semi-shear-strain axis).
2. Draw three lines parallel to the above axis at the appropriate distances representing the values (positive or negative) of ε_a, ε_b and ε_c.
3. On the middle line of these three (representing the middle value of the three strains ε_c) mark a point P representing the origin of the rosette.
4. Draw the rosette configuration at the point P *but lining up gauge c* (in this particular example) *along its vertical ordinate.*
5. Project the directions of gauges *a* and *b* to cut their respective vertical ordinates at Q and R.
6. Construct perpendicular bisectors of PQ and PR; where these intersect is the centre of the strain circle, O.
7. Draw the circle on this centre, which of course should pass through the points P, Q and R. Insert the horizontal strain abscissa through O.
8. Join O to Q, R and S, where S is the other intersection of the circle with the middle vertical line.
9. The lines OQ, OR and OS represent the three gauges *on the circle* where 2α and 2β are the angles between OR and OQ, and OQ and OS, respectively.
10. From the circle read off as required the principal strains ε_1, ε_2 or the chosen co-ordinate direction strains ε_x, ε_y, γ_{xy}.

11. There are six possible circles which can result, depending on the interrelationship of the values of the three strains: e.g. $\varepsilon_a > \varepsilon_b > \varepsilon_c$; $\varepsilon_a > \varepsilon_c > \varepsilon_b$; $\varepsilon_b > \varepsilon_a > \varepsilon_c$; $\varepsilon_b > \varepsilon_c > \varepsilon_a$; $\varepsilon_c > \varepsilon_a > \varepsilon_b$; $\varepsilon_c > \varepsilon_b > \varepsilon_a$.

Two other examples have been given in Fig. 25.13 particularly to emphasize step 4 above.

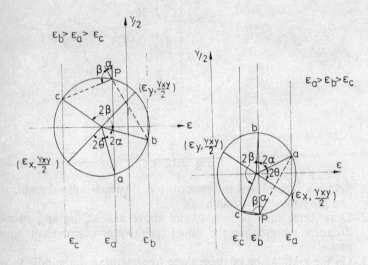

Fig 25.13

Example 25.1. At a point on the surface of a component, a 60° rosette strain gauge positioned as shown in Fig. 25.14 measures strains of $\varepsilon_l = 0.000\,46$, $\varepsilon_m = 0.0002$ and $\varepsilon_n = -0.000\,16$. Use Mohr's strain circle to determine the magnitude and direction of the principal strains and hence the principal stresses. $E = 208\ \text{GN/m}^2$, $\nu = 0.29$.

To construct the strain circle, Fig. 25.14, we use the procedure described above. The principal strain values are represented by TV and TU, and therefore

$$\varepsilon_1 = 0.000\,525 \qquad \text{and} \qquad \varepsilon_2 = -0.000\,19$$

The angle between ε_l and ε_1 on the circle is 34° and between ε_l and ε_2, 214°. Therefore

$$\theta_1 = 17° \qquad \text{and} \qquad \theta_2 = 107°$$

The principal stresses are given by eqns. (25.23); thus

$$\sigma_1 = \frac{208 \times 10^9}{1 - 0.29^2} (0.000\,525 + 0.29 \times -0.000\,19)$$

$$\sigma_2 = \frac{208 \times 10^9}{1 - 0.29^2} (-0.000\,19 + 0.29 \times 0.000\,525)$$

from which

$$\sigma_1 = 106.3\,\text{MN/m}^2$$
$$\sigma_2 = -86.0\,\text{MN/m}^2$$

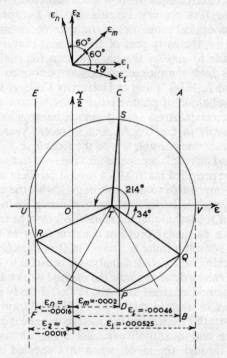

Fig 25.14 60° Rosette Gauge and Mohr's Strain Circle Solution

BIBLIOGRAPHY

Dobie, W. B. and Isaac, P. C., *Electric Resistance Strain Gauges* (E.U.P., 1948).

Perry, C. C. and Lissner, H. R., *The Strain Gauge Primer* (McGraw-Hill, 1955).

Yarnell, J., "Resistance strain gauges; their construction and use." *Electron. Engng*, 1951.

Neubert, H. K. P., *Strain Gauges, Kinds and Uses* (Macmillan, 1967).

Chapter 26

Photoelasticity

26.1. The origin of photoelasticity may be considered as the year 1816 when Sir David Brewster discovered that glass exhibited the optical phenomenon of birefringence when in a stressed condition. During the remainder of the 19th century, investigators in Europe and England studied the stress-optical properties of transparent materials. However, it was only during the early part of the present century that serious efforts were made to employ the phenomenon for engineering stress analysis. In this field, the pioneer work was performed by Professors E. G. Coker and L. N. G. Filon at University College, London. Since that time the technique of photoelasticity has become an established method of experimental stress analysis, much research has been accomplished, particularly in the U.S.A., and in recent years industry has accepted and made considerable use of the technique.

The optical and physical theory and the variety of techniques available in this field have provided material for several excellent textbooks. It is therefore quite impossible to deal thoroughly with the method in one chapter. However, it is hoped that, by explaining the basic principles and the scope of the technique, the reader will then be in a position to decide whether a particular problem could usefully be solved by this method and to consult a fuller treatment of the subject.

Before embarking on a detailed analysis it is useful to have a general picture of what can be done and how the method works. Photoelasticity is an indirect method of stress analysis, in that a study is made of the stress conditions in a scale model made from a transparent plastic material, and from this the stresses in a metal component can be deduced, both qualitative and quantitative results being obtained. Polarized light is passed through the model when in a stressed condition, and changes in the passage of light are thus effected, which may be observed on the image of the model at the end of the optical system. The appearance of an optical stress pattern is illustrated in Plate XIV (*top*) for a pair of mating gear teeth. These optical effects are related to the stress distribution, magnitude and direction in the model; a simple process of scaling provides the actual stresses present in the metal component. The fact that the model is made of plastic and the component of metal is of no significance in the determination of stress in the elastic range, since the widely different elastic constants would only affect the determination of deformations and not stresses.

OPTICAL THEORY

26.2. POLARIZED LIGHT

The condition of the light used in photoelasticity is of some importance and this will be dealt with first. A beam of ordinary light is assumed to travel in wave formation in any number of directions transverse to the axis of propagation. If the wave formation can be confined to one plane, then the light is said to be *plane polarized*.

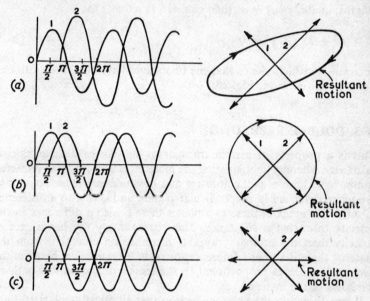

Fig 26.1 Representation of Two Plane-polarized Waves in Perpendicular Planes Forming (a) Elliptical, (b) Circular, (c) Plane Polarization

If two plane-polarized waves having a phase difference α wavelengths vibrate in planes at right angles, Fig. 26.1 (a), then the equations of motion may be represented by

$$x = a_1 \sin \omega t \quad \text{and} \quad y = a_2 \sin (\omega t + \alpha)$$

which gives a resultant motion

$$\frac{y^2}{a_2{}^2} + \frac{x^2}{a_1{}^2} - \frac{2xy \cos \alpha}{a_1 a_2} = \sin^2 \alpha \tag{26.1}$$

which is the equation of an ellipse, and would represent *elliptically-polarized light*.

If the initial phase angle between the two waves is $\alpha = \pm\frac{1}{2}\pi$ or $\frac{1}{4}\lambda$, where λ is the wavelength of the light, and also the amplitudes of the waves are equal, i.e. $a_1 = a_2$, then eqn. (26.1) becomes

$$x^2 + y^2 = a^2 \tag{26.2}$$

which is the equation of a circle, representing *circularly-polarized light*, Fig. 26.1 (b).

A further condition which is of interest is when $\alpha = \pm\pi$ or $\frac{1}{2}\lambda$ wavelengths, and $a_1 = a_2 = a$; then eqn. (26.1) reduces to

$$x^2 + 2xy + y^2 = 0 \tag{26.3}$$

or $x = -y$, which means that the two waves have combined into one plane-polarized wave, Fig. 26.1 (c).

26.3. DOUBLE REFRACTION

This is a property of certain transparent substances such as calcite, mica, etc., whereby light entering the material is split into two refracted components known as the *ordinary* and *extraordinary rays*. These rays are plane polarized in perpendicular planes, and owing to a difference in refractive index in these two planes there is also a difference in the velocity through the substance. The extraordinary ray has a greater velocity than the ordinary ray and therefore on emerging from the material there is a phase difference, or *relative retardation*, between the two rays, which is proportional to the length of the optical path or thickness of the material.

It was shown in the previous section that elliptically-polarized light was the resultant obtained from two plane-polarized rays at right angles having an arbitrary phase difference. Therefore, in general, double refraction produces elliptically-polarized light.

26.4. POLARIZING MEDIUM

Crystals such as tourmaline and herapathite have only one optic axis, that is light can only pass through in one direction. Perpendicular to the optic axis the light is absorbed. Therefore, ordinary light falling on to one of these substances is resolved into components that pass through along the optic axis and components that are stopped, thus resulting in plane-polarized light emerging. In photoelasticity, *Polaroid*, consisting of herapathite crystals embedded in celluloid, is almost universally used as the polarizing medium. It is cheap and can be obtained in large sheets.

26.5. QUARTER-WAVE PLATES

A piece of material which is doubly refracting and has a thickness related to the wavelength of the light used such as to produce a phase difference, or relative retardation, of $\frac{1}{2}\pi$ or $\frac{1}{4}\lambda$ wavelengths between the ordinary and extraordinary ray is known as a *quarter-wave plate*. If the axes in the plate are arranged at 45° to an incoming plane-polarized ray, then the amplitudes of the resolved components which are the ordinary and extraordinary rays will be equal and circularly-polarized light will result. For photoelasticity, mica is the material from which quarter-wave plates are generally made.

26.6. POLARISCOPE

This is the term used to describe the optical equipment required for photoelastic work. A typical arrangement of a *polariscope* is shown in Fig. 26.2. It consists of a light source (1), either white or preferably

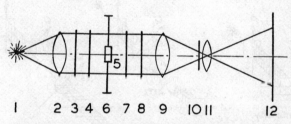

Fig 26.2   Diagrammatic Arrangement of Polariscope

monochromatic (mercury or sodium vapour), followed by a condenser lens (2) to give a parallel beam over the working section. This beam is then plane-polarized by a sheet of Polaroid known as the polarizer (3), from whence, if required, it passes into a mica quarter-wave plate (4) suitably orientated to give circularly-polarized light. The latter would then pass through the stressed model (5) in the loading frame (6) into a second quarter-wave plate (7) and through another sheet of Polaroid termed the analyser (8). Finally, there are collimating (9) and projection lenses (11), a filter (10) to make the light as monochromatic as possible, and at the end of the optical system a viewing screen or camera (12).

26.7. OPTICAL ARRANGEMENTS

Fig. 26.3 shows the basic arrangements of the optical elements of a polariscope for photoelastic work. In Fig. 26.3 (*a*) the quarter-wave plates have been omitted and that set-up is termed a *plane polariscope*.

The optic axes of the polarizer and analyser are at right angles and hence light cannot penetrate beyond the analyser. This is the arrangement used when finding directions of stress in the model. Figs. 26.3 (b) and (c) show the quarter-wave plates included, giving circularly-polarized light and hence the term *circular polariscope*. These arrangements are generally used for finding the distribution and magnitude of stress in the model.

A pair of quarter-wave plates can be arranged in two ways, either crossed, i.e. with the fast (extraordinary) plane in the first opposite the slow (ordinary) plane in the second, or parallel, i.e. the two fast planes coinciding and the two slow planes likewise, Figs. 26.3 (b) and (c).

In the first arrangement the effects of the two plates cancel out and the resultant change is zero, but when the plates are parallel their effects are additive and the resultant change is half a wavelength.

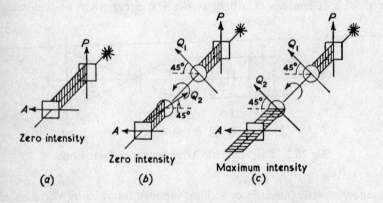

Fig 26.3   Optical Arrangements of Polariscope
(a) Plane polariscope. (b) Crossed-circular polariscope.
(c) Parallel-circular polariscope.

The plane polariscope, that is with crossed polarizer and analyser, causes extinction of the light at the viewing screen. It is therefore evident that insertion of quarter-wave plates in the crossed condition also gives extinction at the viewing screen since they have zero resultant effect. However, in the parallel condition an overall addition of half a wavelength or $\alpha = \pi$ fulfils the condition in Section 26.2 and therefore rotates the resultant plane of vibrations through 90°, thus making it parallel to the optic axis in the analyser, and hence giving maximum brightness at the viewing screen.

PHOTOELASTIC THEORY

26.8. TEMPORARY DOUBLE REFRACTION

As was mentioned previously, it was discovered that certain transparent materials exhibit the property of double refraction or birefringence when in a stressed condition, but return to normal refraction when unstressed. Experimental evidence showed that the extraordinary and ordinary planes of refraction coincide with the planes of principal stress at every point in the material. It was further discovered that the refractive indices and hence the velocities along these planes are related to the magnitude of principal stress. The phase difference or relative retardation between two component rays on emerging from the model was found to be proportional to the difference in principal stresses. Since relative retardation must also be proportional to the thickness of the model, there is the relationship

$$R \propto t(\sigma_1 - \sigma_2)$$

or

$$R = Ct(\sigma_1 - \sigma_2) \tag{26.4}$$

where R is the relative retardation in wavelengths and C is an optical constant. This expression is known as the *stress-optic law* and is the basis of photoelasticity.

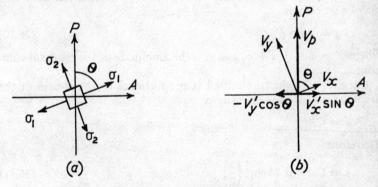

Fig 26.4   Wave Components in a Plane Polariscope on Passing through a Birefringent Model

26.9. ISOCLINICS AND ISOCHROMATICS

Consider the light conditions in a plane polariscope, i.e. no quarter-wave plates; then in Fig. 26.4 (*a*) is shown an element in the model subjected to principal stress σ_1, σ_2 inclined at an angle θ to the incoming plane of polarization. Fig. 26.4 (*b*) represents a view along the axis of the polariscope showing the vector conditions of the components of light passing through the model and resolved by the analyser.

The vector $V_p = a \sin \phi$ represents plane-polarized light from the polarizer. This ray enters the stressed plate and splits into two components:

$$V_x = V_p \cos \theta = a \sin \phi \cos \theta \tag{26.5}$$
$$V_y = V_p \sin \theta = a \sin \phi \sin \theta \tag{26.6}$$

Assuming the x component travels faster through the plate than the y component, then on emerging from the model

$$V_x' = a \cos \theta \sin (\phi + \alpha) \tag{26.7}$$
$$V_y' = a \sin \theta \sin \phi \tag{26.8}$$

where α is the phase difference between the two components. On reaching the analyser, $V_x' \sin \theta$ and $-V_y' \cos \theta$ are allowed to pass through, while $V_x' \cos \theta$ and $V_y' \sin \theta$ are stopped; therefore

$$\begin{aligned} V_A &= V_x' \sin \theta - V_y' \cos \theta \\ &= a \sin \theta \cos \theta [\sin (\phi + \alpha) - \sin \phi] \\ &= a \sin 2\theta \sin \frac{\alpha}{2} \sin \left(\phi + \frac{\alpha}{2} + \frac{\pi}{2} \right) \end{aligned} \tag{26.9}$$

where V_A is the resultant component parallel to the analyser axis. This may be written as

$$V_A = A \sin \phi' \tag{26.10}$$

where $\phi' = \phi + \frac{1}{2}(\alpha + \pi)$ and A is the amplitude of the resultant component.

Intensity of light transmitted is proportional to the square of the amplitude; hence the intensity is

$$I = CA^2$$

Therefore

$$I = Ca^2 \sin^2 2\theta \sin^2 \left(\frac{\alpha}{2} \right) \tag{26.11}$$

i.e. zero intensity and hence extinction at the analyser occurs for $\theta = 0$ and $\alpha = 2m\pi$, where m is 0 or an integer.

The first solution, $\theta = 0$, represents the condition when a plane of principal stress is parallel to the plane of polarization or the optic axis of the polarizer. Hence, at all points in the model where this condition holds, extinction will occur at the analyser, and on the image of the model black bands will be formed which are termed *isoclinics*. An isoclinic may be defined, therefore, as a locus of points in the model along which the directions of principal stress are parallel to the polarizer and analyser optic axes. If the polarizer and analyser are kept crossed and rotated, new isoclinics will appear and previous ones will disappear. A complete set of isoclinics will be obtained by rotating the polarizer and analyser together through 90°. Since isoclinics represent directions of principal stress they are independent of the load value applied to the model. An example of the appearance of isoclinics is given in Plate XIV (*bottom*).

The second condition, $\alpha = 2m\pi$, for zero intensity of light at the analyser, is the relative retardation between an ordinary and extra-ordinary ray emerging from the model. Therefore extinction occurs, and dark bands appear on the image of the model, when the stress conditions produce a relative retardation of $m\lambda$ wavelengths, where $m = 0$ or an integer. These dark bands, which are proportional to the principal stress *difference* at the corresponding points in the model, are termed *isochromatic fringes* or simply fringes. Therefore when $Ct(\sigma_1 - \sigma_2) = \pm m\lambda$ at any point in the model, extinction occurs at the analyser. A fringe represents the locus of points having equal principal stress difference, and may have been formed by 0, 1, 2, 3, etc., wavelengths relative retardation.

26.10. FRINGE IN A CIRCULAR POLARISCOPE

It has already been explained that the insertion of crossed quarter-wave plates into a polariscope has no overall effect, and extinction occurs at the analyser as if it were a plane polariscope. Hence, a stressed model would show a fringe pattern in a crossed-circular polariscope in exactly the same way as in a plane polariscope. However, in the former the isoclinics will no longer appear because with circularly-polarized light it is no longer possible for the light to pass through the model without being resolved into two components at each and every point.

If one of the quarter-wave plates is rotated through 90° so that they are now parallel, there follows an overall change of $\frac{1}{2}\lambda$ in the polariscope, giving a bright field at the viewing screen. Extinction or a fringe now occurs where the conditions of stress in the model are such as to produce $(m + \frac{1}{2})\lambda$ relative retardation. Thus, in the parallel-circular polariscope, fringes are seen for $\frac{1}{2}$, $1\frac{1}{2}$, $2\frac{1}{2}$, etc., wavelengths retardation. An example of half- and whole-order fringe patterns, for a beam fixed at each end and carrying a uniformly distributed load, is shown in Plate XV.

26.11. CALIBRATION

Having obtained a record of the fringe pattern for a particular model, before an analysis of the principal stresses can be made it is necessary to know the optical constant or calibration value for the material, i.e. the amount of principal stress *difference* required to produce one fringe or one wavelength relative retardation.

A calibration can only be made using a known stress field such as simple tension, compression or pure bending, and must be carried out on a specimen cut from the same sheet of material as the model. In the first two, the stress–fringe relationship is given by $R = Ct(\sigma_1 - \sigma_2)$ $= Ct\sigma_1$ or $-Ct\sigma_2$ respectively; hence retardation is proportional to the direct stress on the cross-section of the test piece. Since the stress is uniform over the length of the test piece, likewise the retardation is uniform, and hence a fringe in fact covers the whole of the test piece.

If an increasing load is applied, cycles of alternate brightness and darkness occur on the image, the extinction corresponding to each successively higher fringe order (whole for crossed and half for parallel quarter-wave plates). A plot of load against fringe order gives a straight line, and dividing the slope of this line by the cross-sectional area of the test piece gives a value for stress/fringe order.

This constant is known as the *model fringe value F_m*, in units of stress per fringe, and since retardation is proportional to thickness of model, it only applies for that thickness of test piece. For purposes of comparison or conversion, F_m is multiplied by the thickness of the model to give $F_m t = F_1$, known as the *material fringe value*, which is that stress difference required to produce one fringe order in a 25 mm thickness of the material, i.e. stress per fringe per 25 mm.

Having determined the appropriate fringe value for the model to be analysed, it is then only necessary to multiply the fringe order n at a point by the model fringe value F_m to give the principal stress difference at that point, i.e. $nF_m = (\sigma_1 - \sigma_2)$.

26.12. STRESS TRAJECTORIES

Isoclinics in themselves do not provide a clear picture of direction of stress in the model. However, from them it is possible to construct a set of curves known as *stress trajectories*. The tangent and normal at a point on a stress trajectory give the directions of the two principal stresses at that point. This information obtained from a photoelastic test is in itself quite useful since, with the directions of principal stresses known, for example, a resistance strain gauge analysis of the component can become somewhat simpler.

Of the various ways to construct trajectories one relatively simple and quick approximate method will be described. On the isoclinic diagram draw a reference line parallel to the axis $\theta = 0°$ or the initial setting of the polarizer (or analyser). Draw a series of small parallel guide lines as in Fig. 26.5 at points along an isoclinic at an angle to the reference line equal to θ, the parameter of the isoclinic. Then draw smooth curves to cut the isoclinics at directions parallel to the guide lines as shown. Two orthogonal sets of curves will be obtained represent-ing the σ_1 and σ_2 directions of stress. It is evident that the free boundary of a model will also be a trajectory.

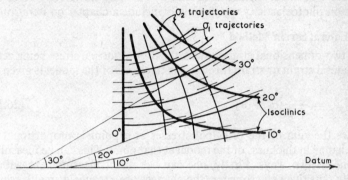

Fig 26.5 Construction of Stress Trajectories

26.13. SEPARATION OF PRINCIPAL STRESSES

The only information obtained directly about the distribution and magnitude of stress from a photoelastic experiment is in terms of the principal stress *difference* at a point. It should also be noted that maxi-mum shear stress is half the principal stress difference and that this is a useful quantity to have directly. It is also important to remember that on a "free", or unloaded, boundary one of the principal stresses must be zero (normal to the boundary), and therefore boundary stresses can be obtained directly from the fringe pattern and model fringe value.

In order to obtain individual principal stresses, normal, or shear stress in the model, it is necessary to "separate" σ_1 from σ_2 in a given fringe pattern. There are many ways of doing this, but it is beyond the scope of this chapter to deal with them all; therefore only a few will be described.

One obvious approach is to find the sum of the principal stresses at a point and combine this with the difference.

(a) Relaxation or Iteration Method

This is a numerical analysis of the problem. It has previously been said that on a free boundary one of the principal stresses is zero. Therefore $(\sigma_1 - \sigma_2) = (\sigma_1 + \sigma_2) = \sigma_1$ say, with σ_2 zero. We can use this condition to solve the Laplace equation

$$\left(\frac{\partial^2}{\partial x^2} + \frac{\partial^2}{\partial y^2}\right)(\sigma_1 + \sigma_2) = 0 \qquad (26.12)$$

by successive approximations of $(\sigma_1 + \sigma_2)$ at the internal points of a grid, given the values of $(\sigma_1 + \sigma_2)$ at the free boundary of the network. Those not familiar with relaxation techniques are advised to consult a detailed work such as that by R. V. Southwell or one of the texts on photoelasticity which usually include a chapter on iteration.

(b) Lateral Strain Method

In a two-dimensional stress system, the third principal stress being zero, the lateral strain or strain through the thickness of the model is given by

$$\varepsilon_3 = -\frac{v}{E}(\sigma_1 + \sigma_2) \tag{26.13}$$

Hence the sum of the principal stresses at a point is proportional to the change in thickness of the model at that point. This can be measured in one of two ways. Firstly, one can use direct measurement with a highly sensitive extensometer (the changes are extremely small), which is a point-by-point method requiring a careful technique. An alternative approach is to use an interferometer. This is an optical device which measures flatness of a surface by the production of an *interference fringe* pattern. These fringes are similar in appearance to isochromatics, but in this case are proportional to $(\sigma_1 + \sigma_2)$, and a set of fringes is termed an *isopachic* pattern. A combination of isopachic and isochromatic fringes permits the separation of σ_1 and σ_2.

(c) Shear Difference Method

For an element of material subjected to direct stress and shear stress, it was shown in Chapter 14 that the equation of equilibrium is

$$\frac{\partial \sigma_x}{\partial x} + \frac{\partial \tau_{xy}}{\partial y} = 0 \tag{26.14}$$

We can make use of this relationship to obtain $(\sigma_1 + \sigma_2)$ and hence separate principal stresses.

Firstly, the stress trajectories must be plotted in the area where stress separation is required. Then on this diagram, Fig. 26.6, draw an arbitrary line from a free boundary and divide it up into increments δx. Next, through points such as B, C and D draw perpendicular lines, and where these lines cut nearby trajectories on either side of A at points such as E and E', determine the maximum shear stress $\frac{1}{2}(\sigma_1 - \sigma_2)$ from the isochromatic fringe pattern. Then

$$\tau_E = \tfrac{1}{2}(\sigma_1 - \sigma_2)_E \sin 2\theta_E \tag{26.15}$$

and

$$\tau_{E'} = \tfrac{1}{2}(\sigma_1 - \sigma_2)_{E'} \sin 2\theta_{E'} \tag{26.16}$$

where θ_E and $\theta_{E'}$ are the angles between EE' and the trajectories. Now,

$$\frac{\partial \tau_{xy}}{\partial y} = \frac{\tau_{xyE} - \tau_{xyE'}}{EE'} \qquad (26.17)$$

But, from eqn. (26.14),

$$\delta \sigma_x = -\frac{\partial \tau_{xy}}{\partial y} \delta x$$

Hence the increments of $\delta \sigma_x$ can be found at points B, C and D, and since $\sigma_x = 0$ at A, the increments can be summed from A to D to give values of σ_x.

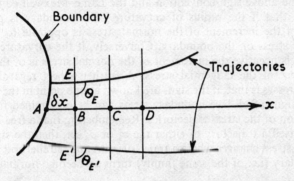

Fig 26.6 Construction for the Shear Difference Method

If principal stresses are also required, then from eqn. (13.25),

$$\sigma_x = \frac{\sigma_1 + \sigma_2}{2} + \frac{\sigma_1 - \sigma_2}{2} \cos 2\theta \qquad (26.18)$$

and with a knowledge of $(\sigma_1 - \sigma_2)$ from the fringe pattern we can obtain $(\sigma_1 + \sigma_2)$ or σ_1 and σ_2.

(d) Integration along a Stress Trajectory

For this method we make use of the Lamé–Maxwell equations of equilibrium:

$$\left. \begin{aligned} \frac{\partial \sigma_1}{\partial S_1} &= \frac{\sigma_2 - \sigma_1}{\rho_2} \\[2mm] \text{and} \quad \frac{\partial \sigma_2}{\partial S_2} &= \frac{\sigma_2 - \sigma_1}{\rho_1} \end{aligned} \right\} \qquad (26.19)$$

where $\partial\sigma_1/\partial S_1$, $\partial\sigma_2/\partial S_2$ are the rates of increase of σ_1, σ_2 along their trajectories and ρ is the radius of curvature of the orthogonal trajectory at the point being analysed. Subscripts 1 and 2 relate the various quantities to the appropriate principal stresses.

Before proceeding further, a sign convention must be adopted for radii of curvature. The one usually associated with the Lamé–Maxwell equations is as follows. Let S_1 and S_2 be two orthogonal trajectories; the positive direction along S_1 may be chosen arbitrarily, but the positive direction along S_2 must be such that, if S_1 were rotated anticlockwise through 90°, it would be in the same positive sense as S_1. The radii of curvature, ρ_1 or ρ_2, are positive if a tangent to the trajectory rotates anticlockwise as the positive distance along S_1 or S_2 increases.

From the above sign convention and the Lamé–Maxwell equations, it follows that if the radius of curvature of a boundary is positive the sign of the increment of the normal stress is opposite to that of tangential stress on the boundary. Conversely, if the curvature of the boundary is negative the increment of the normal stress is of the same sign as that of the tangential stress. A further aspect regarding the signs of stresses is that, if the signs are known at one point in the model, then the signs of all free boundary stress can be determined from an examination of the stress trajectories. Remembering that a free boundary is in itself a trajectory of either the σ_1 or σ_2 set, then the sign of a boundary stress changes where a trajectory running parallel and near to the boundary (i.e. of the same family) turns to become normal to the boundary.

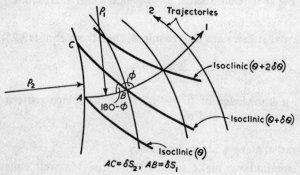

Fig 26.7   Integration along a Stress Trajectory

Referring to Fig. 26.7, the radius of curvature ρ_2 is given by $\delta S_2/\delta\theta$, where $\delta\theta$ is the increment between the θ parameters of two adjacent isoclinics. Eqn. (26.19) becomes

$$\frac{\partial\sigma_1}{\partial S_1} = (\sigma_2 - \sigma_1)\frac{\delta\theta}{\delta S_2} \tag{26.20}$$

or

$$\delta\sigma_1 = (\sigma_2 - \sigma_1)\delta\theta \frac{\delta S_1}{\delta S_2}$$

Now, $\delta S_1/\delta S_2 = AB/AC = -\cot\phi$, Fig. 26.7; therefore

$$\delta\sigma_1 = (\sigma_1 - \sigma_2)\cot\phi \ \delta\theta \qquad (26.21)$$

Starting at a free boundary at various points along a σ_1 trajectory, $(\sigma_1 - \sigma_2)$ is found from the fringe pattern, $\delta\theta$ from the isoclinic pattern, and ϕ from the inclination of the isoclinics with the trajectory. It is then possible to sum the increments of $\delta\sigma_1$ to obtain values of σ_1 along the trajectory.

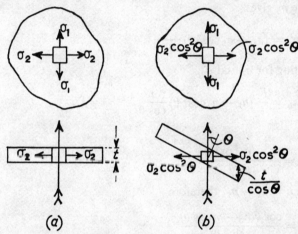

Fig 26.8   Conditions in an Element during Normal and Oblique Incidence

(e) Oblique Incidence Technique

Another most useful technique for separating principal stresses was developed by Drucker. Firstly, the normal fringe pattern and isoclinics are obtained for the model. Then at a point in the model where it is desired to separate the stresses, from a knowledge of the directions of stress from the isoclinic through the point, the model is rotated through a small angle about one of the principal stress axes. A fringe-order reading is made with the polarized light having the oblique incidence from the rotation above. It is necessary to have the model immersed in a fluid of the correct refractive index to avoid dispersion of the light.

The fringe orders obtained from normal and oblique incidence enable the principal stresses to be separated as follows. Under normal incidence, Fig. 26.8 (a), we have the relationship

$$F_1 n = (\sigma_1 - \sigma_2)t \qquad (26.22)$$

where F_1 is the material fringe value. Dividing the retardation into two parts n_1 and n_2, where

$$n_1 = \frac{\sigma_1 t}{F_1} \quad \text{and} \quad n_2 = \frac{\sigma_2 t}{F_1}$$

Then

$$n = n_1 - n_2 \tag{26.23}$$

For rotation about σ_1 and oblique incidence, the component of σ_2 perpendicular to the light is $\sigma_2 \cos^2 \theta$ and the optical path length becomes $t/\cos \theta$, Fig. 26.8 (b). Therefore, the fringe order under oblique incidence n_0 gives

$$F_1 n_0 = (\sigma_1 - \sigma_2 \cos^2 \theta) \frac{t}{\cos \theta} \tag{26.24}$$

Substituting for σ_1 and σ_2,

$$F_1 n_0 = \frac{F_1}{t} (n_1 - n_2 \cos^2 \theta) \frac{t}{\cos \theta}$$

or

$$n_0 = \frac{n_1 - n_2 \cos^2 \theta}{\cos \theta} \tag{26.25}$$

But also $n = n_1 - n_2$; therefore

$$n_1 = \frac{\cos \theta (n_0 - n \cos \theta)}{\sin^2 \theta} \tag{26.26}$$

and

$$n_2 = \frac{n_0 \cos \theta - n}{\sin^2 \theta} \tag{26.27}$$

Hence

$$\sigma_1 = \frac{F_1 \cos \theta (n_0 - n \cos \theta)}{t \sin^2 \theta} \tag{26.28}$$

$$\sigma_2 = \frac{F_1 (n_0 \cos \theta - n)}{t \sin^2 \theta} \tag{26.29}$$

26.14. THREE-DIMENSIONAL PHOTOELASTICITY

So far in this chapter, in explaining the principles and application of photoelasticity, only two-dimensional or plane stress systems have been considered. In practice there are many cases in which a component is subjected to a three-dimensional stress system. It is quite feasible to cast or machine a three-dimensional model out of a photoelastic material, and to load it in the same manner as the component. However, a difficulty arises if polarized light is now passed through the model. Retardation effects will occur and a fringe pattern will be seen, but this would not give a direct relationship for the difference of principal stresses, since it is most unlikely that, all along the path of light through the model, two of the principal stresses would be in the plane of the wavefront and the third coincident with the axis of propagation. It is therefore necessary to resort to a technique known as *stress freezing*.

26.15. FROZEN STRESS TECHNIQUE

It was observed by Solakian in 1935 that a piece of birefringent material loaded when hot and allowed to cool under load retained a fringe pattern at room temperature after removal of load. Further investigators established that the "hot" fringe pattern was of the same distribution and followed the same optical and elastic laws as at room temperature. The only difference was that deformations of the model when hot were very much greater than at room temperature, and the material fringe value was much lower. It was further discovered that, after stress freezing, the model could be sliced without disturbing the fringe pattern and the slice could be analysed optically as if it were a two-dimensional model. The necessary conditions for a slice to be analysed in a simple manner are that it is thin, so that there is little variation in stress through the thickness, and also that one principal stress direction is perpendicular to the plane of the slice. These conditions obtain for slices from the surface of the model, or planes and axes of symmetry. The three principal stress magnitudes and directions can be determined for a general slice in any direction through the model, but a point-by-point technique of observation and calculation is required which is much more complex and lengthy. Isoclinics, or more often the direction of stress at a point, can be determined by the normal plane polariscope procedure, and the oblique incidence technique described earlier is very useful for separating the principal stresses.

It would be too lengthy to describe here the techniques for successfully casting stress-free three-dimensional models, the care necessary in uniform heating, loading and cooling, and slicing methods, and the reader is referred to the several authoritative works on photoelasticity.

26.16. PHOTOELASTIC COATING METHOD

A technique which has attracted some attention in recent years is that in which a thin sheet of photoelastic material is bonded on to a metal surface. Loading of the metal causes the surface strain to be transmitted to the coating, hence inducing birefringence in the latter. If a polari-scope is set up as shown in Fig. 26.9 (*a*) or (*b*) then by reflection from the metal surface, the passage of polarized light through the coating will reveal a fringe pattern proportional to the *principal strain difference* in the metal.

One important application for this technique is the measurement of plastic strains in the metal, since the extreme difference between the elastic moduli of the metal and coating allows the latter to remain elastic both optically and mechanically up to strains of about 4% (photo-stress resins are claimed to go much higher). There is the further possibility of being able to study conditions in a complicated com-ponent or structure for which a normal photoelastic model would be quite inappropriate.

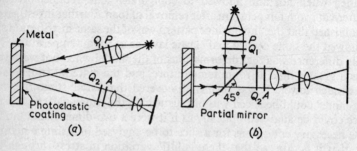

Fig 26.9   Arrangements for the Reflection Type of Polariscope

26.17. PHOTOELASTIC MATERIALS

In this country the most widely used photoelastic material is an epoxy-based resin known as Araldite, which is what is termed a thermosetting resin. In the U.S.A. there are several well-tried and successful materials for both two- and three-dimensional work, namely Bakelite, Fosterite and Kriston; however, these materials are rather expensive and show little advantage over Araldite.

Materials such as celluloid and Perspex, while exhibiting birefringence, are optically too insensitive for normal photoelastic use, although they are sometimes convenient when determining isoclinics.

There are many requirements in the properties of a good photo-elastic material, the more important of which are high optical sensiti-vity, tensile strength and modulus, absence of optical and mechanical creep, and time-edge stress, and the ability of casting and machining.

Although photoelastic properties tend to vary somewhat from one batch of material to another, some typical values for various sub-stances are given in the table below.

Substance		Tensile strength, MN/m²	Modulus of elasticity, GN/m²	Poisson's ratio	Material fringe value, kN/m² per fringe for 25 mm $\lambda = 5,461$ Å
Araldite B (CT.200) (U.K.	20°C	83	3·1	0·3	360
	135°C	2·1	0·013	—	9·5
Bakelite 61-893 (U.S.A.)	20°C	111	4·28	0·36	590
	110°C	2·8	0·0076	0·5	23
Fosterite (U.S.A.) 87°C		3·6	0·0166	0·48	28
Kriston (U.S.A.) 20°C		58	3·73	—	555
Kriston (U.S.A.) 135°C		4·7	0·095	—	43·5
Cellulose nitrate (Xylonite: U.K.)		41 to 55	1·73 to 2·21	0·34	1520
Gelatine* (14% glycerine)		—	0·097 MN/m²	0·5	1·2

* For soil mechanics and body force investigations.

BIBLIOGRAPHY

Coker, E. G. and Filon, L. N. G., *A Treatise on Photoelasticity* (C.U.P., 1931).

Frocht, M. M., *Photoelasticity*, Vols. I and II (Wiley, 1941).

Heywood, R. B., *Designing by Photoelasticity* (Chapman & Hall, 1952).

Jessop, H. T. and Harris, F. C., *Photoelasticity, Principles and Methods* (Cleaver-Hume, 1949).

Chapter 27
Other Methods of Strain Measurement

27.1. The two preceding chapters have dealt in some detail with the two most widely used techniques for experimental stress and strain analysis. However, it is useful to appreciate that there are many other techniques and instruments available which can be very valuable when applied to suitable problems. This chapter attempts no more than to bring to the notice of the reader some of these further techniques, including their main principles and spheres of application. Fuller details may then be obtained if required from the more authoritative texts on the subject.

27.2. BRITTLE LACQUERS

During a tensile test on a hot-rolled steel bar, at the commencement of yielding it is observed that the mill scale cracks in consistent patterns (Lüders lines) before flaking off at larger plastic strains. This is because the scale is brittle and cannot absorb the plastic deformation of the ductile substrate. This is perhaps the most natural example of a strain-sensitive brittle coating.

The above example necessitated a yield strain to be developed in the base metal before the coating would crack, and if yielding could be tolerated at any point in a component or structure, then the phenomenon described might help to indicate points of high strain concentration. However, a brittle coating which could be induced to crack at a strain level within the elastic range of the base material would obviously be of more value.

Lacquers or varnishes are generally of a brittle nature, can be brushed or sprayed on to a surface giving relatively good adhesion, and hence become a possible source for a brittle coating for strain determination.

The first study of a brittle lacquer appears to be that by Dietrich in 1924, followed by Sauerwald and Wieland in 1925. However, it was not until 1932 that Dietrich and Lehr succeeded in making a lacquer crack within the elastic range of a steel, the lacquer consisting of colophony dissolved in benzol. Since then a number of different types of brittle lacquer have been developed and studied. Of these, probably the best known are Stresscoat produced in America and Strainlac developed in this country. These two lacquers are very similar in composition, consisting of wood resin with zinc oxide dissolved in carbon disulphide. Dibutyl phthalate is used as the plasticizer.

The principle of the technique of brittle coatings is very simple.

When the strain transmitted from the surface of the metal to the lacquer reaches a critical value in tension, the lacquer will crack perpendicular to the direction of maximum tensile stress. The first crack will initiate at the point in the component or specimen developing the highest strain, and as loading is continued, more cracks will develop and grow as a larger area of metal reaches the critical strain level. The information provided this far is both qualitative and quantitative, the former because regions of strain concentration are immediately revealed, and the latter because the cracks form an accurate pattern of stress trajectories or *isostatics* (see also Chapter 26). This information on the direction of maximum tensile stress at all points on the surface would then enable strain gauges to be cemented on in the appropriate directions. To obtain quantitative data on the magnitude of strain, and hence stress, it is necessary to determine the cracking strain sensitivity of the lacquer.

This calibration may be accomplished by coating a simple cantilever of the same metal, and applying a deflection to the free end. Cracks will then gradually spread from the root towards the free end as the deflection is increased, and the known strain can then be related to the point of cracking. A cantilever calibration beam is illustrated in Plate XVI (*left*) showing the progress of cracking in the lacquer.

The level of cracking strain sensitivity required in a lacquer for effective use in the elastic range of a metal is generally between 0·0005 and 0·0015. The strain sensitivity of Stresscoat and Strainlac is of the order of 0·0007 but this value can be varied if required from about one-half to double under certain conditions.

A further feature of interest is that the locus of points of equal maximum strain is obtained from a line joining the ends of all the cracks in a pattern. These loci are sometimes termed *isoentatic lines*. An example of three stages of cracking in a notched bar is shown in Plate XVI (*right*).

The accuracy of results depends on a number of factors, but in general is found to be within 10 to 20%. This is not unacceptable when it is considered that this method may provide information unobtainable by theoretical or other experimental methods.

All lacquers are temperature and humidity sensitive, and this can present serious problems in the pursuit of quantitative results. A change in temperature of ±1 to 2°C can markedly alter the strain sensitivity and even make some lacquers craze on occasion. For the best results it is therefore desirable to prepare the specimen, coat, dry and test in a location in which the temperature can be closely controlled. It is also essential that the calibration beam be prepared under exactly the same conditions as the component and kept in close proximity to the latter during testing.

Lacquers are generally applied by spraying, the thickness and evenness of the coating being controlled largely by experience. Thicknesses of coating ranging from 0·05 to 0·2 mm yield a virtually constant strain sensitivity. After coating, a minimum of approximately 24 hours is required for drying. However, if a specimen can also be allowed to age for a few further days the strain sensitivity will increase considerably.

Lacquers exhibit the property of creep at room temperature, and for a lengthy test the results must be corrected to allow for this. The creep behaviour is, however, useful in the study of compressive strain distribution. If the compression is applied slowly and held for a period, the coating absorbs the strain, and stress relaxation occurs owing to creep. On removal of load the release of the compressive strain in the metal produces tensile strain and cracking in the lacquer.

A thorough review of the technique of brittle lacquers has been published by J. R. Linge, *Aircraft Engineering*, April, May and June 1958.

27.3. GRID TECHNIQUES

The deformation of a grid placed on the surface of a component or structure provides a very accurate method of measuring both elastic and plastic strains. In regions where there is a steep strain gradient, such devices as the wire resistance strain gauge and extensometers suffer from the disadvantage that they will only yield an average strain reading over the gauge length which can rarely be made less than 2 mm. There may be difficulties in using photoelasticity because of the geometry of the model required or stresses which enter the plastic range. The grid technique, although perhaps slower and more laborious than the above methods, may be the only answer to the problem.

The principle of the method is simple enough, being that a network of lines is cut into the metal suface, or marked on a thin coating on the metal surface, and the displacement and extension of the lines are measured during an increment of loading, from which the strains and hence stresses can be determined.

Three alternative ways of applying the techniques are by (a) scribed grid, (b) photogrid or (c) replica method. In (a) fine lines are scribed on to the surface using a universal measuring machine. Two mutually perpendicular sets of lines with equal spacing provide a square mesh which is orientated parallel to the principal stress directions if these are known or, if not, some convenient x and y co-ordinate directions. A line thickness of 0·0125 mm and spacing down to a minimum of about 0·25 mm provide a grid suitable for measurement of any conditions of elastic and plastic strain. Unless some form of permanent record can be made of elastic distortion of the grid, measurements must be made with the component under load. If appreciable overall strains are involved these can be measured after removal of load.

The photogrid method employs a photographic emulsion deposited on to the surface of the metal. A fine grid of lines is printed on to the emulsion which is then hardened and will accurately follow the strain in the metal. This method has been used for determining regions of plastic strain.

The replica technique is a quite recent development for highly accurate measurement of elastic or plastic strains. The area of the specimen under examination is first polished, after which an orthogonal grid of scratches of random spacing is put on the specimen by drawing rouge paper across the surface, using a template to obtain straight lines. Datum lines are

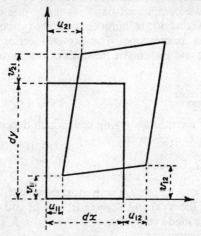

Fig 27.1   Displacement of a Grid Unit

scribed with a sapphire needle. The replica of the scratch pattern is made by applying, under slight impact, a small disc of warm fusible alloy supported by a mushroom-shaped metal platen. The alloy accurately conforms to the scratch pattern and when cool may be levered off the specimen and taken away for measurement under a microscope. The procedure is to take two replicas at the same station under successive loads, which may then be compared to give the strain increments in the required area.

The necessary measurements on an arbitrary rectangle of scratches are indicated in Fig. 27.1. For elastic strains

$$\varepsilon_x = \frac{u_{12} - u_{11}}{dx} \qquad \varepsilon_y = \frac{v_{21} - v_{11}}{dy} \qquad \gamma_{xy} = \frac{u_{21} - u_{11}}{dy} + \frac{v_{12} - v_{11}}{dx}$$

The above correspond to the expressions derived in Chapter 14:

$$\varepsilon_x = \frac{\partial u}{\partial x} \qquad \varepsilon_y = \frac{\partial v}{\partial y} \qquad \gamma_{xy} = \frac{\partial u}{\partial y} + \frac{\partial v}{\partial x}$$

For large (plastic) strains the same measurements of u, v, dx and dy are required, but the above relationships for strain are no longer valid. The way in which the measurements are employed to determine plastic strain is set out in an appendix to the paper on the replica technique by V. M. Hickson, *J. Mech. Engng Sci.*, Vol. 1, 1959.

27.4. DISPLACEMENT GAUGES

The problem of direct measurement of strain in a material or the small displacement of two component parts is mainly one of magnification and recording. The grid techniques described above employ the magnification of a microscope. The present section briefly reviews electrical, acoustic and pneumatic methods of registering and magnifying displacement.

(i) Inductance Gauge

The impedance of a current-carrying coil, which depends on its inductance and resistance, is given by

$$Z = \sqrt{[(2\pi fL)^2 + R^2]}$$

where Z = impedance (ohms), f = frequency of current (hertz), L = inductance of coil (henrys) and R = resistance of coil (ohms).

Usually L can be made very large compared with R, in which case the above expression reduces to

$$Z \approx 2\pi fL$$

so that impedance varies almost proportionately with inductance. If the latter can be made dependent on mechanical displacement then the resulting change in impedance can be measured using an alternating-current bridge circuit and galvanometer, or an oscilloscope where dynamic measurements are required.

Two of the basic methods for changing the inductance of a coil are illustrated in Figs. 27.2 (*a*) and 27.3. In the first, strain in a material, or displacement, causes a change in the small air gap between the coil core and the armature. Unfortunately owing to various factors the single-gap gauge has a linear response only over a very small range. For this reason the double-gap arrangement shown in Fig. 27.3 is more satisfactory, giving much better linearity since the complementary increasing and decreasing air gaps have a compensating effect.

Fig. 27.2 (*b*) shows the other commonly used arrangement for varying inductance by means of a moving-core solenoid. An approximately linear variation of inductance occurs as the core enters the coil.

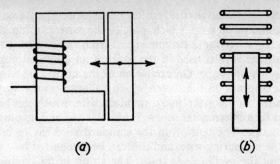

Fig 27.2   Methods of Changing Inductance in a Coil

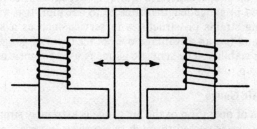

Fig 27.3   Double Gap Change of Inductance

(ii) Capacitance Gauge

This type of gauge operates on the principle of variation in capacitance due to a change in the air gap between parallel plates, or a change in the superimposed area of the plates. The gauge is incorporated as part of an oscillator tuning circuit in a frequency-modulated system, including a frequency discriminator and amplifiers feeding to a cathode-ray oscilloscope.

The care required in the construction of this type of gauge and the associated circuits to obtain sensitivity and linearity, coupled with its size, have limited its use in recent times to certain special applications. The wire resistance strain gauge is now able to cope with the majority of problems, where the capacitance gauge has been employed in the past, more conveniently and accurately. However, the measurement of plastic strains is still a field in which the capacitance gauge has advantages over the wire resistance gauge.

(iii) Acoustic Gauge

This gauge employs the principle of variation in tone of a vibrating wire under tension. It consists of a small frame supporting the knife-edges which bear on the test piece. A wire under initial tension is connected between the knife-edges, such that compressive or tensile strain in the component changes the tension in the wire. The frame also contains a magnetic pick-up which plucks the wire causing a damped oscillation, on receiving a driving signal from the control circuit. The vibration of the wire is used in the form of an electrical impulse to a cathode-ray oscilloscope. On excitation of the measuring wire a Lissajous ellipse appears on the screen which collapses as the amplitude of the wire fades. In the meantime a standard wire, which can be adjusted for tension by a micrometer screw, is also excited and its output fed to the oscilloscope. The tension in the standard wire has to be adjusted to match the measuring wire, and this can be recognized by a particular behaviour of the oscilloscope trace. The strain in the component can then be determined directly from the calibration of the standard wire.

This type of gauge can have quite a rugged construction and is generally used in gauge lengths from 25 to 150 mm. It is very suitable for measuring strains on structures in service such as a ship, bridge or building and has an advantage over other types of gauges owing to its long-term stability. The acoustic gauge is used for measuring strains in the range up to 0·05%.

(iv) Pneumatic Gauge

The principle of operation of the air gauge is extremely simple. If air at constant pressure is passed through two orifices in series, of which the first is fixed in area, then the pressure differential between the two

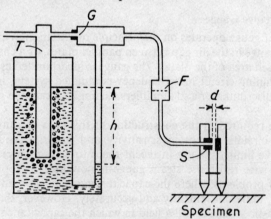

Fig 27.4   Diagrammatic Arrangement of Solex Air Gauge

orifices is a function of the area of the second orifice, or the discharge pressure from the orifice on to a plate. If strain or displacement can be utilized to alter the area of the orifice, or the distance of a plate from the orifice, then there will be a change in the differential pressure between the orifices. This can result in a magnification of strain of up to ×200000.

The principle of the Solex air gauge is illustrated in Fig. 27.4. The input air pressure into the dip tube T is $350 \, \text{kN/m}^2$. Some air passes through the control nozzle G, the excess being bubbled through the water column. From the nozzle G, the flow at $10 \, \text{kN/m}^2$ is via a filter F and thence by a flexible tube to the extensometer measuring nozzle S. At a distance d away from the latter is situated an anvil. The nozzle S and the anvil are located in a pair of knife-edges held on to the specimen. Strain in the latter causes a change in the air flow between the jet and the anvil, and hence varies the air pressure between the two nozzles, which is indicated by the manometer column. The general relationship between the differential pressure h and the distance d is only linear for a very small range and the instrument must be kept within this range. For larger strains a readjustment of the position of the anvil is required.

27.5. MODEL ANALYSIS

Some structures and machines or components are so complex that theoretical analysis is either not possible or very approximate. An experimental approach on a full-size prototype requires it to be built at possibly considerable expense before its behaviour can be studied. As a result a number of changes may be required and a further prototype has to be manufactured. Some of these problems can be overcome by the use of scale models of the structure or machine from which measurements of displacement or strain under load can be made.

For geometrical similarity between model and prototype which are identically loaded and linear elastic in behaviour then the following relations hold between a measured quantity on the model and its equivalent in the prototype:

$$\delta_p = \delta_m \left[\frac{W_p}{W_m} \right] \left[\frac{l_m}{l_p} \right] \left[\frac{E_m}{E_p} \right]$$

$$\sigma_p = \sigma_m \left[\frac{W_p}{W_m} \right] \left[\frac{l_m}{l_p} \right]^2$$

$$\varepsilon_p = \varepsilon_m \left[\frac{W_p}{W_m} \right] \left[\frac{l_m}{l_p} \right]^2 \left[\frac{E_m}{E_p} \right]$$

where W = Applied load
l = Linear dimension
E = Young's modulus
δ = Displacement

and suffixes m and p refer to model and prototype respectively.

The most useful model materials are plastics and rubbers. The former can provide accurate quantitative data while the latter are useful for qualitative studies. PMMA (Perspex) and polystyrene are used for fabricated and machined models or where transparency is an advantage, and epoxy resin (Araldite) is very satisfactory for casting a model. Owing to the much lower modulus of plastics compared with metals, the loads used on the model will develop larger strains and displacements than for the prototype, which assists the measurement sensitivity. As plastics are influenced by temperature and humidity variations it is desirable that these should be kept reasonably constant. Plastics are susceptible to creep at room temperature, and loads must be kept adequately below the proportional limit to minimize this effect.

Measurement of displacement or strain can be made with most of the methods and devices described earlier (brittle lacquer, mechanical gauges, electric strain gauges, etc.). However, care must be taken that the measuring device does not induce some local stiffening or thermal effects into the model.

In summary the principal advantages of model analysis are:

(*a*) Relatively cheap, particularly if various designs have to be considered.
(*b*) Larger elastic displacements for smaller loads.
(*c*) Simpler to modify design and retest.
(*d*) Qualitative analysis rapidly possible.
(*e*) Convenient to store for comparison with future projects.
(*f*) Economical in laboratory space.

27.6. SELECTION OF EXPERIMENTAL TECHNIQUE

The necessity of understanding how to use the methods and instruments for experimental stress analysis described in these last three chapters must also be coupled with a knowledge of which technique to choose for a particular problem. Obviously the apparatus available might largely decide the issue, but it will be assumed in this context that any of the methods described could be employed.

Experimental analysis can be of help at quite early stages in design, particularly where a rigorous theoretical analysis is not possible, or is known to be inexact. Since the component being designed is not in existence, and perhaps a full-scale prototype would be inappropriate or uneconomical at this stage, a model technique is called for. Photoelasticity is the obvious choice, providing boundary stresses (generally of prime interest) and stress direction quickly with little analysis. Design can be varied and studied merely by alterations to one or more models, and even the most complicated three-dimensional frozen stress model and initial analysis can be effected in the space of a week or so. When a component or prototype has already been produced then the full range of techniques can be considered. In this case direct methods of strain measurement are generally preferable to model analysis, with one exception, and that is if stress concentrations are known to exist. Photoelasticity is still probably the most accurate method for determining elastic stress concentration. Brittle lacquers may usefully be applied to the component to discover sources of strain concentration, and to get a general picture of strain distribution and an accurate determination of principal stress direction. The latter can be very useful in enabling strain gauges (electrical, etc.) and extensometers to be attached subsequently in appropriate positions and directions.

Physical size of the problem is an important feature. Small components or areas of a component can be tackled by all methods conveniently. The detailed analysis of a large structure such as an aeroplane, bridge or ship is usually accomplished using wire resistance strain gauges. They are small, relatively inexpensive and suitable for recording a large number of measurements.

The magnitude of strain to be measured will usually be in the elastic range, for which all methods are available with of course various degrees of sensitivity. However, if there is any likelihood of plastic strains being developed at one or more points in the component, only a few of the methods can be used. Although there are special resistance strain gauges capable of measuring up to 10% strain, their gauge length is too great for a sharp strain gradient, and a prior knowledge of the plastic zone is required. The photoelastic coating method is probably the most advantageous, being able to cope with quite large plastic strains and also giving an entire strain field. This method can be backed up, if required, for greater accuracy with one of the grid techniques over a small area. Two most important aspects have not been mentioned so far, and these are dynamic stresses, and conditions which involve temperatures much above atmospheric. Photoelasticity and photo-elastic coatings can be of some help in the former problem, although really good high-speed photographic techniques are required to record fringe patterns. The wire resistance strain gauge coupled to an oscilloscope circuit is probably the most widely used device for recording dynamic strain. At the present time this is also the only accurate way of measuring static or dynamic strains at high temperature, i.e. up to 800°C. All other methods are virtually confined to the range of 0 to 50°C.

It is difficult to summarize all of the above briefly, and so the information has been put in tabular form in Table 27.1.

BIBLIOGRAPHY

Hetenyi, M., *Handbook of Experimental Stress Analysis* (Wiley, 1950).
Lee, G. H., *An Introduction to Experimental Stress Analysis* (Wiley, 1950).

Table 27.1

Method	Uniaxial stress	Biaxial stress	Internal stress	Stress direction	Strain range, %	Gauge length, mm	Point strain	Strain field	Strain gradient or concentration	Static, Dynamic	Temperature limit, °C
Wire resistance strain gauge											
Single	★	—	—	★	2·0	0·8–25	★	—	—	S, D	800
Rosette	—	★	—	★	2·0	0·4–25	★★	—	—	S, D	800
Photoelasticity											
2-dimensional	★	★	—	★	Elastic	—	—	★	★	S, D	50
3-dimensional (frozen stress)	★★	★★	★	★★	Elastic	—	—	★★	★★	S	50
Coating	★	★	—	★	4·0+	—	—	★	★	S, D	50
Brittle lacquer	★	★	—	★	0·3	—	—	★	★	S	30
Grid											
Replica	★	★	—	★	Elastic, plastic	0·08	—	★	★	S	30
Photo- or scribed-	★	★	—	★	Plastic	0·25	—	★	★	S	30
Extensometers and other gauges	★	—	—	—	Varies	2–150	★	—	—	S, D (some)	30

Appendixes

Appendix 1
Sign Conventions in Bending Theory

The sign convention used in this book for the theory of bending is the same as in the authors' previous book and in several others on the mechanics of solids. However, there is no standard convention, and so long as one maintains consistency it is a matter of personal preference.

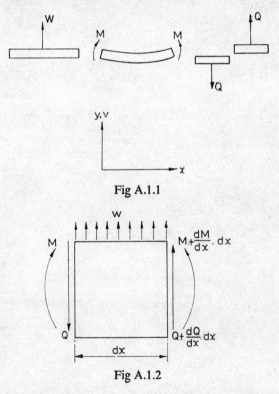

Fig A.1.1

Fig A.1.2

For those who may wish to have an alternative the following is based strictly on mathematical conventions. Distance, y, and deflection, v, are measured positive upwards, distance x is positive from left to right and rotation is positive in the anticlockwise sense.

Fig. A.1.1 illustrates the convention for positive load, bending moment and shear force, and Fig. A.1.2 shows a small length of beam in equilibrium under the above forces and moments.

The more important relationships derived in Chapters 1, 6 and 7 are given below with the appropriate signs for this convention:

$$\frac{dQ}{dx} = -w \qquad \frac{dM}{dx} = -Q$$

$$\sigma = -\frac{My}{I} \qquad \frac{d^2v}{dx^2} = \frac{M}{EI}$$

If it is preferred to have loading and deflection positive downward, the mathematical convention can still be employed in a consistent manner by rotating all the above diagrams through 180°. The equations will then remain the same.

Appendix 2
Properties of Areas

A.2.1. FIRST MOMENT OF AREA

Referring to the plane figure in Fig. A.2.1, the first moment of the element of area dA about the x-axis is $y\,dA$; therefore the first moment of the whole figure is $\int_A y\,dA$ about the x-axis, the suffix A indicating summation over the whole area. The first moment of the whole figure about the y-axis is $\int_A x\,dA$.

Fig A.2.1

A.2.2. POSITION OF CENTRE OF AREA, OR CENTROID

Let the co-ordinates of the centre of area be x and y as shown in Fig. A.2.1. Then the moment of the whole area about an axis is the same as the sum of the moments of all the elements of area about that axis, or

$$A\bar{y} = \int_A y\,dA \qquad \text{so that} \qquad \bar{y} = \frac{1}{A}\int_A y\,dA$$

Similarly,

$$\bar{x} = \frac{1}{A}\int_A x\,dA$$

If either or both axes pass through the centre of area then $\bar{x}$ or $\bar{y}$ or both $= 0$, and

$$\int_A y\,dA = 0 \qquad \text{and/or} \qquad \int_A x\,dA = 0$$

A.2.3. SECOND MOMENT OF AREA

If the first moment of the element dA about an axis is multiplied again by its respective co-ordinate we obtain the second moment of area, namely $y^2 dA$ or $x^2 dA$. The second moment of area of the whole figure about the x-axis is

$$I_{xx} = \int_A y^2 dA$$

and about the y-axis,

$$I_{yy} = \int_A x^2 dA$$

A.2.4. PARALLEL AXIS THEOREM

It is sometimes convenient to determine the second moment of area about axes parallel to the centroidal axes.

Fig A.2.2

Referring to Fig. A.2.2, the second moment of the element dA about the x'-axis is $(y + b)^2 \, dA$, and for the whole figure

$$I_{x'} = \int_A (y + b)^2 \, dA$$

$$= \int_A y^2 \, dA + 2b \int_A y \, dA + \int_A b^2 \, dA$$

but $\int_A y \, dA = 0$ is the first moment about a centroidal axis; therefore

$$I_{x'x'} = I_{xx} + b^2 A$$

and

$$I_{y'y'} = I_{yy} + a^2 A$$

A.2.5. PRODUCT MOMENT OF AREA

Let x and y be the co-ordinates of an element dA about two perpendicular centroidal axes; then $xy\,dA$ is termed the *product moment of area*. For the whole figure,

$$I_{xy} = \int_A xy\,dA$$

This is a very important quantity in the analysis of bending of unsymmetrical cross-sections. It is evident that, if one or both centroidal axes is also an axis of symmetry, then $I_{xy} = 0$.

For the case of parallel axes,

$$I_{x'y'} = I_{xy} + abA$$

A.2.6. POLAR SECOND MOMENT OF AREA

The second moment of an area about an axis perpendicular to the plane of the figure is termed the *polar second moment of area*. Referring

Fig A.2.3

to Fig. A.2.3, the polar second moment of dA is r^2dA, and for the whole figure,

$$J \text{ (or } I_p) = \int_A r^2\,dA$$

Also, since $r^2 = x^2 + y^2$,

$$J = \int_A x^2\,dA + \int_A y^2\,dA$$

$$= I_{yy} + I_{xx}$$

This is known as the *perpendicular axes theorem*.

A.2.7. PRINCIPAL AXES

For any plane figure it is possible to find a pair of perpendicular axes about which the product moment of area is zero. These particular axes are known as *principal axes*, and the second moments of area about them are termed *principal second moments of area*.

Problems

Problems

Problems

CHAPTER 1

1.1 The bracket ABC is freely pivoted on the vertical rod (Fig. P.1.1). Determine the forces transmitted to the rod at A, B and the supports when a 1 kN vertical force is applied at C, and indicate directions on a sketch.

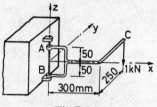

Fig P.1.1

1.2 A rectangular table has sides AB = CD = 2 m and BC = AD = 1 m. Legs are situated at A, B and the mid-point of CD. Loads of 12 kN and 16 kN are applied at the mid-points of AD and BC respectively. Calculate the force set up in each leg.

1.3 Determine the support reactions and the forces in the plane pinjointed framework illustrated in Fig. P.1.2 by (a) resolution at joints, (b) graphical construction.

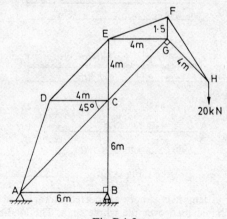

Fig P.1.2

1.4 Find the forces in members CD and EG of Fig. P.1.2 using the method of sections.

1.5 Use the method of tension coefficients to calculate the forces in the pin-jointed space frame shown in plan view in Fig. P.1.3. The pinned supports A, B and C are at the same level, and DE is horizontal and at a height of 4 m above the supports.

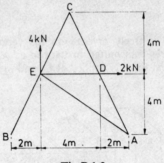

Fig P.1.3

1.6 Sketch the shear force and bending moment diagrams, inserting the principal numerical values, for the beams illustrated in Fig. P.1.4. Also establish any positions of contraflexure.

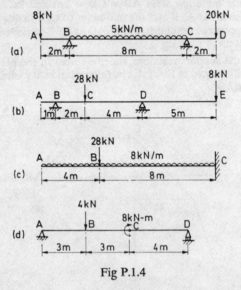

Fig P.1.4

1.7 A beam 6 m in length is simply supported at each end and carries a varying distributed load over the whole span. The loading distribution is represented by the equation $w = ax^2 + bx + c$, where w is the load intensity in kN/m at a distance x along the beam, and a, b and c are constants. The load intensity is zero at each end and has its maximum value of 4 kN/m at mid-span.

Apply the differential equations of equilibrium relating load, shear force and bending moment to obtain the maximum values and distributions of shear force and bending moment.

CHAPTER 2

2.1 A double-acting hydraulic cylinder has a piston 250 mm in diameter and a piston rod 75 mm in diameter. During one stroke the water pressure is 7000 kN/m² on one side of the piston and 280 kN/m² on the other, and on the return stroke the pressures are interchanged. Determine the maximum stress in the rod, allowing for the reduced piston area on the rod side.

2.2 A thin spherical steel vessel is made up of two hemispherical portions bolted together at flanges. The inner diameter of the sphere is 300 mm and the wall thickness is 6 mm. Assuming that the vessel is a homogeneous sphere, what is the maximum working pressure for an allowable tensile stress in the shell of 150 MN/m²?

If 20 bolts are used of 16 mm diameter to hold the flanges together, what is the tensile stress in the bolts when the sphere is under full pressure?

2.3 A suspension bridge is constructed with twin cables and carries a horizontal uniformly distributed loading of 2 kN/m of span, which is 300 m. The lowest point of the cables is 50 m below one support, which is 10 m below the higher support. Determine a suitable cross-sectional area for each cable using a safety factor of 2 and a tensile stress of 300 MN/m².

2.4 A thin-walled steel cylinder of internal diameter 400 mm with closed ends is rotated about a longitudinal axis at a speed of 5000 rev/min. Whilst it is rotating, it is subjected to an internal pressure of 4000 kN/m². If the maximum allowable tensile stress in any direction in the material is 175 MN/m², calculate a suitable shell thickness. Density of steel = 7·83 Mg/m³.

2.5 A conical storage tank has a wall thickness of 20 mm and an apex angle of 60°. If the vessel is filled with water to a depth of 3 m, calculate the maximum meridional and circumferential stresses. The water loading is 9·81 kN/m³.

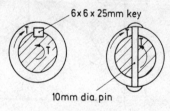

Fig P.2.1

2.6 The hub of a pulley may be fastened to a 25 mm diameter shaft either by a square key or by a pin as shown in Fig. P.2.1. Determine the torque that each connection can transmit if the average shear stress in the key or pin is not to exceed 70 000 kN/m².

2.7 A thin-walled circular tube of 50 mm mean radius is required to transmit 300 kW at 500 rev/min. Calculate a suitable wall thickness so that the shear stress does not exceed 80 MN/m².

CHAPTER 3

3.1 Show that

$$\varepsilon_{vol} = \frac{1 - 2v}{E} (\sigma_x + \sigma_y + \sigma_z)$$

3.2 Express the stresses σ_x, σ_y, σ_z in terms of the three co-ordinate strains and the elastic constants. Obtain similar expressions for the case of plane stress, $\sigma_z = 0$.

3.3 Determine the maximum strain and change in diameter of the cylinder in Problem 2.4 if $E = 208$ GN/m² and $v = 0.3$.

3.4 A copper band 20 mm wide and 2 mm thick is a snug fit on a 100 mm diameter steel bar which may be assumed to be rigid. Determine the stress in the copper if its temperature is lowered by 50°C.

$$\alpha = 18 \times 10^{-6}/°C \qquad E = 105 \text{ GN/m}^2$$

3.5 What are the shear strain and angle of twist per unit length for the tube in Problem 2.7. $G = 85$ GN/m².

CHAPTER 4

The following problems may be used to obtain experience in both the graphical and the virtual work methods.

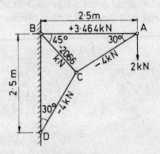

Fig P.4.1

4.1 Determine the vertical and horizontal displacements of joint A of the plane pinjointed framework illustrated in Fig. P.4.1. For all members $AE = 45$ MN.

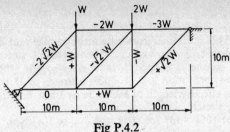

Fig P.4.2

4.2 The plane pinjointed framework shown in Fig. P:4.2 has all members of area 2700 mm². If the movement at the roller support on the inclined surface is limited to 50 mm, determine the maximum value for W. $E = 200\,\text{GN/m}^2$.

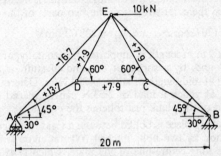

Fig P.4.3

4.3 Determine the horizontal and vertical displacements of joint E in the plane pinjointed truss illustrated in Fig. P.4.3. Some of the member forces are given, and for all members $EA = 100\,\text{MN}$.

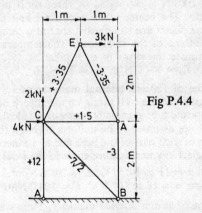

Fig P.4.4

4.4 Obtain the vertical displacement of joint E in the framework shown in Fig. P.4.4. Tension and compression members have cross-sectional areas of 500 and 1000 mm² respectively. $E = 208\,\text{GN/m}^2$.

CHAPTER 5

5.1 A pull of 4·5 kN is resisted jointly by three parallel taut wires with their ends so fixed that *all stretch equally*. The first wire is steel, 15 mm² cross-section; the second is copper, 30 mm² cross-section; the third wire is such that 22·5 kN tension in it produces a stretch of 25 mm.

Taking the effective length of each wire as 2·5 m, find the force in each wire and the stretch.

$E = 207 \times 10^6\,\text{kN/m}^2$ for steel and $100 \times 10^6\,\text{kN/m}^2$ for copper.

5.2 Define the terms elasticity, limit of proportionality, and proof stress.

A composite shaft consists of a brass bar 50 mm in diameter and 200 mm long to each end of which are concentrically friction-welded steel rods of 20 mm diameter and 100 mm length. During a tensile test on the composite bar at a particular stage the overall extension is measured as 0·15 mm. What are then the axial stresses in the two parts of the bar.

$$E_{brass} = 120\,\text{GN/m}^2 \qquad E_{steel} = 208\,\text{GN/m}^2$$

5.3 A rigid plank 2 m in length is supported horizontally, at a height of $\tfrac{1}{3}$ m above the ground, by a spring at one end of stiffness 30 kN/m and a second spring at mid-length of stiffness 20 kN/m. Determine the position on the plank at which a load of 2·5 kN can be placed so that the un-supported end of the plank just touches the ground.

5.4 An elastic packing piece is bolted between a rigid rectangular plate and a rigid foundation by two bolts pitched 300 mm apart and symmetrically placed on the long centre-line of the plate, which is 450 mm long. The tension in each bolt is initially 120 kN, the extension of each bolt is 0·015 mm, and the compression of the packing piece is 0·6 mm. If one bolt is further tightened to a tension of 150 kN, determine the tension in the other bolt.

5.5 A load of 200 kN is supported by three short pillars each of 600 mm² cross-section. The centre pillar is of steel and the two outer ones of copper. The pillars are so adjusted that at a temperature of 15°C each carries one-third of the total load. The temperature is then raised to 115°C. Estimate the stress in each pillar at 15°C and at 115°C. $E_s = 208\,\text{GN/m}^2$, $E_o = 104\,\text{GN/m}^2$, $\alpha_s = 12 \times 10^{-6}/°\text{C}$, $\alpha_o = 18\cdot5 \times 10^{-6}/°\text{C}$.

5.6 A steel tube of 150 mm internal diameter and 8 mm wall thickness is lined internally with a well fitting copper sleeve of 2 mm wall thickness. If the composite tube is initially unstressed, calculate the circumferential stresses set up, assumed to be uniform through the wall thickness, in a unit length of each part of the tube due to an increase in temperature of 100°C. Neglect any temperature effect in the axial direction.

For steel $\alpha = 11 \times 10^{-6}/°\text{C}$; $E = 208\,\text{GN/m}^2$.
For copper $\alpha = 18 \times 10^{-6}/°\text{C}$. $E = 104\,\text{GN/m}^2$.

5.7 A steel tube of 80 mm internal diameter, 2 mm wall thickness and 1·2 m in length is fitted with end plugs and filled with oil at a pressure of 2 MN/m². Determine the volume of oil leakage which would cause the pressure to fall to 1·5 MN/m².

Bulk modulus for the oil $= 2 \cdot 8 \, \text{GN/m}^2$; for the steel $E = 208 \, \text{GN/m}^2$, $v = 0 \cdot 29$.

5.8 A steel bar is subjected to compression and the lateral strains are restrained to one-half and one-third of their normal values. Calculate the slope of the compression stress–strain curve.

$v = 0 \cdot 25$; $E = 208 \, \text{GN/m}^2$.

CHAPTER 6

6.1 A channel which carries water is made of sheet metal 3 mm thick and has a cross-section of 400 mm width and 200 mm depth. Determine the maximum allowable depth of water in the channel for a 7 m length supported at each end. The maximum bending stress (tension or compression) is not to exceed $35 \, \text{MN/m}^2$, and self-weight may be neglected. Loading due to water, $9 \cdot 81 \, \text{kN/m}^3$.

6.2 A cantilevered balcony which projects 2 m out from a wall is constructed of timber joists at $\frac{1}{3}$ m spacing supporting a boarded floor 12 mm thick. The design loading on the floor is $4 \cdot 5 \, \text{kN/m}^2$ and the self-weight of a joist and its associated boarding may be neglected.

If each joist is to be 120 mm deep determine the required width so that the maximum tensile bending stress does not exceed $10 \, \text{MN/m}^2$. For the purpose of calculating the second moment of area, it may be assumed that the neutral axis of the combined joist cross-section and its associated strip of boarding is at the mid-depth of the joist.

6.3 A tapered shaft of length l is built-in at the larger end of diameter d_2, and is free at the smaller end of diameter d_1. A force, W, is applied at the free end perpendicular to the axis of the shaft. Show that the maximum bending stress at any section distant x from the free end of the shaft is given by

$$\frac{Wx}{(\pi/32)[d_1 + (d_2 - d_1)(x/l)]^3}$$

Hence determine the distance x at which point the greatest value of the maximum bending stress occurs.

6.4 A timber beam 80 mm wide by 160 mm deep is to be reinforced with two steel plates 5 mm thick. Compare the resisting moments for the same value of the maximum bending stress in the timber when the plates are: (i) 80 mm wide and fixed to the top and bottom surfaces of the beam, and (ii) 160 mm deep and fixed to the vertical sides of the beam. (E for steel $= 20 \times E$ for timber.)

6.5 A reinforced-concrete beam has a rectangular cross-section 500 mm deep and 250 mm wide. The area of steel reinforcement is 1100 mm, and it is placed at 50 mm above the tension face. Calculate the resisting moment of the section and the stress in the steel if the compressive stress in the concrete is not to exceed $4 \cdot 2 \, \text{MN/m}^2$ and the modular ratio is 15.

6.6 A horizontal beam of rectangular cross-section 100 mm deep and 50 mm wide is simply supported at each end of a 1·5 m span. Vertical loads of 5 kN are applied at $\frac{1}{2}$ m and 1 m from one end, and a horizontal tension of 40 kN is applied at the ends 25 mm below the upper surface. Determine and plot the distribution of longitudinal stress across the section at mid-span.

What eccentricity of end load is required so that there is just no resultant compressive stress at the outer surface?

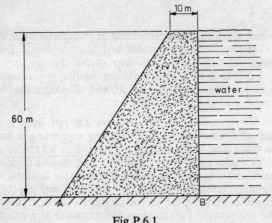

Fig P.6.1

6.7 The cross-section through a concrete dam is illustrated in Fig. P.6.1. Calculate the required width of the base AB so that there is just no tensile stress at B. What is the resultant compressive stress at A? Loading due to water = 9·81 kN/m³. Weight of concrete = 22·7 kN/m³.

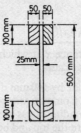

Fig P.6.2

6.8 A beam is made up of four 50 mm × 100 mm pieces of timber glued to a 25 mm × 500 mm web of the same wood as shown in Fig. P.6.2. Calculate the maximum allowable shear force and bending moment that this section can carry. The maximum shearing stresses in the wood and glued joints must not exceed 500 and 250 kN/m² respectively, and the maximum permissible direct stress is 1 MN/m².

6.9 A circular tube of mean radius r and wall thickness t ($\ll r$) is subjected to a transverse shear force Q during bending. Show, by derivation, that the maximum shear stress, τ, occurs at the neutral axis and is equal to $Q/\pi r t$.

6.10 A rolled-steel angle section member illustrated in Fig. P.6.3 is subjected to a bending moment of 1·5 kN-m as shown. Calculate the bending stresses at points A, B and C.

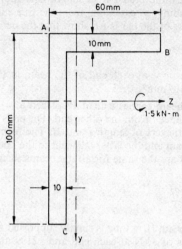

Fig P.6.3

6.11 Determine the location of the shear centre for the beam cross-section shown in Fig. P.6.4. Also calculate maximum values of the horizontal and vertical shear stresses in a flange and the web respectively for a vertical load of 500 kN applied at the shear centre.

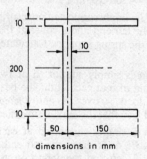

dimensions in mm

Fig P.6.4

6.12 A proving ring, used to calibrate a testing machine, has a mean diameter of 500 mm and a rectangular section 76 mm wide and 12·7 mm thick. If the maximum permitted stress under diametral compression is 55 MN/m², determine the maximum calibration load.

CHAPTER 7

7.1 A beam of length 2·5 L is simply supported at the left end and $2L$ from that end. A uniformly distributed load, w, extends from midway between the supports to the right end. Determine the slope at the right support and the deflection midway between the supports. Flexural stiffness $= EI$.

7.2 A horizontal cantilevered bar of length L is subjected to a uniformly distributed loading, w, extending from the free end to mid-span. If the flexural stiffness of the bar is EI, show that the deflection at the free end is $\dfrac{41}{384}\dfrac{wL^4}{EI}$.

7.3 Determine the slope at each end of the beam in Problem 1.7. $E = 208$ GN/m²; $I = 10^8$ mm⁴.

7.4 A horizontal beam carries a load W placed at a distance $2l$ from one end and a distance l from the other end. The ends of the beam are supported by cantilevers of length l and $2l$. The former is adjacent to the $2l$ length of beam and the latter adjacent to the l portion of beam. Show that if E and I are the same for all the beams then the deflection under the load is

$$v = \frac{14}{9}\frac{Wl^3}{EI}$$

7.5 A horizontal beam 10 m long is simply supported at 1 m from each end. It carries loads of 3 kN at each end and 12 kN at the centre. Find the deflection under the loads and the slope at the supports. $EI = 70$ kN-mm².

7.6 A horizontal beam is simply supported at each end of a 12 m span. It carries a distributed loading varying from 20 kN/m at the left end to 30 kN/m at the right end. Find the position and magnitude of the maximum deflection. $E = 208$ GN/m²; $I = 20 \times 10^{-4}$ m⁴.

7.7 A beam of length l is simply supported on a spring of stiffness K newtons per millimetre at the left end and a roller at the right end. A load W is applied at $\frac{2}{3}l$ from the spring support. Calculate the deflection under the load.

7.8 A horizontal beam is simply supported at two points $2L$ apart, and there is an overhang of a at each end. The beam is subjected to a uniformly distributed load, w, over the entire length. What is the relationship between a and L so that the maximum deflection has the least value?

7.9 A beam of length L simply supported at each end is subjected to a concentrated load, W, at $\frac{2}{3}L$ from the left end. What value of moment must be applied at $\frac{2}{3}L$ so that the deflection at that point is zero?

7.10 A shaft simply supported at each end has a central length of $2a$ of second moment of area $2I$. Two outer sections each of length a have a second moment of area I. Clockwise and anticlockwise moments, M_0, are applied at each change of section. Determine the deflection at either point of application of the moments.

7.11 A shaft is simply supported at each end of a length a having a second moment of area $2I$. There are also overhanging ends each of length a having a second moment of area I. The overhanging ends are subjected to uniform loading, w per unit length. Calculate the deflection at either outer end.

CHAPTER 8

8.1 A bar of length L and cross-sectional area A is required to absorb the same amount of elastic strain energy under axial tension as a bar of area $2A$ for a length of $3L/4$ and area A for a length of $L/4$. Compare the maximum tensile stresses set up in each bar and comment on the result.

8.2 A vertical rod fixed at the upper end has 1 m of 1300 mm² area and 1 m of 2000 mm² area. A load of 100 kg falls onto a collar at the free end through a height of 13·2 mm. Determine the maximum instantaneous stress. $E = 200\,\text{GN/m}^2$.

8.3 Compare the strain energy stored in a beam simply supported at each end and carrying a uniformly distributed load with that of the same beam carrying a concentrated load at mid-span and having the same value of maximum bending stress.

8.4 A horizontal beam of rectangular section 25 mm wide by 50 mm deep is 1 m long and simply supported on (a) rigid rollers, (b) springs of stiffness 300 kN/m. A load of 15 kg falls 50 mm onto the beam at mid-span. Calculate the instantaneous maximum deflections and bending stresses. $E = 208\,\text{GN/m}^2$.

8.5 A bar of rectangular section 1 m in length is simply supported at each end and carries a uniformly distributed load. Determine the maximum depth of section so that the deflection due to shear shall not be greater than 2% of the total deflection.

8.6 A beam of 4 m length carrying a concentrated load W at mid-span is simply supported at the right end and is pin-jointed at the same level at the left end to the free end of a horizontal cantilever of length 2 m. Use Castigliano's theorem to find the deflection under the load. Both beams have a flexural stiffness EI.

8.7 A U-shaped member has a radius R and a leg length L. Show that the deflection caused by forces P applied at the free ends and perpendicular to the legs is

$$\Delta = \frac{P}{EI}\left[\frac{2L^3}{3} + \pi L^2 R + \frac{\pi R^3}{2} + 4LR^2\right]$$

8.8 A split ring of radius R is used as a retainer on a machine shaft, and in order to install it, it is necessary to apply outward tangential forces, F, at the split to open up a gap, δ. If the flexural stiffness of the cross-section is EI, determine the required value of the forces.

CHAPTER 9

9.1 Two shafts are connected end to end by means of a coupling in which there are 12 bolts on a pitch circle diameter of 250 mm. The maximum shear stress is limited to 36 MN/m² in the shafts and 16 MN/m² in the bolts. If one shaft is solid, 50 mm diameter, and the other hollow, 60 mm external diameter, calculate the internal diameter of the latter and the bolt diameter so that both shafts and coupling are equally strong.

9.2 A steel shaft has to transmit 1 MW at 240 rev/min so that the maximum shear stress does not exceed 55 MN/m² and there is not more than 2° twist on a length of 30 diameters. Determine the required diameter of shaft. $G = 80 \, \text{GN/m}^2$.

9.3 A shaft carries five pulleys, A, B, C, D and E, and details of shaft diameters, lengths and pulley torques are given in the following table. Determine in which sections the maximum shear stress and angle of twist occur. $G = 80 \, \text{GN/m}^2$.

Pulley	Torque, N-m	Direction	Shaft	Length, m	Diameter, mm
A	60	Clockwise	AB	1	38
B	900	Anti.	BC	$\frac{1}{2}$	100
C	300	Clockwise	CD	$\frac{1}{2}$	75
D	640	Clockwise	DE	$1\frac{1}{2}$	75
E	100	Anti.			

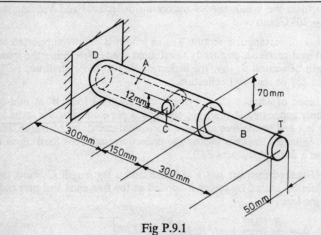

Fig P.9.1

9.4 The inner surface of the aluminium alloy collar A and the outer surface of the steel shaft B of Fig. P.9.1 are smooth. Both the collar and the shaft are rigidly fixed to the wall at D. Pin C fills a hole drilled completely through a diameter of the collar and shaft. The shearing deformation in the pin and the bearing deformation between the pin and shaft can be neglected. Calculate the maximum torque, T, which can be applied to the steel shaft as shown without exceeding an average shearing stress of 8 MN/m² on the cross-sectional area of the pin at E, the interface between the shaft and collar. $G_{steel} = 80 \, GN/m^2$; $G_{aluminium} = 28 \, GN/m^2$.

9.5 A hollow steel shaft is fixed at one end and to the other end is attached a solid circular steel shaft which passes concentrically along the inside of the hollow shaft as shown in Fig. P.9.2. Determine the maximum torque, T, that can be applied to the free end of the solid shaft so that the angle of twist where the torque is applied does not exceed 5°. Local effects where the two parts are connected may be ignored. Shear modulus, $G = 80 \, GN/m^2$.

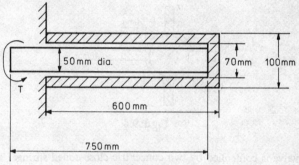

Fig P.9.2

9.6 A cylindrical shaft, 100 mm diameter and 1 m long, is fixed at one end, and a torque of 10 kN-m is applied at the free end. From this end a conical concentric hole of 80 mm diameter extends to a length of $\frac{2}{3}$ m, at which point it has zero diameter. Calculate the angle of twist. $G = 80 \, GN/m^2$.

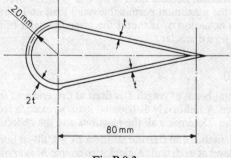

Fig P.9.3

9.7 An aluminium alloy member having the cross-section shown in Fig. P.9.3 is 3 m in length. If the shear stress is not to exceed 30 MN/m² and the applied torque is 134 N-m, determine the required thickness, t, of metal. What is the angle of twist? $G = 28 \, GN/m^2$.

9.8 Compare the torsional strength and stiffness of thin-walled tubes of circular cross-section of mean radius R and thickness t with and without a longitudinal slot.

9.9 A torsional member used for stirring a chemical process is made of a circular tube to which are welded four rectangular strips as shown in Fig. P.9.4. The tube has inner and outer diameters of 94 mm and 100 mm respectively, each strip is 50 mm by 18 mm, and the stirrer is 3 m in length. If the maximum shearing stress in any part of the cross-section is limited to $56\,MN/m^2$, neglecting any stress concentration, calculate the maximum torque which can be carried by the stirrer and the resulting angle of twist over the full length. $\alpha = 0.264$, $\beta = 0.258$, $G = 83\,GN/m^2$.

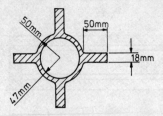

Fig P.9.4

9.10 A valve is controlled by two concentric close-coiled springs. The outer spring has twelve coils of 25 mm mean diameter, 3 mm wire diameter and 5 mm initial compression when the valve is closed. The free length of the inner spring is 6 mm longer than the outer. If the greatest force required to open the valve 10 mm is 150 N, find the stiffness of the inner spring. If the wire diameter for the inner spring is 2 mm, how many coils does it have?

9.11 When an open-coiled spring having ten coils is loaded axially the bending and torsional stresses are $140\,MN/m^2$ and $150\,MN/m^2$ respectively. Calculate the maximum permissible axial load and wire diameter for a maximum extension of 18 mm if the mean diameter of the coils is eight times the wire diameter.

CHAPTER 10

10.1 A horizontal beam of length L is fixed at one end and simply supported at the other. A uniformly distributed load, w, extends from the fixed end to mid-span. Determine all the reactions and the deflection at mid-span.

10.2 A bar of length L and flexural stiffness EI is built-in horizontally and at the same level at each end. A clockwise couple M is applied at mid-span. Find the slope at this point and the deflection curve for the bar.

10.3 An I-section beam of length 5 m is built-in horizontally and at the same level at each end. If the maximum stress is $90\,MN/m^2$, determine what could be the maximum uniformly distributed load. The depth of section is 400 mm, and the second moment of area is $3 \times 10^{-4}\,m^4$.

10.4 A beam of 20 m length is fixed at the left end and simply supported at the right end, where a clockwise couple of 210 kN-m is applied. A uniformly distributed load of 4 kN/m extends from the fixed support to mid-span, and a concentrated load of 60 kN is applied at 5 m from the right end. Calculate the maximum deflection. $EI = 140$ MN-m².

10.5 A beam is fixed horizontally at the left end, A, and is simply supported at the same level at B and C distant 4 m and 6 m from A. A uniformly distributed load of 2·5 kN/m is carried between B and C. Determine the fixing moments and reactions.

10.6 A continuous beam having three spans each of 10 m has the four simple supports A, B, C and D at the same level. Span AB carries a uniform load of 5 kN/m, and a concentrated load of 20 kN acts at 4 m to the left of D. Sketch the shear force and bending moment diagrams.

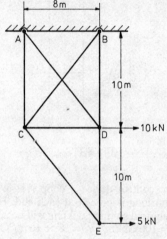

Fig P.10.1

10.7 Determine the forces in the members of the plane pin-jointed redundant framework illustrated in Fig. P.10.1.

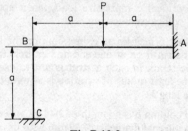

Fig P.10.2

10.8 Find the forces in the members and the principal bending moments for the frame shown in Fig. P.10.2.

CHAPTER 11

11.1 Two rigid bars of equal length are connected by a frictionless hinge at B and are pinned at A and D as illustrated in Fig. P.11.1. A spring of stiffness k is attached to the lower member at C as shown. Determine the critical load for the system.

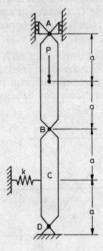

Fig P.11.1

11.2 In a temperature control device a copper strip measuring 8 mm × 4 mm and 100 mm long is pinned at each end. How much axial precompression is required so that buckling will occur after a temperature rise of 50°C. $E = 100 \, \text{GN/m}^2$, $\alpha = 18 \times 10^{-6}/°\text{C}$.

11.3 A vertical compound strut consists of a slender member of length L and stiffness EI, built-in at the lower end and pinned at the upper end to a rigid member of length $L/2$. The upper end of the latter is pinned between rollers which are axially aligned with the whole strut. Determine the critical compressive load, when applied at the roller bearing, which will cause buckling.

11.4 Derive from basic principles an expression for the buckling load of a straight slender strut fixed at one end and entirely free at the other. If the compressive strain in such a strut is not to exceed 0·0008 before buckling and the least radius of gyration is 50 mm, calculate the maximum allowable length.

11.5 A circular steel column has a length of 2·44 m, an external diameter of 101 mm and an internal diameter of 89 mm with its ends position fixed. Assuming that the centre-line is sinusoidal in shape with a maximum displacement at mid-length of 4·5 mm, determine the maximum stress due to an axial compressive load of 10 kN. $E = 205 \, \text{GN/m}^2$.

11.6 A strut of 100 mm diameter with pinned ends has an initial curvature giving a central deflection of 7·5 mm and a slenderness ratio of 100. For a yield stress of 300 MN/m² and a load factor of 2, find the permissible axial load and the maximum stress caused by this load. Area = 1000 mm², $I = 120 \times 10^4$ mm⁴.

11.7 A tubular cast-iron column 5 m long has fixed ends and an external diameter of 250 mm. Calculate a suitable tube thickness if the column supports a load of 1 MN. Assume a constant of 1/6400 in the Rankine formula and a stress of 80 MN/m².

11.8 A beam-column of length L and constant EI is simply supported at each end. It is subjected to a transverse load, W, at mid-span and axial *tensile* end loads, F. Show that the equation for the deflection curve is

$$v = \frac{W}{2\mu F}\left(\operatorname{sech}\frac{\mu L}{2}\right)\sinh \mu x - \frac{W}{2F}x$$

11.9 A column of length L and constant EI has pinned ends subjected to axial compressive loading, P, and a transverse loading of the form $w_0 \sin \pi x/L$. Show that the equation for the deflection curve is

$$v = \frac{w_0 L^4}{\pi^4 EI}\left[\frac{1}{1 - P/P_E}\right]\sin\frac{\pi x}{L}$$

where P_E is the Euler load $\pi^2 EI/L^2$.

11.10 A steel tube having internal and external diameters of 46 and 50 mm respectively is pinned at the ends. It is subjected to axial compression of 5 kN and uniformly distributed transverse loading of 50 N/m. Determine the maximum surface stress. $E = 200$ GN/m².

CHAPTER 12

12.1 At a point in a boiler rivet the material of the rivet is undergoing the action of a shear stress of 50 MN/m² while resisting movement between the boiler plates and a tensile stress of 40 MN/m² due to the extension of the rivet. Find the magnitude of the tensile stresses at the same point acting across two planes making an angle of 80° to the axis of the rivet.

12.2 At a point in the cross-section of a girder there is a tensile stress of 50 MN/m² and a positive shearing stress of 25 MN/m². Find the principal planes and stresses, and sketch a diagram showing how they act.

12.3 The vertical I-section column shown in Figure P.12.1 has a cross-sectional area of 6500 mm² and a second moment of area of 90 × 10⁶ mm⁴. Two strain gauges, A and B, mounted on the web in the position shown, each inclined at 45° to the longitudinal axis give strain readings of 0·0001 and 0·000008 respectively under the action of the given loading system. Determine the value of the transverse load W. $E = 208\,\text{GN/m}^2$; $v = 0·29$.

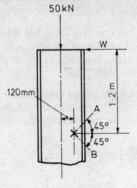

50 kN

W

120mm

1·2 m

A

45°

45°

B

Fig P.12.1

12.4 The loads applied to a piece of material cause a shear stress of 40 MN/m² together with a normal tensile stress on a certain plane. Find the value of this tensile stress if it makes an angle of 30° with the major principal stress. What are the values of the principal stresses?

12.5 At a point in a material the resultant stress on a plane A is 40 MN/m² inclined at 30° to the normal and on a plane B is 10 MN/m² inclined at 45° to the normal. Find the principal stresses and show the position of the two planes A and B relative to the principal planes. Use Mohr's circle only. Planes A and B are not necessarily at right angles to each other.

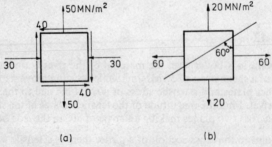

50 MN/m²

40

30   30

40

50

(a)

20 MN/m²

60°

60   60

20

(b)

Fig P.12.2

12.6 Draw the Mohr stress circles for the states of stress at a point given in Figs. P.12.2(a) and (b). For (a) determine and show the magnitude and orientation of the principal stresses. For (b) show the stress components on the inclined plane.

12.7 A pulley of 250 mm diameter is keyed to the unsupported end of a 50 mm diameter shaft which overhangs 200 mm from a bearing. The pulley belt tension on the tight side is three times that on the slack side. Determine the largest values for these tensions if the maximum principal stress in the shaft is not to exceed 150 MN/m².

12.8 At a particular point on the surface of a component the principal strain directions are known but it is not convenient to attach electrical resistance strain gauges in these directions. However, it is possible to cement gauges at 30° and 60° anticlockwise from the major principal strain direction, and the readings from these gauges are +0·0009 and −0·0006 respectively. Construct the Mohr strain circle and find the value of the principal stresses and maximum shear stress. $E = 208 \text{ GN/m}^2$; $v = 0·29$.

12.9 The principal strains, ε_1 and ε_2, are measured at a point on the surface of a shaft which is subjected to bending and torsion. The values are $\varepsilon_1 = +0·0011$ and $\varepsilon_2 = −0·0006$, and ε_1 is inclined at 20° to the axis of the shaft. If the diameter of the shaft is 51 mm and the rigidity modulus for the material is 83 GN/m², determine analytically the applied torque and the maximum shear stress in the material at the point concerned and check graphically.

12.10 A shaft is fitted with a strain gauge the axis of which is inclined at 45° to the axis of the shaft. When calibrated with the shaft in pure torsion the gauge records a strain of 0·0008. In service there is an end load which causes the shaft to extend by 0·228 mm and to contract 0·0127 mm in diameter. What would be the strain gauge record in service if subjected to the same torque as before? Shaft length 152·5 mm, diameter 25·5 mm. Assume that the strain gauge only responds to the component of tensile or compressive strain along its axis.

12.11 Three strain gauges A, B and C are fixed to a point on the surface of a test plate at 120° intervals, and the strains recorded are $\varepsilon_A = +0·001\,08$, $\varepsilon_B = +0·000\,64$, $\varepsilon_C = +0·000\,90$. Draw Mohr's strain circle for this problem and determine the principal strains and the inclination of gauge A to the direction of the greater principal strain.

12.12 At a certain point in a steel structural element the directions of the principal stresses σ_1 and σ_2 are known. Measurements by strain gauges show that there is a tensile strain of 0·000 83 in the direction of σ_1 and a compressive strain of 0·000 52 in the direction of σ_2. Find the magnitudes of σ_1 and σ_2, stating whether tensile or compressive, and the maximum shear stress. $v = 0·25$.

12.13 A rectangular block sustains stresses in three directions at right angles to each other of 50 tensile, 40 compressive and 60 tensile MN/m² respectively. Assuming that Poisson's ratio for this material is 1/3·5 and that E is 201 GN/m², determine the strain in each of the three directions and the values of the volume and torsion moduli.

12.14 A cubical block of material of unit side is strained by direct and shearing strains as follows: a tensile strain of 7×10^{-4} and a compressive strain of 6×10^{-4} on faces perpendicular to those carrying the previous strain, the magnitude of the shear strain being 3×10^{-4}. On faces perpendicular to those carrying the above strains, there is no strain. Determine the principal strains in magnitude and direction, and calculate the magnitude of the single direct stress required to produce the maximum principal strain. Assume $E = 207\,\text{GN/m}^2$.

CHAPTER 13

13.1 For a particular problem the strain displacement equations in cylindrical coordinates are:

$$\varepsilon_r = \frac{\partial u}{\partial r} \qquad \varepsilon_\theta = \frac{u}{r} \qquad \varepsilon_z = \gamma_{r\theta} = \gamma_{\theta z} = \gamma_{zr} = 0$$

Show that the compatibility equation in terms of the stresses σ_r and σ_θ is

$$rv\frac{\partial \sigma_r}{\partial r} - r(1 - v)\frac{\partial \sigma_\theta}{\partial r} + \sigma_r - \sigma_\theta = 0$$

What is the problem?

13.2 Derive compatibility equations from the following strain-displacement relationships:

(a) $\gamma_{xy} = \dfrac{\partial u}{\partial y} + \dfrac{\partial v}{\partial x} \qquad \gamma_{xz} = \dfrac{\partial u}{\partial z} \qquad \gamma_{yz} = \dfrac{\partial v}{\partial z}$

(b) $\varepsilon_z = \dfrac{\partial w}{\partial z} \qquad \gamma_{\theta z} = \dfrac{1}{r}\dfrac{\partial w}{\partial \theta}$

13.3 What is meant by the term *deviatoric strain* as related to a state of strain in three dimensions? Show that the sum of the three deviatoric strains is zero, and also that they can be related to principal strains as follows:

$$\varepsilon_1'^2 + \varepsilon_2'^2 + \varepsilon_3'^2 = \tfrac{1}{3}[(\varepsilon_1 - \varepsilon_2)^2 + (\varepsilon_2 - \varepsilon_3)^2 + (\varepsilon_3 - \varepsilon_1)^2]$$

CHAPTER 14

14.1 A beam of depth d and length l is simply supported at each end and carries a uniformly distributed load over the whole span. Show that the maximum vertical direct stress, σ_y, is $\tfrac{4}{3}(d/l)^2$ times the maximum bending stress, σ_z, at mid-span and therefore in most cases may be considered insignificant in relation to the latter.

14.2 A closed-ended thick-walled cylinder of outer to inner radius ratio k is subjected to internal pressure p. Show that the stress system on an element in the wall is equivalent to pure shear with a superimposed hydrostatic pressure of $p/(k^2 - 1)$. What are the shear stress and the direction of the shear planes?

14.3 Derive expressions for the radial, circumferential and axial strains at the inner and outer surfaces of a thick-walled cylinder of radius ratio k with closed ends subjected to internal pressure p.

14.4 Determine the k ratio for a thick-walled cylinder subjected to an internal pressure of $80\,\text{MN/m}^2$ if the circumferential stress is not to exceed $140\,\text{MN/m}^2$. What are the maximum shear stresses at the inside and outside surfaces?

14.5 One method of determining Poisson's ratio for a material is to subject a cylinder to internal pressure and to measure the axial, ε_z, and circumferential, ε_θ, strains on the outer surface. Show that

$$\nu = \frac{\varepsilon_\theta - 2\varepsilon_z}{2\varepsilon_\theta - \varepsilon_z}$$

The axial and circumferential strains on the outer surface of a closed-ended cylinder of diameter ratio 3 subjected to internal pressure were found to be $1\cdot02 \times 10^{-4}$ and $4\cdot1 \times 10^{-4}$ respectively. Calculate the internal pressure and the hoop strain at the bore. It may be assumed that

$$\sigma_z = \frac{\sigma_r + \sigma_\theta}{2} \qquad E = 207\,\text{GN/m}^2$$

14.6 A solid steel shaft of $0\cdot2\,\text{m}$ diameter has a bronze bush of $0\cdot3\,\text{m}$ outer diameter shrunk onto it. In order to remove the bush the whole assembly is raised in temperature uniformly. After a rise of $100°C$ the bush can just be moved along the shaft. Neglecting any effect of temperature in the axial direction, calculate the original interface pressure between the bush and the shaft.

$$
\begin{aligned}
E \text{ steel} &= 208\,\text{GN/m}^2 \\
\nu \text{ steel} &= 0\cdot29 \\
\alpha \text{ steel} &= 12 \times 10^{-6} \text{ per deg C} \\
E \text{ bronze} &= 112\,\text{GN/m}^2 \\
\nu \text{ bronze} &= 0\cdot33 \\
\alpha \text{ bronze} &= 18 \times 10^{-6} \text{ per deg C}
\end{aligned}
$$

14.7 A compound cylinder is formed by shrinking a tube of 224 mm external diameter and 168 mm internal diameter upon another tube of 126 mm internal diameter. After shrinking, the radial pressure at the common surface is $13\cdot8\,\text{MN/m}^2$. Determine the hoop stresses at the inner and outer surfaces of each tube. Plot diagrams to show the variation of the hoop and radial stresses with radius for both tubes.

14.8 A thick cylindrical sleeve is shrunk on a shaft and the internal diameter is thus extended $0\cdot152\,\text{mm}$ above its original size. The inside and outside diameters of the sleeve are 203 mm and 305 mm respectively. Find (i) the normal pressure intensity between the sleeve and the shaft, (ii) the hoop stresses at the inner and outer surfaces of the sleeve. $E = 207\,\text{GN/m}^2$. Poisson's ratio $= 0\cdot28$.

14.9 A thin disc of inner and outer radii 150 and 300 mm respectively rotates at 150 rad/sec. Determine the maximum radial and hoop stresses. $\nu = 0\cdot304$; $\rho = 7\cdot7\,\text{Mg/m}^3$.

14.10 A thin uniform disc with a central hole is pressed on a shaft in such a manner that when the whole is rotated at n revolutions per minute the pressure at the common surface is p. Derive an expression for the hoop stress in the disc at the periphery, if the inside radius of the disc is r_1 and the outside radius r_2.

14.11 A solid steel disc 457 mm in diameter and of small constant thickness has a steel ring of outer diameter 610 mm and the same thickness shrunk on to it. If the interference pressure is reduced to zero at a rotational speed of 3000 rev/min, calculate the difference in diameters of the mating surfaces of the disc and ring before assembly and the interface pressure. The radial and circumferential stresses at radius r in a ring or disc rotating at ω radians per second are obtained from the following relationships:

$$\sigma_r = A - \frac{B}{r^2} - (3 + v)\frac{\rho}{8g}\omega^2 r^2$$

$$\sigma_\theta = A + \frac{B}{r^2} - (1 + 3v)\frac{\rho}{8g}\omega^2 r^2$$

where A and B are constants, $v = 0.29$ and $\rho = 7.7\,\text{Mg/m}^3$. $E = 207$ GN/m².

14.12 A spherical shell has a k ratio of 1·5. Determine the maximum shear stress at the inner and outer surfaces for an internal pressure of 7 MN/m².

CHAPTER 15

15.1 For a closed-ended thick-walled cylinder of radius ratio k subjected to internal pressure, show that the pressure to cause yielding at the bore according to the shear strain energy criterion is given by

$$p_Y = \frac{k^2 - 1}{k^2\sqrt{3}}\sigma_Y$$

where σ_Y is the yield stress of the material in simple tension. Derive a similar expression for a thick-walled sphere. Compare the yield pressures determined above with those obtained by using the maximum shear stress yield criterion. Which form of thick-walled vessel, cylinder or sphere, having the same k ratio, will withstand the higher pressure before yielding commences?

15.2 A certain material has a yield stress limit in tension of 400 MN/m². The yield limit in compression can be taken as equal to that in tension. The material is subjected to three stresses in mutually perpendicular directions, the stresses being in the ratio $3:2:-1·8$. Determine the stresses which will cause failure according to (a) the maximum shear stress theory, (b) the shear strain energy theory.

15.3 A shaft subjected to pure torsion yields at a torque of 1·2 kN-m. A similar shaft is subjected to a torque of 720 N-m and a bending moment M. Determine the maximum allowable value of M according to (a) maximum shear stress theory, (b) shear strain energy theory.

15.4 A circular steel cylinder of wall thickness 10 mm and internal diameter 200 mm is subjected to a constant internal pressure of 15 MN/m². Determine how much (a) axial tensile load and (b) axial compressive load can be applied to the cylinder before yielding commences according to the maximum shear stress theory. The yield stress of the material in simple tension is 240 MN/m². Assume that the radial stress in the wall of the cylinder is zero. Sketch the plane yield stress locus for the maximum shear stress theory, and show the two points representing the cases above.

15.5 A brass rod of 12 mm diameter is fixed at its lower end and contains a right-angle bend such that it lies in the xy plane shown in Fig. P.15.1. The force P applied at the free end lies in the yz plane inclined at 30° to the z axis. The 0·1% proof stress for yielding in simple tension of the brass is 200 MN/m². Calculate the value of P which will cause yielding at point A on the outer surface according to the shear strain energy criterion.

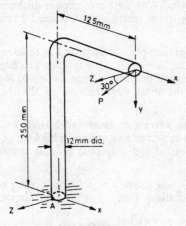

Fig P.15.1

15.6 A steel tube has an internal diameter of 25 mm and an external diameter of 50 mm. Another tube, of the same steel, is to be shrunk over the outside of the first so that the shrinkage stresses just produce a condition of yield at the inner surface of each tube. Determine the necessary difference in diameters of the mating surfaces before shrinking and the required external diameter of the outer tube. Assume that yielding occurs according to the maximum shear stress criterion and that no axial stresses are set up due to shrinking. Yield stress in simple tension or compression = 414 MN/m²; E = 207 GN/m².

15.7 Show that the maximum shear stress in a helical spring subjected either to axial force or axial couple is independent of the helix angle of the coils. An open-coiled helical spring has ten coils of 50 mm pitch and 76 mm mean diameter made from steel wire of 12·7 mm diameter.

If the 0·1% proof stress for the steel is 840 MN/m², determine the corresponding axial load according to the maximum shear stress criterion. $E = 206\,\text{GN/m}^2$; $v = 0\cdot3$.

15.8 A steel rotor disc of uniform thickness 50 mm has an outer rim of diameter 750 mm and a central hole of diameter 150 mm. There are 200 blades each of weight 0·22 kg at an effective radius of 430 mm pitched evenly around the periphery. Determine the rotational speed at which yielding first occurs according to the maximum shear stress criterion. Yield stress in simple tension for the steel is 700 MN/m²; $v = 0\cdot29$; $\rho = 7\cdot3\,\text{Mg/m}^3$; $E = 207\,\text{GN/m}^2$.

CHAPTER 16

16.1 A circular thin steel diaphragm having an effective diameter of 200 mm is clamped around its periphery and is subjected to a uniform gas pressure of 180 kN/m². Calculate a minimum thickness for the diaphragm if the deflection at the centre is not to exceed 0·5 mm. $E = 208\,\text{GN/m}^2$; $v = 0\cdot287$.

16.2 A circular aluminium plate 6 mm thick has an outer diameter of 250 mm and a concentric hole of 50 mm diameter. The edge of the hole is subjected to a bending moment of magnitude 900 N-m/m. Determine the deflection of the inner edge relative to the outer. $E = 70\,\text{GN/m}^2$; $v = 0\cdot3$.

16.2 A circular plate 500 mm diameter and 2·5 mm thick is clamped around its edge and is subjected to a concentrated load of 900 N at its centre. Calculate the radial and tangential bending stresses at the fixed edge. $v = 0\cdot29$.

16.4 Describe briefly the method of solution, stating boundary conditions, and the form of the quantity Q in the governing equation

$$\frac{d}{dr}\left[\frac{1}{r}\frac{d}{dr}\left(r\frac{dw}{dr}\right)\right] = Q/D$$

for the following problems:

1. A piston in a single-acting reciprocating steam engine has a heavy bush and rim and a thin solid web as illustrated in Fig. P.16.1(*a*). It is required to find the deflection of the rim relative to the bush and also the stresses in the web.

2. Part of a ship's derrick is constructed from lengths of thin steel tube. The tubes of radii r_o and r_i are connected by a plate of thickness t as shown in Fig. P.16.1(*b*). If the derrick tubes are subjected to an axial load P, what is the maximum stress due to bending of the interconnecting plate?

16.5 A circular steel plate of 304 mm diameter and 12 mm thick is clamped around the edge. A concentric ring of loading of 20 kN is applied uniformly on a circle of 152 mm diameter. Calculate the deflection at the centre of the plate. $E = 208\,\text{GN/m}^2$; $v = 0\cdot29$.

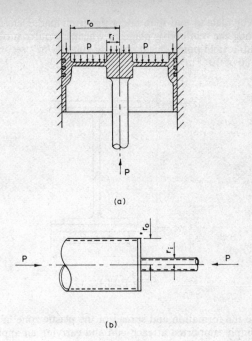

(a)

(b)

Fig P.16.1

CHAPTER 18

18.1 Determine the ratio of the fully plastic to the maximum elastic moment for a beam of elastic-perfectly plastic material subjected to pure bending for the following shapes of cross-section: (a) solid circular, (b) solid square about a diagonal axis, (c) thin-walled circular tube, and (d) thin-walled square tube about a centroidal axis parallel to one of the sides.

18.2 A short column of 0·05 m square cross-section is subjected to a compressive load of 0·5 MN parallel to but eccentric from the central axis. The column is made from elastic-perfectly plastic material which has a yield stress in tension or compression of 300 MN/m². Determine the value of the eccentricity which will result in the cross-section becoming just fully plastic. Also calculate the residual stress at the outer surfaces after elastic unloading from the fully plastic state.

18.3 A steel beam of I-section, as shown in Fig. P.18.1, and of length 5 m is simply supported at each end and carries a uniformly distributed load of 114 kN/m over the full span. Steel reinforcing plates 12 mm thick are welded to each flange and are made of elastic-ideally plastic material. Calculate the plate width such that yielding has just spread through each reinforcing plate at mid-span under the given load. Determine the positions along the reinforcing plates at which the outer surfaces have just reached the yield point. Yield stress = 300 MN/m²; second moment of area = 80 × 10⁻⁶ m⁴.

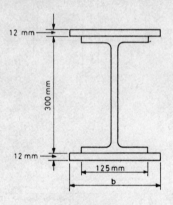

Fig P.18.1

18.4 Examine the formation and spread of the plastic zone in a rectangular beam simply supported at each end and carrying an applied couple at mid-span.

18.5 A solid circular shaft is subjected to pure torsion and the material is elastic-perfectly plastic with a yield stress in shear of 152 MN/m². When the shear stress at one-third of the radius from the centre of the shaft reaches the yield stress, determine the shear strain on the outer surface. Also find the ratio of the torque carried in the above conditions to the maximum elastic torque for the shaft. G = 83 GN/m².

18.6 A solid cylindrical composite shaft 1 m long consists of a copper core of 50 mm diameter surrounded by a well fitting steel sleeve having an external diameter of 62 mm. If the steel has an elastic-perfectly plastic stress–strain relationship, determine the torque that can be applied to the shaft to cause yielding to develop just through to the inner surface of the sleeve. Then calculate the residual shear stress at the outer surface of the steel after removal of the torque. Neglect any stresses set up by the fit between the two parts of the shaft and assume there is no slipping at the copper-steel interface.

G(steel) = 83 GN/m²; G(copper) = 45 GN/m².
τ_Y(steel) = 124 MN/m²; τ_Y(copper) = 76 MN/m².

18.7 A thick-walled cylinder of radius ratio 2:1 and made of an elastic-perfectly plastic material is subjected to internal pressure. Plot a diagram of (p_x/p_{max}) against x, where p_x is the internal pressure to cause yielding to a depth x through the wall, and p_{max} is the pressure which results in yielding right through the wall.

18.8 A thin circular disc with a central hole has inner and outer diameters of 51 and 304 mm respectively. It is required to have a residual compressive hoop stress of 77 MN/m² at the hole when the disc is stationary. Assuming ideal elastic-plastic conditions, yield according to the maximum shear stress theory and elastic unloading, determine the rotational speed necessary to effect the required residual stress. By how much is this speed greater than the speed for initial yielding? Also find the depth of the plastic zone. $\sigma_Y = 340\,MN/m^2$; $w = 7.83\,Mg/m^3$; $v = 0.3$.

18.9 A thin-walled pipe made of polyvinylchloride is subjected to a steady internal pressure of 700 kN/m² at 20°C. If a tensile stress of 17.5 MN/m² is not to be exceeded and the internal radius is 100 mm, determine a suitable wall thickness. What will be the increase in diameter after 1000 hours? The mean creep contraction ratio, v_t, is 0.45, and tensile creep curves provide the following values at 1000 hours:

σ (MN/m²)	6.9	13.8	20.7	27.6	34.5
ε (%)	0.2	0.48	0.97	1.72	3.38

ANSWERS TO PROBLEMS

1.1 $F_{Bz} = 1$; $F_{By} = -F_{Ay} = 2.5$; $F_{Bx} = -F_{Ax} = 3\,kN$.

1.2 $R_A = 5$; $R_B = 9$; $R_{CD} = 14\,kN$.

1.3 $R_A = 22.8$; $R_B = 42.8$; AB = 0; BC = -42.8; AD = $+35.6$; AC = -16.2; DE = $+48.3$; CD = -22.8; EC = -22.8; CG = -48.3; EF = $+40.5$; EG = $+4$; FG = -76; FH = $+73$; GH = $-57.2\,kN$.

1.4 CD = -22.8; EG = $+4\,kN$.

1.5 Reactions: $Ax = 1.2$; $Ay = -0.8$; $Az = -0.8$; $Bx = -0.6$; $By = -1.2$; $Bz = -1.2$; $Cx = -0.6$; $Cy = -1.2$; $Cz = 1.2\,kN$. Members: DE = 2; DA = CD = 0; EA = 1.65; EB = 1.8; EC = $-1.8\,kN$.

1.6 (a) S.F.: A, -8; B, $+17$; linear to C, -23; D, $+20\,kN$. B.M.: A, 0; linear to B, -16; parabolic (max. $+13$) to C, -40; linear to D, 0 kN-m. Contraflexure at 3.16 and 7.64 m.

   (b) S.F.: A, 0 to B, $+12$; C, -16; D, $+8$; E, $+8\,kN$. B.M.: A, 0 to B, 0; C, $+24$; linear to D, -40; linear to E, 0 kN-m. Contraflexure at 4.5 m.

   (c) S.F.: A, 0; linear to B, -32, -60; linear to C, $-124\,kN$. B.M.: A, 0; parabolic to B, -64; another parabola to C, $-800\,kN$-m.

   (d) S.F.: A, $+2$ const.; B, -2 const.; D, $-2\,kN$. B.M.: A, 0; linear to B, $+6$; linear to C, 0, $+8$; linear to D, 0 kN-m.

1.7 $Q_{max} = 8$; $M_{max} = 15\,kN$-m.

654 PROBLEMS

2.1 75 MN/m².
2.2 $p = 12$; $\sigma = 210$ MN/m².
2.3 1710 mm².
2.4 9 mm.
2.5 Meridional, 550; circumferential, 735 kN/m².
2.6 131·2; 137·5 N-m.
2.7 4·56 mm.

3.2 (i) $\quad \sigma_z = \dfrac{E}{(1 + \nu)(1 - 2\nu)} [(1 - \nu)\varepsilon_x + \nu\varepsilon_y + \nu\varepsilon_z]$; etc.

(ii) $\quad \sigma_z = \dfrac{E}{1 - \nu^2} (\varepsilon_x + \nu\varepsilon_y) = \dfrac{E}{1 + \nu} (\varepsilon_x - \varepsilon_z)$; etc.

3.3 0·000 763; 0·305 mm.
3.4 94·5 MN/m².
3.5 0·000 94; 0·019 rad/m.

4.1 Horizontal, 0·196; vertical, 1·045 mm.
4.2 $W = 300$ kN.
4.3 Horizontal, 7·306; vertical, 8·148 mm.
4.4 0·114 mm.

5.1 1·67, 1·61, 1·21 kN; 1·35 mm.
5.2 Steel; 123; brass, 19·7 MN/m².
5.3 1·25 m from raised end.
5.4 124·3 kN.
5.5 111; 144·8 (copper); 178·6 MN/m² (steel).
5.6 16·2 (steel); −64·8 MN/m² (copper).
5.7 1640 mm³.
5.8 230 GN/m².

6.1 99·5 mm.
6.2 32 mm.
6.3 $x = ld_1/2(d_2 - d_1)$.
6.4 M(i) = 1·424M(ii).
6.5 38 kN-m; 89 MN/m².
6.6 −10 to +26 MN/m²; 46 mm.
6.7 51 m; 1·635 MN/m².
6.8 S.F., 4·83 kN; B.M., 4·3 kN-m.
6.10 A, 75·7; B, −35·2; C, −87·5 MN/m².
6.11 43·4 mm to left of web centre-line. Flange, 155; web, 256 MN/m².
6.12 7 kN.

7.1 $\dfrac{5}{48} \dfrac{wL^3}{EI}$; $\dfrac{7}{96} \dfrac{wL^4}{EI}$

7.3 ±0·0016 rad.
7.5 −0·5 mm; +1·48 mm; 0·514 × 10⁻³ rad.
7.6 6·03 m from left end; 16 mm.

7.7 $\dfrac{P}{9K} + \dfrac{4}{243}\dfrac{PL^3}{EI}$

7.8 $L = 1.24a$

7.9 $\frac{2}{5}WL$ clockwise.

7.10 $\dfrac{23}{24}\dfrac{M_0 a^2}{EI}$

7.11 $\dfrac{wa^4}{4EI}$

8.1 Stress in stepped bar is 1.265 times that in uniform bar.

8.2 $50\,\text{MN/m}^2$.

8.3 1.6.

8.4 (a) $2.43\,\text{mm}$; $152\,\text{MN/m}^2$. (b) $7.99\,\text{mm}$; $51.5\,\text{MN/m}^2$.

8.5 $9.2\,\text{mm}$.

8.6 $2P/EI$.

8.8 $F = \delta EI/3\pi R^3$.

9.1 Shaft, $45\,\text{mm}$; bolts, $6.85\,\text{mm}$.

9.2 $163\,\text{mm}$.

9.3 $\tau_{CD} = 6.5\,\text{MN/m}^2$; $\theta_{AB} = 0.003\,67\,\text{rad}$.

9.4 $61.1\,\text{N-m}$.

9.5 $5.35\,\text{kN-m}$.

9.6 $\theta = 0.004\,2 + 0.006\,14$

$$\left[0.439\,\log_e\left(\frac{0.83 + x}{0.83 - x}\right) + 0.878\,\tan^{-1}\left(\frac{x}{0.83}\right)\right]_{x=0}^{x=2/3}$$

9.7 $1\,\text{mm}$; $0.165\,\text{rad}$.

9.8 $3R/t$; $t^2/3R^2$.

9.9 $2.83\,\text{kN-m}$; $0.004\,05\,\text{rad}$.

9.10 $4\,\text{kN/m}$; 10 coils.

9.11 $125\,\text{N}$; $3.9\,\text{mm}$.

10.1 $\dfrac{21}{384}wL$; $\dfrac{171}{384}wL$; $\dfrac{9}{128}wL^2$; $\dfrac{3}{51}\dfrac{wL^4}{EI}$

10.2 $\theta = \dfrac{\bar{M}L}{16}$; $v = \dfrac{1}{EI}\left\{\dfrac{\bar{M}}{4L}x^3 - \dfrac{\bar{M}}{8}x^2 - \dfrac{\bar{M}}{2}\left[x - \dfrac{L}{2}\right]^2\right\}$

10.3 $65\,\text{kN/m}$.

10.4 $15.1\,\text{mm}$.

10.5 $R_A = -187.5$, $R_B = 2938$, $R_C = 2250\,\text{N}$; $M_A = -250\,\text{N-m}$.

10.6 $R_A = 22.12$, $R_B = 29.8$, $R_C = 10.88$, $R_D = 11.04\,\text{kN}$.
$M_B = 28.8$, $M_C = 9.6\,\text{kN-m}$.

10.7 AC, $+15.4$; BD, -15.85; AD, $+12.3$; BC, -11.7; DE, -6.25; CE, $+8.0$; CD, $+2.31\,\text{kN}$.

10.8 $F_{AB} = -\dfrac{P}{4}$, $F_{BC} = -\dfrac{7}{16}P$; $M_A = \dfrac{7}{24}Pa$; $M_B = \dfrac{Pa}{6}$; $M_C = \dfrac{Pa}{12}$

11.1 $ak/3$.
11.2 1·34 kN.
11.3 $P = 0·235EI/L^2$.
11.4 2·77 m.
11.5 6·3 MN/m².
11.6 70 kN; 105 MN/m².
11.7 30 mm.
11.10 38 MN/m².

12.1 55·8; 21·7 MN/m².
12.2 60, −10 MN/m²; $22\frac{1}{2}°$, $112\frac{1}{2}°$ to the neutral axis.
12.3 24·6 kN.
12.4 46·3, 69·3, −23·1 MN/m².
12.5 49, 6 MN/m²; 34°30′, 80°.
12.6 (a) 67·5, −47·5 MN/m²; $112\frac{1}{2}°$, $22\frac{1}{2}°$. (b) Tension 39, shear 35 MN/m².
12.7 2·25, 6·75 kN.
12.8 278, −190, 234 MN/m².
12.9 141 MN/m²; 2·32 kN-m.
12.10 0·0013.
12.11 0·001 135, 0·000 615; 161°.
12.12 154·5, −69, 111·7 MN/m².
12.13 0·000 341, 0·000 548, 0·000 44; 156, 78 GN/m².
12.14 7·09 × 10⁻⁴; −6·09 × 10⁻⁴; 146·7 MN/m².

14.2 $\tau = pr_0^2/(k^2 - 1)r^2$.
14.4 1·92; 110 and 30 MN/m².
14.5 396 MN/m²; 28·7 × 10⁻⁴.
14.6 20·2 MN/m².
14.7 −63·5, −49·5, +49·1, +35·2 MN/m².
14.8 −54, 140, 86·2 MN/m².
14.9 1·6, 13·6 MN/m².

14.10 $\sigma_\theta = \dfrac{2pr_1^2}{r_2^2 - r_1^2} + \dfrac{w}{g}\left(\dfrac{\pi n}{60}\right)^2 [3r_1^2 + r_2^2 + v(r_1^2 - r_2^2)]$

14.11 0·1315 mm; 12·8 MN/m².
14.12 14·7, 4·36 MN/m².

15.1 Sphere; $P_Y = \dfrac{2}{3}\dfrac{(k^3 - 1)}{k^3}\sigma_Y$.

15.2 (a) 250, 166·5, −150 MN/m²; (b) 274, 182·5, −164 MN/m².
15.3 (a) 960 N-m; (b) 830 N-m.
15.4 (a) 1037 kN; (b) 1037 kN.
15.5 139·5 N.
15.6 100, 0·127 mm.
15.7 8·9 kN.
15.8 7300 rev/min.

16.1 3·1 mm.
16.2 1·14 mm.
16.3 66·7, 19·35 MN/m².
16.5 0·113 mm.

18.1 (a) 1·7, (b) 2, (c) 1·272, (d) 1·125.
18.2 10·33 mm; 252·8 and 147·2 MN/m².
18.3 175 mm; 1·92 m from each end.
18.5 0·0055; 1·321.
18.6 5320 N-m; 17 MN/m².
18.8 15 900 rev/min; 10·6%; 6·85 mm approximately.
18.9 4 mm; 1·09 mm.

Index

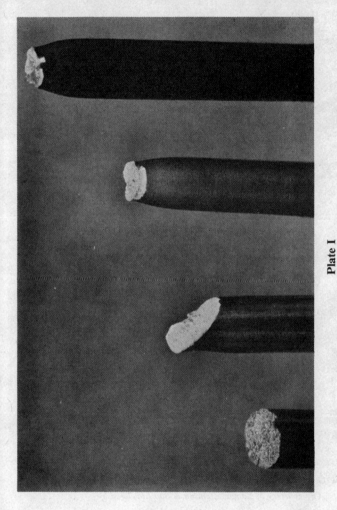

Plate I

Typical Tensile Fractures

Left to right: cast iron—transverse plane without necking; aluminium alloy—single shear plane and slight necking; mild steel—"cup" end of a "cup and cone" fracture; alloy steel (hardened and tempered)—irregular cup and cone failure with marked necking

Plate II
Serrated Wedge Grips for Tension Tests on Flat Bar
(*By courtesy of Samuel Denison & Son Ltd, Leeds*)

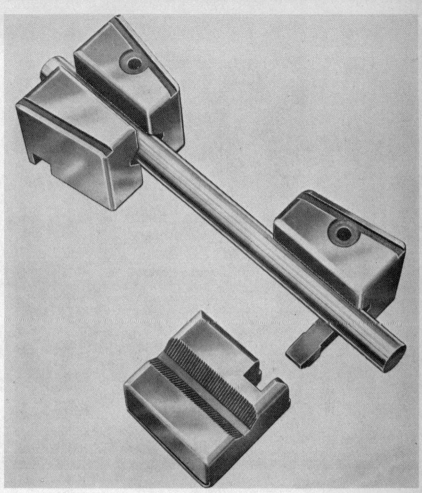

Plate III
Serrated Wedge Grips for Tension Tests on Round Bar
(*By courtesy of Samuel Denison & Son Ltd, Leeds*)

Plate IV

250 kN Capacity Universal Testing Machine for Tension, Compression and Bend Tests

Showing on the left the hydraulic straining frame and the weighing
and recording cabinet on the right

(*By courtesy of W. & T. Avery Ltd, Birmingham*)

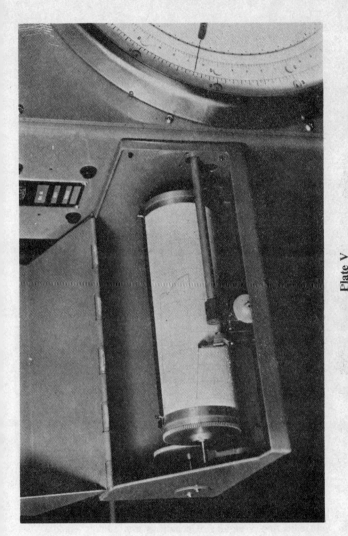

Plate V

Autographic Recording of a Load–Extension Diagram

Horizontal movement of the pen proportional to load and rotation of the drum
proportional to extension

(By courtesy of W. & T. Avery Ltd, Birmingham)

Plate VI
**1.7 kN-m Torsion Testing Machine for Manual or Powered Straining
and with Self-indicating Weighing Mechanism and Sliding Saddle**
(*By courtesy of W. & T. Avery Ltd, Birmingham*)

Plate VII
Above: **Brittle Fracture of a Chain Link**

Below: **Brittle Fracture of the Oil Tanker "World Concord"**
(*By courtesy of the Belfast News-Letter Ltd*)

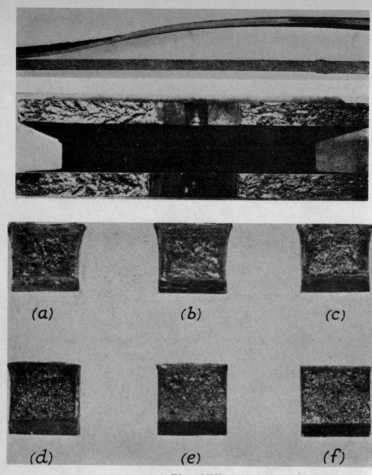

Plate VIII

Top: **Fractures in Plates from an Oil Storage Tank**
Shear fracture above; brittle (cleavage) fracture below—chevron
markings are just visible

(*By courtesy of the Institution of Mechanical Engineers*)

Centre: **Chevron Patterns on a Brittle Fracture Surface, Pointing to the
Source of Initiation at the Notch**

Bottom: **A Set of Charpy Impact Test Fractures of Mild Steel at the
Following Temperatures**
(*a*) 70°C, (*b*) 53°C, (*c*) 48°C, (*d*) 40°C, (*e*) 25°C, (*f*) 18°C

Plate IX
A Machine for Carrying Out Charpy and Izod Impact Tests
By courtesy of W. & T. Avery Ltd, Birmingham)

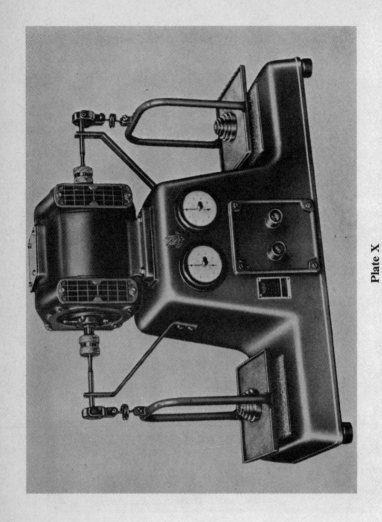

Plate X

A Fatigue Testing Machine for Rotating Cantilever Bending Tests

(By courtesy of W. & T. Avery Ltd, Birmingham)

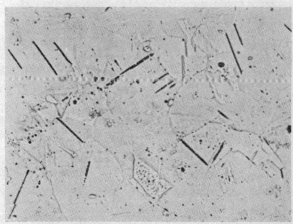

Plate XI
Fatigue Slip Bands in Pure Copper
Above: after 2×10^6 cycles
Below: electro-polished after 4×10^6 cycles, showing persistent slip bands (magnification $\times 200$)
(*By courtesy of Dr. N. Thompson, Bristol University*)

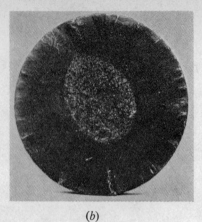

(a)   (b)

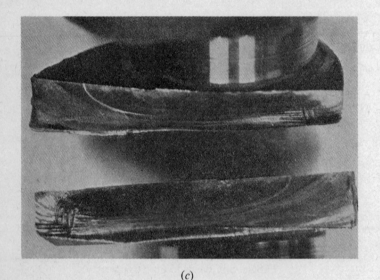

(c)

Plate XII

Typical Fatigue Fractures

(a) Mild steel laboratory specimen. The darker flat area on the fracture surface to the left is due to fatigue. The remainder is the final rupture by shear

(b) Hardened steel wagon axle. The fatigue crack area surrounds the central portion of final failure

(c) Automobile crankshaft web. Note the concoidal markings emanating from the source of fracture

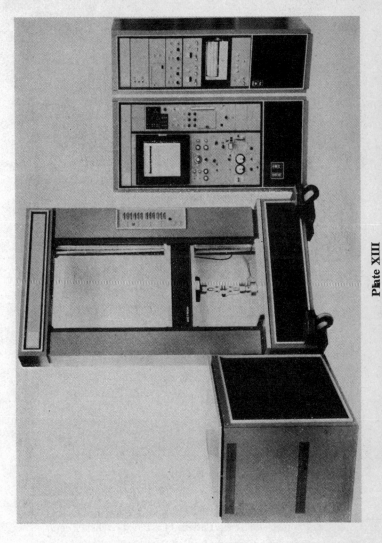

Plate XIII

50 kN Capacity Universal Testing Machine with Load–Strain Control System

(By courtesy of Instron Ltd.)

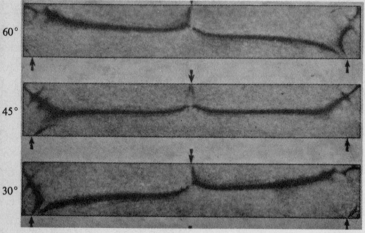

60°

45°

30°

Plate XIV

Above: **Photoelastic Stress Pattern of a Pair of Wheel Teeth**

Below: **Isoclinics for a Centrally Loaded Beam in Bending, shown for 60°, 45° and 30° Orientation of the Polarizer and Analyser with Respect to a Vertical Datum**

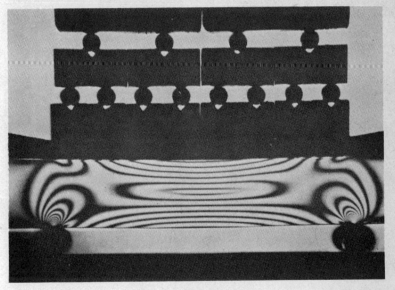

Plate XV

Fringe Patterns for a Beam Built-in at Each End and Carrying a Uniformly Distributed Load

Above: whole-order fringes
Below: half-order fringes

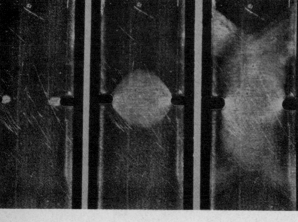

Plate XVI

Cantilever Beam for Calibration of Brittle Lacquer

(From J. R. Linge, "Some developments and applications of brittle lacquers," *Aircraft Engineering*, April 1958; *by courtesy of Westland Aircraft Ltd, Fairey Aviation Division, Middlesex*)

Three Stages of Cracking of Brittle Lacquer Coating on a Notched Bar

(From J. R. Linge, "Some developments and applications of brittle lacquers," *Aircraft Engineering*, May 1958; *by courtesy of Westland Aircraft Ltd, Fairey Aviation Division, Middlesex*)